Introduction to Continuous Symmetries

Introduction to Continuous Symmetries

From Space–Time to Quantum Mechanics

Franck Laloë

Translated by Nicole and Dan Ostrowsky

WILEY-VCH

Author

Prof. Franck Laloë
Laboratoire Kastler Brossel (ENS)
24 rue Lhomond
75231 Paris Cedex 05
France

Cover: © agsandrew/Shutterstock

Library of Congress Card No.: applied for

British Library Cataloguing-in-Publication Data
A catalogue record for this book is available from the British Library.

Bibliographic information published by the Deutsche Nationalbibliothek
The Deutsche Nationalbibliothek lists this publication in the Deutsche Nationalbibliografie; detailed bibliographic data are available on the Internet at <http://dnb.d-nb.de>.

Print ISBN: 978-3-527-41416-1
ePDF ISBN: 978-3-527-84054-0
ePub ISBN: 978-3-527-84053-3

Instructor Material (= password protected)
→ Supplementary material for and instructors available for download
www.wiley-vch.de/9783527414161

Contents

Preface

The birth of Quantum Mechanics has often been compared to that of Relativity, initially introduced by Albert Einstein in 1905. The force of Einstein's reasoning was to start from physical ideas based on the invariance of light velocity, come up with a very general symmetry principle (the equivalence of the inertial reference frames), translate these ideas into equations, and then construct a theory that was the necessary consequence of these equations. This led to the theory of relativity in its so-called "special" form. This deductive construction gave to the theories of relativity, both special and general (also based on a physical principle, the equivalence principle), a very convincing character. By contrast, Quantum Mechanics was not established as the consequence of such a superb abstract reasoning but rather as the result of a collection of experimental challenges: how can one explain black body radiation, atomic spectra, or the photoelectric effect?

After 25 years of efforts, from 1900 to 1925, the solution appeared almost as a magical algorithm, offering results explaining all the experiments. Within a few years, it was also shown that this formalism had a mathematical coherence, even though it could be written in different forms, either as wave equations or as matrix equations. It was a fantastic success for physics, subject however to one enormous reservation: whereas the equations were simple and coherent, and the predictions always verified, the physical objects themselves remained poorly defined. Numerous contradictory interpretations were proposed, desperately trying to reconstruct objects and properties based on these extraordinary equations. The lack of success of these attempts may have led to the conclusion that they were bound to fail and that there was nothing, or just something senseless, between the experimental data and the mathematical formalism.

Now that Quantum Mechanics is almost a century old, has the situation changed? Is it possible to identify some milestones along this rough road that might give a sense to this entire construction? Can we come back to the simple idea underlying physics that it does describe objects and their properties in the real world?

I believe this book is such a milestone along that path. I discovered these ideas while being a graduate student in the "DEA Brossel" at the beginning of the 1980's, and taking Franck's course that gave rise to this book. I was immediately fascinated by discovering that, instead of using mysterious and almost magical canonical quantization rules, most of quantum mechanics' formalism could be obtained from much more down-to-earth considerations. You simply had to start from the rules governing classical physics and incorporate them into symmetry principles. These rules include geometric transformations, space and time translations, rotations, and may be extended to changes of reference frame in Galilean or Einstein relativity. Taking for granted that the description of quantum states requires a Hilbert space (this point will be discussed below), it can be shown that all the continuous symmetry transformations must be represented, in the mathematical sense, by unitary (or anti-unitary) transformations acting in the Hilbert space. Starting from the structure itself of these generalized geometrical transformations, and making a small (but crucial) detour via Lie algebra to write the commutation relations between the infinitesimal generators of the symmetry operations, numerous essential characteristics of quantum mechanics "spontaneously" appear: the Schrödinger equation in the case of Galilean relativity, the Klein–Gordon and Dirac equations in the case of Einstein relativity, and (surprisingly!) the spin in both cases. These reconstructions are very clearly and convincingly presented in this work.

In the last part of this introduction, I would like to come back to the previous statement that "the description of a quantum system requires a Hilbert space": why is this necessary? This question puzzled me for many years, but I believe I now have an answer, even though it is not (yet?) accepted by Franck. What is needed before this book is some form of a non-classical probability theory based on projections (in a Hilbert space) rather than partitions (in a Borel set) as would be the case in classical physics. The basic idea of this new probability theory is to associate mutually orthogonal projectors to mutually exclusive measurement results. This is in agreement with a basic idea of the quantum world: mutually exclusive events corresponding to measurement results cannot be subdivided, and when performing feasible measurements on a given system there will never be more than N mutually exclusive results. Using Hilbert basis and distributions, this contextual quantization can be extended – with some mathematical precautions – to cases where N goes to infinity while remaining countable.

Accepting this fundamentally quantum idea, and a few simple other arguments, there is no longer a choice: very powerful mathematical theorems show that quantum mechanics is the only possible theory. More precisely, Uhlhorn's theorem, discussed in one of the complements of the book, shows that, in each context, unitary transformations between projectors are necessary to conserve the mutually exclusive character of the events. Gleason's theorem also shows that Born's law is necessary to ensure the general structure of a probability law. Once this probabilistic framework is established, one has reached the starting point of this book, and real physics – the continuous symmetry transformations – may come into play.

I am not sure Franck is willing to embark on this adventurous path, though it appeared to me as a consequence of the developments presented here. In any case, this book presents very exciting ideas for the reader to discover (or rediscover).

Philippe Grangier
CNRS - Institut d'Optique Graduate School - Ecole Polytechnique,
Palaiseau, March 2021

Introduction

This book started as a course given to students for several years, with its corresponding teaching notes. There were also a number of presentations given by the students to illustrate some specific points. At the very beginning, it was part of a master's course in "Quantum Physics" given at the end of the 1970's at the Physic's Laboratory of the ENS in Paris. The main objective was to familiarize the graduate students with the computation techniques based on rotational invariance, irreducible tensor operators, the Wigner–Eckart theorem, etc. These techniques, often imported from nuclear physics, were becoming a basic tool in quantum optics, optical pumping theory, relaxation, etc. With that goal in mind, I still thought it useful to expand the context of the course and teach Wigner's ideas on the essential role of the Poincaré group generators. This aspect turned out to be particularly interesting for the students (and the teacher!), and the initial two-hour course rapidly expanded. After a few years, this more fundamental aspect became a good half of the course. The present book naturally reflects this duality in the teaching objectives.

A reader may wish for example to rapidly master the technical tools of the course. After a general introduction, the reader can directly go to chapters VII and VIII for the principal results on rotational symmetries, or to chapter IX for particle exchange symmetries. With the same objective of offering practical useful tools, complement D_V as well as the appendix, include forays into the domain of discrete symmetries: space parity and time reversal. Instead of considering, as is often the case, this latter as a symmetry in itself, we shall view it in the general frame of the space–time symmetries.

On the other hand, another reader may wish to explore the more fundamental aspects. The reader should then go to the chapters V and VI where we present the approach of Wigner [4] in the context of Einstein's special relativity, and that of Levy-Leblond [29, 30] for the Galilean case. These approaches show how "the quantum world arises from the classical one" using very general hypotheses. We simply assume that classical space–time remains, in quantum mechanics, the general framework for describing the evolution of physical systems, and that this description is done in a linear (and complex) state space. There is no need to use more or less artificial and sometimes ambiguous "quantization rules" to go from a classical to a quantum description of a physical system. Using the sole properties of classical space–time one can predict the existence of linear quantum operators acting in the space state, and obeying very precise commutation rules. One can then build various relatively simple state spaces. Without adding any other hypothesis, one predicts the existence of several operators, one (diagonal) for the mass, one for the momentum, another one for the angular momentum, etc. In addition there appears, even for a point particle, a spin operator associated to an internal rotation, unimaginable in classical physics (a point object cannot rotate around itself). In a way, we can say that physical considerations (space–time has a classical structure imposed by special relativity) lead to mathematical results on possible quantum descriptions for the simplest objects (irreducible representations). These considerations enable us to construct various quantum wave equations: Schrödinger, Klein–Gordon, or Dirac equations.

Note that smaller characters are used for passages that can be skipped in a first reading. We also kept, whenever possible, the notations of reference [9]. After some hesitation, we did not use, in general, the relativity covariant notations. These notations are practically indispensable in field theory, and the reader should be familiar with them. They are also handy when treating, for example, the six generators of the Lorentz group as the components of a single second-order tensor, or when building the Pauli–Lubanski generator, etc. On the other hand, they prevent us from retaining our goal: conduct parallel lines of reasoning for the Galilean and Poincaré groups, identify the additional $1/c$ terms appearing in the second case, and highlight the origin of the relativistic effects. We thus kept the more elementary notation, which, in the end, does not lead to longer computations when each steps of the reasoning is detailed.

Many colleagues and students of this course have given useful advice on the various versions of the teaching notes and the present manuscript. I cannot thank them all, and will only cite two, who later became well-known researchers (and friends): Dominique Delande, one of the first who took the time to read in detail the initial teaching notes and proposed very useful corrections; Philippe Grangier, who has always been an enthusiastic supporter of the methods using symmetries for the "construction" of quantum mechanics, and has used them in his own research. The appendix on time reversal has greatly benefited from Guy Fishman's very pertinent remarks. Very special thanks are due to Michel Le Bellac who, after carefully reading several chapters, made several very interesting suggestions and encouraged me to add a few developments that were effectively lacking.

One of the best things that can happen to a book is to benefit from outstanding translators. They can bring out and correct weaknesses and missing elements of the texts ... as well as eliminate typographic errors. This book largely benefited from questions and remarks of its translators. Nicole and Dan Ostrowsky have, as they previously did for the three volumes of Quantum Mechanics with Claude Cohen-Tannoudji and Bernard Diu, based their translation on an attentive reading, which led to numerous improvements. As for Carsten Henkel, the German translator, he never hesitated to deeply analyze the subjects treated in the book, which led to several important additions. I wish to express both my friendship and gratitude to all the translators for their valuable contributions.

Chapter I

Symmetry transformations

A. Basic symmetries

A-1. Definition

Consider a physical system which, at time t_0, is in the state $S(t_0)$. For a classical system containing N particles for example, $S(t_0)$ could be defined by the $2N$ values of the vectors giving their positions and velocities at time t_0. Once it has evolved, the system is at time t in the state $S(t)$.

Introduction to Continuous Symmetries: From Space–Time to Quantum Mechanics, First Edition. Franck Laloë.

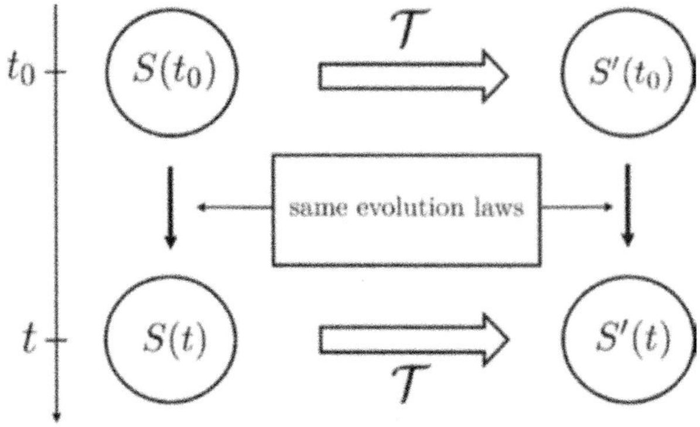

Figure 1: Consider a transformation $\mathcal{T}$ that changes an arbitrary physical system state $S(t)$ into another state $S'(t)$. If the sequence of the states $S(t)$ and that of the states $S'(t)$ follow the same evolution laws, $\mathcal{T}$ is said to be a symmetry transformation.

Imagine now a transformation $\mathcal{T}$ that changes a system in a given state S into another system in a state S' (figure 1). Obviously, the kind of transformations $\mathcal{T}$ one can think of are numerous: translation of the positions, rotation through a constant or time-dependent angle, multiplication by a factor 2 of the inter-particle distances, changing the sign of the electric charges, etc.

Applying this transformation $\mathcal{T}$ to a sequence of states $S(t)$ that describe a possible motion of the system yields another sequence of states $S'(t)$. By definition, $\mathcal{T}$ is said to be a symmetry transformation if that latter sequence of states $S'(t)$ also describes a possible motion of the system, following the same evolution laws. This condition must be satisfied for any initial state $S(t_0)$. Therefore:

> A transformation $\mathcal{T}$ is said to be a symmetry transformation if it transforms any possible motion into another possible motion.

This means that in figure 1 one can, at any instant t and for any motion, "finish closing the square" with a transformation $\mathcal{T}$.

The definition of a symmetry transformation $\mathcal{T}$ not only concerns a particular state of the system $S(t)$ at a given instant (as, for example, when a given geometrical figure is said to be symmetrical), but all its history as well, i.e. the whole set of states the system successively goes through as time passes.

Comment:

One can also consider $\mathcal{T}$ transformations that are not instantaneous, such as translations or expansions of the time scale, or Lorentz transformations of a field that is extended in space (its different points then undergo different time transformations), etc. In the left and right parts of figure 1, different time scales should then be used. In fact, for an extended system, a Lorentz transformation is not a succession of transformations, each acting at a single time t, but a global transformation acting on the entire time evolution of the system.

A-2. Examples

A transformation $\mathcal{T}$ is not necesssarily a symmetry transformation. For instance, space dilation by a factor 2 is not in general a symmetry transformation in classical mechanics, when the forces between particles depend on their distance. Neither is the one associated with the rotation of the system by an angle proportional to time (change to a non Galilean reference frame, where inertia effects are different). On the other hand, and this will be explained in detail, the translation or rotation by a fixed quantity of an *isolated* system is a symmetry operation (space homogeneity and isotropy).

Here are a number of so-called fundamental symmetries:

- space translations ;
- space rotations ;
- time translations ;
- Lorentz (or Galilean) "relativistic" transformations;
- P (parity, meaning space symmetry with respect to the origin), C (charge conjugation), and T (time reversal);
- exchange between identical particles.

Among these transformations, all are at the present considered as symmetry transformations for the whole set of physical laws governing isolated systems[1], with the exception of P, C, and T. These latter are symmetry transformations if the interactions within the system are electromagnetic (or strong), but no longer if weak interactions play a role (the product CPT nevertheless remains a symmetry operation).

Note that the translation invariance of the evolution of an isolated system is a concept that would be hard to abandon completely as it is almost the definition of a so-called "isolated physical system". As for the time translations, the basis of physics or even of the scientific method would be destroyed if they were not, at least approximately[2], symmetry transformations for isolated systems: the same experiment performed today or tomorrow would yield different results.

As it is considered at present that all the physical laws known or to be discovered must satisfy all these symmetries, these latter can be viewed as fundamental "*superlaws*" (Wigner), well worth studying.

[1]Obviously, Galilean transformations must also be excluded since they only qualify as symmetry transformations when they are approximations of the Lorentz transformations, in the "non-relativistic" limit (all the velocities $\ll$ light velocity, all the distances $\Delta x \ll c\Delta t$).

[2]In certain cosmological theory, "fundamental" physical constants change (Dirac) as the Universe expands. This changes the group of transformations for which the physical laws are invariant.

Comments:

(i) If both $\mathcal{T}$ and $\mathcal{T}'$ are symmetry transformations, so is their product $\mathcal{T}'\mathcal{T}$ (transformation obtained by applying $\mathcal{T}$ first, and then $\mathcal{T}'$): a group structure is thus expected for the set of symmetry transformations.

(ii) In the case of the time-reversal symmetry, one must actually change in figure 1 $S'(t)$ into $S'(-t)$ and invert the sense of the vertical arrow on the right, which represents the system evolution (*cf.* appendix and its figure 2).

(iii) For an isolated physical system, all the translations and rotations are symmetry transformations. If the physical system is subjected to an external potential (and hence no longer isolated), some of these transformations may still keep their symmetry character. This happens, for example, for the rotations around the origin O of a system subjected to a central potential around O.

A-3. Active and passive points of views

A transformation $\mathcal{T}$ can be defined from two points of views. The first one concerns a single observer in a given reference frame. To any motion of the physical system, the observer associates another motion obtained by a transformation that can be a translation, a rotation, a time delay, etc. As we saw above, $\mathcal{T}$ is a symmetry transformation if the observer can describe both motions with the same dynamical equations. The two motions only differ by their initial conditions. This is the so-called "active" point of view that gives the observer the role of applying the transformation.

One can also adopt another "passive" point of view where a single motion of the system is described by two observers, each using its own reference frame deduced from one another by the transformation $\mathcal{T}$. Each observer will give the system, for example, a different position, or orientation, or velocity, etc. and hence a different mathematical description. $\mathcal{T}$ is a symmetry transformation if the evolutions of the coordinates in the two reference frames are solutions of the same dynamical equations.

In short, the system changes in the active point of view, whereas the reference frame (the axes) changes in the passive point of view. Depending on the case, one or the other point of view will seem more natural. For example, for a time translation it is easy to imagine two different motions delayed in time as in the active point of view. On the other hand, when studying relativity where several Galilean reference frames are used by different observers, the passive point of view is often preferred.

Comments:

(i) As the definition of a translation, rotation, etc. amounts to a change of the space (or time) coordinates of the physical system, and as these coordinates define the relative position of the physical system with respect to the reference frame, it is clear that the active and passive points of view are in fact equivalent. Mathematically, the operations that must be performed on the equations to account for the effects of the transformation $\mathcal{T}$ are identical in both cases. From a physics point of view, one can describe the transformation in the passive point of view with a single motion of the system, but seen from two different reference frames; but one can also switch to the active point of view and introduce a new motion that is seen, in the first reference frame, as the initial motion in the second.

(ii) The physical distinction between these two points of views becomes meaningful if one of the reference frames is prominent with respect to the other. This happens, for example, if there implicitly exists a third reference frame, independent both from $Oxyz$ and from the system under study, like the laboratory reference frame ($Oxyz$ could then be the reference frame of the measuring devices, which, like the system, are mobile with respect to the laboratory). The difference between the two points of views then becomes clear: one moves either the system or the measuring devices.

Another example where the two points of view are not equivalent is when $Oxyz$ is an inertial reference frame, but not its transformed $O'x'y'z'$ since the transformation $\mathcal{T}$ is time-dependent (for instance, a rotation at constant angular velocity). In such a case, the passive point of view is to be preferred in quantum mechanics[3], and we shall use it most of the time.

B. Symmetries in classical mechanics

Let us show that, in classical mechanics, symmetries enforce certain forms for the physical laws, and, in addition, impose constants of motion. We begin with a particularly simple example using Newton's equation, that is, $\boldsymbol{F} = m\boldsymbol{\gamma}$.

[3]A simple example shows the difficulties of the active point of view in that case. It is known that the circulation of an electron's "velocity" (probability current) in a central potential is quantized. An arbitrary increase of the electron's angular velocity in a hydrogen atom is therefore forbidden in quantum mechanics (no quantum family of states exist where this velocity varies continuously).

B-1. Newton's equations

Consider two particles with mass m_1 and m_2, positions $\boldsymbol{r}_1$ and $\boldsymbol{r}_2$, and interacting through a potential $U(\boldsymbol{r}_1, \boldsymbol{r}_2\,; t)$. The equations of motion read:

$$\begin{cases} m_1\,\ddot{\boldsymbol{r}}_1 = \boldsymbol{f}_1\,(\boldsymbol{r}_1, \boldsymbol{r}_2\,; t) = -\boldsymbol{\nabla}_{\boldsymbol{r}_1}\,U\,(\boldsymbol{r}_1, \boldsymbol{r}_2\,; t) \\ m_2\,\ddot{\boldsymbol{r}}_2 = \boldsymbol{f}_2\,(\boldsymbol{r}_1, \boldsymbol{r}_2\,; t) = -\boldsymbol{\nabla}_{\boldsymbol{r}_2}\,U\,(\boldsymbol{r}_1, \boldsymbol{r}_2\,; t) \end{cases} \tag{I-1}$$

where $\ddot{\boldsymbol{r}}_1$ represents the second derivative of $\boldsymbol{r}_1$ and $\boldsymbol{\nabla}_{\boldsymbol{r}_1}$ the gradient with respect to the coordinates $\boldsymbol{r}_1$.

• **Translational invariance**

Let $\boldsymbol{b}$ be an arbitrary constant vector. Consider the transformation of the positions at every time t:

$$\begin{cases} \boldsymbol{r}_1 \stackrel{\mathcal{T}}{\Longrightarrow} \boldsymbol{r}_1 + \boldsymbol{b} \\ \boldsymbol{r}_2 \stackrel{\mathcal{T}}{\Longrightarrow} \boldsymbol{r}_2 + \boldsymbol{b} \end{cases} \tag{I-2}$$

(the two velocities are then unchanged). It transforms a possible motion into another possible motion (with the same potential U). Since the transformation does not change the accelerations, we have:

$$\begin{cases} \boldsymbol{f}_1\,(\boldsymbol{r}_1 + \boldsymbol{b}, \boldsymbol{r}_2 + \boldsymbol{b}\,; t) = \boldsymbol{f}_1\,(\boldsymbol{r}_1, \boldsymbol{r}_2\,; t) \\ \boldsymbol{f}_2\,(\boldsymbol{r}_1 + \boldsymbol{b}, \boldsymbol{r}_2 + \boldsymbol{b}\,; t) = \boldsymbol{f}_2\,(\boldsymbol{r}_1, \boldsymbol{r}_2\,; t) \end{cases} \tag{I-3}$$

This means that, when the two variables increase by a quantity $\boldsymbol{b}$, the gradients of the potential function U with respect to these two variables remains constant. It follows that, in this change of the two variables, U only varies by a constant. This constant could be time-dependent, but has no consequence on the particle's motion since it does not depend on their positions. If we furthermore require U to go to zero at infinity, this constant is necessarily equal to zero. This means that the potential U is invariant in the translation of the two variables, and is therefore only a function of $\boldsymbol{r}_1 - \boldsymbol{r}_2$:

$$U\,(\boldsymbol{r}_1 + \boldsymbol{b}, \boldsymbol{r}_2 + \boldsymbol{b}\,; t) \equiv U\,(\boldsymbol{r}_1, \boldsymbol{r}_2\,; t) \tag{I-4}$$

This restricts the possible potentials, thus imposing a constraint on the form of the physical laws:

$$U \equiv U\,(\boldsymbol{r}_1 - \boldsymbol{r}_2\,; t) \tag{I-5}$$

In addition, it is easy to show that $\boldsymbol{f}_1 = -\boldsymbol{f}_2$, which leads to:

$$\frac{\mathrm{d}}{\mathrm{d}t}\,(m_1\,\dot{\boldsymbol{r}}_1 + m_2\,\dot{\boldsymbol{r}}_2) = 0 \tag{I-6}$$

Accordingly, when translations are symmetry transformations, the total momentum is a constant of the motion.

• Rotational invariance

If, in addition, rotations are symmetry transformations, other properties appear. Following the same line of reasoning as above, we see that the field of forces $\boldsymbol{f}_1$ (or $\boldsymbol{f}_2$), considered as a function of $\boldsymbol{r} = \boldsymbol{r}_1 - \boldsymbol{r}_2$, is invariant under any rotation of the vector $\boldsymbol{r}$. It is thus a field such that $\boldsymbol{f}_1$ and $\boldsymbol{f}_2$ are parallel to $\boldsymbol{r}_1 - \boldsymbol{r}_2$ and have a modulus that only depends on $|\boldsymbol{r}|$. It follows:

$$U \equiv U\left(|\boldsymbol{r}|\,;t\right) \tag{I-7}$$

and:

$$\begin{aligned}\frac{\mathrm{d}}{\mathrm{d}t}\left[m_1\, \boldsymbol{r}_1 \times \dot{\boldsymbol{r}}_1 + m_2\, \boldsymbol{r}_2 \times \dot{\boldsymbol{r}}_2\right] &= \boldsymbol{r}_1 \times \boldsymbol{f}_1 + \boldsymbol{r}_2 \times \boldsymbol{f}_2 \\ &= \left(\boldsymbol{r}_1 - \boldsymbol{r}_2\right) \times \boldsymbol{f}_1 = 0\end{aligned} \tag{I-8}$$

The conservation over time of the total angular momentum thus comes from the rotational invariance of all the possible motions.

• Time translation

The correspondence between the motions is given by:

$$\begin{cases} \boldsymbol{r}_1(t) \stackrel{\mathcal{T}}{\Longrightarrow} \boldsymbol{r}_1(t+\tau) \\ \boldsymbol{r}_2(t) \stackrel{\mathcal{T}}{\Longrightarrow} \boldsymbol{r}_2(t+\tau) \end{cases} \tag{I-9}$$

where τ is an arbitrary constant (the new motion has a temporal advance of $+\tau$ compared to the initial motion). This time translation is a symmetry transformation if:

$$\begin{cases} \boldsymbol{f}_1\left(\boldsymbol{r}_1, \boldsymbol{r}_2\,; t+\tau\right) = \boldsymbol{f}_1\left(\boldsymbol{r}_1, \boldsymbol{r}_2\,; t\right) \\ \boldsymbol{f}_2\left(\boldsymbol{r}_1, \boldsymbol{r}_2\,; t+\tau\right) = \boldsymbol{f}_2\left(\boldsymbol{r}_1, \boldsymbol{r}_2\,; t\right) \qquad \forall\, \tau \end{cases} \tag{I-10a}$$

The forces depend explicitly on the position but not on the time. It follows that:

$$U\left(\boldsymbol{r}_1, \boldsymbol{r}_2\,; t+\tau\right) = U\left(\boldsymbol{r}_1, \boldsymbol{r}_2\,; t\right) + G(t,\tau) \tag{I-10b}$$

The function G does not have any physical meaning since it does not depend on the positions, and does not change the accelerations. We shall ignore it and, since τ is arbitrary in (I-10b), choose a time-independent potential energy W that describes the possible motions:

$$W\left(\boldsymbol{r}_1, \boldsymbol{r}_2\right) = U\left(\boldsymbol{r}_1, \boldsymbol{r}_2\,; t=0\right) \tag{I-11}$$

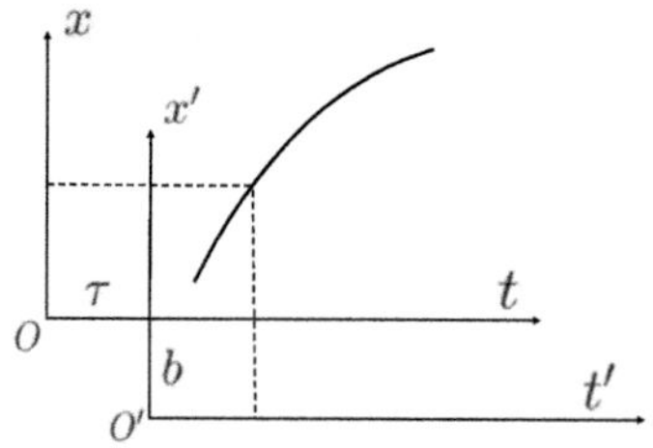

Figure 2: The trajectories of a physical system are symbolized in this figure by the trajectory of a single position as a function of time. Two space–time reference systems Oxt and $O'x't'$ are used to describe this motion; they have a space offset b and a time offset τ. The figure shows a motion described in the second reference system as being in advance, in both time and space, with respect to the description in the first reference system. Nevertheless, the origin O has been moved to O' by a positive amount along the time axis, but by a negative amount along the position axis.

We now compute:

$$\frac{\mathrm{d}}{\mathrm{d}t}\left[m_1\,\dot{\boldsymbol{r}}_1^2 + m_2\,\dot{\boldsymbol{r}}_2^2\right] = 2\left(\dot{\boldsymbol{r}}_1\cdot\boldsymbol{f}_1 + \dot{\boldsymbol{r}}_2\cdot\boldsymbol{f}_2\right) = -2\left(\dot{\boldsymbol{r}}_1\cdot\boldsymbol{\nabla}_{\boldsymbol{r}_1}W + \dot{\boldsymbol{r}}_2\cdot\boldsymbol{\nabla}_{\boldsymbol{r}_2}W\right)$$
$$= -2\frac{\mathrm{d}}{\mathrm{d}t}\,W \qquad \text{(I-12)}$$

which leads to:

$$\frac{\mathrm{d}}{\mathrm{d}t}\left[\frac{m_1}{2}\,\dot{\boldsymbol{r}}_1^2 + \frac{m_2}{2}\,\dot{\boldsymbol{r}}_2^2 + W\right] = 0 \qquad \text{(I-13)}$$

The total energy (kinetic and potential) is thus a constant of motion.

Comment:

Space and time translations belong to the same category of transformations, even though they show some differences. In the preceding equations, time was not considered as a system's dynamical variable like the position $\boldsymbol{r}(t)$, but as a parameter the dynamical variables depend on. In relativity, however, time also plays the role of a coordinate that is needed, in addition to the three components of $\boldsymbol{r}$, to define an event in space–time. One should then be careful about signs, which may differ in both cases.

To see why in a simple case, let us consider a one-dimensional space. Figure 2 shows the motion of a particle seen by two different observers placed in reference frames Oxt and $O'x't'$. These reference frames have a space offset b and a time offset τ. The two offsets are counted positive if the second observer sees a motion that has progressed in space and time with respect to the first observer – this is the case shown in the figure. The same event is then described, either by the coordinates (x, t), or by the coordinates (x', t') given by:

$$\begin{aligned} x' &= x + b \\ t' &= t - \tau \end{aligned} \tag{I-14a}$$

We notice an opposite sign between the space and time variations. The minus sign for the time is not surprising: if the event "the train arrives at the station" occurs at ten to twelve instead of twelve, the train is ten minutes early.

The sign of τ is also different in (I-9) and (I-14a). The reason lies in the difference between the two points of view mentioned above: in the first case, time was considered as a parameter the dynamical variables depend on, in the second case, as the coordinate of a space–time event.

Starting from equation (I-14a), we can recover the plus sign of (I-9) . The same motion is described in the first reference frame by a function $x(t)$ of time t, and in the second reference frame by a different function $x'(t')$. If times t and t' are such that $t' = t - \tau$, they relate to the same event, so that $x' = x + b$, and therefore:

$$x'(t') = x(t) + b = x(t' + \tau) + b \tag{I-14b}$$

Time τ now follows a plus sign. The conclusion is that, when time plays the role of a parameter the position depends on, one has to add (and not subtract) τ to obtain a time progression, in agreement with (I-9).

We shall go no further in the symmetry studies around Newton's equations as these are not a convenient starting point for the quantization of a physical system. It is best to use either the Lagrangian or the Hamiltonian formalism.

B-2. Lagrange's equations

B-2-a. General formalism

The system is described[4] by a set of generalized coordinates q_i, with $i = 1, 2, \ldots, N$, which define its configuration $\mathscr{C}$ (distinct from its state S, which also contains velocities); it can also depend on a number of parameters λ_α (particles' masses, charges, etc.). One associates with that system a function called

[4] A more detailed presentation can be found, for example, in the appendix III of reference [9], or in a standard book on analytical mechanics.

"*Lagrangian*", which depends on all the coordinates q_i and q_i':

$$L \equiv L(q_i, \dot{q}_i\,; t\,; \lambda_\alpha) \tag{I-15}$$

and such that the equations of motion are given by the Lagrangian equations:

$$\boxed{\frac{\mathrm{d}}{\mathrm{d}t}\left(\frac{\partial L}{\partial \dot{q}_i}\right) = \frac{\partial L}{\partial q_i} \qquad [i = 1, 2, \ldots, N]} \tag{I-16}$$

In these equations, $\dot{q}_i$ stands for the time derivative of q_i. Remember that the total time derivative of a function $K(q_i, \dot{q}_i\,; t)$ is given by:

$$\frac{\mathrm{d}}{\mathrm{d}t} K(q_i, \dot{q}_i\,; t) = \frac{\partial K}{\partial t} + \sum_i \dot{q}_i \frac{\partial K}{\partial q_i} + \sum_i \ddot{q}_i \frac{\partial K}{\partial \dot{q}_i} \tag{I-17}$$

Like Newton's equations, the Lagrangian equations are of second-order with respect to the time.

They are equivalent to a principle of least action that yields, in a global way (rather than local in time), the possible motions of the physical system. This principle states that, among all the possible motions leading the system at time t_1 from the configuration $\mathscr{C}_1$ (symbolizing the set of all q_i) to the configuration $\mathscr{C}_2$ at time t_2, the only realizable motion (that which satisfies the equations of motion) must have a stationary action $\mathscr{A}$:

$$\mathscr{A} = \int_{t_1}^{t_2} \mathrm{d}t\; L\left[q_i(t), \dot{q}_i(t)\,; t\right] \tag{I-18}$$

Figure 3, where the set of coordinates q_i is symbolized by a single ordinate axis q, shows with dashed lines several a priori possible paths, and with a solid line the path actually followed by the system, as it minimizes $\mathscr{A}$.

The primary advantage of the Lagrangian point of view is its great generality, as, with a proper choice of the function L, the equations of motion of a large number of physical systems can be put in the form (I-16). Furthermore, whatever the variables q_i chosen to describe the system (for example, taking spherical instead of Cartesian coordinates, etc.), the system's equations of motion keep the same form (I-16), which is not the case for Newton's equations.

For a system of particles of mass m_n, with an interaction potential $V(\boldsymbol{r}_n\,; t)$, the Lagrangian can be written:

$$L = T\left(\dot{\boldsymbol{r}}_1, ..., \dot{\boldsymbol{r}}_n, ...\right) - V(\boldsymbol{r}_1, ..., \boldsymbol{r}_n\, ...; t) \tag{I-19}$$

Here, the 3 components of the particles' position vectors $\boldsymbol{r}_n$ play the role of the q_i, and T is the kinetic energy:

$$T\left(\dot{\boldsymbol{r}}_1, ..., \dot{\boldsymbol{r}}_n, ...\right) = \frac{1}{2} \sum_n m_n\, \dot{\boldsymbol{r}}_n^2 \tag{I-20}$$

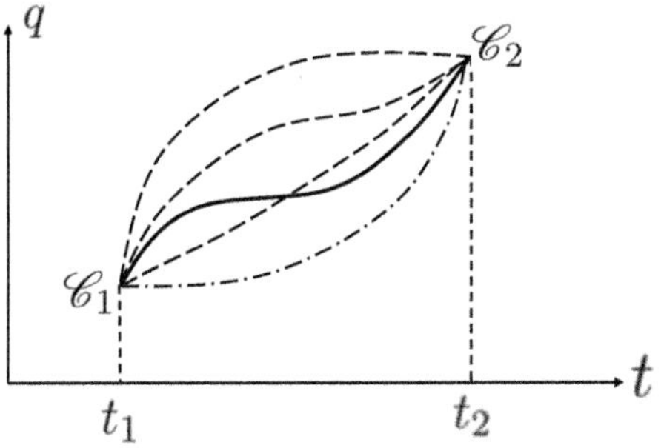

Figure 3: Two successive configurations of a physical system, $\mathscr{C}_1$ and $\mathscr{C}_2$, are represented in a symbolic way by the value of a single position variable q on the vertical axis. These states being fixed, various "paths" (set of states at all the intermediary times) going from $\mathscr{C}_1$ to $\mathscr{C}_2$ are shown with dashed lines. None of these virtual paths will be followed by the system for real, except the one, drawn with a solid line, that makes action stationary.

In this example, the forces can be obtained by taking the derivative of a potential, but Lagrangian formalism can be used in more general cases. One can, for example, compute the evolution of a system of particles interacting with a given electromagnetic field described by a scalar potential V and a vector potential $\boldsymbol{A}$. A possible Lagrangian is then (*cf.*, for example, Appendix III, § 4.b, of [9]):

$$L = \sum_n \left\{ \frac{1}{2} m_n \dot{\boldsymbol{r}}_n^2 + q_n^{\mathrm{e}} \boldsymbol{A}(\boldsymbol{r}_n, t) \cdot \dot{\boldsymbol{r}} - q_n^{\mathrm{e}} V(\boldsymbol{r}_n, t) \right\} \tag{I-21}$$

(where q_n^{e} is the electric charge of the nth particle). As mentioned above, many other equations of motion (where the values of the fields at each point of space become dynamical variables, playing the role of the q_i and $\dot{q}_i$) can be obtained from Lagrange's equations and a variational principle. This is the case, for example, of Maxwell's equations.

Comments:

(i) It should not be assumed that a unique Lagrangian corresponds to the equations of motion of a given physical system. There are actually many "equivalent" Lagrangians. For example, if Λ is an arbitrary function $\Lambda(q_i, t)$, one can, starting from a Lagrangian L obtain another[5] Lagrangian L':

$$L'(q_i, \dot{q}_i\,; t) = L(q_i, \dot{q}_i; t) + \frac{\mathrm{d}}{\mathrm{d}t}\,\Lambda(q_i\,; t) \tag{I-22}$$

The variation δL of the Lagrangian is:

$$\delta L = L' - L = \frac{\partial \Lambda}{\partial t} + \sum_j \dot{q}_j \frac{\partial \Lambda}{\partial q_j} \tag{I-23}$$

and:

$$\begin{cases} \dfrac{\mathrm{d}}{\mathrm{d}t}\left(\dfrac{\partial}{\partial \dot{q}_i}\delta L\right) = \dfrac{\mathrm{d}}{\mathrm{d}t}\dfrac{\partial \Lambda}{\partial q_i} = \dfrac{\partial^2 \Lambda}{\partial q_i \partial t} + \displaystyle\sum_j \dot{q}_j \dfrac{\partial^2 \Lambda}{\partial q_i \partial q_j} \\ \dfrac{\partial}{\partial q_i}\delta L = \dfrac{\partial^2 \Lambda}{\partial q_i \partial t} + \displaystyle\sum_j \dot{q}_j \dfrac{\partial^2 \Lambda}{\partial q_i \partial q_j} \end{cases} \tag{I-24}$$

This means that the contributions of δL to each side of equation (I-16) are identical and hence compensate each other. Lagrangians that lead to the same differential equations of motion are called "equivalent Lagrangians".

Another way to check that L and L' are equivalent is to note that the corresponding variation of the action is written:

$$\delta \mathscr{A} = \int_{t_1}^{t_2} \mathrm{d}t\,\frac{\mathrm{d}\Lambda}{\mathrm{d}t} = \Lambda\Big[q_i(t_2)\,; t_2\Big] - \Lambda\Big[q_i(t_1)\,; t_1\Big] \tag{I-25}$$

The variation $\delta\mathscr{A}$ only depends on S_1, S_2 and t_1, t_2, and not on the path followed by the system between S_1 and S_2. It follows that $\mathscr{A}$ and $\mathscr{A} + \delta\mathscr{A}$ will always be stationary for the same paths.

(ii) Reciprocally, it is sometimes suggested that the difference between two equivalent Lagrangians must be a total derivative with respect to time. This is simply not true.

A first simple counter-example is to multiply the Lagrangian by an arbitrary constant. Actually, the ensemble of the equivalent Lagrangians is in general much larger. For example, for a free particle, one can choose either L or L' given by:

$$L = \dot{x}^2 + \dot{y}^2 + \dot{z}^2$$
$$L' = \alpha\dot{x}^2 + \beta\dot{y}^2 + \gamma\dot{z}^2$$

where α, β, and γ are arbitrary constants.

It might be interesting to establish a general procedure to find all the Lagrangians

[5] Note that Λ should not depend on the $\dot{q}_i$ or else second derivatives of the q_i would appear, which is not possible for a Lagrangian.

equivalent to a given Lagrangian. This would enable one to find the necessary and sufficient conditions for a given transformation $\mathcal{T}$ to be a symmetry transformation. Furthermore, it would allow seeing if the quantization obtained from these Lagrangians leads to the same physical results. This general problem does not seem to have been solved at the present.

B-2-b. Constants of motion; Noether's theorem

α. *Simple cases*

The invariance properties of L can lead to the existence of constants of motion. For example, imagine the Lagrangian (I-19) is invariant under the translation of the whole set of particles by a given value $\boldsymbol{b}$ (substitution $\boldsymbol{r}_n \Longrightarrow \boldsymbol{r}_n + \boldsymbol{b}$). In that case:

$$\sum_n \boldsymbol{\nabla}_{\boldsymbol{r}_n} L = 0 \tag{I-26}$$

leads to, according to (I-16):

$$\frac{\mathrm{d}}{\mathrm{d}t} \sum_n \boldsymbol{p}_n(t) = 0 \tag{I-27}$$

where $\boldsymbol{p}_n$ is defined as:

$$\boldsymbol{p}_n(t) = \boldsymbol{\nabla}_{\dot{\boldsymbol{r}}_n} L = m_n \dot{\boldsymbol{r}}_n(t) \tag{I-28}$$

Conservation of the total momentum is thus a consequence of space translation invariance of the Lagrangian[6].

Similarly, if L is invariant under a time translation:

$$\frac{\partial L}{\partial t} = 0 \tag{I-29}$$

Defining the function H as:

$$H(t) = \sum_n \dot{\boldsymbol{r}}_n(t) \cdot \boldsymbol{p}_n(t) - L\Big(\boldsymbol{r}_1(t), ..., \boldsymbol{r}_n(t) ...; \dot{\boldsymbol{r}}_1(t), ..., \dot{\boldsymbol{r}}_n(t), ...\Big) \tag{I-30}$$

it is easy to show, using (I-16) and (I-28) that, as the system evolves, H remains constant:

$$\frac{\mathrm{d}}{\mathrm{d}t} H(t) = \sum_n \left\{ \ddot{\boldsymbol{r}}_n \cdot \boldsymbol{p}_n + \dot{\boldsymbol{r}}_n \cdot \boldsymbol{\nabla}_{\boldsymbol{r}_n} L - \dot{\boldsymbol{r}}_n \cdot \boldsymbol{\nabla}_{\boldsymbol{r}_n} L - \ddot{\boldsymbol{r}}_n \cdot \boldsymbol{p}_n \right\} = 0 \tag{I-31}$$

[6] The total momentum conservation results from the invariance of L when all the interacting particles undergo the same translation. An invariance of L under the translation of a single particle would lead to the conservation of that particle's momentum (particle without interaction).

Time translation invariance thus leads to the energy H being a constant of motion.

In the two examples above, we have assumed that L is invariant, but the invariance of the equations of motion in a transformation $\mathcal{T}$ does not forcibly imply that the Lagrangian itself is invariant under this transformation. Another possibility (*cf.* comment above) is that the transformation $\mathcal{T}$ adds to L a total derivative with respect to time. Emily Noether has shown in 1918 that, in this case as well, the invariance of the transformation $\mathcal{T}$ also leads to the existence of a constant of motion.

β. *General demonstration of the theorem*

Consider a transformation $\mathcal{T}$ of the generalized coordinates q_i of an arbitrary system, that can be written as:

$$q_i \stackrel{\mathcal{T}}{\Longrightarrow} q_i + \delta q_i \tag{I-32}$$

The transformation $\mathcal{T}$ is supposed to be infinitesimal and:

$$\delta q_i = \delta\varepsilon \; f_i\,(q_j, \dot{q}_j\,; t) \tag{I-33a}$$

where $\delta\varepsilon$ is an infinitely small quantity. The variations $\delta\dot{q}_i$ of the time derivatives of the q_i are given by:

$$\delta\dot{q}_i = \delta\varepsilon \; g_i\,(q_j, \dot{q}_j, \ddot{q}_j\,; t) \tag{I-33b}$$

Note that g_i, contrary to f_i, may depend on $\ddot{q}_j$, as can be seen from the definition of $\delta\dot{q}_i$:

$$\frac{\mathrm{d}}{\mathrm{d}t}\,(q_i + \delta q_i) = \dot{q}_i + \delta\dot{q}_i \tag{I-34}$$

which leads to:

$$g_i = \frac{\mathrm{d}}{\mathrm{d}t}\,f_i = \frac{\partial f_i}{\partial t} + \sum_j \left(\dot{q}_j\,\frac{\partial f}{\partial q_j} + \ddot{q}_j\,\frac{\partial f}{\partial \dot{q}_j}\right) \tag{I-35}$$

In the transformation $\mathcal{T}$, the infinitesimal variation δL of the Lagrangian is written:

$$\delta L = \sum_i \left(\frac{\partial L}{\partial q_i}\delta q_i + \frac{\partial L}{\partial \dot{q}_i}\delta\dot{q}_i\right) = \delta\varepsilon\,\sum_i \left(f_i\frac{\partial L}{\partial q_i} + g_i\frac{\partial L}{\partial \dot{q}_i}\right) \tag{I-36}$$

If the transformation $\mathcal{T}$ is chosen in such a way that δL is proportional to the total time derivative of a function Λ:

$$\delta L = \delta\varepsilon\frac{\mathrm{d}}{\mathrm{d}t}\,\Lambda \tag{I-37}$$

Noether's theorem states that the function:

$$\boxed{F = \sum_i f_i \frac{\partial L}{\partial \dot{q}_i} - \Lambda} \tag{I-38}$$

is a constant of motion; $\mathrm{d}F/\mathrm{d}t$ is zero along all the possible trajectories of the physical system.

Demonstration: the total time derivative of $\sum_i f_i \, \partial L / \partial \dot{q}_i$ is:

$$\begin{aligned}\frac{\mathrm{d}}{\mathrm{d}t} \sum_i \left[f_i \frac{\partial L}{\partial \dot{q}_i} \right] &= \sum_i \left[g_i \frac{\partial L}{\partial \dot{q}_i} + f_i \frac{\mathrm{d}}{\mathrm{d}t} \left(\frac{\partial L}{\partial \dot{q}_i} \right) \right] \\ &= \sum_i \left(g_i \frac{\partial L}{\partial \dot{q}_i} + f_i \frac{\partial L}{\partial q_i} \right)\end{aligned} \tag{I-39}$$

(the second equality, obtained from the equations of motion, is valid along any possible trajectory of the system). It follows, using (I-36) and (I-37):

$$\frac{\mathrm{d}}{\mathrm{d}t} \left[\sum_i f_i \frac{\partial L}{\partial \dot{q}_i} \right] = \frac{\delta L}{\delta \varepsilon} = \frac{\mathrm{d}}{\mathrm{d}t} \Lambda \tag{I-40}$$

Inserting this result into the time derivative of (I-38) leads to $\mathrm{d}F/\mathrm{d}t = 0$, which proves the theorem.

Comments:

(i) If Λ depends only on the coordinates q_i and of the time t, the variation δL of the Lagrangian depends on the q_i, the $\dot{q}_i$, and of time; $L + \delta L$ then provides a Lagrangian that is equivalent to L. But, if Λ depends on the $\dot{q}_i$, the function $L+\delta L$ also depends on the $\ddot{q}_i$, which is no longer compatible with the standard definition of a Lagrangian. The Noether theorem nevertheless remains valid in this case; see, for instance, the first example below.

(ii) One may have to use Lagrange's equations to go from (I-36) to (I-37). As an example, they can be used to express the $\ddot{q}_j$ appearing in general in (I-36), as a function of the q_i and $\dot{q}_i$; one can then look for a function $\Lambda(q_i, t)$ independent of the $\dot{q}_i$ (and whose total time derivative does not contain the $\ddot{q}_i$).

(iii) Obviously, Noether's theorem is only interesting if it yields a non-trivial constant of motion F, and not, for example, a zero or a constant independent of the generalized coordinates!

If one accepts functions such as $\Lambda(q_i, \dot{q}_i, t)$, there is a greater chance of obtaining zero information. A trivial case is to take completely arbitrary functions

f_i , and write δL in the form (I-37) choosing the function $\Lambda = \sum_i f_i \, \partial L/\partial \dot{q}_i$ (this is always possible since (I-40) simply states that, along the trajectories, δL is always proportional to the total derivative of that function Λ). We then get the trivial result $F \equiv 0$.

γ. *Examples*

Here are a few simple examples of the application of Noether's theorem; the theorem is also valid in field theory, where it has important applications – *cf.* complement B_I.

• Consider first an arbitrary system whose Lagrangian is not explicitly time-dependent. As a transformation law, we choose:

$$q_i(t) \overset{\mathcal{T}}{\Longrightarrow} q_i(t + \delta t) \tag{I-41a}$$

where δt is a constant that plays the role of $\delta\varepsilon$. Operation $\mathcal{T}$ shifts the time evolution of the system; we are thus dealing with a time translation. As δt is infinitesimal, we can write:

$$\delta q_i(t) = \dot{q}_i \delta t \tag{I-41b}$$

so that:

$$\begin{cases} f_i = \dot{q}_i \\ g_i = \ddot{q}_i \end{cases} \tag{I-42}$$

On the other hand:

$$\delta L = \delta t \sum_i \left[\dot{q}_i \frac{\partial L}{\partial q_i} + \ddot{q}_i \frac{\partial L}{\partial \dot{q}_i} \right] = \delta t \frac{\mathrm{d}L}{\mathrm{d}t} \tag{I-43}$$

(since by hypothesis $\partial L/\partial t = 0$). We find that the function Λ is none other than the Lagrangian itself. According to (I-38), the constant of motion is:

$$F = \sum_i \dot{q}_i \frac{\partial L}{\partial \dot{q}_i} - L \tag{I-44}$$

which is simply the usual definition of the Hamiltonian H (energy).

• Let us now go back to the example discussed in § B-2-b above, and find which constants of motion derive from the invariance under space translation, or under a change of Galilean reference frame. The Lagrangian is written as:

$$L = \sum_n \frac{m_n}{2} \dot{\boldsymbol{r}}_n^2 - V(\boldsymbol{r}_1, \ldots, \boldsymbol{r}_n \ldots) \tag{I-45}$$

where the interaction potential V is invariant under the translation of all the positions (as usual, we have replaced the q_i by the $\boldsymbol{r}_n$, or more precisely by the three components of these vectors). The invariance of L under the transformation $\boldsymbol{r}_n \Longrightarrow \boldsymbol{r}_n + \delta\boldsymbol{b}$ readily yields the conservation of the total momentum. The 3 components of the infinitesimal vector $\delta\boldsymbol{b}$ now play the role of 3 infinitely small $\delta\varepsilon$; setting $f_i = 1$, $g_i = 0$, $\delta L = 0$, yield as constants of motion the 3 components of the vector:

$$\boldsymbol{P} = \sum_n \boldsymbol{\nabla}_{\dot{\boldsymbol{r}}_n} L = \sum_n \boldsymbol{p}_n \tag{I-46}$$

- We now introduce a change of Galilean reference frame by:

$$\boldsymbol{r}_n \Longrightarrow \boldsymbol{r}_n + t\,\delta\boldsymbol{v} \tag{I-47}$$

Relations (I-33) then become:

$$\delta\boldsymbol{r}_n = t\delta\boldsymbol{v} \qquad\qquad \delta\dot{\boldsymbol{r}}_n = \delta\boldsymbol{v} \tag{I-48}$$

so that we have $f_i = t$ and $g_i = 1$ for $i = 1, 2, 3$; the 3 components of the vector $\delta\boldsymbol{v}$ now play the role of $\delta\varepsilon$. In this case, δL is written:

$$\delta L = \delta\boldsymbol{v} \cdot \sum_n \left(m_n\,\dot{\boldsymbol{r}}_n - t\,\boldsymbol{\nabla}_{\boldsymbol{r}_n} V\right) \tag{I-49}$$

If V is translation invariant as in (I-26), the sum of all its gradients is equal to zero, and:

$$\delta L = \delta\boldsymbol{v} \cdot \frac{\mathrm{d}}{\mathrm{d}t}\boldsymbol{G} \tag{I-50}$$

where:

$$\boldsymbol{G}(t) = \sum_n m_n\,\boldsymbol{r}_n(t) \tag{I-51}$$

$\boldsymbol{G}$ is the position of all the particles' center of mass, multiplied by the sum of masses. Replacing Λ by $\boldsymbol{G}$ in (I-38) yields the (vectorial) constant of the motion:

$$\boldsymbol{G}_0 = t\sum_n m_n\,\dot{\boldsymbol{r}}_n - \sum_n m_n\,\boldsymbol{r}_n = \boldsymbol{P}t - \boldsymbol{G}(t) \tag{I-52}$$

Hence:

$$\boldsymbol{G}(t) = \boldsymbol{P}t - \boldsymbol{G}_0 \tag{I-53}$$

Dividing this equation by the sum of masses we check that, as expected, the particles' center of mass moves at constant velocity.

Finally, imagine that V is not translation invariant, but that the sum of the (outside) forces acting on the particles is a constant vector $\boldsymbol{F}$. The variation of L is then:

$$\delta L = \delta L = \delta \boldsymbol{v} \cdot \Big(\sum_n m_n \, \dot{\boldsymbol{r}}_n + t \, \boldsymbol{F} \Big) = \delta \boldsymbol{v} \cdot \frac{\mathrm{d}}{\mathrm{d}t} \boldsymbol{\Lambda} \tag{I-54a}$$

with:

$$\boldsymbol{\Lambda}(t) = \boldsymbol{G}(t) + \frac{1}{2} t^2 \, \boldsymbol{F} \tag{I-54b}$$

The constant of motion is now:

$$\boldsymbol{G}_0 = \boldsymbol{P} t - \boldsymbol{G}(t) - \frac{1}{2} \boldsymbol{F} \, t^2 = \boldsymbol{P}_0 \, t + \frac{1}{2} \boldsymbol{F} \, t^2 - \boldsymbol{G}(t) \tag{I-55}$$

where, in the second equality, we have taken into account the linear time variation of the total momentum $\boldsymbol{P} = \boldsymbol{P}_0 + \boldsymbol{F} t$. As a result, we have to add a term $(t^2/2)\boldsymbol{F}$ to the right-hand side of (I-53), which shows that the center of mass now moves with a constant acceleration.

Exercise:

Consider a particle of mass m, position $\boldsymbol{r}$, momentum $\boldsymbol{p} = m\dot{\boldsymbol{r}}$, and angular momentum $\boldsymbol{\ell} = \boldsymbol{r} \times \boldsymbol{p}$. This particle is subjected to a central potential $V = -\alpha/r^n$ (n is a positive integer). Introducing the transformation

$$\delta \boldsymbol{r} = \delta \boldsymbol{\varepsilon} \times \boldsymbol{\ell} \tag{I-56}$$

show that:

$$\delta L = -m \frac{n\alpha}{r^{n+2}} \left[\boldsymbol{r}^2 \left(\delta \boldsymbol{\varepsilon} \cdot \dot{\boldsymbol{r}} \right) - \left(\delta \boldsymbol{\varepsilon} \cdot \boldsymbol{r} \right) \left(\boldsymbol{r} \cdot \dot{\boldsymbol{r}} \right) \right] \tag{I-57}$$

In the case of a Coulomb or Newton potential ($n = 1$), show that:

$$\delta L = \frac{\mathrm{d}}{\mathrm{d}t} \left[-m\alpha \, \delta \boldsymbol{\varepsilon} \cdot \frac{\boldsymbol{r}}{r} \right] \tag{I-58}$$

and that, consequently, the vector $\boldsymbol{M}$ (Runge–Lenz vector):

$$\boldsymbol{M} = \boldsymbol{p} \times (\boldsymbol{r} \times \boldsymbol{p}) - m\alpha \frac{\boldsymbol{r}}{r} \tag{I-59}$$

is a constant. Physical interpretation: the points of the particle's (plane) trajectory where its velocity is perpendicular to $\boldsymbol{r}$ are fixed. Instead of following a "rosette pattern", the trajectory is a closed curve. It is actually an ellipse whose major axis is parallel to $\boldsymbol{M}$ and whose eccentricity equals $|\boldsymbol{M}|/\alpha m$.

Comment:

In (I-33), we assumed that the transformation only concerned the q_i (and the $\dot{q}_i$), but not the time. This restriction can be lifted by introducing, in addition to the variations δq_i and $\delta \dot{q}_i$ written in (I-33), a time variation:

$$\delta t = \delta\varepsilon\, h\,(q_j, \dot{q}_j\,; t) \tag{I-60}$$

Relation (I-35) is no longer valid in this case. Writing:

$$(\dot{q}_i + \delta\dot{q}_i)\ \mathrm{d}\,(t + \delta t) = \mathrm{d}\,[q_i + \delta q_i] \tag{I-61}$$

we readily obtain, to first-order in $\delta\varepsilon$:

$$g_i\, \mathrm{d}t + \dot{q}_i\, \mathrm{d}h = \mathrm{d}f_i \tag{I-62}$$

meaning:

$$g_i = \frac{\mathrm{d}f_i}{\mathrm{d}t} - \dot{q}_i\,\frac{\mathrm{d}h}{\mathrm{d}t} \tag{I-63}$$

We also have:

$$\delta L = \delta\varepsilon \left\{ \sum_i \left(f_i\,\frac{\partial L}{\partial q_i} + g_i\,\frac{\partial L}{\partial \dot{q}_i} \right) + h\,\frac{\partial L}{\partial t} \right\} \tag{I-64}$$

Writing δL as:

$$\delta L = \delta\varepsilon \left[\frac{\mathrm{d}}{\mathrm{d}t}\Lambda + L\,\frac{\mathrm{d}h}{\mathrm{d}t} \right] \tag{I-65a}$$

it can be shown that the function:

$$F = \sum_i \frac{\partial L}{\partial \dot{q}_i}\,[f_i - h\dot{q}_i] + hL - \Lambda \tag{I-65b}$$

is a constant of motion. Using Lagrange's equation to get the total time derivative of $\partial L/\partial \dot{q}_i$ and (I-65a) to get that of Λ, one shows that:

$$\begin{aligned}\frac{\mathrm{d}F}{\mathrm{d}t} &= \sum_i \frac{\partial L}{\partial q_i}\,[f_i - h\dot{q}_i] + \sum_i \frac{\partial L}{\partial \dot{q}_i} \left[\frac{\mathrm{d}f_i}{\mathrm{d}t} - \dot{q}_i\,\frac{\mathrm{d}h}{\mathrm{d}t} - h\ddot{q}_i \right] \\ &\quad + L\,\frac{\mathrm{d}h}{\mathrm{d}t} + h \left[\frac{\partial L}{\partial t} + \sum_i \left(\dot{q}_i\,\frac{\partial L}{\partial q_i} + \ddot{q}_i\,\frac{\partial L}{\partial \dot{q}_i} \right) \right] \quad - \frac{\delta L}{\delta\varepsilon} - L\,\frac{\mathrm{d}h}{\mathrm{d}t}\end{aligned} \tag{I-66}$$

Using relation (I-64) to compute $\delta L/\delta\varepsilon$, and finally replacing g_i by its value (I-63), all the terms in the right-hand side of this equality cancel each other two by two. A is indeed a constant of motion.

Exercise: Setting $f_i = g_i = 0$, $h = 1$, show, as in (I-44), that H is a constant of the motion if $\partial L/\partial t = 0$.

B-3. Hamilton's equations

Before leaving classical mechanics, let us quickly review the Hamiltonian formalism, which is often used for the "quantization" of a physical system. Note that another quantization procedure starts from the Lagrangian formalism [10] (Feynman's postulates), and is often used, especially in field theory.

B-3-a. General formalism

To each generalized coordinate of the system q_i we associate the conjugate momentum:

$$p_i = \frac{\partial L}{\partial \dot{q}_i} \tag{I-67}$$

We assume that all the $\dot{q}_i$ can be expressed in terms of the p_i and q_i, so that the dynamical state of the system can be defined either by the set of all the q_i and $\dot{q}_i$, or by the set of q_i and p_i. In Hamilton's point of view, the q_i and p_i are chosen as the independent variables. They are the coordinates of a point defining the state of the physical system in a space called "phase space", with a large dimension if there are a large number of particles. We introduce the Hamiltonian:

$$H(p_i, q_i) = \sum_i p_i \, \dot{q}_i(q_j, p_j) - L[q_i, \dot{q}_i(q_j, p_j)] \tag{I-68}$$

where, in the right-hand side, the notation $\dot{q}_i(q_j, p_j)$ explicits the fact that the derivatives of the q_i are functions of the variables q_j and p_j for any j. From now on we shall simply write $\dot{q}_i$. The differential of H is written:

$$\mathrm{d}H(p_i, q_i) = \sum_i [\dot{q}_i \, \mathrm{d}p_i + p_i \, \mathrm{d}\dot{q}_i] - \frac{\partial L}{\partial q_i}\mathrm{d}q_i - \frac{\partial L}{\partial \dot{q}_i}\mathrm{d}\dot{q}_i \tag{I-69a}$$

Using definition (I-67) for p_i, we see that the $\mathrm{d}\dot{q}_i$ terms on the right-hand side cancel out. Using Lagrange's equations, we can replace $\partial L/\partial q_i$ by $\dot{p}_i$ and write:

$$\mathrm{d}H(p_i, q_i) = \sum_i [\dot{q}_i \, \mathrm{d}p_i - \dot{p}_i \mathrm{d}q_i] \equiv \sum_i \left[\frac{\partial H}{\partial p_i}\mathrm{d}p_i + \frac{\partial H}{\partial q_i}\mathrm{d}q_i\right] \tag{I-69b}$$

Identifying the terms in $\mathrm{d}p_i$ and $\mathrm{d}q_i$ leads to the Hamilton's equations of motion:

$$\boxed{\begin{cases} \dot{q}_i = \dfrac{\partial H}{\partial p_i} \\ \dot{p}_i = -\dfrac{\partial H}{\partial q_i} \end{cases}} \tag{I-70}$$

Instead of N second-order differential equations (Lagrangian point of view), we now have $2N$ equations between $2N$ variables, but these are first-order differential equations.

B-3-b. Poisson bracket

Consider two physical quantities $A(q_i, p_i)$ and $B(p_i, q_i)$ defined using Hamilton's point of view. The Poisson bracket $\{A, B\}$ is defined as:

$$\{A,B\} = \sum_i \left[\frac{\partial A}{\partial q_i}\frac{\partial B}{\partial p_i} - \frac{\partial B}{\partial q_i}\frac{\partial A}{\partial p_i}\right] \tag{I-71}$$

It's easy to show that:

$$\begin{aligned}\{q_i, p_j\} &= \delta_{ij} \\ \{q_i, q_j\} &= \{p_i, p_j\} = 0\end{aligned} \tag{I-72}$$

as well as:

$$\{A, A\} = 0 \tag{I-73a}$$

$$\{A, B\} = -\{B, A\} \tag{I-73b}$$

$$\{A, BC\} = \{A, B\}\, C + B\,\{A, C\} \tag{I-73c}$$

(they follow the same rules as the commutators).

According to (I-70), the time evolution of any given physical quantity $F(q_i, p_i\,; t)$ can be written as:

$$\frac{\mathrm{d}F}{\mathrm{d}t} = \frac{\partial F}{\partial t} + \sum_i \left(\dot{q}_i\,\frac{\partial F}{\partial q_i} + \dot{p}_i\,\frac{\partial F}{\partial p_i}\right) = \frac{\partial F}{\partial t} + \{F, H\} \tag{I-74}$$

The Poisson bracket thus yields directly the time evolution of any function F of the position and momentum. A constant of motion is therefore described by a function F whose Poisson bracket $\{F, H\}$ with the Hamiltonian vanishes.

In certain cases, the Poisson brackets also yield new constants of motion: using (I-73c) and (I-74) we see that, if both F and G are constants of the motion, their Poisson bracket is also a constant of the motion (Poisson's theorem)[7].

Exercise: Show that expression:

$$\{A, \{B, C\}\} + \{B, \{C, A\}\} + \{C\,\{A, B\}\} = 0 \tag{I-75}$$

equals zero (Jacobi identity). Infer from this result Poisson's theorem.

[7]This Poisson bracket may simply be equal to zero or to a constant independent of the q and p, or else be proportional to F or G; in such cases, Poisson's theorem does not bring any new information.

B-3-c. Infinitesimal transformations

The Poisson brackets can also yield the transformation of a system's motion in an infinitesimal operation $\mathcal{T}$. Let us show this on the concrete example of a particle of mass m whose motion is given by the values of its 4 space–time coordinates as a function of a parameter u: $x = f_1(u)$, $y = f_2(u)$, $z = f_3(u)$, $t = f_4(u)$.

• Spatial translation by an infinitesimal constant vector $\delta\boldsymbol{b}$. By definition of this translation:

$$x' = x + \delta b_x \qquad y' = y + \delta b_y \qquad z' = z + \delta b_z \qquad t' = t \qquad \boldsymbol{p}' = \boldsymbol{p} \tag{I-76}$$

Now:

$$\{p_x, x\} = -1 \qquad \{p_y, x\} = \{p_z, x\} = 0 \tag{I-77}$$

and consequently:

$$\{\delta\boldsymbol{b}\cdot\boldsymbol{p}, x\} = -\delta b_x \tag{I-78}$$

We can thus write:

$$\begin{cases} \boldsymbol{r}' = \boldsymbol{r} - \{\delta\boldsymbol{b}\cdot\boldsymbol{p}, \boldsymbol{r}\} \\ \boldsymbol{p}' = \boldsymbol{p} - \{\delta\boldsymbol{b}\cdot\boldsymbol{p}, \boldsymbol{p}\} = \boldsymbol{p} \end{cases} \tag{I-79}$$

which shows that the Poisson brackets of $\delta\boldsymbol{b}\cdot\boldsymbol{p}$ with the dynamical variables yield infinitesimal translations.

• Spatial rotation of an infinitesimal angle $\delta\varphi$ around the unit vector $\boldsymbol{u}$. This rotation is associated with an infinitesimal (time-independent) vector $\delta\boldsymbol{a}$:

$$\delta\boldsymbol{a} = \boldsymbol{u}\,\delta\varphi \tag{I-80}$$

We can write:

$$\begin{cases} \delta\boldsymbol{r} = \boldsymbol{r}' - \boldsymbol{r} = \delta\boldsymbol{a}\times\boldsymbol{r} + \ldots \\ \delta\boldsymbol{p} = \boldsymbol{p}' - \boldsymbol{p} = \delta\boldsymbol{a}\times\boldsymbol{p} + \ldots \end{cases} \tag{I-81}$$

We now introduce the (angular momentum) vector:

$$\boldsymbol{\ell} = \boldsymbol{r}\times\boldsymbol{p} \tag{I-82}$$

This yields:

$$\begin{aligned} \{\delta\boldsymbol{a}\cdot\boldsymbol{\ell}\ ,\ x\} &= -\delta a_y\, z + \delta a_z\, y \\ \{\delta\boldsymbol{a}\cdot\boldsymbol{\ell}\ ,\ p_x\} &= -\delta a_y\, p_z + \delta a_z\, p_y \end{aligned} \tag{I-83}$$

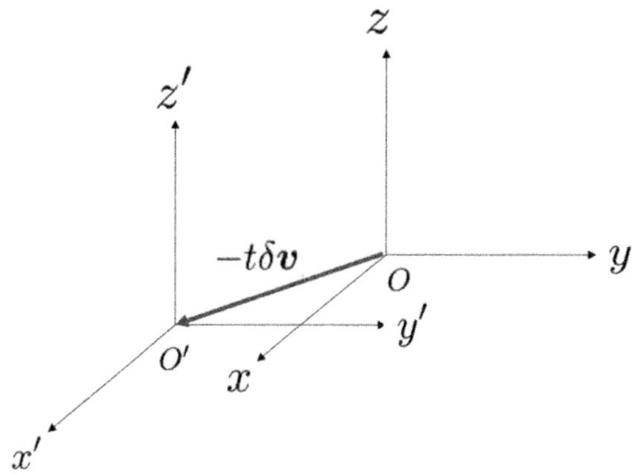

Figure 4: Two Galilean reference frames $Oxyz$ and $O'x'y'z'$ have parallel axes, and their origins O and O' are linked by a vector equal to $-t\delta\boldsymbol{v}$, where $\delta\boldsymbol{v}$ is an infinitesimal constant vector.

and, consequently:

$$\begin{cases} \boldsymbol{r}' = \boldsymbol{r} - \{\delta\boldsymbol{a}\cdot\boldsymbol{\ell}\,,\,\boldsymbol{r}\} + \ldots \\ \boldsymbol{p}' = \boldsymbol{p} - \{\delta\boldsymbol{a}\cdot\boldsymbol{\ell}\,,\,\boldsymbol{p}\} + \ldots \end{cases} \tag{I-84}$$

The angular momentum $\boldsymbol{\ell}$ thus plays for the rotations the same role as $\boldsymbol{p}$ for the translations.

• Time translation

Consider two motions defined by $\boldsymbol{r}(t)$ and $\boldsymbol{r}'(t)$, the second being ahead of the first one by a time interval δt (figure 2), as in relation (I-9). We then have:

$$\begin{cases} \boldsymbol{r}'(t) = \boldsymbol{r}(t+\delta t) \simeq \boldsymbol{r}(t) + \delta t\,\dot{\boldsymbol{r}}(t) \\ \boldsymbol{p}'(t) = \boldsymbol{p}(t+\delta t) \simeq \boldsymbol{p}(t) + \delta t\,\dot{\boldsymbol{p}}(t) \end{cases} \tag{I-85}$$

According to (I-74), and to first-order in δt:

$$\begin{cases} \boldsymbol{r}' = \boldsymbol{r} - \{\delta t\,H, \boldsymbol{r}\} \\ \boldsymbol{p}' = \boldsymbol{p} - \{\delta t\,H, \boldsymbol{p}\} \end{cases} \tag{I-86}$$

H is thus associated with time translations

• Pure Galilean transformation

Consider (figure 4) two Galilean reference frames $Oxyz$ and $O'x'y'zy$ with parallel axes, and such that the vector $\overrightarrow{OO'}$ linking their origins is equal to $-t\delta\boldsymbol{v}$ ($\delta\boldsymbol{v}$ is an infinitesimal constant vector). If the position vector of a given particle is $\boldsymbol{r}(t) = \overrightarrow{OM}(t)$ in the first reference frame, it will become $\boldsymbol{r}'(t) = \overrightarrow{O'M}(t)$ in the second (passive point of view, § A-3), with:

$$\boldsymbol{r}' = \boldsymbol{r} + t\delta\boldsymbol{v} \tag{I-87}$$

In a similar way:

$$\boldsymbol{p}' = \boldsymbol{p} + m\delta\boldsymbol{v} \tag{I-88}$$

As in (I-52), we introduce the vector:

$$\boldsymbol{G}_0(t) = t\boldsymbol{p} - m\boldsymbol{r} \tag{I-89}$$

We get:

$$\begin{cases} \{x\ ,\ \boldsymbol{G}_0 \cdot \delta\boldsymbol{v}\} = t\delta v_x \\ \{p_x\ ,\ \boldsymbol{G}_0 \cdot \delta\boldsymbol{v}\} = m\delta v_x \end{cases} \tag{I-90}$$

and hence:

$$\begin{aligned} \boldsymbol{r}' &= \boldsymbol{r} - \{\delta\boldsymbol{v} \cdot \boldsymbol{G}_0, \boldsymbol{r}\} \\ \boldsymbol{p}' &= \boldsymbol{p} - \{\delta\boldsymbol{v} \cdot \boldsymbol{G}_0, \boldsymbol{p}\} \end{aligned} \tag{I-91}$$

Using Poisson brackets, the vector $\boldsymbol{G}_0$ generates infinitesimal Galilean transformations. For an ensemble of several particles, we can reason in a similar way by introducing the position of the system's center of mass – *cf.* (I-52).

B-3-d. Symmetry transformations and constants of motion

Consider an infinitesimal transformation generated by Poisson brackets containing a function $G(t)$:

$$q_i' = q_i - \{G(t), q_i\}\,\delta\varepsilon \qquad p_j' = p_j - \{G(t), p_j\}\,\delta\varepsilon \tag{I-92}$$

This transformation remains the same if one adds to $G(t)$ any function of time t, leading to a whole family of equivalent functions $G'(t)$, associated with the same transformation. We now show that:

The transformation generated by Poisson brackets containing a function $G(t)$ is a symmetry transformation provided either $G(t)$, or one of the equivalent functions $G'(t)$, is a constant of motion.

As we will see in § C-2, this result has a direct analogue in quantum mechanics, where Poisson brackets are replaced with commutators between operators.

To prove this result, we come back to the scheme of figure 1, and take as an example the transformation associated with changing the system's description from reference frame $Oxyzt$ to another frame $O'x'y'z't'$. In a first case, the dynamical variables of the physical system are changed at an initial time t_0 by the transformation associated with $G(t_0)$. The system then evolves during an infinitesimal time δt under the effect of its Hamiltonian H. In the second case, we first let the system evolve during the same time δt, and then change the reference frame at time $t_0 + \delta t$. Since the relative position of the two reference frames may have changed during δt, a different transformation has to be applied, associated with $G(t_0 + \delta t)$. If the same final physical state is obtained in both cases, the function $G(t)$ is associated with a symmetry transformation.

In the first case, the successive variations of the dynamical variables are:

$$\begin{aligned} q_i &\Rightarrow q_i - \{G(t_0),\, q_i\}\,\delta\varepsilon \\ &\Rightarrow q_i - \{G(t_0), q_i\}\,\delta\varepsilon - \{H,\, q_i\}\,\delta t + \{H, \{G(t_0),\, q_i\}\}\,\delta t\delta\varepsilon + \ldots \\ p_j &\Rightarrow p_j - \{G(t_0),\, p_j\}\,\delta\varepsilon \\ &\Rightarrow p_j - \{G(t_0), p_j\}\,\delta\varepsilon - \{H,\, p_j\}\,\delta t + \{H, \{G(t_0),\, p_j\}\}\,\delta t\delta\varepsilon + \ldots \end{aligned} \tag{I-93a}$$

while, in the second case, they become:

$$\begin{aligned} q_i &\Rightarrow q_i - \{H,\, q_i\}\,\delta t \\ &\Rightarrow q_i - \{H,\, q_i\}\,\delta t - \{G(t_0), q_i\}\delta\varepsilon + \left[- \{\frac{\partial G}{\partial t}, q_i\} + \{G(t_0), \{H,\, q_i\}\} \right] \delta t\delta\varepsilon + \ldots \\ p_j &\Rightarrow p_j - \{H,\, p_j\}\,\delta t \\ &\Rightarrow p_j - \{H,\, p_j\}\,\delta t - \{G(t_0), p_j\}\delta\varepsilon + \left[- \{\frac{\partial G}{\partial t}, p_i\} + \{G(t_0), \{H,\, p_j\}\} \right] \delta t\delta\varepsilon + \ldots \end{aligned} \tag{I-93b}$$

The transformation is a symmetry transformation if, at time t_0:

$$\{H, \{G(t),\, q\}\} + \{\frac{\partial G}{\partial t}, q\} + \{G(t), \{q,\, H\}\} = 0 \tag{I-94}$$

where q is any of the dynamical variables, position or momentum. The Jacobi identity (I-75) then allows us to write this condition as:

$$-\{q, \{H,\, G(t)\}\} + \{q, \frac{\partial G}{\partial t}\} = 0 \tag{I-95}$$

It follows that the function $\{H, G(t)\} - \partial G/\partial t$ has a zero Poisson bracket with all the dynamical variables, which means that this function is independent of all of them. It is therefore a function of time only:

$$-\{H,\, G(t)\}\} + \frac{\partial G}{\partial t} = F(t) \tag{I-96}$$

The left-hand side of this equation is the total time derivative $\mathrm{d}G/\mathrm{d}t$ of G – *cf.* (I-74). We therefore have:

$$\frac{\mathrm{d}}{\mathrm{d}t}\left[G(t) - \int \mathrm{d}t\; F(t)\right] = 0 \tag{I-97}$$

The function between brackets, which is equivalent to $G(t)$ since it defines the same transformation when inserted in (I-92), is a constant of motion.

C. Symmetries in quantum mechanics

In quantum mechanics, symmetry transformations play at least as big a role as in classical mechanics, as will be shown below.

C-1. Quantization standard procedure

In quantum mechanics, the set of q_i and p_i are replaced by a ("ket") vector $|\psi\rangle$ belonging to the state space $\mathscr{E}$ of the system. For a spinless particle, one can choose one of two possible "bases" , either the basis $\{|\boldsymbol{r}\rangle\}$ of the eigenvectors of the position operator $\boldsymbol{R}$, or the basis $\{|\boldsymbol{p}\rangle\}$ of the eigenvectors of the momentum $\boldsymbol{P}$. To each of these bases correspond the functions:

$$\psi(\boldsymbol{r}) = \langle \boldsymbol{r}|\psi\rangle \tag{I-98a}$$

$$\overline{\psi}(\boldsymbol{p}) = \langle \boldsymbol{p}|\psi\rangle \tag{I-98b}$$

called the wave functions in the $\boldsymbol{r}$ and $\boldsymbol{p}$ representations respectively. For a particle with non-zero spin, several wave functions are necessary to characterize a quantum state $|\psi\rangle$.

The classical quantities $\mathscr{A}(q_i, p_i)$ become linear Hermitian operators acting in $\mathscr{E}$:

$$\mathscr{A}(q_i, p_i) \Longrightarrow A \tag{I-99}$$

In general, any ket $|\psi\rangle$ may be expanded on eigenvectors of A; A is then called an "observable". To the classical Poisson brackets correspond commutators between observables. Relations (I-72) become:

$$[R_i, P_j] = i\hbar\, \delta_{ij} \tag{I-100}$$

(where i and j stand for x, y or z).

Comment:

Quantization rules are often defined by a replacement of classical functions $\mathscr{A}$ by operators A, obtained by simply substituting an operator Q for every variable q, an operator P for every variable p, and finally performing a symmetrization of the products of operators to ensure the hermiticity of A. The operator value of a commutator is obtained in the same way from the Poisson bracket (Dirac rule [37]). This method is nevertheless rather ill-defined over the ensemble of classical quantities, and may even sometimes lead to contradictory results [38, 39]. Moreover, if one changes the classical variables in the Lagrangian (for instance, in order to use angle and distance variables), the method does not provide a unique and well-defined quantization. In this book, we proceed differently, and start only from general considerations on space–time and symmetries.

The time evolution of $|\psi\rangle$ is the Schrödinger equation:

$$i\hbar \frac{\mathrm{d}}{\mathrm{d}t} |\psi(t)\rangle = H(t)\,|\psi(t)\rangle \tag{I-101}$$

where the Hamiltonian $H(t)$ is the quantum observable associated with the classical Hamiltonian function.

The equation is equivalent to:

$$|\psi(t)\rangle = U(t, t_0)\,|\psi(t_0)\rangle \tag{I-102}$$

In this equation, $|\psi(t)\rangle$ represents the state vector at time t, $|\psi(t_0)\rangle$ its value at the initial time t_0, and $U(t, t_0)$ is the unitary operator obeying:

$$i\hbar \frac{\mathrm{d}}{\mathrm{d}t} U(t, t_0) = H(t)\, U(t, t_0) \tag{I-103a}$$

with:

$$U(t_0, t_0) = \mathbb{1} \tag{I-103b}$$

When H is time-independent, we simply have:

$$U(t, t_0) = \exp\left\{-\frac{i}{\hbar} H \times (t - t_0)\right\} \tag{I-104}$$

The operator $U(t)$ is called an "evolution operator". More details on its properties can be found for instance in complement $\mathrm{F_{III}}$ and in § A of chapter XX of [9].

C-2. Symmetry transformations

Consider a transformation $\mathcal{T}$ of the physical system. Before the transformation, the system is described at time t by a ket $|\psi(t)\rangle$, and after the transformation by a ket $|\psi'(t)\rangle$. We define the operator T (acting in the state space $\mathscr{E}$) that transforms $|\psi\rangle$ into $|\psi'\rangle$:

$$|\psi'(t)\rangle = T(t)\,|\psi(t)\rangle \tag{I-105}$$

(in many cases, $T(t)$ is a linear unitary operator). The diagram of figure 1 now becomes that of figure 5. $\mathcal{T}$ will be a symmetry transformation if, at any time t, we have:

$$U(t, t_0)T(t_0)|\psi(t_0)\rangle = T(t)|\psi(t)\rangle = T(t)U(t, t_0)|\psi(t_0)\rangle \tag{I-106}$$

meaning, since this is true for any $|\psi(t_0)\rangle$:

$$\boxed{U(t, t_0)\,T(t_0) = T(t)\,U(t, t_0)} \tag{I-107}$$

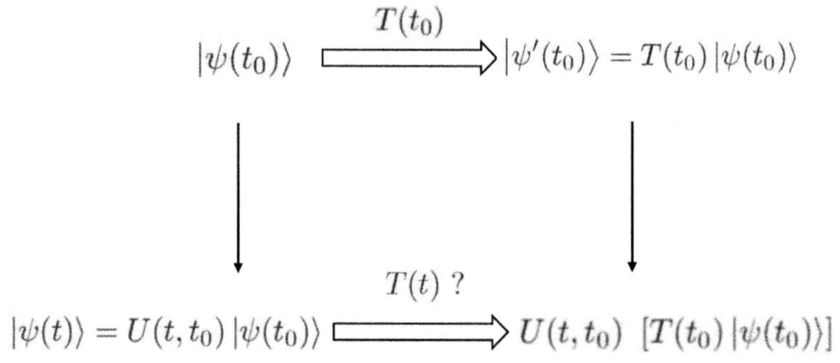

Figure 5: Diagram expliciting the condition for the transformation $\mathcal{T}$, represented in quantum mechanics by the operator $T(t)$ acting in the state space, to be a symmetry transformation. Arrows pointing down represent the time propagation due to the evolution operator $U(t,t_0)$, the double horizontal arrows the effect of $T(t)$. The transformation $\mathcal{T}$ is a symmetry transformation if $T(t)$ transforms a possible evolution of the system into another possible evolution. This is indeed the case if one can "close the square" with the horizontal arrow at the bottom of the figure, just below the question mark.

If this condition is met, we can write:

$$U^\dagger(t,t_0)\,T(t)\,U(t,t_0) = T(t_0) \tag{I-108}$$

This equality means that the operator $T(t)$ corresponds to a time-independent operator in Heisenberg's point of view. T is thus a constant of motion.

If T is time-independent, (I-107) becomes:

$$\boxed{[T\,,\,U(t,t_0] = 0} \tag{I-109}$$

which shows that, at any time, T and the evolution operator U commute. If, in addition, H is also time-independent (conservative system), U is given by (I-104), and (I-109) is written as:

$$\boxed{[T,H] = 0} \tag{I-110}$$

This shows again that T is a constant of motion.

Comment:

Two kets that differ by a global phase factor correspond to identical physical states. Relation (I-107) may thus be replaced by the more general equality:

$$U(t,t_0)\,T(t_0)|\psi(t_0)\rangle = e^{i\alpha(t)}T(t)\,U\,(t,t_0)\,|\psi(t_0)\rangle \qquad \forall\;|\psi(t_0)\rangle \tag{I-111a}$$

where $\alpha(t)$ is a function of time, which could, a priori, also depend on $|\psi(t_0)\rangle$. This is actually not the case as we now show. Setting:

$$T(t)\,U(t,t_0)|\psi(t_0)\rangle = |\chi\rangle \tag{I-111b}$$

relation (I-111a) can be written (operator T can be inverted):

$$U(t,t_0)\,T(t_0)\,U^{-1}\,(t_0,t)\;T^{-1}(t)\;|\chi\rangle = e^{i\alpha(t)}\;|\chi\rangle \tag{I-111c}$$

where α could be a function of $|\chi\rangle$. But the operator acting on the left of this equality is a linear operator having as eigenvector any ket $|\chi\rangle$, which means it corresponds to a diagonal matrix in any base; this is possible only if all its eigenvalues are equal (if two of them were different, one can easily see that the sum of the two corresponding eigenvectors would not be an eigenvector of the operator). As a result, α is necessarily independent of the ket.

Consequently we have:

$$U(t,t_0)\,T(t_0) = e^{i\alpha(t)}\;T(t)\,U(t,t_0) \tag{I-111d}$$

where $\alpha(t)$ is a real function of time, equal to zero for $t = t_0$. Setting:

$$T'(t) = e^{i\alpha(t)}\;T(t) \tag{I-112}$$

we find for T' an equality of the type (I-107). We shall assume that T' is defined [8] in such a way that $\alpha(t) = 0$.

C-3. General consequences

Relations (I-107), (I-109), or (I-110) imply that, if $\mathcal{T}$ is a symmetry transformation, operators U and H cannot take an arbitrary form.

For example, we cannot have a Hamiltonian proportional to $\boldsymbol{L}\cdot\boldsymbol{R}$ (with $\boldsymbol{L} = \boldsymbol{R}\times\boldsymbol{P}$) if the geometrical parity operation is a symmetry transformation. Similarly, a time-reversal symmetry does not allow a Hamiltonian proportional to $\boldsymbol{R}\cdot\boldsymbol{P}$, etc.

We already noted that (I-108) or (I-110) implies the existence of *constants of motion*. For example, (I-110) shows that the average values of T, T^2, T^3, etc. are constant. Hence, wherever they appear, all the T operators, their products, etc. yield constants of motion. For *conservative* systems, the existence of such constants simplifies the *search for stationary states* (eigenkets of H): one looks for a basis of eigenkets common to H and T.

If H and T commute, we have:

$$\langle\theta_1|H|\theta_2\rangle = 0 \quad \text{if} \quad \theta_1 \neq \theta_2 \tag{I-113}$$

[8] We are discussing here a single transformation $\mathcal{T}$. To eliminate the phase factors for a group of transformations is more delicate, and will be discussed later, in particular in chapter IV.

where $|\theta_1\rangle$ and $|\theta_2\rangle$ are eigenvectors of T, with eigenvalues θ_1 and θ_2. This type of relation yields "*selection rules*" for H; these rules are actually valid for all the *invariant observables* T.

One can also obtain some information concerning the *degeneracy*. Consider for example $|E_0\rangle$, an eigenket of H, with eigenvalue E_0. Relation (I-110) shows that if:

$$|E_0'\rangle = T|E_0\rangle \tag{I-114}$$

$|E_0'\rangle$ is an eigenket of H with the same eigenvalue E_0, since:

$$\begin{aligned} H\,|E_0'\rangle &= H\,T\,|E_0\rangle = T\,H\,|E_0\rangle = E_0\,T\,|E_0\rangle \\ &= E_0\,|E_0'\rangle \end{aligned} \tag{I-115}$$

Unless $|E_0\rangle$ and $|E_0'\rangle$ are proportional, the eigenvaleue E_0 is at least doubly degenerate.

If both T_1 and T_2 obey (I-108), so does their product. The operators T associated with the symmetry transformations form a group. The T do not necessarily commute with each other (non-commutative group), and we shall see the importance of the commutation relations between these operators.

Complement A_I

Eulerian and Lagrangian points of view in classical mechanics

The purpose of this complement is to present two different points of views used in classical statistical mechanics to describe an ensemble of physical systems moving in the phase space. In fluid mechanics, Eulerian or Lagrangian points of view can be used. In the first (Eulerian), one focuses on the time evolution of the fluid flow at each fixed position in space, computing for example the fluid flow rate at every point in space and as a function of time. In other words, one studies successive "snapshots" of the fluid. In the second point of view (Lagrangian), one follows the motion of each fluid element along its own trajectory. To study the motion of these fluid elements, one must define total time derivatives along each of these trajectories.

Extending these ideas from ordinary space to a larger space, that we call "phase space", we now use Eulerian and Lagrangian points of view to describe in classical statistics an ensemble of physical systems. These two points of view present a certain analogy with those of Schrödinger and Heisenberg in quantum mechanics, respectively. Just as Heisenberg's point of view, that of Lagrange is better adapted for the computation of two-time average values (correlation functions).

Consider an ensemble of physical systems, each being described, in the Hamiltonian formalism, by N coordinates q_i and N conjugated momenta p_i. Each dynamical state is defined by the coordinates q_i and p_i of a point that belongs to a $2N$ dimensional space, the so-called "phase space" (in the case of a single particle, the phase space has 6 dimensions). This point's motion obeys Hamilton's equations:

$$\begin{cases} \dot{q}_i(t) = \dfrac{\partial}{\partial p_i} H\left[q_j(t), p_j(t)\,;\, t\right] \\ \dot{p}_i(t) = -\dfrac{\partial}{\partial q_i} H\left[q_j(t), p_j(t)\,;\, t\right] \end{cases} \tag{1}$$

where $H[q_i, p_i\,;\, t]$ is the Hamiltonian of each system, supposed to be the same for all of them.

1. Eulerian point of view

Consider now a statistical ensemble of such systems, define at time $t = t_0$ by a probability density $\rho(q_i, p_i, t_0)$ in the phase space. A time t, this ensemble is described by a probability density $\rho(q_i, p_i, t)$ that yields the probability $\mathrm{d}\mathcal{P}(t)$:

$$\mathrm{d}\mathcal{P}(t) = \rho\,(q_1, q_2 \ldots q_N, p_1, p_2 \ldots p_N\,;\,t) \times \mathrm{d}q_1 \ldots \mathrm{d}q_N\;\mathrm{d}p_1 \ldots \mathrm{d}p_N \tag{2}$$

for the point characterizing the system to be in the volume element:

$$\mathrm{d}q_1\;\mathrm{d}q_2 \ldots \mathrm{d}q_N\;\;\mathrm{d}p_1\;\mathrm{d}p_2 \ldots \mathrm{d}p_N \tag{3}$$

centered at:

$$q_1, q_2, \ldots, q_N,\;\; p_1, p_2 \ldots p_N \tag{4}$$

We obviously have:

$$\int \mathrm{d}^N q \int \mathrm{d}^N p\; \rho\,(q_1, \ldots, q_N,\; p_1, \ldots, p_N\,;\,t) = 1 \tag{5}$$

where $\int \mathrm{d}^N q$ and $\int \mathrm{d}^N p$ are simplified notations for the pluri-dimensional integrals $\int \mathrm{d}q_1 \int \mathrm{d}q_2 \ldots \int \mathrm{d}q_N$ and $\int \mathrm{d}p_1 \int \mathrm{d}p_2 \ldots \int \mathrm{d}p_N$, respectively. We shall also simplify the notation for the whole set of coordinates (4) into $q_1, ..., p_N$. Remember that the system's evolution is fully determined (the function H is not random), so that probabilities only come into play in the choice of the system's initial state.

In the $2N$-dimensional phase space, the probability density ρ can be assimilated to the density of a fluid; the associated flow density will then be a $2N$-component vector:

$$\boldsymbol{J}(q_1, ..., p_N) \begin{cases} \rho\,(q_j, p_j\,;\,t)\;\dot{q}_i = \rho\,(q_j, p_j\,;\,t)\;\dfrac{\partial}{\partial p_i} H\,(q_j, p_j\,;\,t) \\ \rho\,(q_j, p_j\,;\,t)\;\dot{p}_i = -\rho\,(q_j, p_j\,;\,t)\;\dfrac{\partial}{\partial q_i} H\,(q_j, p_j\,;\,t) \end{cases} \tag{6}$$

where the two i-ranking components of that vector are written on the right-hand side of the bracket. This allows computing the number of systems of the ensemble that enter or exit each volume of the phase space, hence obtaining the local variation $\partial\rho/\partial t$ of the density ρ. The divergence theorem shows:

$$-\frac{\partial \rho}{\partial t} = {}^{2N}\boldsymbol{\nabla} \cdot \boldsymbol{J} \tag{7}$$

where ${}^{2N}\boldsymbol{\nabla} \cdot \boldsymbol{J}$ symbolizes the divergence of the vector $\boldsymbol{J}$ in the phase space:

$$\begin{aligned} {}^{2N}\boldsymbol{\nabla} \cdot \boldsymbol{J} = \sum_i \bigg\{ &\frac{\partial}{\partial q_i} \left[\rho\,(q_j, p_j\,;\,t)\;\frac{\partial}{\partial p_i}\;H\,(q_j, p_j\,;\,t)\right] \\ &-\frac{\partial}{\partial p_i} \left[\rho\,(q_j, p_j\,;\,t)\;\frac{\partial}{\partial q_i}\;H\,(q_j, p_j\,;\,t)\right]\bigg\} \end{aligned} \tag{8}$$

The $\partial^2 H/\partial q_i\, \partial p_i$ terms cancel out from this expression and, using definition (I-71) for the Poisson brackets, we have:

$$ {}^{2N}\boldsymbol{\nabla} \cdot \boldsymbol{J} = \Big\{ \rho\,(q_j, p_j\,;\,t)\,, H\,(q_j, p_j\,;\,t)\,\Big\} \tag{9} $$

so that relation (7) becomes:

$$ \frac{\partial \rho}{\partial t} = \{ H\,(q_j, p_j\,;\,t)\,, \rho\,(q_j, p_j\,;\,t)\,\} \tag{10} $$

Consider now a quantity $\mathscr{A}$ that depends on the q_i and p_i, and possibly on time:

$$ \mathscr{A}\,(q_i, p_i\,;\,t) \tag{11} $$

$\mathscr{A}$ could be, for example, the kinetic energy of an ensemble of particles, their interaction potential energy, etc. The average value of $\mathscr{A}$ at any given time t is written:

$$ \overline{\mathscr{A}}(t) = \int \mathrm{d}^N q \int \mathrm{d}^N p\;\rho\,(q_i, p_i\,;\,t)\;\;\mathscr{A}\,(q_i, p_i\,;\,t) \tag{12} $$

Using (10), we can compute the time evolution of this average value:

$$ \frac{\mathrm{d}}{\mathrm{d}t}\,\overline{\mathscr{A}}(t) = \int \mathrm{d}^N q \int \mathrm{d}^N p\;\left[\{H, \rho\}\,\mathscr{A} + \rho \frac{\partial}{\partial t} \mathscr{A} \right] \tag{13} $$

Slightly anticipating on the Lagrangian point of view, we can also compute the total time derivative of ρ. We "go along" with the particles and measure at each instant their "density" in the phase space:

$$ \begin{aligned} \frac{\mathrm{d}}{\mathrm{d}t}\,\rho &= \frac{\partial}{\partial t}\,\rho + \sum_i \left(\dot{q}_i \frac{\partial \rho}{\partial q_i} + \dot{p}_i \frac{\partial \rho}{\partial p_i} \right) = \frac{\partial}{\partial t}\,\rho + \sum_i \left(\frac{\partial H}{\partial p_i}\frac{\partial \rho}{\partial q_i} - \frac{\partial H}{\partial q_i}\frac{\partial \rho}{\partial p_i} \right) \\ &= \frac{\partial}{\partial t}\,\rho + \{\rho, H(t)\} \end{aligned} \tag{14} $$

Since $\{\rho, H\} = -\{H, \rho\}$, we are left with:

$$ \frac{\mathrm{d}}{\mathrm{d}t}\,\rho = 0 \tag{15} $$

This result is called Liouville's theorem. It states that the "fluid" particles flow along their trajectories keeping their "density" constant (which can vary at a fixed point of the phase space). They occupy, as time goes by, a constant volume of the phase space.

2. Lagrangian point of view

In the point of view of § 1 above, the integrals are performed with probability distributions that are time-dependent. It is sometimes easier, in particular when calculating two-time average values, to compute integrals with a sole distribution, the one describing the system at time $t = t_0$, chosen as the initial time. In the Lagrangian point of view, one works with a single probability density ρ^L in the phase space, which is time-independent and equal to:

$$\rho^L\left(q_i^0, p_i^0\right) = \rho\left(q_i = q_i^0, p_i = p_i^0\,;\, t = t_0\right) \tag{16}$$

where ρ is the function already defined in § 1.

This is reminiscent of the Heisenberg point of view in quantum mechanics, where the state vector $|\psi_H\rangle$ is constant, and it is the operators A_H that show the Hamiltonian time dependence. In a similar way, we define a classical function $\mathscr{A}^L(q_i^0, p_i^0\,;\, t)$ that shows the time dependence induced by the Hamiltonian. At $t = t_0$, this function has the same value as in the Eulerian point of view. At time t, it yields by definition the contribution to the quantity $\overline{\mathscr{A}}$ from all the points in the phase space that were at time t_0 at q_i^0 and p_i^0. In other words, if the solutions of equations (1) with the initial conditions $q_j(t = t_0) = q_j^0$ and $p_j(t = t_0) = p_j^0$ are written as:

$$\begin{cases} q_j\left(q_i^0, p_i^0\,;\, t\right) \\ p_j\left(q_i^0, p_i^0\,;\, t\right) \end{cases} \tag{17}$$

the function $\mathscr{A}^L\left(q_i^0, p_i^0\,;\, t\right)$ is defined by:

$$\mathscr{A}^L\left(q_i^0, p_i^0\,;\, t\right) = \mathscr{A}\left[q_i\left(q_i^0, p_i^0\,;\, t\right),\ p_i\left(q_i^0, p_i^0\,;\, t\right)\,;\, t\right] \tag{18}$$

Using the conservation of the probability density (15), we can write:

$$\overline{\mathscr{A}}(t) = \int \mathrm{d}^N q^0 \int \mathrm{d}^N p^0\ \rho^L\left(q_i^0, p_i^0\right)\ \mathscr{A}^L\left(q_i^0, p_i^0\,;\, t\right) \tag{19}$$

We see that $\mathscr{A}^L$ is explicitly time-dependent, even if this is not the case for $\mathscr{A}$, since according to (18) and (1):

$$\begin{aligned} \frac{\partial}{\partial t}\mathscr{A}^L\left(q_i^0, p_i^0\,;\, t\right) &= \sum_i \left\{\frac{\partial \mathscr{A}}{\partial q_i}\frac{\partial H}{\partial p_i} - \frac{\partial \mathscr{A}}{\partial p_i}\frac{\partial H}{\partial q_i}\right\} + \frac{\partial \mathscr{A}}{\partial t} \\ &= \{\mathscr{A}(t), H(t)\} + \frac{\partial \mathscr{A}}{\partial t} \end{aligned} \tag{20}$$

In the right-hand side of the equality, the q_i^0 and p_i^0 dependence is explicit when replacing the q_j and p_j by their expressions (17).

A two-time average value is easy to compute:

$$\overline{\mathscr{A}(t)\mathscr{B}(t')} = \int \mathrm{d}^N q^0 \int \mathrm{d}^N p^0\ \rho^L\left(q_i^0, p_i^0\right) \times \mathscr{A}^L\left(q_i^0, p_i^0\,;\, t\right) \mathscr{B}^L\left(q_i^0, p_i^0\,;\, t'\right) \tag{21}$$

Comment:

Relation (20) is a "mixed" equality as it yields the time variation of $\mathscr{A}^L$ as a function of a Poisson bracket involving the variables q_i and p_i, introduced in the Eulerian point of view. It is also possible to only use, in the Lagrangian point of view, the variables q_i^0 and p_i^0, and introduce the corresponding Poisson brackets:

$$\left\{\mathscr{A}^L(t), \mathscr{B}^L(t)\right\}_L = \sum_n \left(\frac{\partial \mathscr{A}^L}{\partial q_n^0} \frac{\partial \mathscr{B}^L}{\partial p_n^0} - \frac{\partial \mathscr{A}^L}{\partial p_n^0} \frac{\partial \mathscr{B}^L}{\partial q_n^0} \right) \tag{22}$$

Equality (20) can then be written, as will be shown below:

$$\begin{aligned} \frac{\partial}{\partial t} \mathscr{A}^L\left(q_i^0, p_i^0\,; t\right) = &\left\{\mathscr{A}^L(t), H^L(t)\right\}_L \\ &+ \frac{\partial \mathscr{A}}{\partial t} \left[q_j\left(q_i^0, p_i^0\,; t\right),\; p_j\left(q_i^0, p_i^0\,; t\right)\,;\, t\right] \end{aligned} \tag{23a}$$

or, if $\mathscr{A}$ (in the Eulerian point of view) is time-independent:

$$\frac{\partial}{\partial t} \mathscr{A}^L\left(q_i^0, p_i^0\,; t\right) = \left\{\mathscr{A}^L(t), H^L(t)\right\}_L \tag{23b}$$

Demonstration: Let us first compute the simplest Poisson brackets:

$$\{q_i(t), q_j(t)\}_L \qquad \{p_i(t), p_j(t)\}_L \qquad \{q_i(t), p_j(t)\}_L \tag{24}$$

At time $t = t_0$, they are the usual Poisson brackets. At a later time t, one must bear in mind that the $q_i(t)$ and $p_i(t)$ a priori depend on the q_k^0 and p_k^0 for all values of k. After an infinitesimal time dt, we have:

$$\begin{cases} q_i(t_0 + \mathrm{d}t) = q_i^0 + \dfrac{\partial H}{\partial p_i}(q_k^0, p_k^0, t_0)\ \mathrm{d}t \\ p_j(t_0 + \mathrm{d}t) = p_j^0 - \dfrac{\partial H}{\partial q_j}(q_k^0, p_k^0, t_0)\ \mathrm{d}t \end{cases} \tag{25}$$

In these relations, we can replace the partial derivatives of H with respect to q_i and p_i by those with respect to q_i^0 and p_i^0. A derivation then provides:

$$\begin{aligned} \frac{\partial q_i(t_0 + \mathrm{d}t)}{\partial q_l^0} &= \delta_{il} + \frac{\partial^2 H}{\partial q_l^0 \partial p_i^0}(q_k^0, p_k^0, t_0)\ \mathrm{d}t \\ \frac{\partial q_i(t_0 + \mathrm{d}t)}{\partial p_l^0} &= \frac{\partial^2 H}{\partial p_l^0 \partial p_i^0}(q_k^0, p_k^0, t_0)\ \mathrm{d}t \\ \frac{\partial p_j(t_0 + \mathrm{d}t)}{\partial q_l^0} &= -\frac{\partial^2 H}{\partial q_l^0 \partial q_j^0}(q_k^0, p_k^0, t_0)\ \mathrm{d}t \\ \frac{\partial p_j(t_0 + \mathrm{d}t)}{\partial p_l^0} &= \delta_{jl} - \frac{\partial^2 H}{\partial p_l^0 \partial q_j^0}(q_k^0, p_k^0, t_0)\ \mathrm{d}t \end{aligned} \tag{26}$$

We take the product of the first and fourth of these equalities, minus the product of the second and third, and sum over the index l. We then obtain, to first-order in dt:

$$\{q_i(t_0+\mathrm{d}t), p_j(t_0+\mathrm{d}t)\}_L = \delta_{ij} - \frac{\partial^2 H}{\partial p_i^0 \partial q_j^0}\mathrm{d}t + \frac{\partial^2 H}{\partial q_j^0\, \partial p_i^0}\mathrm{d}t = \delta_{ij} \tag{27a}$$

As for the first two equalities (26), they lead to:

$$\{q_i(t_0+\mathrm{d}t), q_j(t_0+\mathrm{d}t)\}_L = \frac{\partial^2 H}{\partial p_i^0 \partial p_j^0}\mathrm{d}t - \frac{\partial^2 H}{\partial p_j^0 \partial p_i^0}\mathrm{d}t = 0 \tag{27b}$$

A similar calculation starting from the two last lines of (26) shows that the Poisson-Lagrange brackets $\{p_i(t_0+\mathrm{d}t), p_j(t_0+\mathrm{d}t)\}_L$ also vanish. The three brackets (24) therefore take the same value at times t_0 and $t_0+\mathrm{d}t$.

In the above equations, we can replace t_0 by t_0+t', q_i^0 and p_i^0 by $q_i(t_0+t')$ and $p_i(t_0+t')$, and see by the same reasoning that the Poisson–Lagrange brackets are still stationary at any time. These brackets are therefore constant, and keep the value of the usual Poisson brackets:

$$\{q_i(t), p_j(t)\}_L = \delta_{ij} \tag{28a}$$
$$\{q_i(t), q_j(t)\}_L = 0 \tag{28b}$$
$$\{p_i(t), p_j(t)\}_L = 0 \tag{28c}$$

We can now evaluate the quantity $\{\mathscr{A}^L(t), \mathscr{B}^L(t)\}_L$ appearing in (22). According to definition (18) and the chain rule for taking the derivative of the composition of two functions, we get:

$$\begin{aligned}\{\mathscr{A}^L(t), \mathscr{B}^L(t)\}_L = \sum_{ijn} \Bigg[&\left[\frac{\partial \mathscr{A}}{\partial q_i}\frac{\partial q_i}{\partial q_n^0} + \frac{\partial \mathscr{A}}{\partial p_i}\frac{\partial p_i}{\partial q_n^0}\right]\left[\frac{\partial \mathscr{B}}{\partial q_j}\frac{\partial q_j}{\partial p_n^0} + \frac{\partial \mathscr{B}}{\partial p_j}\frac{\partial p_j}{\partial p_n^0}\right] \\ &- \left[\frac{\partial \mathscr{A}}{\partial q_i}\frac{\partial q_i}{\partial p_n^0} + \frac{\partial \mathscr{A}}{\partial p_i}\frac{\partial p_i}{\partial p_n^0}\right]\left[\frac{\partial \mathscr{B}}{\partial q_j}\frac{\partial q_j}{\partial q_n^0} + \frac{\partial \mathscr{B}}{\partial p_j}\frac{\partial p_j}{\partial q_n^0}\right]\Bigg] \\ = \sum_{ij} \Bigg[&\frac{\partial \mathscr{A}}{\partial q_i}\frac{\partial \mathscr{B}}{\partial q_j}\{q_i(t),\, q_j(t)\}_L + \frac{\partial \mathscr{A}}{\partial p_i}\frac{\partial B}{\partial q_j}\{p_i(t),\, q_j(t)\}_L \\ &+ \frac{\partial \mathscr{A}}{\partial q_i}\frac{\partial \mathscr{B}}{\partial p_j}\{q_i(t),\, p_j(t)\}_L + \frac{\partial \mathscr{A}}{\partial p_i}\frac{\partial \mathscr{B}}{\partial p_j}\{p_i(t),\, p_j(t)\}_L\Bigg]\end{aligned} \tag{29}$$

Using the general relations (28), we obtain:

$$\begin{aligned}\{\mathscr{A}^L(t), \mathscr{B}^L(t)\}_L &= \sum_i \left[\frac{\partial \mathscr{A}}{\partial q_i}\frac{\partial \mathscr{B}}{\partial p_i} - \frac{\partial \mathscr{A}}{\partial p_i}\frac{\partial \mathscr{B}}{\partial q_i}\right] \\ &= \{\mathscr{A}, \mathscr{B}\}\end{aligned} \tag{30}$$

where, on the right-hand side, the q_j and p_j must be replaced by their expressions (17) as a function of the q_n^0 and p_n^0.

A simple example: One-dimensional particle subjected to a force $mg(t)$:

- Eulerian point of view:

$$H(t) = \frac{p^2}{2m} - mg(t)x$$
$$\frac{\mathrm{d}}{\mathrm{d}t}x = \frac{p}{m} \qquad \frac{\mathrm{d}}{\mathrm{d}t}p = mg(t)$$
$$\begin{cases} p(t) = m\displaystyle\int_0^t \mathrm{d}t'\ g(t') + p_0 \\ x(t) = \displaystyle\int_0^t \mathrm{d}t' \int_0^{t'} \mathrm{d}t''\ g(t'') + \frac{p_0}{m}t + x_0 \end{cases}$$

- Lagrangian point of view:

$$\begin{aligned} H^L(x_0, p_0\,;\,t) &= \frac{1}{2m}\left[p_0 + m\int_0^t \mathrm{d}t'\ g(t')\right]^2 \\ &\quad - mg(t)\left[\int_0^t \mathrm{d}t' \int_0^{t'} \mathrm{d}t''\ g(t'') + \frac{p_0}{m}t + x_0\right] \\ &= \frac{p_0^2}{2m} - mg(t)\,x_0 + p_0\left[\int_0^t \mathrm{d}t'\ g(t') - t\,g(t)\right] \\ &\qquad + F(t) \end{aligned}$$

where :

$$F(t) = m\left[\frac{1}{2}\left(\int_0^t \mathrm{d}t'\ g(t')\right)^2 - \int_0^t \mathrm{d}t' \int_0^{t'} \mathrm{d}t''\ g(t)\ g(t'')\right]$$

In the Lagrangian point of view, the particle's momentum and position are:

$$p^L(p_0\,;\,t) = p_0 + m\int_0^t \mathrm{d}t'\ g(t')$$
$$x^L(x_0, p_0\,;\,t) = x_0 + \frac{t}{m}p_0 + \int_0^t \mathrm{d}t' \int_0^{t'} \mathrm{d}t''\ g(t'')$$

It's easy to show that $\left\{x^L(t), p^L(t)\right\}_L = 1$, and we can compute:

$$\begin{aligned} \left\{x^L(t), H^L(t)\right\}_L &= \frac{p_0}{m} + \int_0^t \mathrm{d}t'\ g(t') - t\ g(t) - \left(\frac{t}{m}\right)(-mg(t)) \\ &= \frac{p_0}{m} + \int_0^t \mathrm{d}t'\ g(t') = \frac{p(t)}{m} \end{aligned}$$
$$\left\{p^L(t), H^L(t)\right\}_L = m\ g(t) = -\frac{\partial H(t)}{\partial x}$$

which shows that equality (30) is satisfied.

Complement B$_{\text{I}}$

Noether's theorem for a classical field

Noether's theorem was introduced in § B-2-b of chapter I for the case where the physical system is described by dynamical variables q_i whose index i takes a finite number of discrete values $i = 1, 2, \ldots, N$. Actually, the main interest of the theorem occurs when the system under study is a classical or quantum field. As a field varies in space, the $q_i(t)$ are replaced by the components $\phi_j(\boldsymbol{r}, t)$ of the field, where $\boldsymbol{r}$ labels the dynamical variables: this continuous index replaces the discrete index i of the previous case. The discrete index j labels the components of the field: a scalar field has only one component (and the index j is not needed): a vector field in 3-dimensional space has three components, $j = 1, 2, 3$; one can also define a second-order tensor field for which j takes on 9 values, etc.

For the sake of simplicity, we assume in this complement that the field components are real; the case of complex components will be studied at the beginning of complement C$_{\text{VI}}$.

1. Lagrangian density and Lagrange equations for continuous variables

The Lagrangian L depends on all the dynamical variables of the physical system, and is written:

$$L = \int \mathrm{d}^3r \; \mathcal{L}(\boldsymbol{r}, t) \tag{1}$$

where $\mathcal{L}(\boldsymbol{r}, t)$ is the "Lagrangian density":

$$\mathcal{L} = \mathcal{L}\Big(\phi_j(\boldsymbol{r}, t), \dot{\phi}_j(\boldsymbol{r}, t), \partial_k\phi_j(\boldsymbol{r}, t); \boldsymbol{r}, t\Big) \tag{2}$$

We use the same notation as in complement A$_{\text{XVIII}}$ of reference [9]: $\dot{\phi}_j$ is the time derivative of the field ϕ_j, and $\partial_k\phi_j$ its spatial derivative with respect to the k component of its position. The Lagrangian density $\mathcal{L}$ depends on all these functions for various values of j and k and, eventually, directly on the position $\boldsymbol{r}$ in space, as well as on the time t (for example, when describing the effect of an outside time-dependent potential applied to the system).

Consider a possible history of the field, i.e. a path Γ for the field going from the value $\phi_j(\boldsymbol{r}, t_1)$ at an initial time t_1 to the final value $\phi_j(\boldsymbol{r}, t_2)$ at a final time t_2. The action $\mathscr{A}[\Gamma]$ associated with this path is written[1]:

$$\mathscr{A}[\Gamma] = \int_{t_1}^{t_2} \mathrm{d}t \int \mathrm{d}^3r \, \mathcal{L}\left(\phi_j(\boldsymbol{r}, t), \dot{\phi}_j(\boldsymbol{r}, t), \partial_k \phi_j(\boldsymbol{r}, t); \boldsymbol{r}, t\right) \tag{3}$$

The principle of least action postulates that among all possible paths starting from the same initial state and ending at the same final state, the path(s) actually followed by the system is the one (or are those) for which $\mathscr{A}[\Gamma]$ presents an extremum. To obtain this (or those) path(s), we compute the variation $\delta\mathscr{A}$ of the action for an infinitesimal variation of the path characterized by the infinitesimal variations $\delta\phi_j(\boldsymbol{r}, t)$, $\delta\left(\partial\phi_j(\boldsymbol{r}, t)/\partial t\right)$, and $\delta\left(\partial\phi_j(\boldsymbol{r}, t)/\partial r_k\right)$:

$$\delta\mathscr{A} = \int_{t_1}^{t_2} \mathrm{d}t \int \mathrm{d}^3r \sum_j \left[\delta\phi_j(\boldsymbol{r}, t)\frac{\partial\mathcal{L}}{\partial\phi_j} + \delta\dot{\phi}_j(\boldsymbol{r}, t)\frac{\partial\mathcal{L}}{\partial\dot{\phi}_j} + \sum_k \delta\left(\partial_k\phi_j(\boldsymbol{r}, t)\right)\frac{\partial\mathcal{L}}{\partial(\partial_k\phi_j)}\right] \tag{4}$$

Since:

$$\delta\dot{\phi}_j(\boldsymbol{r}, t) = \frac{\partial}{\partial t}\delta\phi_j(\boldsymbol{r}, t) \qquad \text{and:} \qquad \delta\left(\partial_k\phi_j(\boldsymbol{r}, t)\right) = \partial_k\left(\delta\phi_j(\boldsymbol{r}, t)\right) \tag{5}$$

we can perform an integration by parts of the terms containing the time and space derivatives. We find that the integrated terms are zero, because of the boundary conditions for $\delta\phi_j(\boldsymbol{r}, t)$ at the initial and final times, and for $\boldsymbol{r} \to \infty$. The remaining terms are all proportional to $\delta\phi_j(\boldsymbol{r}, t)$, and we get[2]:

$$\delta\mathscr{A} = \int_{t_1}^{t_2} \mathrm{d}t \int \mathrm{d}^3r \sum_j \delta\phi_j(\boldsymbol{r}, t)\left[\frac{\partial\mathcal{L}}{\partial\phi_j} - \frac{\mathrm{d}}{\mathrm{d}t}\frac{\partial\mathcal{L}}{\partial\dot{\phi}_j} - \sum_k \partial_k\frac{\partial\mathcal{L}}{\partial(\partial_k\phi_j)}\right] \tag{6}$$

As $\delta\mathscr{A}$ must be zero for any temporal or spatial variations of $\delta\phi_j(\boldsymbol{r}, t)$, we deduce the Lagrange equations for the field:

$$\boxed{\frac{\mathrm{d}}{\mathrm{d}t}\frac{\partial\mathcal{L}}{\partial\dot{\phi}_j} = \frac{\partial\mathcal{L}}{\partial\phi_j} - \sum_k \partial_k\frac{\partial\mathcal{L}}{\partial(\partial_k\phi_j)} = 0} \tag{7}$$

[1]When $\boldsymbol{r}$ goes to infinity, we assume the Lagrangian density is zero or goes to zero sufficiently rapidly for the integral on the right-hand side of (3) to be convergent.

[2]Even when the function $\mathcal{L}$ does not directly depend on time, it does so indirectly when we replace in (2) the fields and their derivatives by their values for a given history of the field. The notation $\frac{\mathrm{d}}{\mathrm{d}t}\frac{\partial\mathcal{L}}{\partial\dot{\phi}_j}$ in (6) and (7) designates the derivative of this function with respect to t in the course of this history. This total derivative includes the contributions of all the partial derivatives with respect to all the variables of $\mathcal{L}$ appearing in the right-hand side of (2): the fields, their derivatives, and possibly time.

When dealing with continuous variables, and when the Lagrangian density depends on the spatial derivatives of the fields, we note the presence of the term in $\sum_k$; this term does not appear when the system is described by discrete variables. This equation is written in a more condensed form in the relativistic notations with $\mu = 0, 1, 2, 3$ labeling the space–time components:

$$\frac{\partial \mathcal{L}}{\partial \phi_j} = \partial_\mu \frac{\partial \mathcal{L}}{\partial(\partial_\mu \phi_j)} \tag{8}$$

where $\partial_0 = \mathrm{d}/\mathrm{d}(ct)$ and $\partial_k = \mathrm{d}/\mathrm{d}x^k$ with $x^k = x, y, z$; on the right-hand side the index μ is summed from 0 to 3, following Einstein's convention.

Comment:

Adding to $\mathcal{L}$ the total time derivative of any function of the fields does not change the equations of motion; it simply adds to the action a term that depends only on the initial and final states of the system, but not on the actual path leading from one to the other. One can also add any spatial derivative without changing the Lagrangian, after integration over $\mathrm{d}^3 r$ (provided the functions go to zero at infinity).

2. Symmetry transformations and current conservation

Consider now variations of the ϕ_j components having the form:

$$\delta\phi_j(\boldsymbol{r}, t) = \delta\varepsilon\, f_j\Big(\phi_l(\boldsymbol{r}, t), \dot{\phi}_l(\boldsymbol{r}, t), \partial_m \phi_l(\boldsymbol{r}, t); \boldsymbol{r}, t\Big) \tag{9a}$$

where f_j depends on all the variables inside the parentheses, for any values of l and m. It follows that:

$$\begin{aligned} \delta\dot{\phi}_j(\boldsymbol{r}, t) &= \delta\varepsilon\, \dot{f}_j\Big(\phi_l(\boldsymbol{r}, t), \dot{\phi}_l(\boldsymbol{r}, t), \partial_m \phi_l(\boldsymbol{r}, t); \boldsymbol{r}, t\Big) \\ \delta[\partial_k \phi_j(\boldsymbol{r}, t)] &= \delta\varepsilon\, \partial_k f_j\Big(\phi_l(\boldsymbol{r}, t), \dot{\phi}_l(\boldsymbol{r}, t), \partial_m \phi_l(\boldsymbol{r}, t); \boldsymbol{r}, t\Big) \end{aligned} \tag{9b}$$

The variation of the Lagrangian density is written:

$$\delta\mathcal{L} = \delta\varepsilon \sum_j \left[f_j \frac{\partial \mathcal{L}}{\partial \phi_j} + \dot{f}_j \frac{\partial \mathcal{L}}{\partial \dot{\phi}_j} + \sum_k (\partial_k f_j) \frac{\partial \mathcal{L}}{\partial(\partial_k \phi_j)} \right] \tag{9c}$$

As in relation (I-37) of chapter I, we assume that $\delta\mathcal{L}$ is proportional to a total time derivative[3] of a function Λ:

$$\delta\mathcal{L} = \delta\varepsilon\, \frac{\mathrm{d}}{\mathrm{d}t} \Lambda\Big(\phi_l(\boldsymbol{r}, t), \dot{\phi}_l(\boldsymbol{r}, t), \partial_m \phi_l(\boldsymbol{r}, t); \boldsymbol{r}, t\Big) \tag{10}$$

[3]When $\boldsymbol{r}$ goes to infinity, we assume that Λ goes to zero fast enough for the integral (1) yielding the Lagrangian L to remain finite. When Λ does not depend on $\dot{\phi}_j$, L and $L + \delta L$ are equivalent Lagrangians – *cf.* note 5 page 12.

The field transformations (9a) are then said to be symmetry transformations.

We now compute the time variation of the function $\sum_j f_j \, \partial\mathcal{L}/\partial\dot{\phi}_j$. Equation (7) for the time evolution of the field yields:

$$\begin{aligned}\frac{\mathrm{d}}{\mathrm{d}t}\Big[\sum_j f_j \frac{\partial\mathcal{L}}{\partial\dot{\phi}_j}\Big] &= \sum_j \Big[\dot{f}_j \frac{\partial\mathcal{L}}{\partial\dot{\phi}_j} + f_j \frac{\mathrm{d}}{\mathrm{d}t}\Big(\frac{\partial\mathcal{L}}{\partial\dot{\phi}_j}\Big)\Big] \\ &= \sum_j \Big[\dot{f}_j \frac{\partial\mathcal{L}}{\partial\dot{\phi}_j} + f_j \Big(\frac{\partial\mathcal{L}}{\partial\phi_j} - \sum_k \partial_k \frac{\partial\mathcal{L}}{\partial(\partial_k\phi_j)}\Big)\Big] \\ &= \frac{\delta\mathcal{L}}{\delta\varepsilon} - \sum_{j,k} \partial_k\Big(f_j \frac{\partial\mathcal{L}}{\partial(\partial_k\phi_j)}\Big) = \frac{\mathrm{d}}{\mathrm{d}t}\Lambda - \sum_{j,k} \partial_k\Big(f_j \frac{\partial\mathcal{L}}{\partial(\partial_k\phi_j)}\Big)\end{aligned} \tag{11}$$

or:

$$\frac{\mathrm{d}}{\mathrm{d}t}\Big[\sum_j f_j \frac{\partial\mathcal{L}}{\partial\dot{\phi}_j} - \Lambda\Big] + \sum_{j,k} \partial_k\Big(f_j \frac{\partial\mathcal{L}}{\partial(\partial_k\phi_j)}\Big) = 0 \tag{12}$$

This result has the form of a local conservation equation. Defining a local density $\rho(\boldsymbol{r},t)$ and the k component of its associated current $\boldsymbol{J}(\boldsymbol{r},t)$:

$$\boxed{\rho(\boldsymbol{r},t) = \sum_j f_j \frac{\partial\mathcal{L}}{\partial\dot{\phi}_j} - \Lambda \quad ; \quad \big(\boldsymbol{J}(\boldsymbol{r},t)\big)_k = \sum_j f_j \frac{\partial\mathcal{L}}{\partial(\partial_k\phi_j)}} \tag{13}$$

we get[4]:

$$\frac{\mathrm{d}}{\mathrm{d}t}\rho(\boldsymbol{r},t) + \boldsymbol{\nabla}\cdot\boldsymbol{J}(\boldsymbol{r},t) = 0 \tag{14}$$

This equation, similar to the one expressing charge conservation, means that the local increase of the density (of a physical quantity to be determined) in a small volume is equal to the incoming flux of a current $\boldsymbol{J}$ across the surface of that small volume. In the discrete case studied in chapter I, Noether's theorem yielded only one constant of motion; in our present case with continuous field variables, there are an infinite number of them as there are an infinite number of points $\boldsymbol{r}$ in space. This is because assuming the invariance of $\mathcal{L}$ at each point in space is much stronger than assuming the invariance of its spatial integral L.

3. Generalization, relativistic notation

Time plays a prominent role in relation (10). It can be modified to become more symmetric with respect to time and space, and we set:

$$\delta\mathcal{L} = \delta\varepsilon\Big[\frac{\mathrm{d}}{\mathrm{d}t}\Lambda + \sum_k \mathrm{d}_k\Lambda_k\Big] = \delta\varepsilon \; \partial_\mu\Lambda^\mu \tag{15}$$

[4]This equation is generally written with a partial time derivative $\partial\rho/\partial t$ (instead of a total derivative) when one adopts a point of view where ρ is a function of only $\boldsymbol{r}$ and t, ignoring its possible dependence on the fields ϕ_j.

where the index k is summed from 1 to 3, and μ from 0 to 3. This expression involves four functions Λ^μ similar[5] to Λ, and we set $\Lambda^0 = c\Lambda$ and $\Lambda^{\mu=k} = \Lambda_k$; the ∂_μ have been defined in relation (8). We can follow a reasoning almost the same as in § 2, but we end up with a modified expression for the current:

$$\boxed{(\boldsymbol{J}(\boldsymbol{r},t))_k = \sum_j f_j \frac{\partial \mathcal{L}}{\partial(\partial_k \phi_j)} - \Lambda_k} \qquad \text{with: } k = 1,2,3 \tag{16}$$

This is a generalization of Noether's theorem to the case where the variation of the Lagrangian density $\mathcal{L}$ also contains spatial derivatives. In relativistic notation, we write J^μ the components of the four-vector $(c\rho, \boldsymbol{J})$, and relation (14) simply becomes:

$$\partial_\mu J^\mu = 0 \tag{17}$$

Noether's theorem has numerous applications in field theory, including in quantum theory where the conserved quantities and their associated currents play a central role; the interested reader may consult references [24, 25] for example. The applications to the Schrödinger, Klein–Gordon, and Dirac equations will be discussed in complement C_{VI}. We shall only mention here a general and very simple example, the local conservation of energy and momentum.

4. Local conservation of energy

We assume that the Lagrangian density is not explicitly time-dependent, and introduce a variation of the field $\phi(\boldsymbol{r},t)$ that is a shift in time by a quantity $\delta\varepsilon$. For the sake of simplicity, we assume that the field is a scalar so that the index j is no longer needed. We have $\delta\phi = \delta\varepsilon\,\dot{\phi}$, and hence $f = \dot{\phi}$. As a result, $\mathcal{L}$ will vary since it depends on the field and its derivatives; as we assume that $\mathcal{L}$ does not directly depend on time, we have $\delta\mathcal{L} = \delta\varepsilon\ \text{d}\mathcal{L}/\text{d}t$. This yields $\Lambda = \mathcal{L}$ (and $\Lambda^{1,2,3} = 0$), and relations (13) lead to a local energy density ρ_E and an energy current $\boldsymbol{J}_E$ written as:

$$\rho_E(\boldsymbol{r},t) = \dot{\phi}\frac{\partial\mathcal{L}}{\partial\dot{\phi}} - \mathcal{L} \quad ; \quad (\boldsymbol{J}_E(\boldsymbol{r},t))_k = \dot{\phi}\frac{\partial\mathcal{L}}{\partial(\partial_k\phi)} \tag{18}$$

Integrating over space $\rho_E(\boldsymbol{r},t)$ yields the definition of the Hamiltonian; ρ can thus be identified to the Hamiltonian density.

If, for example, the Lagrangian density is written:

$$\mathcal{L} = \frac{1}{2}\left[\dot{\phi}^2 - c^2(\boldsymbol{\nabla}\phi)^2 - \omega^2\phi^2\right] = \frac{1}{2}\left[c^2(\partial^\mu\phi)(\partial_\mu\phi) - \omega^2\phi^2\right] \tag{19}$$

[5] As before, we assume that, when $\boldsymbol{r}$ goes to infinity, the Λ^μ go to zero fast enough for the spatial integral yielding L to remain finite.

where c and $\omega = mc^2/\hbar$ are constants, we get:

$$\rho_E(\boldsymbol{r},t) = \frac{1}{2}\Big[\dot{\phi}^2 + \omega^2\phi^2 + c^2(\boldsymbol{\nabla}\phi)^2\Big] \quad ; \quad \boldsymbol{J}_E(\boldsymbol{r},t) = -c^2\dot{\phi}\,\boldsymbol{\nabla}\phi \tag{20}$$

The energy density ρ_E is the sum of the kinetic energy density of the field (term in $\dot{\phi}^2$) and its potential energy density (terms in ϕ^2 and $(\boldsymbol{\nabla}\phi)^2$); the energy current includes temporal and spatial derivatives of ϕ.

In electromagnetism, the field has 6 components, those of the electric and magnetic fields $\boldsymbol{E}$ and $\boldsymbol{B}$, often arranged in a second-order antisymmetric tensor. The energy density at each point is proportional to $(\boldsymbol{E}^2 + c^2\boldsymbol{B}^2)$ and the current $\boldsymbol{J}_E$ to the Poynting vector, proportional to $\boldsymbol{E}\times\boldsymbol{B}$.

Exercise: We assume that the Lagrangian density does not depend directly on the position but only through the values of the field and its derivatives. Show that this yields 3 local conservation laws, with the following densities and currents:

$$\rho_{x_i}(\boldsymbol{r},t) = \partial_{x_i}\phi\,\frac{\partial\mathcal{L}}{\partial\dot{\phi}} \quad ; \quad \big(\boldsymbol{J}_{x_i}(\boldsymbol{r},t)\big)_k = \partial_{x_i}\phi\,\frac{\partial\mathcal{L}}{\partial(\partial_k\phi)} - \delta_{k,x_i}\,\mathcal{L} \tag{21}$$

where $x_i = x, y, z$. These relations express a local conservation of the three components of the field's momentum. The various values of i and k yield the 9 components of the current, which form a 3×3 tensor, the so-called "constraint tensor".

Chapter II

Some ideas about group theory

We mentioned in the previous chapter the importance of the group theory concept in symmetry problems. We present in this chapter some ideas about group theory without trying to give a general course on the subject. We underline the important concepts, introducing the vocabulary useful for reading books on symmetries in physics. The reader who wishes more details is referred to the bibliography.

Introduction to Continuous Symmetries: From Space–Time to Quantum Mechanics, First Edition. Franck Laloë.

A. General properties of groups

A-1. Definitions; examples

A group G is a set of elements, noted g_1, g_2, g_3, ..., g_i, g_j, ..., equipped with an internal composition law (multiplication):

$$g_i, g_j \Longrightarrow \text{product, noted } g_i\, g_j \tag{II-1}$$

This law obeys the following rules:

- it must be associative:

$$g_i(g_j\, g_k) = (g_i\, g_j)\, g_k \qquad\qquad \forall\, i, j, k \tag{II-2}$$

- it must have an identity element, noted e, such that:

$$eg = ge = g \qquad\qquad \forall\, g \tag{II-3}$$

- any element g must have an inverse, noted g^{-1}, with which it commutes:

$$gg^{-1} = g^{-1}g = e \tag{II-4}$$

(it follows that the "right" and "left" inverses are identical).

The multiplication law is not necessarily commutative, and we can have:

$$g_i\, g_j \neq g_j\, g_i \tag{II-5}$$

If the law is commutative, meaning if $g_i\, g_j = g_j\, g_i$ for any i and j, we have a *commutative*, or *abelian* group.

A finite group contains a finite number N of elements; N is called the order of the group. An infinite group contains an infinite number of elements (which can be countable or not).

The inverse of $g_1 g_2$ is $g_2^{-1} g_1^{-1}$:

$$(g_1\, g_2)^{-1} = g_2^{-1}\, g_1^{-1} \tag{II-6}$$

(easy to prove).

The group element :

$$g_i\, g_j\, g_i^{-1}\, g_j^{-1} \tag{II-7}$$

is called the "*commutator*" of g_i and g_j. If g_i and g_j commute ($g_i\, g_j = g_j\, g_i$), this commutator is obviously the identity element e, and conversely.

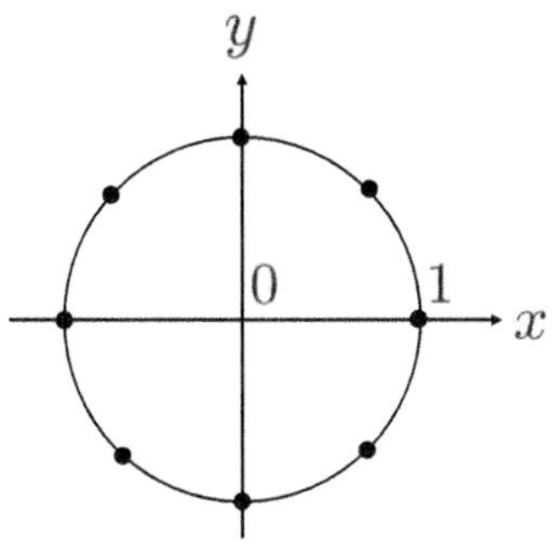

Figure 1: The set of complex numbers $e^{2i\pi p/8}$ *(*$p = 0, 1, \ldots, 7$*) form the cyclic group* Z_8*; as the multiplication between two complex numbers is commutative, this is an abelian group.*

> ***Comment:***
>
> This definition of the commutator between elements of a group (where an addition–subtraction rule has not necessarily been defined) should not be confused with the completely different definition of the commutator between operators.

A finite group obeys the often-useful "rearrangement lemma": if we multiply (on the left or on the right) all the elements of G by an arbitrary element g_0 of G, we get once and only once all the elements of G, but in a different order (if $g_0 \neq e$). This is because the product $g_0 g$ is equal to a given element g' of the group for a single value of $g = g_0^{-1} g'$.

Examples of groups

- The set of the two numbers $+1$ and -1, equipped with the usual multiplication law, constitute a particularly simple group (commutative and of order 2).
- Similarly, the set of the complex numbers $e^{2i\pi p/N}$ (N is fixed, $p = 0, 1, \ldots, N-1$) form an abelian group, of order N, represented by points regularly spaced on a circle in the complex plane (figure 1). This is the so-called Z_n (or sometimes C_N) *cyclic group*.
- The group of all the positive or negative integers (equipped with the usual addition law) is noted Z_∞. The set of positive integers does not form a group, since it does not contain the inverse (here the opposite) of each of its elements.

Below are some examples of non-abelian groups.

- The permutation group[1] S_N of N objects, of order $N!$ This is a non-abelian group if $N \geq 3$. Take for example S_3, and note $P_{2\,3\,1}$ the permutation that puts the object labeled 2 in the place of object 1, object 3 in the place of object 2, and object 1 in that of object 3. Another common notation is:

$$\begin{pmatrix} 2 & 3 & 1 \\ 1 & 2 & 3 \end{pmatrix} \equiv P_{231}$$

It's easy to show that the action of $P_{2\,3\,1}$ followed by $P_{2\,1\,3}$ on the objects 123, yields 321, whereas if the same permutations are applied on 123 but starting with $P_{2\,1\,3}$, we get 132. It thus follows that:

$$P_{2\,1\,3}\, P_{2\,3\,1} \neq P_{2\,3\,1}\, P_{2\,1\,3}$$

which shows that S_3 is non-commutative.

- The group of geometric operations that preserve a given figure. As the product of two elements of the group is defined by the successive applications of the two operations, these operations do indeed form a group.

Take as an example the set of transformations on the plane that preserve distance and an equilateral triangle (figure 2). It includes:

  - the rotations around the triangle's center O , by the angles 0, $2\pi/3$, and $4\pi/3$; these operations are noted I, C_3, ${C_3}^2$;
  - the symmetries with respect to the three straight lines going through O and the summits of the triangle, noted V, V', and V''.

It's easy to show that $VC_3 = V'$ and $C_3V = V''$: the transformations form a non-commutative group, called C_{3v}.

This is a first example of the strong relation between physical invariance and the concept of group.

- The abelian group of the translations $T_{(n)}$ in an n-dimensional space, or of the rotations $R_{(n)}$ around a fixed point O in that space. One can also

[1] Consider N objects $a_1, a_2, \ldots, a_N$ arranged in a predetermined order. Following standard usage in physics, we call permutation $P_{ijk\ldots}$ of these N objects the operation where object i is moved to the first place, object j to the second, etc. This operation changes the order of the objects. Another possible definition would be to consider that the permutation concerns the nature of the objects, transforming object a_1 into a_i, etc. This would lead to a different multiplication table.

In mathematics, such a bijection of an ensemble onto itself is generally called a "substitution", keeping the word permutation for a given order of the objects.

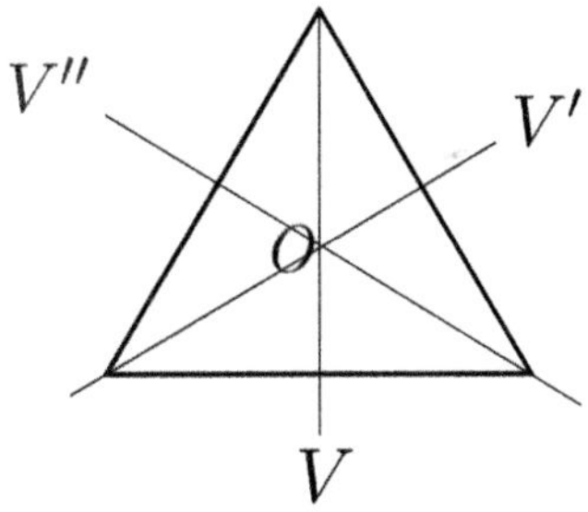

Figure 2: The C_{3v} group is the set of the transformations in the plane that preserve the distances as well as an equilateral triangle. It includes the rotations around the triangle's center O , by the angles O, $2\pi/3$, and $4\pi/3$, as well as the symmetries with respect to the three straight lines going through O and the summits of the triangle, noted V, V', and V''.

introduce the group of displacements whose elements are the product of an element of $R_{(n)}$ by an element of $T_{(n)}$.

They are examples of infinite continuous groups, which we will return to in the next chapter.

- Groups of matrices, with the usual matrix multiplication law. These are important groups in physics, but have been labeled somewhat differently throughout the literature:

  - GL(n, C): (linear) group of the $n \times n$ complex matrices that are regular (with a non-zero determinant, hence invertible).
  - GL(n, R): the same as above, but including only real matrices.
  - SL(n, C) and SL(n, R): groups of matrices with the same properties as above, but with the additional constraint of having a determinant equal to 1 (unimodular matrices). The S in the group names stands for "Special" as opposed to the previous G for "General".
  - U(n): $n \times n$ unitary matrices (preserve the norm and the scalar product)[2].
  - SU(n): $n \times n$ unitary matrices with a determinant equal to 1. These groups play an important role in particle physics.
  - O(n): $n \times n$ orthogonal matrices (unitary and real).

[2]Hermitian matrices do not form a group since, even if M_1 and M_2 are both Hermitian, their product $M_1 M_2$ is not necessarily Hermitian (unless $M_1 M_2 = M_2 M_1$).

- SO(n): $n \times n$ orthogonal matrices with a determinant equal to 1. For example, SO(3) is the set of rotation matrices about a point in the usual 3-dimensional space (as they are unimodular, the symmetries with respect to a point or a plane have been eliminated). SO(3) and $R_{(3)}$ (the rotation group in the usual 3-dimensional space) are in practice two names for the same group.

A-2. Multiplication table – Isomorphic groups

The multiplication law in a group G can be defined by a double entry table (multiplication table) where the product $g_i\, g_j$ is placed at the intersection of the ith row and the jth column:

	g_1	g_2 ·····	g_j ·······	g_N
g_1	$(g_1)^2$	$g_1\, g_2$ ··	$g_1\, g_j$ ··	$g_1\, g_N$
g_2	$g_2\, g_1$	$(g_2)^2$ ··	$g_2\, g_j$ ··	$g_2\, g_N$
.	.	.	.	.
.	.	.	.	.
.	.	.	.	.
g_i	$g_i\, g_1$	$g_i g_2$ ··	$g_i\, g_j$ ··	$g_i\, g_N$
.	.	.	.	.
.	.	.	.	.
.	.	.	.	.
g_N	$g_N\, g_1$	$g_N g_2$ ··	$g_N\, g_j$ ··	$(g_N)^2$

This table specifies the *structure* of the group. It obeys specific rules: for example, each element g appears once and only once in each row and each column. If G is commutative, the table is symmetric (with respect to the principal diagonal).

Exercise: Write the multiplication table for the C_{3v} group.

A subset $g_\alpha, g_\beta, \ldots, g_\gamma$ of the group's elements may exist such that all the elements of G may be obtained by taking the product of elements of that subset. The $g_\alpha, g_\beta, \ldots, g_\gamma$ are then called the group "generators" .

Examples:

- Z_N: a single generator, $e^{2i\pi/N}$
- S_N: the subset of transpositions
- C_{3v}: C_3 and V

A group G may be defined by a certain number of relations between its generators, without having to specify the entire multiplication table. For example, for C_{3v}:

$$V^2 = I \ ; \ (C_3)^3 = I \ ; \quad C_3 V C_3 = V \ ; \quad C_3 \text{ and } V \neq I \quad ; \ C_3 \neq V$$

Homomorphism: $G \Longrightarrow G'$

There is a group homomorphism from G to G' if, to each element $g \in G$, there corresponds an element:

$$g' = f(g) \in G'$$

that obeys the groups' internal composition law:

$$g'_1 \, g'_2 = f(g_1) \; f(g_2) = f(g_1 \, g_2) \tag{II-8}$$

(the product of the transformed g'_1 and g'_2 is equal to the transformed of the product $g_1 \, g_2$).

Isomorphisme: $G \Longleftrightarrow G'$

If there is a one-to-one correspondence between G and G' (bijection), i.e. when:

$$g_1 \neq g_2 \Longrightarrow g'_1 \neq g'_2$$

and each element g' of G' corresponds to one element g of G, the homomorphism becomes an isomorphism.

The two groups G and G' have the same multiplication table. They only differ by the labeling of their elements. They are often considered as two different realizations of the same *abstract group*.

Examples:

(i) SO(3) and $R_{(3)}$.

(ii) The group of transformations in the 3-dimensional space (figure 3) that includes:

- 3 rotations, around an Oz axis, through the angles 0, $2\pi/3$, $4\pi/3$
- 3 symmetries with respect to three planes passing through Oz and at 120 degrees from each other.

This group is isomorphic to C_{3_v}.

(iii) the two previous groups are isomorphic to S_3 (permutations of the summits of an equilateral triangle).

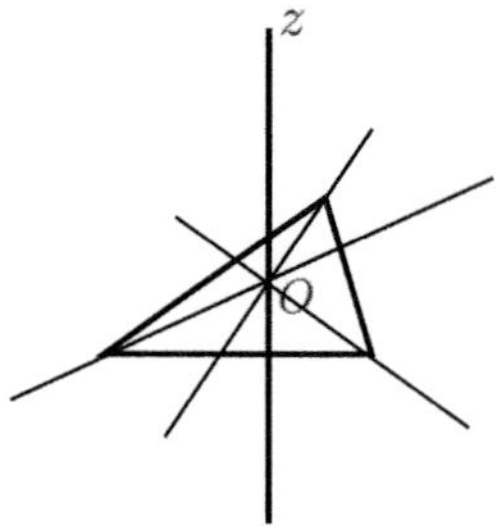

Figure 3: In three dimensions, a group includes three rotations around the Oz axis, through the angles 0, $2\pi/3$, $4\pi/3$, as well three symmetries with respect to three planes passing through Oz and at 120 degrees from each other. This group is isomorphic to the C_{3_v} group of figure 2.

A way to check the isomorphism between two groups is to ensure they have the same number of generators with the same relations in their multiplication table.

A-3. Subgroups – Center

Let H be a subset of G. If H also forms a group (under the same operation as for G), H is a subgroup of G. Any subset K of the group G does not necessarily form a subgroup: the subset must be closed under the multiplication law (meaning the product of two elements of K belongs to K), and K must also contain the inverse of any of its elements.

Examples:

(i) I, C_3, ${C_3}^2$ all together form a subgroup of C_{3_v} (isomorphic to Z_3), but not just I and C_3.

(ii) $R_{(3)}$ is a subgroup of the set of displacements in 3-dimensional space.

(iii) $R_{(2)}$ is a subgroup of $R_{(3)}$.

Cayley's theorem:

Any finite group of order N is isomorphic to a subgroup of S_N (group of permutations of N objects).

This property may be understood by looking at the multiplication table of the group: any element of the group is associated to a row that includes a permutation of the N elements of the group (rearrangement lemma). It is easy to show the isomorphism applying the multiplication law in the initial group and in that of the permutations.

Invariant subgroup:

A subgroup H of G is invariant if:

$$g\,H\,g^{-1} \subset H \qquad \forall\, g \in G \tag{II-9a}$$

which means that, for any g belonging to G and h to H, the element:

$$h' = g\,h\,g^{-1} \tag{II-9b}$$

also belongs[3] to H.

This is equivalent to saying that, for any $g \in G$ and $h \in H$, there exists $h' \in H$ such that:

$$g\,h = h'\,g \tag{II-10}$$

(or the opposite, interchanging the roles of h and h').

The set of the elements c of G that commute with all the elements of G obviously form an invariant abelian subgroup, called the *center* C of the group G. It may happen that $C = G$ (if G is abelian) or, on the contrary, that C consists only of the identity element (example: C_{3_v}). A group that contains no invariant subgroup other than the trivial subgroup (identity element e) is called a "simple group".

A-4. Conjugacy classes

Consider two elements g and g' of G. g is said to be conjugate of g' (through the element $f \in G$) if:

$$g = f^{-1}\,g'\,f \tag{II-11}$$

It follows that:

- g is conjugate of itself ($f = e$)
- g' is conjugate of g (through f^{-1}). g and g' are thus conjugate of each other.
- If g_1 is conjugate of g_2 (through f) and g_2 of g_3 (through j), g_1 is conjugate of g_3, since:

$$g_1 = f^{-1}\,g_2\,f = \left(f^{-1} j^{-1}\right) g_3 \left(j f\right) \tag{II-12}$$

[3]If G is a finite group, equality (II-9a) is equivalent to $g\,H\,g^{-1} = H$ (the demonstration is easy and uses the fact that, if $h_1 \neq h_2$, then $g\,h_1\,g^{-1} \neq g\,h_2\,g^{-1}$).

Conjugacy is an equivalence relation between elements of G. The set of all the elements that are conjugate of an element g form the class of that element. All the elements of a class are thus conjugate two by two.

It is easy to show that two classes are either disjoint (no common elements) or identical. In other words, the classes form a *partition* of G, which can be viewed as the reunion of disjoint classes.

- An invariant subgroup may be viewed as the reunion of complete equivalence classes.

Comments:

(i) The identity element is always the only element in its class. This class is the only one to be a subgroup of G.

(ii) In a commutative group, the elements themselves are the classes (each containing a single element).

(iii) A "rearrangement lemma" also exists for the classes C of a group: when an element g spans all the values of its class C, the set of $g_0^{-1} g g_0$ (where g_0 is an arbitrary element of G), also yields all the elements of C, but in a different order (the $g_0^{-1} g g_0$ obviously belong to C and are all different).

(iv) Equality (II-10) shows that h' and h are conjugate. Therefore, an invariant subgroup can be defined as a subgroup containing entire classes.

Exercises:

(i) Show that the inverse of the elements of a class forms a class, not necessarily identical to the former.

(ii) Let g be an element of a class of a finite group G of order N, and let m be the number of elements of G that commute with g. Show that g is the conjugate of each element of its class (in particular of itself) exactly m times (meaning through m different elements of G). Infer from this result that $N = mn$, where n is the number of elements in the class of g: the order of a class is a divisor of that of the group.

(iii) Let C and C' be two classes of G. The product of these two classes is the set of the elements of G that can be written gg' with $g \in C$, $g' \in C'$. Show that this product is the reunion of a certain number of classes of G.

Example: geometric transformations

Imagine the elements of G are invertible geometric transformations A, B, ... acting on vectors $\boldsymbol{x}, \boldsymbol{y}$, ... As an example, $G = C_{3_v}$ or $G = R_{(3)}$. The concept of class then has a very simple geometrical interpretation. We note $\boldsymbol{y}$ the transform of $\boldsymbol{x}$ through operation A:

$$\boldsymbol{y} = A\boldsymbol{x} \tag{II-13}$$

Choosing an arbitrary transformation C of the group, we set:

$$\begin{cases} \boldsymbol{x}' = C\boldsymbol{x} \\ \boldsymbol{y}' = C\boldsymbol{y} \end{cases} \tag{II-14}$$

The operation that transforms $\boldsymbol{x}'$ into $\boldsymbol{y}'$ is the conjugate of A through the element C^{-1}:

$$\boldsymbol{y}' = \tilde{A}\,\boldsymbol{x}' \tag{II-15}$$

where:

$$\tilde{A} = CAC^{-1} \tag{II-16}$$

For linear transformations, formulas (II-14) correspond to a change of axes: the components of $\boldsymbol{x}$ and $\boldsymbol{x}'$ (or those of $\boldsymbol{y}$ and $\boldsymbol{y}'$) correspond to the same vector but on different axes (the components of the new basis vectors on the old ones appear in the columns of the matrix C). In this point of view, A and its conjugate correspond to the same geometric operation, but with different axes.

For example, if $G = C_{3_v}$, we have:

$$C_3^{-1}\,V\,C_3 = V'' \tag{II-17}$$

The conjugate of V through C_3 is the symmetry with respect to an axis whose direction has undergone the rotation C_3^{-1} with respect to its initial direction. In a similar way:

$$\left[C_3{}^2\right]^{-1}\,V\,[C_3]^2 = V' \tag{II-18}$$

Likewise, if $G = R_{(3)}$, the set of all the elements of a class includes the rotations through a given angle, around an arbitrary axis.

A-5. Tensor product (or direct product) group

Consider two different groups G and G', with elements g_i and g'_k. By definition, the elements of the tensor product $G \otimes G'$ are the pairs (g_i, g'_k). It's easy to show that the composition law:

$$(g_i, g'_k)\,(g_j, g'_e) = (g_i g_j, g'_k g'_e) \tag{II-19}$$

yields a group structure to the product $G \otimes G'$.

Comment:

If H is an invariant subgroup of G, we can define the concept of quotient group G/H, similar in certain case to the opposite of the product's concept (see complement $\mathrm{A_{II}}$).

Examples:

(i) If G and G' are the groups of $N \times N$ and $N' \times N'$ matrices respectively, with elements M_i and M'_k, their direct product yields the set of tensor product matrices:

$$\mathscr{M}_{ik} = M_i \otimes M'_k \tag{II-20a}$$

whose elements are, by definition:

$$\langle n, p|\mathscr{M}_{ik}|n', p'\rangle = \langle n|M_i|n'\rangle \times \langle p|M'_k|p'\rangle \tag{II-20b}$$

These are simply products of matrix elements (we are using here the Dirac notations, standard in quantum mechanics).

(ii) Consider $G = T_{(3)}$ and $G' = R_{(3)}$. In the group of displacements (Euclidean group $\mathscr{E}_{(3)}$), each transformation can be obtained by the product of a rotation and a translation. This could lead us to think that $\mathscr{E}_{(3)}$ is the direct product of $T_{(3)}$ and $R_{(3)}$. But this is wrong: the two groups act on the same objects and combine their effects in a non-independent way, as opposed to what should happen with the tensor product rule (II-19).

If g_1 is a rotation $\mathscr{R}_1$ and g'_1 a translation $\mathcal{T}(\boldsymbol{b}_1)$ defined as $\boldsymbol{r} \Rightarrow \boldsymbol{r}+\boldsymbol{b}_1$, their product $h_1 = \mathcal{T}(\boldsymbol{b}_1)\mathscr{R}_1$ is a displacement $\boldsymbol{r} \Rightarrow \mathscr{R}_1\boldsymbol{r}+\boldsymbol{b}_1$. Another displacement h_2 is defined in a similar way with a rotation $\mathscr{R}_2$ and a translation of vector $\boldsymbol{b}_2$. The product h_2h_1 of the two displacements then acts according to $\boldsymbol{r} \Rightarrow \mathscr{R}_2\mathscr{R}_1\boldsymbol{r} + \mathscr{R}_2\boldsymbol{b}_1 + \boldsymbol{b}_2$; but this contradicts relation (II-19), which would lead to $\mathscr{R}_2\mathscr{R}_1\boldsymbol{r} + \boldsymbol{b}_1 + \boldsymbol{b}_2$.

The conjugate of a translation $\mathcal{T}(\boldsymbol{b})$ by a rotation $\mathscr{R}$ is $\mathscr{R}\,\mathcal{T}(\boldsymbol{b})\,\mathscr{R}^{-1}$, corresponding to the sequence $\boldsymbol{r} \Rightarrow \mathscr{R}^{-1}\boldsymbol{r} \Rightarrow \mathscr{R}^{-1}\boldsymbol{r} + \boldsymbol{b} \Rightarrow \boldsymbol{r} + \mathscr{R}\boldsymbol{b}$; it is therefore another translation $\mathcal{T}(\mathscr{R}\boldsymbol{b})$. This shows that the translations form an invariant subgroup of the displacement group $\mathscr{E}_{(3)}$. The displacement group is then called the "semidirect product " of $T_{(3)}$ by $R_{(3)}$, and is noted $\mathscr{E}_{(3)} = R_{(3)} \ltimes T_{(3)}$ (with a vertical bar added to the product sign).

B. Linear representations of a group

B-1. Definition and properties

Consider a group G with elements noted g, and assume we know a set of square $q \times q$ matrices M that are *regular* ($\det M \neq 0$, M^{-1} exists). This matrix set will be noted ${}^{(q)}R$:

$$M \in {}^{(q)}R \tag{II-21}$$

To define a linear representation of G by the set of M we need to have a correspondence between each element g and a matrix M, noted $M(g)$, such that:

$$M(gg') = M(g) \times M(g') \tag{II-22}$$

(the matrix associated with the product must be the product of the matrices). In other words, as the internal composition law is obeyed, a linear representation can be viewed as an homomorphism from G to ${}^{(q)}R$.

The number q is called the "dimension" (or sometimes the "degree") of the representation. If all the matrices M are unitary, we are then dealing with a "unitary representation".

Comments:

(i) Choosing a regular square $q \times q$ matrix amounts to choosing a linear invertible operator $\mathcal{M}$, acting in a q-dimensional vector space $\mathscr{E}_q$. A representation can thus be viewed as an homomorphism from G to a set of linear invertible operators acting in $\mathscr{E}_q$.

(ii) If G itself is defined as a set of invertible matrices, $\mathrm{SO}(n)$ for example, a representation of G is readily found with those matrices. This does not mean it is the only possible representation. For example, one could associate with each element of the group the (same) $q \times q$ identity matrix, which leads to a so-called *trivial* representation.

(iii) The matrices M are supposed to be invertible. This excludes correspondences that associate the elements g with the same matrix:

$$N(g) = \begin{pmatrix} 0 & 0 \\ 1 & 1 \end{pmatrix} \tag{II-23}$$

even though this correspondence obeys the internal composition law [it's easy to show that N is a (non-orthogonal) projector, meaning $N = N^2 = N^3 = \ldots$ (idempotent matrix)].

(iv) The same word "representation" will mean either the correspondence $G \Longrightarrow {}^{(q)}R$, or the set ${}^{(q)}R$ of matrices.

B-1-a. Identity element and inverse

If e is the identity element of G, and g^{-1} the inverse of g, equation (II-22) shows that:

$$M(ge) = M(g) = M(g)\, M(e) \tag{II-24}$$

Since the M are invertible, we can write:

$$M(e) = [M(g)]^{-1} M(g) = (\mathbb{1}) \qquad \text{(identity matrix)} \tag{II-25a}$$

In a similar way:

$$M(gg^{-1}) = M(e) = M(g)\, M(g^{-1}) \tag{II-25b}$$

which, after a multiplication by $[M(g)]^{-1}$ on the left, leads to:

$$M(g^{-1}) = [M(g)]^{-1} \tag{II-26}$$

The matrix associated with e is thus always the identity matrix $P \times P$; two inverse matrices are associated with two inverse elements.

B-1-b. Faithful representation

If, when g_1 and g_2 are different, $M(g_1)$ and $M(g_2)$ are also always different, the representation is said to be "*faithful*":

$$g_1 \neq g_2 \Longrightarrow M(g_1) \neq M(g_2) \tag{II-27}$$

The correspondence $G \Longrightarrow {}^{(q)}R$ then becomes a bijection (isomorphism).

B-1-c. Projective representations

As in quantum mechanics, the multiplication of the state vector $|\psi\rangle$ by a global phase factor does not change the physical state. One often introduces the concept of "*projective representation*" where (II-22) is replaced by:

$$M(gg') = \mathrm{e}^{i\theta(g,g')}\, M(g)\, M(g') \tag{II-28}$$

with $\theta(g, g')$ being a real function of g and g'. This last equality imposes less constraints than (II-22); projective representations are *not* in general representations in the strict sense of the term.

B-1-d. Examples

(i) One can say that $R_{(3)}$ is represented by SO(3). Many other representations of $R_{(3)}$ will be mentioned later (chapter VII).

(ii) Group of the translations $T_{(3)}$ in 3-dimensional space. These translations depend on 3 parameters, the components b_x, b_y, b_z of the translation vector $\boldsymbol{b}$. The corresponding transformations of the coordinates are written:

$$x' = x + b_x \qquad y' = y + b_y \qquad z' = z + b_z \tag{II-29}$$

Unlike the preceding example, the present transformation is not linear. Nevertheless, we can write:

$$\begin{pmatrix} x' \\ y' \\ z' \\ 1 \end{pmatrix} = M(\boldsymbol{b}) \begin{pmatrix} x \\ y \\ z \\ 1 \end{pmatrix} \tag{II-30}$$

with :

$$M(\boldsymbol{b}) = \begin{pmatrix} 1 & 0 & 0 & b_x \\ 0 & 1 & 0 & b_y \\ 0 & 0 & 1 & b_z \\ 0 & 0 & 0 & 1 \end{pmatrix} \tag{II-31}$$

It's easy to show that:

$$M(\boldsymbol{b}_1 + \boldsymbol{b}_2) = M(\boldsymbol{b}_1)\ M(\boldsymbol{b}_2) \tag{II-32}$$

which means that the M are a 4-dimensional representation of $T_{(3)}$.

(iii) As an example, the S_N group may be represented as follows:

$$\begin{aligned} \text{even permutation} &\Longrightarrow \text{matrix } (\mathbb{1}) \\ \text{odd permutation} &\Longrightarrow \text{matrix } (-\mathbb{1}) \end{aligned} \tag{II-33}$$

This is a 1-dimensional representation, the simplest one after the trivial representation. It plays a fundamental role in quantum mechanics for a system of N identical fermions (antisymmetrization postulate), whereas it is the trivial representation that plays that role for identical bosons systems (symmetrization postulate).

There exist representations of S_N with higher dimensions that are non-commutative $[M(g)\ M(g') \neq M(g')\ M(g)]$. For example, consider N objects $a_1, a_2, \ldots, a_N$, arranged in a determined order, that a permutation P_α changes into $a'_1, a'_2, \ldots, a'_N$. We associate with this permutation the square $N \times N$ matrix $M(P_\alpha)$ defined as:

$$\begin{pmatrix} a'_1 \\ a'_2 \\ \vdots \\ a'_N \end{pmatrix} = M(P_\alpha) \begin{pmatrix} a_1 \\ a_2 \\ \vdots \\ a_N \end{pmatrix} \tag{II-34}$$

Matrix $M(P_\alpha)$ has a single non-zero element on each of its rows or columns: the element $(M)_{ij}$ of row i and column j equals 1 if the permutation replaces object a_i by object a_j. For example, in S_3:

$$M(P_{231}) = \begin{pmatrix} 0 & 1 & 0 \\ 0 & 0 & 1 \\ 1 & 0 & 0 \end{pmatrix} \tag{II-35}$$

It is easy to show using matrix products that the product of two M matrices is the matrix associated with the product of the corresponding permutations, meaning that the matrices are a representation of S_N.

For systems of identical particles, rather than limit the symmetrization postulate to 1-dimensional representations of S_N, it has been suggested that quantum mechanics requires introducing representations with a higher dimension (parastatistics)[4] for certain types of identical particles. However, the existence of elementary particles demanding such a generalization of the postulates has not been proven.

(iv) It is clear that having a representation of an arbitrary group, we also have a representation of any subgroup. We mentioned above that any finite group of order N is isomorphic to a subgroup of S_N. Consequently, using representation (II-34), we can obtain a representation of any finite group, the so-called "*regular representation*". To each element g of the group (in which a predetermined order $g_1, g_2, \ldots, g_N$ has been defined), we associate the square $N \times N$ matrix $M(g)$ such that:

$$\begin{pmatrix} gg_1 \\ gg_2 \\ gg_3 \\ \vdots \\ gg_N \end{pmatrix} = M(g) \begin{pmatrix} g_1 \\ g_2 \\ g_3 \\ \vdots \\ g_N \end{pmatrix} \tag{II-36}$$

Multiplying the relation on the left by $(M(g'))$ shows that the matrix product $(M(g))(M(g'))$ is associated in the group with the product element gg'; we have thus constructed a representation of the group.

Exercise: Construct the regular representation of C_{3v}.

B-2. Equivalent representations

Let $M_1(g)$ and $M_2(g)$ be two representations of the group G, with the same dimension n. They are said to be *equivalent* if there exists an invertible matrix S such that:

$$M_2(g) = S^{-1}\, M_1(g)\, S \qquad \forall\, g \tag{II-37}$$

[4]In a space limited to two dimensions, one can define quasiparticles called "anyons" that obey such parastatistics.

Comments:

(i) $M_1(g)$ and $M_2(g)$ can be considered to be associated, in two different bases of a vector space, with the same operator $\mathcal{M}(g)$. S is then the change-of-basis matrix. In terms of operators, two equivalent representations are actually the same representation.

(ii) It can be shown (*cf.* [15], p. 74) that any matrix representation of a finite group is equivalent to a representation constructed with unitary matrices.

B-3. Characters

The *characters* of a representation are the set of numbers:

$$\chi(g) = \mathrm{Tr}\{M(g)\} \tag{II-38}$$

Comments:

(i) For a given representation, the characters of all the elements of the same class are equal:

$$\mathrm{Tr}\{N^{-1} M N\} = \mathrm{Tr}\{N N^{-1} M\} = \mathrm{Tr}\{M\} \tag{II-39}$$

(ii) It is easy to show that two equivalent representations have the same characters $\chi(g)$ for all the elements $g \in G$.

(iii) An important theorem states that any representation (or its equivalent) is uniquely determined by the set of its characters (those of the different classes of G). Characters are particularly important in the study of irreducible representations, which will be introduced below.

B-4. Sum and product of two representations

• The (direct) sum of two representations $M_1(g)$ and $M_2(g)$ of the *same* group G, with respective dimensions q_1 and q_2, is given by the $n \times n$ matrices (with $n = q_1 + q_2$):

$$M(g) = \left(\begin{array}{c|c} M_1(g) & 0 \\ \hline 0 & M_2(g) \end{array}\right) \tag{II-40}$$

It's easy to show that the matrices $M(g)$ yield a different representation of G. If $M_1(g)$ and $M_2(g)$ are matrices associated with operators $\mathcal{M}_1(g)$ and $\mathcal{M}_2(g)$

acting on vector spaces $\mathscr{E}_{q_1}$ and $\mathscr{E}_{q_2}$ respectively, $M(g)$ can be associated with an operator acting in the direct sum space:

$$\mathscr{E}_n = \mathscr{E}_{q1} \oplus \mathscr{E}_{q2} \tag{II-41}$$

Starting from two representations (or a single one if $M_1 \equiv M_2$), the sum operation allows building a new one with a larger dimension, and whose characters are the sums:

$$\chi(g) = \chi_1(g) + \chi_2(g) \tag{II-42}$$

• The (tensor) product of representations M_1 and M_2 (with matrix elements $\langle n_1|M_1|n_1'\rangle$ and $\langle n_2|M_2|n_2'\rangle$) is defined by the tensor product matrices:

$$M(g) = M_1(g) \otimes M_2(g) \tag{II-43}$$

that is:

$$\langle n_1, n_2|M(g)|n_1', n_2'\rangle = \langle n_1|M_1(g)|n_1'\rangle \; \langle n_2|M_2(g)|n_2'\rangle \tag{II-44}$$

The $M(g)$ matrices can be associated with operators acting in a tensor product space $\mathscr{E}_{p_1} \otimes \mathscr{E}_{p_2}$; their dimension is $P = p_1 \times p_2$, and the characters are given by:

$$\chi(g) = \chi_1(g) \times \chi_2(g)$$

Comment:

One can perform the product of representations of two different groups G_1 and G_2. This yields a representation of the (tensor) product $G_1 \otimes G_2$.

B-5. Reducible or irreducible representations

A representation is said to be "reducible" if it is equivalent to a representation that is a direct sum of representations of the same group. This is the case if we can find in relation (II-37) a matrix S such that, for any g:

$$M_2(g) = \left(\begin{array}{c|c} M_a(g) & 0 \\ \hline 0 & M_b(g) \end{array} \right) \tag{II-45}$$
$$\qquad\qquad \underbrace{\qquad\quad}_{q_a} \; \underbrace{\qquad\quad}_{q_b}$$

where the dimensions q_a and q_b of each block are the same for all g (with of course $q_a + q_b = q$).

Finding a matrix S that obeys equality (II-45) for all g, means that the initial representation can be *decomposed* into two representations with lower dimensions q_a and q_b. The problem is to find a *single* matrix S for *all* the matrices $M(g)$. For example, it would not be sufficient to diagonalize each matrix $M(g)$ in its own eigenbasis to completely decompose the representation, unless all the eigenbases were the same.

The representation is said to be "*irreducible*" if no invertible matrix S can be found that obeys relation (II-45) for all the $M(g)$.

Comments:

(i) The concept of reducibility can be used for any set of matrices, whether or not they represent a group G.

(ii) Saying that the representation is reducible is equivalent to saying that the operators $\mathcal{M}(g)$ act in a space $\mathscr{E}_q$ that is the direct sum of two subspaces $\mathscr{E}_{q1}$ and $\mathscr{E}_{q2}$, each of them being (globally) invariant under the action of all the $\mathcal{M}(g)$.

(iii) The reducibility of a set of matrices is sometimes defined with less constraints, imposing only the global invariance of a single subspace $\mathscr{E}_{q1}$ under the action of the $\mathcal{M}(g)$. Relation (II-45) must then be replaced by:

$$M_2(g) = S^{-1} M_1(g)\; S = \left(\begin{array}{c|c} M_a(g) & M_c(g) \\ \hline 0 & M_b(g) \end{array} \right) \tag{II-46}$$

which is obviously a weaker condition. In this context, if the stronger condition (II-45) is met, the representation is said to be "completely reducible". This distinction is only pertinent for infinite sets of matrices, as it can be proven that if a *finite* set of matrices is reducible, it must be completely reducible.

We shall use the word "reducible" in its full sense as in (II-45); it amounts to considering that the word "completely" is always implied.

An important task is the decomposition of a given representation of a group into irreducible representations. This involves finding a suitable matrix S' to transform the representation into the sum of irreducible blocs:

$$M_2(g) = (S')^{-1} M_1(g)\; S' = \begin{pmatrix} M'_a(g) & & & \\ & M'_b(g) & & 0 \\ & & M'_c(g) & \\ & 0 & & M'_d(g) \end{pmatrix} \tag{II-47}$$

Here are, without demonstration, a certain number of results:

- The irreducible representations (or their equivalent) obtained through the decomposition of a given representation, do not depend on the way the decomposition was performed (unicity theorem).
- The irreducible representations of an abelian group all have the dimension 1. Conversely, any non-commutative group must have at least one irreducible representation with a dimension larger than or equal to two.
- The number of non-equivalent irreducible representations of a finite group must be equal to the number of its classes.

The theory of characters is useful in the search for irreducible representations of a finite group. In particular, it yields an *irreducibility criterion* of a representation.

There are many physical applications of finite group theory, which use the theory of characters: examples include the crystallographic point groups and the group of molecular vibrations. The infinite groups are also of very great importance, and will be studied next.

Exercise: Symmetric and alternating square of a representation

The tensor product of an irreducible representation $M_1(g)$ by itself is written:

$$M(g) = M_1(g) \otimes M_1(g) \tag{II-48}$$

The operators $M(g)$ act in the space:

$$\mathcal{E}_{q^2} = \mathcal{E}_q \otimes \mathcal{E}_q \tag{II-49}$$

where we define the orthonormal basis:

$$|v_{i,j}\rangle = |u_i\rangle \otimes |u_j\rangle \tag{II-50}$$

(the $|u_i\rangle$ are an orthonormal basis of $\mathcal{E}_q$).

Introduce the exchange operator P_{exc} defined as:

$$P_{\text{exc}}|v_{i,j}\rangle = |v_{j,i}\rangle \tag{II-51}$$

and show that it is unitary and commutes with all the $M(g)$. Infer from this that the $M(g)$ representation is reducible and can be decomposed into two representations with dimensions $q(q+1)/2$ and $q(q-1)/2$.

Complement A_{II}

Left coset of a subgroup; quotient group

Let H (with elements noted h) be a subgroup of G (with elements noted g).

1. Left cosets

With each subgroup H of G, we can associate a way of regrouping the elements of G into classes $C_0, C_1, C_2, \ldots$, called "left cosets" of H, whose reunion forms the entire group G.

Consider two elements g and g' and assume we can find two elements h and h' of H such that:

$$g\,h = g'\,h' \tag{1}$$

As we saw in chapter II, this defines an equivalence relation between the elements g (it's easy to show that since H is a subgroup, this is a transitive relation: g being equivalent to g' and g' to g'' implies that g is equivalent to g''). The classes C are then the sets of all the elements of G that are equivalent with each other.

Class C_0 will gather all the elements of G equivalent to e, meaning all the elements g such that:

$$g\,h = e\,h' \tag{2}$$

or:

$$g = h'\,h^{-1} \in H \tag{3}$$

Conversely, if $g \in H$, we can write:

$$g\,g^{-1} = e \tag{4}$$

which is equality (1) with $h = g$, $g' = e$, $h' = e$. Consequently, C_0 is simply the subgroup H itself. The coset C_1 will be obtained by choosing any element of G outside of C_0 and associating with it all the equivalent elements, and so on for C_2, C_3, etc.

Relation (1) can be written:

$$g' = g\, h'' \tag{5}$$

with:

$$h'' = h\, h'^{-1} \in H \tag{6}$$

Conversely, relation (5) yields an equality similar to (1) in the particular case where $h' = e$. This means that g and g' are equivalent through H if there exists an element $h'' \in H$ such that (5) is satisfied.

Consequently, each element g can be associated, through (5), to as many distinct equivalent elements g' as they are elements in H. If H is finite and of order p, each class C contains the same number p of distinct elements.

Let q be the number of classes and N the order of the finite group G. As the latter is the reunion of the (disjoint) classes $C_0, C_1, C_2, \ldots$ we must have:

$$N = p\, q \tag{7}$$

This leads to the following result (Lagrange's theorem): the order of a subgroup of a group of finite order N is a divisor of the integer N.

Comments:

(i) One can define in a similar way right cosets of H. They are in general different from the left cosets (unless, obviously, G is abelian).

(ii) If H is an invariant subgroup, it is easy to show that the right and left cosets are the same [*cf.* relation (II-10) of chapter II].

(iii) The cosets associated with H are not to be confused with the conjugacy classes (§ A-4 of chapter II).

2. Quotient group

Imagine that H is an invariant subgroup of G. We can define a multiplication law between classes, calling "product class" C_iC_j the set of all the elements that can be written as g_ig_j, where $g_i \in C_i$, $g_j \in C_j$. Let us check that this product is indeed another class C_k (thanks to the fact that H is invariant). Noting g'_i and g'_j two other elements of C_i and C_j respectively, we have, using (5):

$$\begin{aligned} g'_i &= g_i\, h_1 \\ g'_j &= g_j\, h_2 \end{aligned} \tag{8}$$

which yields, using relation (II-10) of chapter II:

$$\begin{aligned} g'_i g'_j &= g_i \, h_1 \, g_j \, h_2 = g_i \, g_j \, h'_1 \, h_2 \\ &= g_i \, g_j \, h'' \end{aligned} \tag{9}$$

This equality shows that the product $g'_i g'_j$ is equivalent to $g_i g_j$. Conversely, proceeding with the same computation backwards, we can obtain that way all the elements of C_k.

Using this composition law, the inverse of the class C_k, noted C_k^{-1}, will contain the inverse of all the elements of C_k (it can be shown that if H is invariant, it is itself a class C). The identity element is the C_0 class, i.e. H itself.

The ensemble of the C thus forms a group, which by definition is the quotient group G/H.

Exercises:

(i) Consider the tensor product $\mathscr{G} = G_1 \otimes G_2$. Show that $\mathscr{H} = G_1 \otimes e_2$ is an invariant subgroup of $\mathscr{G}$ (e_2 is the identity element of G_2). Show that the equivalence relation (1) becomes particularly simple for the elements of $\mathscr{G}$, and that the quotient group $\mathscr{G}/\mathscr{H}$ is isomorphic to G_2.

(ii) Consider the group $\mathscr{E}_{(3)}$ of the displacements in the real three-dimensional space. $T_{(3)}$ is the set of translations in real space. Show that $T_{(3)}$ is an invariant subgroup of $\mathscr{E}_{(3)}$. Show that $\mathscr{E}_{(3)}/T_{(3)}$ is isomorphic to the rotation subgroup $R_{(3)}$.

Chapter III

Introduction to continuous groups and Lie groups

This chapter explains a few concepts related to infinite groups (groups containing an infinite number of elements), in particular those having the cardinality of the continuum (uncountable sets[1]). Take, for example, the group of translations along the Ox axis (with unit vector $\boldsymbol{e}_x$) by a quantity $q\boldsymbol{e}_x$, where q is a positive, negative, or zero integer. This group is infinite, but countable (q varies by steps). On the other hand, if you consider the group $T_{(1)}$ of the translations by a quantity $a\boldsymbol{e}_x$, where a is any real number, this infinite group is uncountable.

[1]An infinite set is countable if a one-to-one correspondence can be established between its elements and the natural integers 1, 2, 3, ... Otherwise, the set is said to have the cardinality of the continuum. Example: the set of rational numbers (fractions) is countable, but the set of real (rational and irrational) numbers in the interval [0,1] has the cardinality of the continuum.

Introduction to Continuous Symmetries: From Space–Time to Quantum Mechanics, First Edition. Franck Laloë.

We first introduce (§ A) some general notions concerning continuous groups, and a few ideas about topology. We then study simple examples (§ B) before focusing on two very important special cases: the Galilean and the Poincaré groups (§ C).

A. General properties

A-1. Continuous groups

A-1-a. Definitions

Consider a group G whose elements are identified by a finite number n of real parameters $a_1, a_2, \ldots a_n$. These parameters can be thought of as the components of a vector $\boldsymbol{a} \in R^n$:

$$g = g(a_1, a_2, \ldots, a_n) = g(\boldsymbol{a}) \qquad \boldsymbol{a} \in R^n \tag{III-1}$$

To span the entire group, $\boldsymbol{a}$ must vary within a certain domain D of R^n, which can be finite or infinite, and has the cardinality of the continuum. The parameters are supposed to be *essential* (non-redundant): the group cannot be described in a continuous way[2] by a smaller number of parameters. As an example, for the group $T_{(1)}$ of the translations in a one-dimensional space, it is natural to use, as a parameter b, the algebraic length of the translation vector $b\boldsymbol{e}_x$. This single parameter is then essential. On the other hand, if two parameters b_1 and b_2 were introduced to define the translation by a vector $(b_1 + b_2)\boldsymbol{e}_x$, these parameters would not be essential.

We assume there is a one-to-one correspondence between the elements of G and those of D (there are no two points $\boldsymbol{a}_1$ and $\boldsymbol{a}_2$ in D that correspond to the same element g of G). It is thus equivalent to define an element $g \in G$ or a vector $\boldsymbol{a}$ with components $a_1, a_2, \ldots, a_n$. It will often be easier to reason with the vectors $\boldsymbol{a}$ rather than with the elements g. The integer n is called the *dimension* of the group.

Product:

If $g(\boldsymbol{a})$ and $g(\boldsymbol{b})$ are two elements of G, so is the product $g(\boldsymbol{a})\, g(\boldsymbol{b})$, which can then be described by a value $\boldsymbol{c}$ of the parameter in D:

$$g(\boldsymbol{c}) = g(\boldsymbol{a})\, g(\boldsymbol{b}) \tag{III-2}$$

The value of $\boldsymbol{c}$ is a function of $\boldsymbol{a}$ and $\boldsymbol{b}$:

$$\boldsymbol{c} = \boldsymbol{\Phi}_2\,(\boldsymbol{a}, \boldsymbol{b}) \tag{III-3}$$

[2] We somewhat anticipate the definition of topologies in the group and the parameter's space to introduce the concept of continuity.

The function $\boldsymbol{\Phi}_2$ reflects in D the internal composition law in G. It cannot just be any function. As the multiplication law in G is associative, it follows that:

$$\boldsymbol{\Phi}_2\left(\boldsymbol{a}, \boldsymbol{\Phi}_2\left(\boldsymbol{b}, \boldsymbol{c}\right)\right) = \boldsymbol{\Phi}_2\left(\boldsymbol{\Phi}_2\left(\boldsymbol{a}, \boldsymbol{b}\right), \boldsymbol{c}\right) \tag{III-4}$$

Inverse:

Element $[g(\boldsymbol{a})]^{-1}$, the inverse of $g(\boldsymbol{a})$, is described by the value $\boldsymbol{a}'$ of the parameter in D:

$$[g(\boldsymbol{a})]^{-1} = g(\boldsymbol{a}') \tag{III-5}$$

with:

$$\boldsymbol{a}' = \boldsymbol{\Phi}_1(\boldsymbol{a}) \tag{III-6}$$

Example: For the group $T_{(1)}$, or more generally for the group of translations defined by their vector $\boldsymbol{a}$:

$$\begin{cases} \boldsymbol{\Phi}_2\left(\boldsymbol{a}, \boldsymbol{b}\right) = \boldsymbol{a} + \boldsymbol{b} \\ \boldsymbol{\Phi}_1(\boldsymbol{a}) = -\boldsymbol{a} \end{cases} \tag{III-7}$$

A-1-b. Intuitive notions of topology

It is easy to define a topology in R^n, such as the distance δ between two points $\boldsymbol{a}$ and $\boldsymbol{b}$:

$$\delta\left(\boldsymbol{a}, \boldsymbol{b}\right) = \left\{\sum_{i=1}^{n} \left(a_i - b_i\right)^2\right\}^{1/2} \tag{III-8}$$

This leads to the concept of the neighborhood of a point $\boldsymbol{a}$ (open set), set of all the points $\boldsymbol{b}$ such that:

$$\delta\left(\boldsymbol{a}, \boldsymbol{b}\right) < \varepsilon \tag{III-9}$$

and allows us to talk about points $\boldsymbol{a}'$ approaching a point $\boldsymbol{a}$, about limits and continuity, etc.

We will assume that a topology has also been introduced in the group G itself, for example by defining a "distance" between two of its elements $g(\boldsymbol{a})$ and $g(\boldsymbol{b})$.

Examples:

(i) In the translation group $T_{(3)}$, the distance between two translations can be defined as the modulus of the difference of their vectors $\boldsymbol{\ell}$. It is simply the distance (in the usual sense of the word in ordinary R^3 space) between the images of a given point through the two operations.

(ii) In the rotation group $R_{(3)}$, the distance between two rotations $\mathscr{R}'$ and $\mathscr{R}''$ can be defined as the upper value of the modulus of the vector $\boldsymbol{e}'' - \boldsymbol{e}' = \mathscr{R}''\boldsymbol{e} - \mathscr{R}'\boldsymbol{e}$ as the direction of a unit vector $\boldsymbol{e}$ changes. For two rotations around the same axis [in SO(2) for example], through the angles φ_1 and φ_2, the distance is $2|\sin[(\varphi_1 - \varphi_2)/2]|$.

On can then define, as in (III-9), the neighborhood of each element g. The group G is then said to be a "topological group".

Such a group where the product gg' and the inverse g^{-1} are continuous functions of g and g' is called a "continuous group". Furthermore, we shall assume that $\boldsymbol{a}$ varies "in a continuous way" in D and that the functions $\boldsymbol{\Phi}_1$ and $\boldsymbol{\Phi}_2$ are continuous functions of their variables (except, eventually, on the boundary of the domain D).

An important concept, *homeomorphism*, is a bicontinuous, one-to-one correspondence between two topological sets D_1 and D_2. When such a correspondence exists, D_1 and D_2 are said to be homeomorphic.

Examples:

(i) Homeomorphic sets (see figure 1):

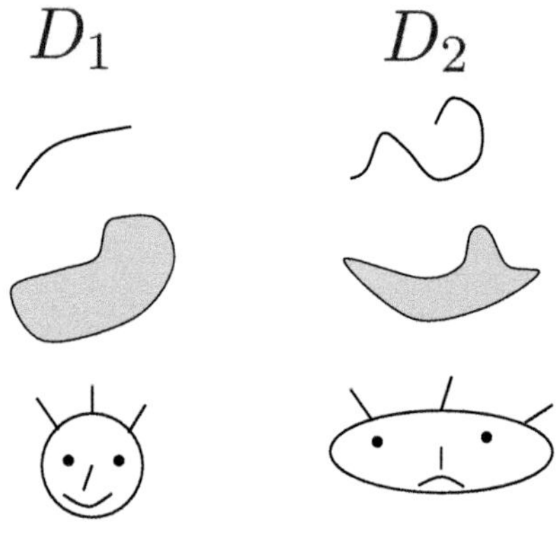

Figure 1: Schematic representations of homeomorphic sets. Each set on the left is homeomorphic to the corresponding one on the right.

(ii) Non-homeomorphic sets (see figure 2):

If the volume of D is finite and contains its boundary (D is a closed set), the group is said to be "*compact*"[3]. If this volume is composed of "a single piece", the domaine D and the group G are said to be "*connected*"[4].

[3]For the study of Lie groups, this simple definition is sufficient. More generally, a set (metric space) is said to be compact if one can extract from any infinite sequence of its elements a converging sequence.

[4]The general definition of a connected topological set is that it is not the union of two open non-empty separated sets.

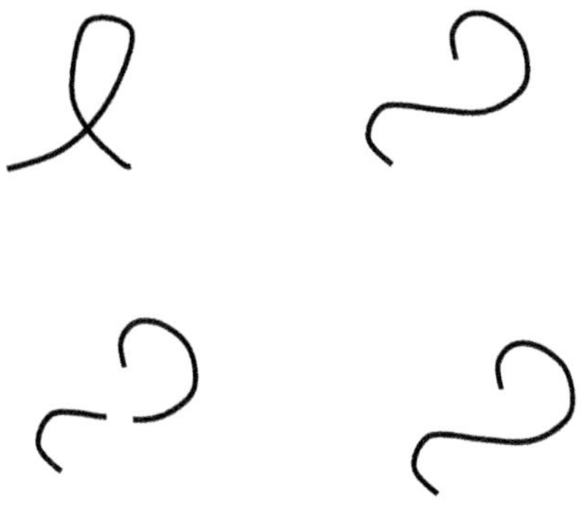

Figure 2: Schematic representations of non-homeomorphic sets. Bicontinuitiy is prevented on the first line by the intersection in the figure on the left. On the second line, it is prevented by the gap in the figure on the left.

(iii) Figure 3 symbolizes a disconnected domain (in two pieces):

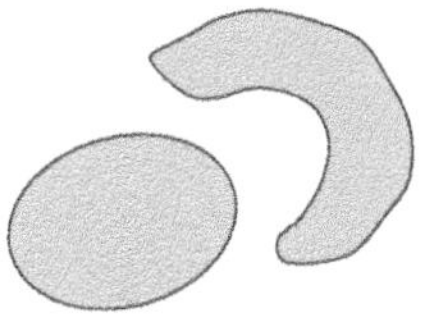

Figure 3: Schematic representation of a "disconnected" set, composed of two pieces.

The notion of "homotopic" (or more simply "equivalent") paths is a useful concept. A path in D is defined as the trajectory of a point M in D whose position depends in a continuous way on a parameter (such as the time) that varies continuously within a finite interval. More precisely, a path is the image in D, through a continuous correspondence f, of a closed interval $[t_0, t_1]$ of a real straight line (the time axis for example). The points M along the path are given by $M = f(t)$ with $t_0 \leq t \leq t_1$ (if $t_0 < t_1$). One can choose once and for all $t_0 = 0$ and $t_1 = 1$. Points $M_0 = f(t_0)$ and $M_1 = f(t_1)$ are the endpoints of the paths (figure 4). When M_0 and M_1 are the same (the path closes onto itself), the path is said to be a "loop" or a "closed path".

Imagine that, in a set, any two points are always the endpoints of one or several paths (one can always go continuously from one point to the other): the set is then connected.

Two paths having the same endpoints M_0 and M_1 are said to be "homotopic" if the first one can be continuously deformed into the second (or con-

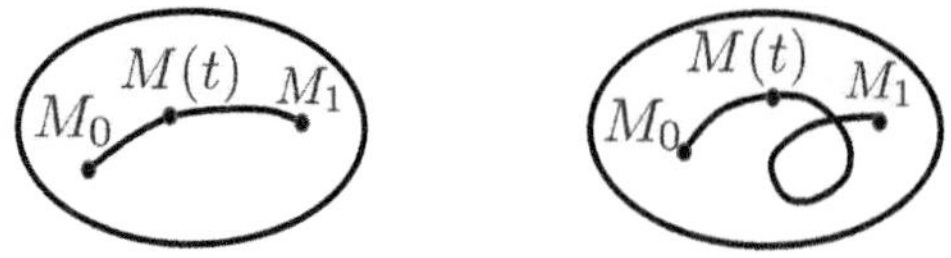

Figure 4: Two possible paths in a set, with endpoints M_0 and M_1 and followed by the point $M(t)$, where, by convention, $0 \leq t \leq 1$.

versely). Starting from the first path $M(t)$ going from M_0 to M_1, one should be able to define at each time a continuous transformation of the points $M(t)$ that brings them onto the points of the second path (figure 5).

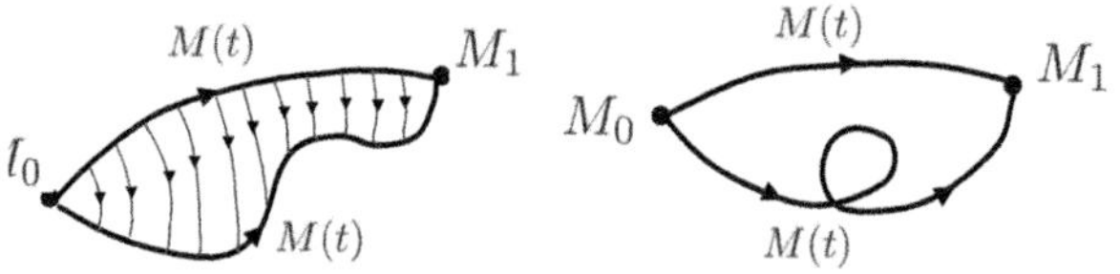

Figure 5: Two paths having the same endpoints M_0 and M_1 are homotopic if a continuous deformation can change one into the other. The figure shows two examples of such paths.

This concept of homotopic paths applies to closed paths. If a closed path can be continuously deformed into a "zero path", where all the points $M(t)$ coincide with M_0 for all t (the point M does not move), this path is said to be "homotopic to zero".

Examples:

(i) In the set of all the points inside a planar circle all the closed paths are homotopic to zero (*cf.* figure 6).

The same is true for the set of all the points on the surface of a sphere (in the three-dimensional space).

(ii) On the other hand, this is no longer true if we exclude from the preceding set all the points included in a small circle concentric to the first one, no matter how small it is. One cannot go continuously from a closed path that goes q times around the origin to another one that goes around q' times (if $q' \neq q$). Consequently there exist in such a set non-homotopic paths, as well as loops that are not homotopic to zero.

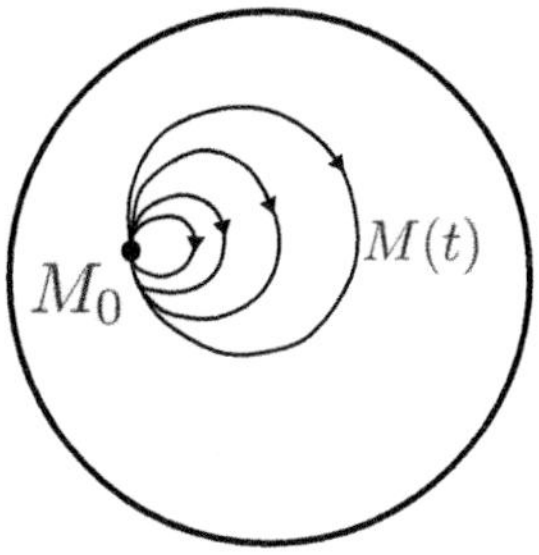

Figure 6: In the set of all the points inside a planar circle all the closed paths are homotopic to zero.

(iii) Consider the set of points inside a sphere through which passes, along one of its diameters, a wire carrying an electrical current i . Ampere's law shows that the line integrals of the magnetic field $\boldsymbol{B}$ along homotopic paths are equal, and reduce to zero for paths homotopic to zero.

(iv) In the complex plane, the integral of a complex function along a closed path is zero if that path belongs to a domain Δ where the function is differentiable (Cauchy conditions), and if it is homotopic to zero in Δ.

If all the closed paths of the topological set are homotopic to zero, the set is said to be "simply connected"[5]; otherwise, it is "non-simply connected".

Example:

A set that includes "holes", even though it may be "in one piece" (connected), is not simply connected. This is the case of the previous example (ii), or of a set such as the one shown in figure 7:

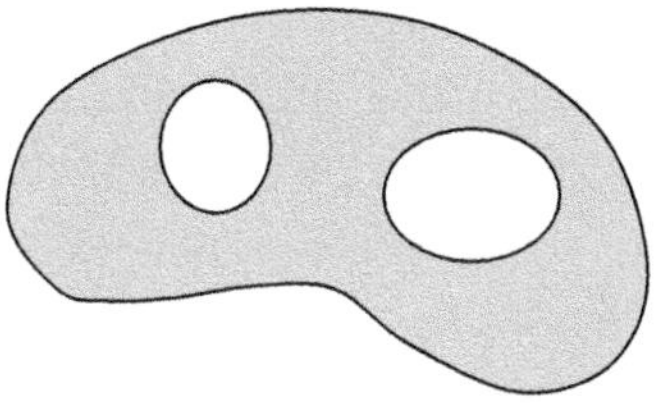

Figure 7: In a set that includes "holes", all the closed paths are not homotopic to zero. The set is not simply connected.

[5]We shall call p-connected, or class p connected, a set having a maximum number p of non-homotopic paths. For a set including "holes" as in figure 7, p is infinite. In the case of the rotation group $R_{(3)}$, we shall see in chapter VI that $p = 2$.

Consider finally the set of closed paths based at a given point M_0. This set can be divided into a certain number of equivalence classes, each containing all the closed paths that are homotopic with each other. The ensemble of all these classes is called $\Pi_1(M_0)$. The product of two closed paths based at the same point M_0 can be defined (intuitively, when the mobile point goes along the first path from M_0 to M_0, then along the second one, this yields a new closed path called product). It can be shown that this product defines a composition law in $\Pi_1(M_0)$, that each class has an inverse (paths taken in the opposite direction), that there exists an identity element (path homotopic to zero). The ensemble $\Pi_1(M_0)$ is thus a group, called the "homotopy group". For a connected group, it can be shown that the structure of the group Π_1 does not depend on the choice of point M_0.

Examples:

(i) For a simply connected group, Π_1 is the Z_1 group, with one element. If the group is p-connected, Π_1 has p distinct elements.

(ii) In the cases discussed above where the set is the inside of a circle minus its center, or the inside of a sphere minus a diameter, Π_1 is isomorphic to the commutative group Z_∞.

(iii) In the case of a torus, in 3-dimensional space, Π_1 is a group isomorphic to $Z_\infty \otimes Z_\infty$.

(iv) For a set including r "holes", the group Π_1 is generated by r elements (associated with a path going around each hole) and has an infinite number of elements. Π_1 is not a commutative group (if $r > 1$): going q times around a first hole and q' times around a second hole is not the same as inverting these two operations.

A-2. Lie groups

A-2-a. Definitions

In addition to the conditions stated above for the group G to be continuous, we add that $\boldsymbol{\Phi}_1$ and $\boldsymbol{\Phi}_2$ must be differentiable. The group is then a *Lie group.*

In practice, the Lie groups considered in physics are often groups of regular $N \times N$ matrices[6] $M(\boldsymbol{a})$. They are often matrices coming from the representation of a group of transformations, such as geometric transformations, and this will be the only case we shall study. For convenience, we will assume that the $N \times N$ identity matrix is described by the parameter $\boldsymbol{a} = \mathbf{0}$.

$$M(\mathbf{0}) = \mathbb{1} \tag{III-10}$$

If this were not the case, a simple translation of domain D in R^n would make this assumption true. We start with a few simple examples of such groups.

[6]Do not confuse N with the dimension n of the group (see examples below).

Examples:

(i) Group of the one-dimensional homotheties defined by $x' = ax$ $(a > 0)$, and hence by 1×1 matrices. This is an abelian group of dimension 1, which is not compact since a can span the entire positive axis.

(ii) Group GL$(2, C)$ of regular complex 2×2 matrices:

$$M(\boldsymbol{a}) = \begin{pmatrix} 1 + a_1 & a_3 \\ a_4 & 1 + a_2 \end{pmatrix} \quad \text{with} \quad a_1 + a_2 + (a_1\, a_2 - a_3\, a_4) \neq -1 \quad \text{(III-11)}$$

We have:

$$M(\boldsymbol{a})\ M(\boldsymbol{b}) = \begin{pmatrix} (1 + a_1)(1 + b_1) + a_3\, b_4 & (1 + a_1)b_3 + a_3(1 + b_2) \\ a_4(1 + b_1) + (1 + a_2)b_4 & a_4 b_3 + (1 + a_2)(1 + b_2) \end{pmatrix} \quad \text{(III-12)}$$

which yields the $\boldsymbol{\Phi}_2$ function. This is a non-abelian group, with dimension $n = 8$ (since defining a complex number is equivalent to defining two real numbers), while $N = 2$.

(iii) Group $R_{(2)}$, or SO(2), of the matrices associated with the rotation around a fixed axis. We shall call φ the single parameter $(n = 1)$:

$$M(\varphi) = \begin{pmatrix} \cos\varphi & -\sin\varphi \\ \sin\varphi & \cos\varphi \end{pmatrix} \qquad 0 \leq \varphi < 2\pi \quad \text{(III-13)}$$

We can simply write here:

$$\begin{cases} \Phi_1(a = \varphi) = -\varphi \\ \Phi_2(a = \varphi,\, b = \varphi') = \varphi + \varphi' \end{cases} \quad \text{(III-14)}$$

This is an abelian group, whose domain D can be represented in the plane by a circumference (figure 8). It is better to take into account the topology of the group and represent the domain D by a circumference rather than by a line segment of length 2π, whose two extremities would correspond to infinitely close rotations. This is a compact group, connected but not simply connected [group $\Pi_1 = Z_\infty$]. It is homeomorphic to the group of complexe numbers of modulus 1.

(iv) Group O(2) of the matrices associated with the rotations in a plane around a fixed point O, or to their product through a symmetry with respect to an axis passing through O:

$$M(\varphi) = \begin{pmatrix} \cos\varphi & -\varepsilon\sin\varphi \\ \sin\varphi & \varepsilon\cos\varphi \end{pmatrix} \qquad 0 \leq \varphi < 2\pi\ ;\quad \varepsilon = \pm 1 \quad \text{(III-15a)}$$

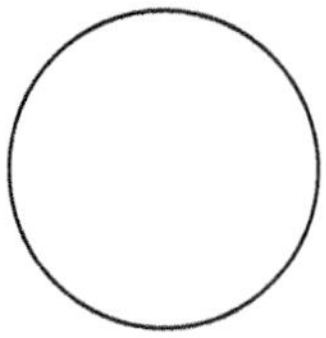

Figure 8: A simple circumference is the variation domain of the parameter φ for the group of rotations around a fixed axis.

We have:

$$\det M(\varphi) = \varepsilon \tag{III-15b}$$

The domain D associated with this group has two components, associated with the matrices having determinants $+1$ and -1 (one cannot go continuously from $+1$ to -1). It is represented by two circumferences (figure 9). This group is not connected, but it is compact.

A-2-b. Infinitesimal operators

Consider a group of $N \times N$ matrices $M(\boldsymbol{a})$, depending on the (vector) parameter $\boldsymbol{a}$. We assume that the matrices $M(\boldsymbol{a})$ form a Lie group and that their matrix elements are continuous and differentiable functions of the $\boldsymbol{a}$ components. Remember that, by hypothesis:

$$M(\mathbf{0}) = (\mathbb{1}) \tag{III-16}$$

We shall focus on transformations, close to the identity, for which the parameter $\boldsymbol{a}$ takes on an infinitesimal value. A first-order Taylor expansion of

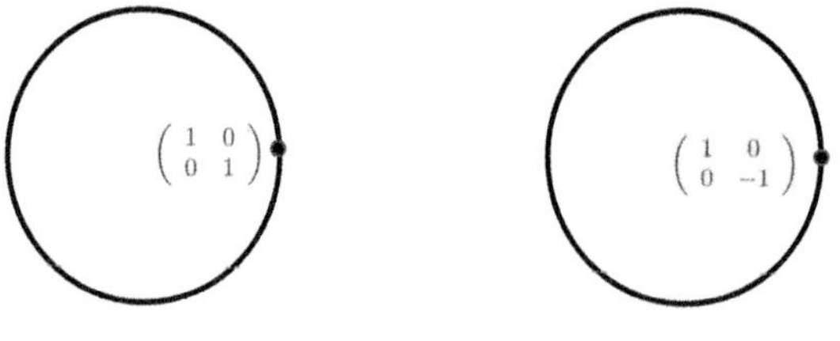

Figure 9: Variation domain of the parameters for the group O(2), which has two "components", one for the determinant $+1$, the other for the determinant -1.

$M(\delta\boldsymbol{a})$ yields:

$$M(\delta\boldsymbol{a}) = (\mathbb{1}) + \sum_{i=1}^{n} \delta a^i\, X_i + \ldots \tag{III-17}$$

where X_i is an $N \times N$ matrix:

$$X_i = \frac{\partial}{\delta a^i}\, M(\boldsymbol{a})|_{\boldsymbol{a}=0} \tag{III-18}$$

From now on, we shall often use an upper index for the δa^i components of $\delta\boldsymbol{a}$ (or the a^i components of $\boldsymbol{a}$).

There is no reason, a priori, for the X_i to belong to the group G, as we see in the following examples. We shall assume that the X_i matrices are non-zero and independent (any one of them is not a linear combination of the others). Relation (III-17) shows that, when $\delta\boldsymbol{a}$ is infinitesimal, the difference between $M(\delta\boldsymbol{a})$ and the identity matrix is a linear combination of the X_i (meaning it belongs to a vector space having the X_i as a basis). The X_i are called the "infinitesimal generators " of the group.

Consider now the "commutator" of two infinitesimal operators $M(\delta\boldsymbol{a})$ and $M(\delta\boldsymbol{b})$. It belongs to the group G and can be written $M(\delta\boldsymbol{c})$ with:

$$M(\delta\boldsymbol{c}) = M(\delta\boldsymbol{a})\; M(\delta\boldsymbol{b})\; [M(\delta\boldsymbol{a})]^{-1}\; [M(\delta\boldsymbol{b})]^{-1} \tag{III-19}$$

Let us compute this matrix to second-order in $\delta\boldsymbol{a}$ and $\delta\boldsymbol{b}$, i.e. including either the "square" terms in $\delta a_i\; \delta a_j$ and $\delta b_i\; \delta b_j$, or the "cross" terms in $\delta a_i\; \delta b_j$ $(i, j = 1, 2, \ldots, n)$. A simplification occurs as all the square terms are equal to zero as we now show. If $\delta\boldsymbol{b} = 0$, $M(\delta\boldsymbol{b})$ is the identity matrix and so is $M(\delta\boldsymbol{c})$ [using relation (III-19)] . This matrix does not depend on $\delta\boldsymbol{a}$, and all the terms of its expansion in $\delta\boldsymbol{a}$ (except the 0th-order) must be zero. In a similar way, if $\delta\boldsymbol{a} = 0$, we also have $M(\delta\boldsymbol{c}) = (\mathbb{1})$, and all the square terms in $\delta\boldsymbol{b}$ must be zero. We thus know from the very beginning that we can ignore all the second-order terms that are not cross terms in $\delta\boldsymbol{a}$ and $\delta\boldsymbol{b}$, as they will cancel out in the end. This is why, for example, formula (III-17) can be used as such for each matrix in (III-19), without developing the expansion further. Similarly, we can write:

$$[M(\delta\boldsymbol{a})]^{-1} = (\mathbb{1}) - \sum_{i=1}^{n} \delta a^i\, X_i + \ldots \tag{III-20}$$

ignoring the following terms of the expansion [one can check that multiplying together the right-hand sides of (III-19) and (III-20), we do get $(\mathbb{1})$, neglecting second-order terms].

We thus get:

$$M(\delta\boldsymbol{a})\, M(\delta\boldsymbol{b}) = (\mathbb{1}) + \sum_{i=1}^{n} \delta a^i\, X_i + \sum_{j=1}^{n} \delta b^j\, X_j + \sum_{ij} \delta a^i\, \delta b^j\, X_i\, X_j \tag{III-21}$$

Multiplying on the right both sides of the relation by $(\mathbb{1} - \sum_i \delta a^i X_i)$, we get:

$$\begin{aligned} M(\delta\boldsymbol{a})\, M(\delta\boldsymbol{b})\, [M(\delta\boldsymbol{a})]^{-1} = (\mathbb{1}) + \sum_{i=1}^{n} \delta a^i\, X_i + \sum_{j=1}^{n} \delta b^j\, X_j + \sum_{ij} \delta a^i\, \delta b^j\, X_i\, X_j + \\ - \sum_{i=1}^{n} \delta a^i\, X_i - \sum_{ij} \delta a^i\, \delta b^j\, X_j\, X_i + \dots \\ = (\mathbb{1}) + \delta b^j\, X_j + \sum_{i,j=1}^{n} \delta a^i\, \delta b^j\, [X_i, X_j] + \dots \end{aligned} \tag{III-22}$$

where on the right-hand side appears the commutator of the two X_i and X_j matrices:

$$[X_i, X_j] = X_i\, X_j - X_j\, X_i \tag{III-23}$$

We finally multiply on the right both sides of this equation by $(1 - \sum_j \delta b^j\, X_j)$. This amounts, within our approximation, to subtracting $\sum_j \delta b^j\, X_j$ from the right-hand side of (III-22). We thus obtain:

$$M(\delta\boldsymbol{a})\, M(\delta\boldsymbol{b})\, [M(\delta\boldsymbol{a})]^{-1}\, [M(\delta\boldsymbol{b})]^{-1} = M(\delta\boldsymbol{c}) = (\mathbb{1}) + \sum_{i,j=1}^{n} \delta a^i\, \delta b^j\, [X_i, X_j] + \dots \tag{III-24}$$

(where we have neglected the third or higher-order terms).

Let us now see what these operations mean in terms of the group parameters. Using the functions $\boldsymbol{\Phi}_1$ and $\boldsymbol{\Phi}_2$, we can write:

$$\begin{aligned} &M(\delta\boldsymbol{a})\, M(\delta\boldsymbol{b}) \Longleftrightarrow \delta\boldsymbol{c}_1 = \boldsymbol{\Phi}_2(\delta\boldsymbol{a}, \delta\boldsymbol{b}) \\ &M(\delta\boldsymbol{a})\, M(\delta\boldsymbol{b})\, [M(\delta\boldsymbol{a})]^{-1} \Longleftrightarrow \delta\boldsymbol{c}_2 = \boldsymbol{\Phi}_2(\delta\boldsymbol{c}_1, \boldsymbol{\Phi}_1(\delta\boldsymbol{a})) \end{aligned} \tag{III-25}$$

and finally:

$$M(\delta\boldsymbol{c}) \Longleftrightarrow \delta\boldsymbol{c} = \boldsymbol{\Phi}_2(\delta\boldsymbol{c}_2, \boldsymbol{\Phi}_1(\delta\boldsymbol{b})) \tag{III-26}$$

Within our approximation, we saw above that $M(\delta\boldsymbol{a})$ and $M(-\delta\boldsymbol{a})$ can be considered as the inverse of each other. We thus have:

$$\boldsymbol{\Phi}_1(\delta\boldsymbol{a}) = -\delta\boldsymbol{a} + \dots \tag{III-27}$$

To obtain $\delta\boldsymbol{c}$, our computation will be similar to the one that led us to (III-24). We first notice that, if $\delta\boldsymbol{a}$ is zero, $\delta\boldsymbol{c}_1$ equals $\delta\boldsymbol{b}$; in the same way, if $\delta\boldsymbol{b}$

is zero, $\delta \boldsymbol{c}_1 = \delta \boldsymbol{a}$. Consequently, the only second-order terms are the cross terms in $\delta \boldsymbol{a}$ and $\delta \boldsymbol{b}$, which leads to:

$$\delta \boldsymbol{c}_1 = \boldsymbol{\Phi}_2(\delta \boldsymbol{a}, \delta \boldsymbol{b}) = \delta \boldsymbol{a} + \delta \boldsymbol{b} + \sum_{i,j=1}^{n} \delta a^i \, \delta b^j \, \boldsymbol{F}_{ij} + \ldots \tag{III-28}$$

with:

$$\boldsymbol{F}_{ij} = \frac{\partial^2}{\partial a^i \, \partial b^j} \boldsymbol{\Phi}_2(\boldsymbol{a}, \ \boldsymbol{b})|_{\boldsymbol{a}=\boldsymbol{b}=0} \tag{III-29}$$

We now evaluate $\delta \boldsymbol{c}_2$:

$$\delta \boldsymbol{c}_2 = \boldsymbol{\Phi}_2(\delta \boldsymbol{c}_1, -\delta \boldsymbol{a}) = \delta \boldsymbol{a} + \delta \boldsymbol{b} + \sum_{ij} \delta a^i \, \delta b^j \, \boldsymbol{F}_{ij} - \delta \boldsymbol{a} + \sum_{ij} (\delta a^i + \delta b^i)(-\delta a^j) \boldsymbol{F}_{ij} + \ldots \tag{III-30}$$

where the $\delta \boldsymbol{a}$ terms cancel each other, and the $\delta a_i \, \delta a_j$ terms can be ignored. Finally:

$$\delta \boldsymbol{c} = \boldsymbol{\Phi}_2(\delta \boldsymbol{c}_2, -\delta \boldsymbol{b}) = \delta \boldsymbol{b} + \sum_{ij} \delta a^i \, \delta b^j \, (\boldsymbol{F}_{ij} - \boldsymbol{F}_{ji}) - \delta \boldsymbol{b} + \sum_{ij} \delta b^i \, \delta b^j \, \boldsymbol{F}_{ij} + \ldots \tag{III-31}$$

which simply yields, ignoring the $\delta b_i \, \delta b_j$ terms:

$$\delta \boldsymbol{c} = \sum_{ij} \delta a^i \, \delta b^j \, [\boldsymbol{F}_{ij} - \boldsymbol{F}_{ji}] \tag{III-32}$$

Calling f^k_{ij} the kth component of $\boldsymbol{F}_{ij}$:

$$\boldsymbol{F}_{ij} = \sum_{k=1}^{n} f^k_{ij} \, \boldsymbol{e}_k \tag{III-33}$$

(the $\boldsymbol{e}_k$ are the unit vectors of the R^n basis) we can also write:

$$\delta c^k = \sum_{i,j=1}^{n} \delta a^i \, \delta b^j (f^k_{ij} - f^k_{ji}) \tag{III-34}$$

Equality (III-20) gives an approximation of any matrix close to the identity, with an infinitesimal parameter. Applied to $\delta \boldsymbol{c}$, it yields:

$$M(\delta \boldsymbol{c}) = (\mathbb{1}) + \sum_{i,j,k=1}^{n} \delta a^i \, \delta b^j (f^k_{ij} - f^k_{ji}) \, X_k \tag{III-35}$$

Comparing this relation to equality (III-24) and using the fact that the X_k are linearly independent matrices, we obtain by identification:

$$\boxed{[X_i, X_j] = \sum_{k=1}^{n} C_{ij}^k \; X_k} \tag{III-36}$$

with:

$$\boxed{C_{ij}^k = f_{ij}^k - f_{ji}^k} \tag{III-37}$$

Consequently, the commutator of any X_i and X_j matrices is still a linear combination of n matrices X.

A-2-c. Structure constants; Lie algebra

The numbers C_{ij}^k, which according to (III-37) are related to the second derivative at the origin of the function $\boldsymbol{\Phi}_2$, are called the "structure constants" of the group. They are real numbers since the parameters $\boldsymbol{a}$ and the function $\boldsymbol{\Phi}_2$ are real. Because of the antisymmetric relation:

$$C_{ij}^k = -C_{ji}^k \tag{III-38}$$

there are actually $n^2(n-1)/2$ structure constants in a group of dimension n.

These constants have been introduced while considering only matrices $M(\delta\boldsymbol{a})$ infinitely close to the identity matrix ($\mathbb{1}$). This could lead us to believe that the C_{ij}^k reflect properties of the group only in the neighborhood of the identity element. This is actually not the case. The product:

$$M(\boldsymbol{a}_0)\, M(\delta\boldsymbol{a}) \tag{III-39}$$

allows obtaining matrices infinitely close to any element $M(\boldsymbol{a}_0)$ of the group. The operators X_i, the structure constants, etc. reflect properties of the group that are valid close to any element of the group.

The constants C_{ij}^k play a very important role in the study of Lie groups. They give the *local properties* of the group composition law, indicate if it is commutative, etc.. If all the structure constants of a group are zero, the group is abelian.

In a continuous group, the structure constants play a role similar to that of the multiplication table for a finite group. They do *not*, however, completely determine the structure of the group, i.e. its multiplication law (the function $\boldsymbol{\Phi}_2$). It can be shown (and we shall admit it without further proof) that two Lie groups, having the same dimension n and the same structure constants, are *locally*, but not necessarily globally, *isomorphic*. These two groups are then homomorphic to the same group[7]. An important example we shall return to are the groups SO(3) and SU(2), which have the same structure constants but are only locally isomorphic.

As shown by relations (III-17) and (III-39), the infinitesimal variations of any matrix M, in the neighborhood of a given element $M(\boldsymbol{a}_0)$, are given by a linear combination of the X_i:

$$X = \Sigma_i \lambda^i X_i \tag{III-40}$$

The set $\mathscr{L}$ of the operators X thus defined (where the λ_i can take any real values) form a vector space called the "*Lie algebra*" of the group. By definition, the X_i constitute a basis of $\mathscr{L}$. As we already noted, the elements of this algebra do not generally belong to G.

A-2-d. Algebra structure

In a general way, an algebra is a vector space equipped with an internal bilinear composition law (multiplication) that is distributive with respect to the addition. A total of 3 composition laws (of which 2 are internal) are defined in an algebra: multiplication by a scalar λ, internal addition, internal multiplication. In the case of a Lie algebra of operators, the second law is simply the sum $X + X'$ of the two operators, and the third one associates X and X' with their commutators:

$$X, X' \Longrightarrow [X, X'] \tag{III-41}$$

This last law is not necessarily associative (the algebra is said to be non-associative) since we have the Jacobi equality:

$$[X, [X', X'']] + [X', [X'', X]] + [X'', [X, X']] = 0 \tag{III-42}$$

This equality establishes certain constraints satisfied by the structure constants. Inserting (III-36) into (III-42), and taking into account the linear independence of the X_i, we have :

$$\sum_\ell \left\{ C_{ij}^\ell \, C_{\ell k}^m + C_{jk}^\ell \, C_{\ell i}^m + C_{ki}^\ell \, C_{\ell j}^m \right\} = 0 \quad \forall\, m \ ; \ \ \forall\, i, j, k \tag{III-43}$$

where the indices i, j, k undergo a circular permutation, while m does not change. Another demonstration of this relation is given in comment (i) below.

---

[7]Given a Lie algebra, it can correspond to several groups G, G', ... that are not necessarily simply connected. One of them, G_u, is simply connected and called the "universal covering group" of the G. One can transform G_u into G by a discrete kernel homomorphism, isomorphic to the homotopy group Π_1 of G.

Conversely, it can be shown that if we are given arbitrary constants, anti-symmetric $[C_{ij}^k = -C_{ji}^k]$ and satisfying Jacobi relation (III-43), we can define (at least) one composition law $\boldsymbol{\Phi}_2(\boldsymbol{a},\boldsymbol{b})$ with the corresponding Lie group G.

Using relations (III-38) and (III-42) to construct possible Lie algebras for any given dimension n is an interesting mathematical problem. At the end of the 19th century, Elie Cartan and Wilhelm Killing carried out major work on this subject, which we shall not discuss here.

We shall call Lie operator any matrix of $\mathscr{L}$, i.e. any matrix X in the form (III-40). We shall also use the name "infinitesimal generator", in particular when these same operators are multiplied by the imaginary constant i, which makes them Hermitian (and not anti-Hermitian) in several interesting physical cases.

Comments:

(i) Associativity of the internal composition law in the group G, expressed by relation (III-4) for the function $\boldsymbol{\Phi}_2$, also leads to equality (III-43). Consider the product $M(\delta\boldsymbol{h}) = M(\delta\boldsymbol{a})\, M(\delta\boldsymbol{b})\, M(\delta\boldsymbol{c})$ of 3 matrices of the group. We have:

$$\delta\boldsymbol{h} = \boldsymbol{\Phi}_2\left[\boldsymbol{\Phi}_2(\delta\boldsymbol{a},\delta\boldsymbol{b}),\delta\boldsymbol{c}\right] = \boldsymbol{\Phi}_2\left[\delta\boldsymbol{a},\boldsymbol{\Phi}_2(\delta\boldsymbol{b},\delta\boldsymbol{c})\right] \tag{III-44}$$

Let us compute the $\delta a\,\delta b\,\delta c$ terms that appear in $\delta\boldsymbol{h}$. According to (III-28):

$$\begin{cases} \boldsymbol{\Phi}_2(\delta\boldsymbol{a},\delta\boldsymbol{b}) = \delta\boldsymbol{a} + \delta\boldsymbol{b} + \sum\limits_{ij} \boldsymbol{F}_{ij}\,\delta a^i\,\delta b^j + \ldots \\ \boldsymbol{\Phi}_2(\delta\boldsymbol{b},\delta\boldsymbol{c}) = \delta\boldsymbol{b} + \delta\boldsymbol{c} + \sum\limits_{jk} \boldsymbol{F}_{jk}\,\delta b^j\,\delta c^k + \ldots \end{cases} \tag{III-45}$$

Using again (III-28), the terms we are looking for are:

$$\sum_{\ell k} \boldsymbol{F}_{\ell k}\left(\sum_{ij} f_{ij}^{\ell}\,\delta a^i\,\delta b^j\right)\delta c^k = \sum_{i\ell} \boldsymbol{F}_{i\ell}\left(\sum_{jk} f_{jk}^{\ell}\,\delta b^j\,\delta c^k\right)\delta a^i \tag{III-46}$$

Identifying the corresponding terms, we get:

$$\sum_{\ell} f_{ij}^{\ell}\,\boldsymbol{F}_{\ell k} = \sum_{\ell} f_{jk}^{\ell}\,\boldsymbol{F}_{i\ell} \tag{III-47a}$$

or:

$$\sum_{\ell} f_{ij}^{\ell}\,f_{\ell k}^{m} = \sum_{\ell} f_{jk}^{\ell}\,f_{i\ell}^{m} \tag{III-47b}$$

(for any i, j, k, and m). Now, according to the definition of C_{ij}^k, the left-hand side of (III-43) equals:

$$\sum_{\ell} (f_{ij}^{\ell} - f_{ji}^{\ell})\,(f_{\ell k}^{m} - f_{k\ell}^{m}) + \sum_{\ell} (f_{jk}^{\ell} - f_{kj}^{\ell})\,(f_{\ell i}^{\ell} - f_{i\ell}^{\ell}) \tag{III-47c}$$

$$+ \sum_{\ell} (f_{ki}^{\ell} - f_{ik}^{\ell})\,(f_{\ell j}^{m} - f_{j\ell}^{m}) \tag{III-47d}$$

Using 6 times equality (III-47b) (and changing the names of the indices), we can show that this quantity is zero, thus proving again relation (III-43).

(ii) If one has:

$$\mathrm{Tr}\{X_i X_j\} = \delta_{ij}$$

the C_{ij}^k are completely symmetric. This is because:

$$\begin{aligned}\mathrm{Tr}\{X_k[X_i,X_j]\} = C_{ij}^k = \mathrm{Tr}\{X_k X_i X_j - X_k X_j X_i\} = \mathrm{Tr}\{[X_k,X_i]X_j\} \\ = C_{ki}^j = -C_{ik}^j \end{aligned} \qquad \text{(III-48)}$$

A-3. Compact groups

A compact group was defined in § A-1-b as a group for which the volume of the domain D in which the parameters a_1, a_2, ..., a_n vary is finite and contains its boundary (finite and closed domain). Let us mention, without demonstration, some general properties of the representations of compact groups.

• Any representation of a compact group is equivalent to a unitary representation.

• Any irreducible representation (§ B-5 of chapter II) of a compact group has a finite dimension.

• Conversely, for non-compact simple groups (§ A-3 of chapter II), there is no faithful and unitary representation with a finite dimension.

These theorems will not be directly used in our computations, but will be illustrated in multiple examples: finite unitary representation of the compact rotation group, finite but non-unitary representation of the non-compact Lorentz group by the $\mathrm{SL}(2, C)$ matrices, representations of the Galilean and Poincaré groups by 5×5 non-unitary matrices, etc.

B. Examples

B-1. SO(2) group

Its matrices are given in (III-13). We have:

$$M(\mathrm{d}\varphi) = \begin{pmatrix} 1 & -\mathrm{d}\varphi \\ \mathrm{d}\varphi & 1 \end{pmatrix} \qquad \text{(III-49)}$$

and thus:

$$X = \begin{pmatrix} 0 & -1 \\ 1 & 0 \end{pmatrix} = -i\sigma_y \qquad \text{(III-50)}$$

where σ_y is the second Pauli matrix:

$$\sigma_y = \begin{pmatrix} 0 & -i \\ i & 0 \end{pmatrix} \tag{III-51}$$

Its Lie algebra is of dimension 1, and the only associated infinitesimal generator is σ_y. There are no structure constants (abelian group).

To go from the infinitesimal generator to the finite transformations we can write:

$$\begin{aligned} M(\varphi + \mathrm{d}\varphi) &= M(\mathrm{d}\varphi)\, M(\varphi) \\ &= [1 - i\sigma_y \mathrm{d}\varphi]\, M(\varphi) \end{aligned} \tag{III-52}$$

and:

$$\mathrm{d}M(\varphi) = -i\,\sigma_y\, \mathrm{d}\varphi\, M(\varphi) \tag{III-53}$$

Since we are dealing with a single operator (that obviously commutes with itself), we can integrate the differential equation:

$$\frac{\mathrm{d}M(\varphi)}{\mathrm{d}\varphi} = -i\,\sigma_y\, M(\varphi) \tag{III-54}$$

as an ordinary differential equation. This leads to:

$$M(\varphi) = \mathrm{e}^{-i\varphi\,\sigma_y} \tag{III-55}$$

Note that, since $\sigma_y^2 = 1$, we have:

$$\begin{aligned} \mathrm{e}^{-i\varphi\,\sigma_y} &= 1 - i\varphi\,\sigma_y + \frac{1}{2}(i\varphi)^2 + \dots \\ &= \cos\varphi - i\sin\varphi\,\sigma_y \end{aligned} \tag{III-56}$$

Taking (III-51) into account, this again yields (III-13).

As will often be the case, the matrices of the group are all the exponentials of the matrices of the Lie algebra (*exponential group*). We have:

$$M(\varphi) = \lim_{p\to\infty} \left(1 + \varphi\frac{X}{p}\right)^p \tag{III-57}$$

B-2. One-dimensional Galilean transformation

Consider two Galilean reference frames. The origin of the second starts from an initial position b; it has a velocity $-v$ in the first reference frame (which means that the velocities in the second reference frame are higher by $+v$). The time origin of the second reference frame corresponds, in the first, to a time τ. The transformations of the coordinates of an event associated with a one-dimensional Galilean transformation are written[8]:

$$\begin{cases} x' = x + vt + b \\ t' = t - \tau \end{cases} \tag{III-58}$$

Let us define a 3-component vector $\boldsymbol{a}$:

$$\boldsymbol{a} \begin{cases} b \\ v \\ \tau \end{cases} \tag{III-59}$$

Introducing an homogeneous coordinate, we can write:

$$\begin{pmatrix} x' \\ t' \\ 1 \end{pmatrix} = M(\boldsymbol{a}) \begin{pmatrix} x \\ t \\ 1 \end{pmatrix} \tag{III-60a}$$

with:

$$M(\boldsymbol{a}) = \begin{pmatrix} 1 & v & b \\ 0 & 1 & -\tau \\ 0 & 0 & 1 \end{pmatrix} \tag{III-60b}$$

Taking the product of two matrices $M(\boldsymbol{a})$ and $M(\boldsymbol{a}')$, we easily find the composition law of the group (function $\boldsymbol{\Phi}_2$):

$$M(\boldsymbol{a}'') = M(\boldsymbol{a})\, M(\boldsymbol{a}') \Longleftrightarrow \begin{cases} v'' = v + v' \\ b'' = b + b' - v\,\tau' \\ \tau'' = \tau + \tau' \end{cases} \tag{III-61}$$

This is a non-abelian group. For example:

$$M(0, v, 0)\, M(0, 0, \tau') = \begin{pmatrix} 1 & v & -v\tau' \\ 0 & 1 & -\tau' \\ 0 & 0 & 1 \end{pmatrix} \tag{III-62a}$$

[8]In most books on relativity, the Lorentz transformation is characterized by the opposite parameter $-v$. The convention we have chosen here (and in the rest of the book) is better adapted for discussing symmetries. It is coherent with relation (I-87) of chapter I, as well as its figure 4.

As for the minus sign in front of τ, it comes from the fact that time is considered here as the coordinate of an event, rather than a parameter on which the dynamical variables of the system depend [*cf.* comment in page 8 and relation (I-14a)].

whereas:

$$M(0,0,\tau')\,M(0,v,0) = \begin{pmatrix} 1 & v & 0 \\ 0 & 1 & -\tau' \\ 0 & 0 & 1 \end{pmatrix} \tag{III-62b}$$

The two products differ by a $v\tau'$ term, corresponding to a residual translation. Imagine that, at a station, you jump on a moving train and wait for a time τ ; or else you first wait for a time τ and then jump on the train. In the two cases you will not arrive at the same place in the train.

Let us build the Lie algebra:

$$X_1 = X_K = \begin{pmatrix} 0 & 1 & 0 \\ 0 & 0 & 0 \\ 0 & 0 & 0 \end{pmatrix} \qquad X_2 = X_P = \begin{pmatrix} 0 & 0 & 1 \\ 0 & 0 & 0 \\ 0 & 0 & 0 \end{pmatrix}$$

$$X_3 = X_H = \begin{pmatrix} 0 & 0 & 0 \\ 0 & 0 & -1 \\ 0 & 0 & 0 \end{pmatrix} \tag{III-63}$$

We have called X_K the operator associated with the pure Galilean transformations; X_P, the space translation operator; X_H, the time translation operator (the reason for these names will appear in the following chapters). All the products of these matrices are zero, except for $X_K X_H = X_P$.

It is easy to show that:

$$\begin{cases} [X_H, X_P] = 0 \\ [X_K, X_P] = 0 \\ [X_K, X_H] = -X_P \end{cases} \tag{III-64}$$

The group is thus "almost" abelian: the only non-zero structure constant is C_{13}^2.

Let us go from the X to the finite transformations. It is easy to show that $(X_{1,2,3})^2 = 0$ (nilpotent matrices). Consequently:

$$\mathrm{e}^{aX_i} = 1 + aX_i + \frac{a^2}{2}X_i^2 + \ldots = 1 + aX_i \tag{III-65}$$

Thus:

$$\mathrm{e}^{vX_K} = \begin{pmatrix} 1 & v & 0 \\ 0 & 1 & 0 \\ 0 & 0 & 1 \end{pmatrix} \quad \mathrm{e}^{bX_P} = \begin{pmatrix} 1 & 0 & b \\ 0 & 1 & 0 \\ 0 & 0 & 1 \end{pmatrix} \quad \mathrm{e}^{\tau X_H} = \begin{pmatrix} 1 & 0 & 0 \\ 0 & 1 & -\tau \\ 0 & 0 & 1 \end{pmatrix} \tag{III-66a}$$

which leads to:

$$M(b,v,t) = \mathrm{e}^{\tau X_H}\,\mathrm{e}^{bX_P}\,\mathrm{e}^{vX_K} \tag{III-66b}$$

B-3. SU(2) group

This is the group of the 2×2 complex unitary ($M^\dagger M = MM^\dagger = \mathbb{1}$) matrices, which have a determinant equal to 1 and are hence unimodular.

B-3-a. Form of the matrices and Lie algebra

The eigenvectors of any unitary matrix, associated with different eigenvalues, are orthogonal. They form an orthonormal basis in which the matrix can be diagonalized. In this basis, M becomes M', which depends on two real quantities, φ_1 and φ_2:

$$M' = \begin{pmatrix} e^{-i\varphi_1} & 0 \\ 0 & e^{-i\varphi_2} \end{pmatrix} \tag{III-67}$$

Setting:

$$N' = \begin{pmatrix} \varphi_1 & 0 \\ 0 & \varphi_2 \end{pmatrix} \tag{III-68}$$

readily yields $M' = e^{-iN'}$. Coming back to the initial basis, N' becomes N, and we have:

$$M = e^{-iN} \tag{III-69}$$

where N is a Hermitian matrix [according to (III-68), it is symmetric and real in an orthonormal basis]. As the determinant does not depend on the basis:

$$\det M = \det M' = e^{-i(\varphi_1+\varphi_2)} = e^{-i\,\mathrm{Tr}\{N'\}} \tag{III-70}$$

which, for the unimodular matrices of SU(2), is equal to 1. Consequently, as the trace does not depend on the basis:

$$\mathrm{Tr}\{N'\} = \mathrm{Tr}\{N\} = 2n\pi \tag{III-71}$$

where n is an integer. We can choose $n = 0$ as subtracting from φ_1 (or from φ_2) any multiple of 2π does not change the M matrix. Calling σ_x, σ_y, and σ_z the Pauli matrices:

$$\sigma_x = \begin{pmatrix} 0 & 1 \\ 1 & 0 \end{pmatrix} \qquad \sigma_y = \begin{pmatrix} 0 & -i \\ i & 0 \end{pmatrix} \qquad \sigma_z = \begin{pmatrix} 1 & 0 \\ 0 & -1 \end{pmatrix} \tag{III-72}$$

we can write:

$$N = \frac{a_1}{2}\sigma_x + \frac{a_2}{2}\sigma_y + \frac{a_3}{2}\sigma_z \tag{III-73}$$

since condition (III-71) with $n = 0$ shows that N cannot have a component on the 2×2 identity matrix. Finally:

$$\begin{aligned} M(\boldsymbol{a}) &= \mathrm{e}^{-i[a_1\sigma_x + a_2\sigma_y + a_3\sigma_z]/2} \\ &= \mathrm{e}^{-i\boldsymbol{a}\cdot\boldsymbol{\sigma}/2} = \mathrm{e}^{-ia\,\sigma_u/2} \end{aligned} \tag{III-74}$$

where:

$$\begin{aligned} &a = \sqrt{a_1^2 + a_2^2 + a_3^2} \\ &\boldsymbol{u} = \boldsymbol{a}/a \text{ (unit vector parallel to the vector } \boldsymbol{a}\text{, see figure 10)} \\ &\sigma_u = \boldsymbol{\sigma}\cdot\boldsymbol{u} = [a_1\sigma_x + a_2\sigma_y + a_3\sigma_z]/a \end{aligned} \tag{III-75}$$

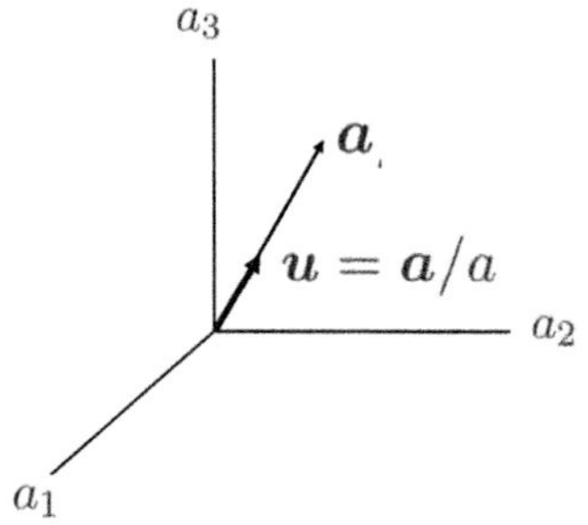

Figure 10: The matrices of the SU(2) group depend on three parameters, which can be represented in the ordinary (three-dimensional) space by a vector $\boldsymbol{a}$ *of length* a. *We note* $\boldsymbol{u} = \boldsymbol{a}/a$ *the unit vector parallel to the vector* $\boldsymbol{a}$.

As the square of any Pauli matrix is the identity matrix, (III-56) can be generalized:

$$\mathrm{e}^{-ia\sigma_u/2} = \cos\frac{a}{2} - i\sin\frac{a}{2}\sigma_u \tag{III-76}$$

which yields:

$$M(\boldsymbol{a}) = \begin{pmatrix} \cos\frac{a}{2} - i\,u_z\sin\frac{a}{2} & (-i\,u_x - u_y)\sin\frac{a}{2} \\ (-i\,u_x + u_y)\sin\frac{a}{2} & \cos\frac{a}{2} + i\,u_z\sin\frac{a}{2} \end{pmatrix} \tag{III-77}$$

(where u_x, u_y, and u_z are the 3 components a_1/a, a_2/a, and a_3/a of the vector $\boldsymbol{u}$). The product of two unitary unimodular matrices has the same property.

The dimension of the SU(2) group is $n = 3$. Taking the derivative of (III-74) or (III-77), we obtain the three operators that generate the Lie algebra:

$$\begin{cases} X_1 = \dfrac{\partial}{\partial a_1} M(\boldsymbol{a})|_{\boldsymbol{a}=0} = -i\,\sigma_x/2 \\ X_2 = \dfrac{\partial}{\partial a_2} M(\boldsymbol{a})|_{\boldsymbol{a}=0} = -i\,\sigma_y/2 \\ X_3 = \dfrac{\partial}{\partial a_3} M(\boldsymbol{a})|_{\boldsymbol{a}=0} = -i\,\sigma_z/2 \end{cases} \tag{III-78}$$

As we know that:

$$[\sigma_x, \sigma_y] = 2i\sigma_z \qquad [\sigma_y, \sigma_z] = 2i\sigma_x \qquad [\sigma_z, \sigma_x] = 2i\sigma_y \tag{III-79}$$

we can write:

$$[X_1, X_2] = X_3 \qquad [X_2, X_3] = X_1 \qquad [X_3, X_1] = X_2 \tag{III-80}$$

These relations readily yield the 9 structure constants C_{ij}^k of the group, 6 of which are zero[9], those for which $k = i$ or j. We will come back to this Lie algebra, which is also that of SO(3).

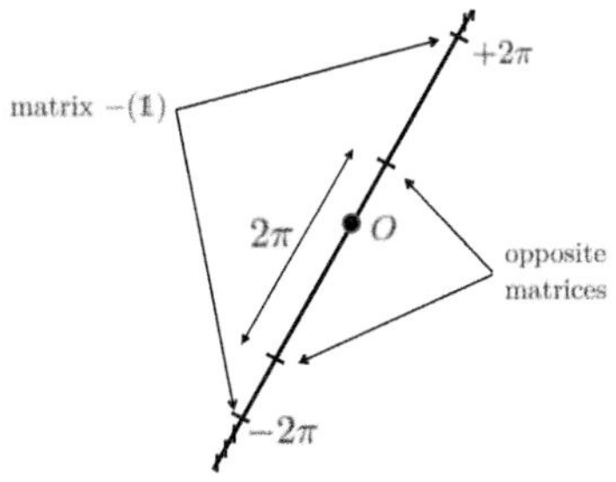

Figure 11: In each direction of the three-dimensional space, the vector $\boldsymbol{a}$ *must vary from* $+2\pi$ *to* -2π *to cover all the matrices of the* $SU(2)$ *group. The identity matrix* $(\mathbb{1})$ *is obtained at the center* O *of the interval* $[-2\pi, +2\pi]$*, whose extremities correspond to the same matrix* $-(\mathbb{1})$*. Two matrices whose parameters differ by* 2π *are opposite. As the direction of* $\boldsymbol{a}$ *changes, its variation domain becomes a sphere of radius* 2π*.*

[9] We have $C_{ij}^k = \varepsilon_{ijk}$, where ε_{ijk} is the Levi-Civita symbol. That symbol is zero if any two i, j, k indices are the same, equal to $+1$ if ijk is a circular permutation of the three indices 123, and to -1 if it is a transposition of two of those indices.

B-3-b. General properties of the group

Consider the variation domain D of $\boldsymbol{a}$ in real three-dimensional space. Choosing the basis where σ_u is diagonal, we can write:

$$e^{-ia\,\sigma_u/2} = \begin{pmatrix} e^{-ia/2} & 0 \\ 0 & e^{ia/2} \end{pmatrix} \tag{III-81}$$

This means that, along the axis of vector $\boldsymbol{u}$, a must vary between -2π and $+2\pi$ to be able to obtain any unitary diagonal matrix (with a determinant equal to 1). Along this axis, two vectors $\boldsymbol{a}$ whose extremities differ by 2π correspond to opposite matrices (figure 11).

When the unit vector $\boldsymbol{u}$ can take any direction, the domain accessible to the parameter $\boldsymbol{a}$ becomes the inside of a sphere of diameter 4π. The center of the sphere corresponds to the identity matrix. The points on the surface all correspond to the same matrix, the opposite of the identity matrix. We shall thus gather all these points together, imagining a sort of space "folding" onto itself (impossible in reality) where all the points on the surface of the sphere come together [in a way similar to what we did for SO(2), in § A-2; or as the operation of folding a cylinder into a torus by bringing its two extremities into contact]. As a result, a path such as the one represented in figure 12 is a closed path.

Another way to represent this group, giving the $(\mathbb{1})$ and $-(\mathbb{1})$ matrices a more symmetric role, would be to use two spheres, each having a diameter 2π, the first one centered at O (identity matrix), the second at O' (matrix opposed to the identity matrix), and identifying the points A and A' on their surfaces when

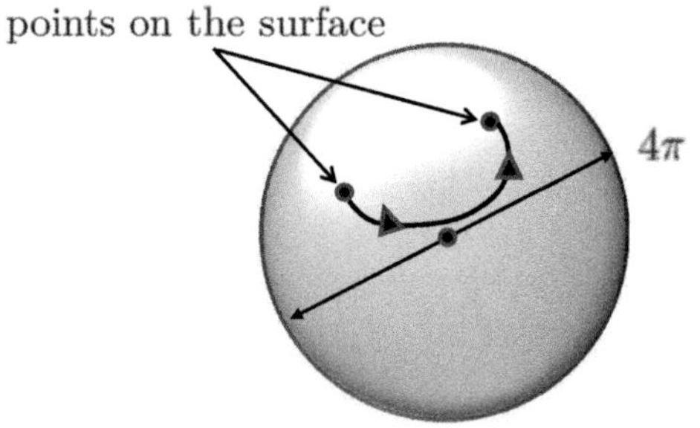

Figure 12: Representation of a closed path in $SU(2)$. It appears on the figure as an open line having its extremities at two different points on the surface of the sphere of radius 2π. Nevertheless, these two points are associated with the same $-(\mathbb{1})$ matrix of $SU(2)$, which means that the path is actually closed.

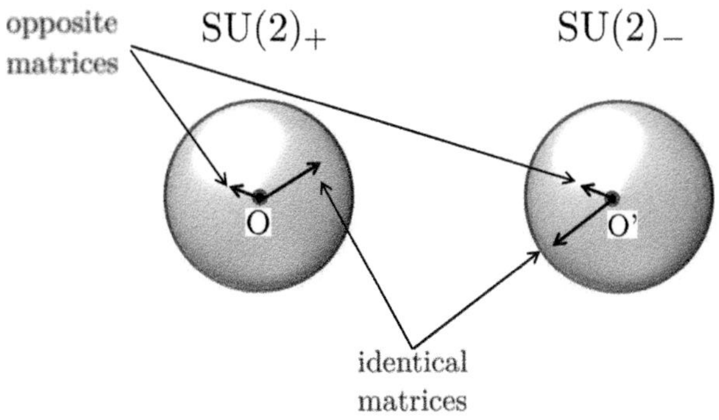

Figure 13: Another representation of the SU(2) group, different from the one shown in figure 12. We separate the two subsets $SU(2)_+$ (matrices with a positive trace) and $SU(2)_-$ (matrices with a negative trace), associated with two different spheres of radius π. We shall consider as identical the matrices on the surface of each spheres, described by opposite vectors starting from the centers O and O' of each spheres, and ending at points A and A'. Two matrices described by the same vector (regardless of its length) but starting from each center O and O', are opposite matrices.

$\overrightarrow{OA} = -\overrightarrow{O'A'}$ (figure 13). More precisely, the points $\boldsymbol{a}$ of the initial sphere of diameter 4π correspond, either to a point of the sphere on the left, or to a point of the one on the right. If $0 \leq a \leq \pi$ ("heart" of the initial sphere), it is simply a point A of the sphere on the left, such that $\overrightarrow{OA} = \boldsymbol{a}$. On the other hand, if $\pi \leq a \leq 2\pi$ ("external crown" of the initial sphere), one chooses a point A' of the sphere on the right, such that:

$$\overrightarrow{O'A'} = \boldsymbol{a} - 2\pi\frac{\boldsymbol{a}}{a} = \boldsymbol{a}' \tag{III-82}$$

($\boldsymbol{a}'$ is the parameter corresponding to the opposite matrix). It is easy to check that:

$$\overrightarrow{O'A'} \to \mathbf{0} \quad \text{if} \quad a \to 2\pi \tag{III-83}$$

This point of view amounts to decomposing SU(2) into two subsets, $\mathrm{SU}(2)_+$ and $\mathrm{SU}(2)_-$, each associated with one of the two spheres of figure 13, and hence to opposite matrices. These subsets can be characterized by noting that, according to relation (III-77), the trace of each SU(2) matrix is given by:

$$\mathrm{Tr}\,\{M(\boldsymbol{a})\} = 2\cos\frac{a}{2} \tag{III-84}$$

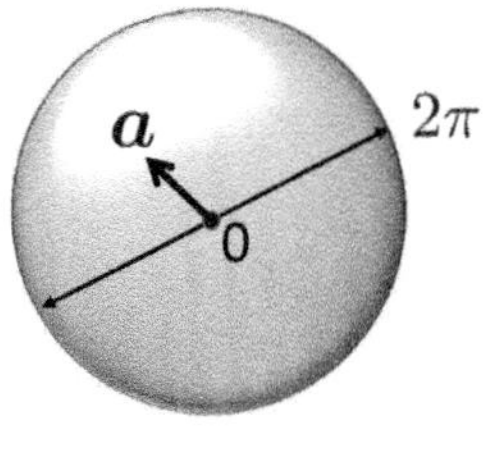

Figure 14: Rotations are determined by a parameter $\boldsymbol{a}$ *whose extremity lies inside a sphere with a* 2π *diameter. A zero rotation corresponds to a zero vector* $\boldsymbol{a} = 0$*. Diametrically opposed points on the surface of the sphere correspond to the same rotation.*

Consequently, the $\mathrm{SU}(2)_+$ matrices have a positive trace, whereas those of $\mathrm{SU}(2)_-$ have a negative trace. Neither the set $\mathrm{SU}(2)_+$, nor the set $\mathrm{SU}(2)_-$ are subgroups of SU(2): these subsets are not closed with respect to the matrix multiplication law. This point is easy to see on relation (III-81). We will discuss it in more detail in § A-3-d of chapter VII.

Looking at figure 13, one might believe that SU(2) is not connected. This is not true as corresponding points on the surfaces of the two spheres must actually be considered as the same point. The fact that SU(2) is connected is obvious in figure 12. In both figures, one sees that any closed path can be contracted into an arbitrarily small loop, which shows that SU(2) is simply connected. We shall see in § B-4 of complement $\mathrm{C_V}$ that this is not the case for SO(3), though this group has the same Lie algebra as SU(2).

B-4. Rotation group and its topology

A rotation is defined by a rotation axis and an angle. Both parameters can be summarized by a vector $\boldsymbol{a}$, parallel to the axis and whose length equals the rotation angle (expressed in radians). The rotation is clockwise with respect to $\boldsymbol{a}$, meaning that if $\boldsymbol{a}$ points along the Oz axis of an $Oxyz$ coordinate system, the rotation will bring Ox onto Oy. The ensemble of rotations forms a non-commutative group $R_{(3)}$.

The variation domain D of the vector $\boldsymbol{a}$ is the inside of a sphere with diameter 2π (figure 14). Two rotations with different $\boldsymbol{a}$ vectors are always different, except if the vectors' extremities are diametrically opposed on the surface of the sphere (two rotations around the same axis but with angles $+\pi$ and $-\pi$ yielding the same result). Two opposite vectors $\boldsymbol{a}$ such that $|\boldsymbol{a}| = \pi - \varepsilon$, where $\varepsilon \to 0$, correspond to infinitely close rotations. To preserve the topology of the rotation group, we must identify two by two all the points diametrically opposed on the surface of the sphere. This is an important difference with the SU(2) group, where we had to identify together *all* the points on the surface of the boundary sphere.

B-4-a. Commutation relations and Lie algebra

A 3×3 orthogonal matrix, with a $+1$ determinant, is associated with each rotation. The set of these matrices form the SO(3) group, isomorphic to $R_{(3)}$. For example, a rotation around the O_z axis through an angle φ is associated with the matrix:

$$\mathscr{R}_{\mathbf{e}_z}(\varphi) = \begin{pmatrix} \cos\varphi & -\sin\varphi & 0 \\ \sin\varphi & \cos\varphi & 0 \\ 0 & 0 & 1 \end{pmatrix} \tag{III-85}$$

More generally, the matrix $\mathscr{R}(\boldsymbol{a})$ is a function of the three components of the vector $\boldsymbol{a}$:

$$\mathscr{R}(\boldsymbol{a}) = \mathscr{R}(a_x, a_y, a_z) \tag{III-86}$$

A more detailed study of the rotation matrices will be given in chapter VII. All we need here are the infinitesimal rotation matrices. Using (III-85), we can write (with $\delta a_z = \delta\varphi$):

$$\mathscr{R}\left(0, 0, \delta a_z\right) = (\mathbb{1}) + \delta a_z \begin{pmatrix} 0 & -1 & 0 \\ 1 & 0 & 0 \\ 0 & 0 & 0 \end{pmatrix} = (\mathbb{1}) + \delta a_z \mathscr{M}_z \tag{III-87}$$

with:

$$\mathscr{M}_z = \begin{pmatrix} 0 & -1 & 0 \\ 1 & 0 & 0 \\ 0 & 0 & 0 \end{pmatrix} \tag{III-88}$$

A circular permutation of the Ox, Oy, and Oz axes yields:

$$\begin{aligned} \mathscr{R}\left(\delta a_x, 0, 0\right) &= (\mathbb{1}) + \delta a_x \mathscr{M}_x \\ \mathscr{R}\left(0, \delta a_y, 0\right) &= (\mathbb{1}) + \delta a_y \mathscr{M}_y \end{aligned} \tag{III-89}$$

with :

$$\mathscr{M}_x = \begin{pmatrix} 0 & 0 & 0 \\ 0 & 0 & -1 \\ 0 & 1 & 0 \end{pmatrix} \qquad \mathscr{M}_y = \begin{pmatrix} 0 & 0 & 1 \\ 0 & 0 & 0 \\ -1 & 0 & 0 \end{pmatrix} \tag{III-90}$$

It is easy to show that:

$$\mathscr{M}_x\mathscr{M}_y = \begin{pmatrix} 0\,0\,0 \\ 1\,0\,0 \\ 0\,0\,0 \end{pmatrix} \qquad \mathscr{M}_y\mathscr{M}_x = \begin{pmatrix} 0\,1\,0 \\ 0\,0\,0 \\ 0\,0\,0 \end{pmatrix} \tag{III-91}$$

and thus:

$$[\mathscr{M}_x\,,\mathscr{M}_y] = \mathscr{M}_z \tag{III-92}$$

as well as the other two relations obtained by a circular permutation of x, y, and z. We obtain the same commutation relations as in (III-80), for the SU(2) group, and thus the same Lie algebra. The two groups $R_{(3)}$ and SU(2) have identical structure constants.

B-4-b. Topology

The rotation group is compact and connected. it is *not* simply connected, but 2-connected: its homotopy group Π_1 is homomorphic to the cyclic abelian group Z_2. A path as shown in figure 15, with its extremities at two diametrically opposed points A and A' on the surface of the sphere, must be considered as closed (A and A' correspond to the same element of $R_{(3)}$). It is clear, however, that this path cannot be continuously reduced to an infinitesimal path since A and A' must always remain diametrically opposed.

However, if the path goes through two couples of points diametrically opposed, A and A', B and B', it can be continuously reduced to a zero path: this closed path is then homotopic to zero. To see this, imagine in $R_{(3)}$ a closed path represented on the left-hand side of figure 16. This path penetrates the sphere through two couples of diametrically opposed points, A and A' on one side, B and B' on the other. We can change the directions of the AA' and BB' axes,

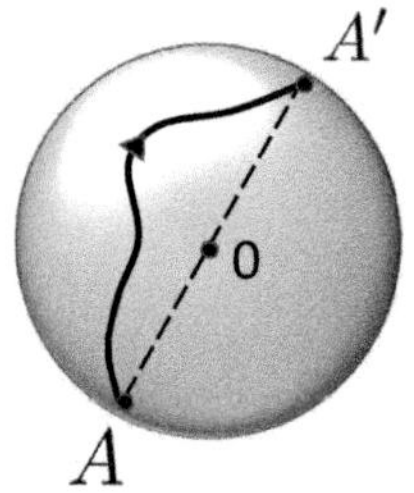

Figure 15: A path whose extremities are two diametrically opposed points on the sphere of radiums π is actually a closed path (a loop).

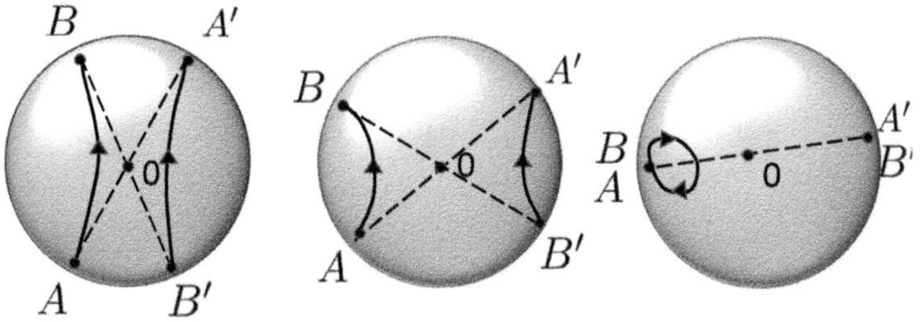

Figure 16: A path that penetrates the sphere through two couples of opposite points A and A′, B and B′, is equivalent to a closed path, homotopic to zero: continuously moving the points on the sphere, one can eliminate two by two the parts of the path that go through the sphere.

so as to move B closer to A, and B' closer to A' (middle figure). When those points coincide, the path is now formed of two loops, one going through A, the other through A'. Following the closed path, one will successively go around a loop starting and ending in A, then around another one starting and ending in A'. One can continuously reduce one of the loops to a point (zero path), point A' for example. As a result, after going around the first non-zero loop, one remains immobile in A' (figure on the right). This amounts to saying that one remains stationary at A, and we can ignore the second loop. It is then obvious that the remaining loop is homotopic to zero.

In a more general way, this process enables canceling by pairs the couples of points where the path penetrates the surface of the sphere (it goes through a rotation of an angle π). This does not apply to an isolated couple. Consequently, the closed paths having an even number of these couples are homotopic to zero, those having an odd number are homotopic to the loop shown in figure 15. In particular, for rotations around a fixed axis, a path where the object goes around once (2π rotation) is non-homotopic to zero, whereas if it goes around twice (4π), it will be homotopic to zero, etc.

This reveals an important property of $R_{(3)}$ [or $\mathrm{SO}_{(3)}$], which has a more complex topological structure than a simply connected group. Compare, for example, $R_{(3)}$ to SU(2), a group with the same dimension $n = 3$ and the same structure constants (see § B-4-a). The $R_{(3)}$ and SU(2) groups are locally, but not globally, isomorphic.

We can define a homomorphism between SU(2) and $R_{(3)}$ (this point will be more explicit in complement $\mathrm{A_{VII}}$) as there is a correspondence between any matrix of SU(2) defined by the parameter $\boldsymbol{a}$ as in § B-3, and the rotation $\mathscr{R}$

associated with the same value of the parameter if $|\boldsymbol{a}| < \pi$; or with the value of $\boldsymbol{a}$ defining the opposite matrix of SU(2). SU(2) appears "twice as large" as $R_{(3)}$, as can be seen by comparing[10] figure 14 with figure 13.

Unlike $R_{(3)}$, the SU(2) group is simply connected, as we saw in § B-3. The SU(2) group is thus the "universal covering group" of $R_{(3)}$ [*cf.* note (7)] The general space property whereby the rotation group is 2-connected has important consequences: in quantum mechanics, it allows the existence of particles with half-integer spins. We will return to this point in chapter VII.

C. Galilean and Poincaré groups

Galilean and Poincaré groups play a particularly fundamental role in physics. They express our essential conceptions of space and time, and constitute a framework for all physical theories ("superlaws").

Historically, physicists started by accepting (more or less explicitly) a first sort of relativity, that of Galileo[11]. In this case, the space–time transformation laws are simple and intuitive, as will be explained. This will allow us to clearly point out the modifications imposed by Einstein's relativity (Lorentz transformations, Poincaré group).

C-1. Galilean group

The Galilean group is formed by the set of transformations that change the space–time coordinates of an event from one Galilean reference frame to another. They are linear space–time transformations that, for any two events, preserve their time interval $t_2 - t_1$, as well as the distance $|\boldsymbol{r}_2 - \boldsymbol{r}_1|$ of all the simultaneous events. There are several subgroups of the Galilean group:

(i) The $R_{(3)}$ group of rotations around a fixed point O. We obtained in § B-4-a the following commutation rules for the elementary rotation operators:

$$[\mathscr{M}_x\,, \mathscr{M}_y] = \mathscr{M}_z \qquad [\mathscr{M}_y\,, \mathscr{M}_z] = \mathscr{M}_x \qquad [\mathscr{M}_z\,, \mathscr{M}_x] = \mathscr{M}_y \qquad \text{(III-93)}$$

[10]The center of SU(2) [i.e. the invariant subgroup that commutes with all the elements of SU(2)] consists of the 2×2 identity matrix and its opposite. This is a group with 2 elements, isomorphic to Z_2. The rotation group is isomorphic to the quotient group $\mathrm{SU}(2)/Z_2$ (*cf.* complement $\mathrm{A_{II}}$).

[11]A physical law invariant in a Galilean transformation is often called "non-relativistic". This is a poor denomination as, theoretically, Galilean transformations are no "less relativistic" than Lorentz transformations. We now know that nature obeys the Einstein relativistic invariance laws, and not those of Galileo. This evidently limits the impact of Galilean transformations. However, in many interesting physical situations, they offer good approximations to the Lorentz transformations [*cf.* note (1) of chapter I].

(ii) Translation group.

– Space translations, the $T_{(3)}$ group. There are three infinitesimal translation operators, which commute with each other as opposed to the infinitesimal rotation operators. This is because translations commute with each other (this group will be studied in more detail in § 1-a of complement C_V).

– Time translations.

Introducing a homogeneous coordinate, we can write the matrices associated with all infinitesimal space–time translations, and verify that they all commute with each other.

(iii) Pure Galilean transformations. In three-dimensional space, the coordinate transformations are written[12]:

$$\begin{cases} x' = x + v_x\, t \\ y' = y + v_y\, t \\ z' = z + v_z\, t \\ t' = t \end{cases} \tag{III-94}$$

where the space and time coordinates are mixed together.

(iv) We could add the space symmetry:

$$\begin{cases} \boldsymbol{r}' = -\boldsymbol{r} \\ t' = t \end{cases} \tag{III-95a}$$

or the time symmetry:

$$\begin{cases} \boldsymbol{r}' = \boldsymbol{r} \\ t' = -t \end{cases} \tag{III-95b}$$

but these transformations do not depend on a continuous parameter, and will not be discussed here (*cf.* complement D_V and appendix).

Combining these sets, we get other subgroups of the Galilean group. For example, rotations and translations (i.e. the continuous geometrical operations) yield the displacement group $\mathscr{E}_{(3)}$.

Galilean transformations can be written with 4×4 matrices, acting on the 4 space–time coordinates x, y, z, and t. The translations, however, do not correspond to linear transformations of these coordinates. To be able to represent

[12]The sign we choose for the velocity parameter $\boldsymbol{v}$ is the same as in § B-2, *cf.* note 8: the origin O' of the second reference frame moves with velocity $-\boldsymbol{v}$ with respect to the first reference frame.

them with matrices, we add a fifth coordinate, equal to one (homogeneous coordinate). As a result, with each transformation of the group is associated a 5×5 matrix such that:

$$\begin{pmatrix} x' \\ y' \\ z' \\ t' \\ 1 \end{pmatrix} = M \begin{pmatrix} x \\ y \\ z \\ t \\ 1 \end{pmatrix} \qquad \text{(III-96)}$$

where :

$$M = \left(\begin{array}{ccc|cc} & & & v_x & b_x \\ & \mathscr{R}(\boldsymbol{a}) & & v_y & b_y \\ & & & v_z & b_z \\ \hline 0 & 0 & 0 & 1 & -\tau \\ 0 & 0 & 0 & 0 & 1 \end{array}\right) \qquad \text{(III-97)}$$

Each transformation depends on a number of parameters given by: 3 (rotations) + 3 (translations) + 3 (pure Galilean transformations) + 1 (time translation) = 10. We shall write:

$$M \equiv M\left(\boldsymbol{a}, \boldsymbol{b}, \boldsymbol{v}, \tau\right) \qquad \text{(III-98)}$$

To each parameter corresponds an infinitesimal operator:

$$\begin{aligned} &\boldsymbol{a} : a_x, a_y, a_z \Longrightarrow X_{J_x, J_y, J_z} \\ &\boldsymbol{b} : b_x, b_y, b_z \Longrightarrow X_{P_x, P_y, P_z} \\ &\boldsymbol{v} : v_x, v_y, v_z \Longrightarrow X_{K_x, K_y, K_z} \\ &\tau \Longrightarrow X_H \end{aligned} \qquad \text{(III-99)}$$

(the letters chosen for the various indices will be explained later). They correspond to the following 5×5 matrices [the vertical and horizontal lines delimit two sectors, one corresponding to the 3 space coordinates, and one to the coordinate t and unity; a zero in one of the two sectors means that all the matrix elements of that sector are zero, for instance 4 in the bottom right square of III-100a)]:

$$X_{J_{x,y,z}} = \left(\begin{array}{ccc|cc} & & & 0 & 0 \\ & \mathscr{M}_{x,y,z} & & 0 & 0 \\ & & & 0 & 0 \\ \hline 0 & 0 & 0 & 0 & 0 \\ 0 & 0 & 0 & 0 & 0 \end{array}\right) \qquad \text{(III-100a)}$$

$$X_{P_x} = \left(\begin{array}{ccc|cc} 0&0&0&0&1 \\ 0&0&0&0&0 \\ 0&0&0&0&0 \\ \hline 0&0&0&0&0 \\ 0&0&0&0&0 \end{array}\right) \quad X_{P_y} = \left(\begin{array}{ccc|cc} 0&0&0&0&0 \\ 0&0&0&0&1 \\ 0&0&0&0&0 \\ \hline 0&0&0&0&0 \\ 0&0&0&0&0 \end{array}\right) \quad X_{P_z} = \left(\begin{array}{ccc|cc} 0&0&0&0&0 \\ 0&0&0&0&0 \\ 0&0&0&0&1 \\ \hline 0&0&0&0&0 \\ 0&0&0&0&0 \end{array}\right) \tag{III-100b}$$

$$X_{K_x} = \left(\begin{array}{ccc|cc} 0&0&0&1&0 \\ 0&0&0&0&0 \\ 0&0&0&0&0 \\ \hline 0&0&0&0&0 \\ 0&0&0&0&0 \end{array}\right) \quad X_{K_y} = \left(\begin{array}{ccc|cc} 0&0&0&0&0 \\ 0&0&0&1&0 \\ 0&0&0&0&0 \\ \hline 0&0&0&0&0 \\ 0&0&0&0&0 \end{array}\right) \quad X_{K_z} = \left(\begin{array}{ccc|cc} 0&0&0&0&0 \\ 0&0&0&0&0 \\ 0&0&0&1&0 \\ \hline 0&0&0&0&0 \\ 0&0&0&0&0 \end{array}\right) \tag{III-100c}$$

$$X_H = \left(\begin{array}{ccc|cc} 0&0&0&0&0 \\ 0&0&0&0&0 \\ 0&0&0&0&0 \\ \hline 0&0&0&0&-1 \\ 0&0&0&0&0 \end{array}\right) \tag{III-100d}$$

Comment:

The -1 appearing in X_H has the same origin as the minus sign in the second equation (I-14a) of chapter I: to move forward in time a space–time event, one must decrease its temporal coordinate.

From now on, and to avoid triple index notations such as $X_{J_{x_i}}$, we shall note X_{J_i}, X_{K_i}, etc. the X operators associated with J_{x_i}, K_{x_i}, etc. . The symbol ε_{ijk} equals 0 if two of the indices i, j, k are the same, $+1$ if i, j, k are an even permutation of xyz and -1 they are an odd permutation. Finally, we shall imply a summation over a repeated index k (Einstein summation convention). The following summations have only one non-zero term (if $i \neq j$), or even none (if $i = j$). This leads to the following commutation relations:

$$\left[X_{J_i}, X_{J_j}\right] = \varepsilon_{ijk}\, X_{J_k} \tag{III-101a}$$

$$\left[X_{J_i}, X_{P_j}\right] = \varepsilon_{ijk}\, X_{P_k} \tag{III-101b}$$

$$\left[X_{P_i}, X_{P_j}\right] = 0 \qquad \text{or} \quad [X_{\boldsymbol{P}}, X_{\boldsymbol{P}}] = 0 \tag{III-101c}$$

$$\left[X_{J_i}, X_{K_j}\right] = \varepsilon_{ijk}\, X_{K_k} \tag{III-101d}$$

$$\left[X_{P_i}, X_{K_j}\right] = 0 \qquad \text{or} \quad [X_{\boldsymbol{P}}, X_{\boldsymbol{K}}] = 0 \tag{III-101e}$$

$$\left[X_{K_i}, X_{K_j}\right] = 0 \qquad \text{or} \quad [X_{\boldsymbol{K}}, X_{\boldsymbol{K}}] = 0 \tag{III-101f}$$

$$[X_{J_i}, X_H] = 0 \quad \text{or} \quad [X_{\boldsymbol{J}}, X_H] = 0 \tag{III-101g}$$

$$[X_{P_i}, X_H] = 0 \quad \text{or} \quad [X_{\boldsymbol{P}}, X_H] = 0 \tag{III-101h}$$

$$[X_{K_i}, X_H] = -X_{P_i} \quad \text{or} \quad [X_{\boldsymbol{K}}, X_H] = -X_{\boldsymbol{P}} \tag{III-101i}$$

These relations yield the structure constants of the Galilean group, many of which are zero. They can be interpreted as follows:

- (III-101a): non-commutativity of infinitesimal rotations (*cf.* chapter VII);
- (III-101b): non-commutativity of rotations and translations in space;
- (III-101d): non-commutativity of rotations and changes of Galilean frame.

The similarity between relations (III-101a, b, d) arises from the fact that $\boldsymbol{J}$, $\boldsymbol{P}$ and $\boldsymbol{K}$ are vector quantities (*cf.* § A of chapter VIII).

Relations (III-101g) and (III-101h) express the commutativity between either purely spatial operations (translations, rotations), or time-dependent operations. As an example, relation (III-101h) expresses the commutativity of space and time translations in the Galilean group. As for the last relation (III-101i), it has already been discussed (§ B-2) in a one-dimensional case.

Note that, so far, we have not used any quantum mechanics postulates (quantization rules, etc.). Nevertheless, commutation relations appear, similar to those of angular momenta: $[J_x, J_y] = i\hbar J_z$, etc. or to that of the free particle Hamiltonian H with its momentum: $[\boldsymbol{P}, H] = 0$, (the definition of the quantum operators associated with the X will be given in later chapters). This shows the very fundamental character of these commutation relations, which are more closely linked to the very structure of space–time rather than to the type of quantum approach in use, or to the nature of the system under study. For example, the commutation relations will be the same for a field or for a system of particles.

Comment:

We noted, on several occasions (*cf.* for instance chapter I, third example of § B-2-b-γ), the link between pure Galilean transformations and the position of the system's center of mass. We thus expect a relation of the type:

$$R_i = -X_{K_i}/M \tag{III-102}$$

(where M is the system's mass, such as the mass of an isolated particle). This relation does indeed exist, and equation (III-101i) simply states that the time variations of a system's position are proportional to its momentum ($\frac{\mathrm{d}\boldsymbol{r}}{\mathrm{d}t} = \frac{\boldsymbol{p}}{M}$). Now how should we understand (III-101e) since we know that, in quantum mechanics, position and momentum do not commute?

This difficulty is solved, in the case of the Galilean group, by the need to use projective representations whose phase factors change the commutation rules (III-101e) – *cf.* chapter IV, § E-1-b. When these phase factors are properly taken into account, we shall see in § B of chapter V that we easily find all the usual quantum mechanical relations, including those concerning the particle's spin.

C-2. Poincaré group

The Poincaré group concerns space–time transformations in the framework of Einstein's relativity. It is the set of linear space–time transformations that preserve the space–time interval:

$$c^2\left(t_2-t_1\right)^2-\left(\boldsymbol{r}_1-\boldsymbol{r}_2\right)^2 \tag{III-103}$$

between any two events. Examples of such transformations are:

(i) Rotations (already mentioned in the preceding paragraph);

(ii) Space and time translations (also mentioned earlier);

(iii) *Special* (or *pure*) Lorentz transformations. These transformations depend on a vector parameter $\boldsymbol{v}$ (i.e. on 3 parameters, like the Galilean transformations). We set:

$$\beta=v/c \qquad \gamma=\left(1-\beta^2\right)^{-1/2} \tag{III-104}$$

When $\boldsymbol{v}$ is parallel to Ox, these transformations are written:

$$\begin{cases} x'=\dfrac{x+vt}{\sqrt{1-\beta^2}}=\gamma(x+\beta ct) \\ y'=y \\ z'=z \\ t'=\dfrac{t+vx/c^2}{\sqrt{1-\beta^2}}=\gamma\left(t+\dfrac{\beta x}{c}\right) \end{cases} \tag{III-105}$$

These are homogeneous transformations between the 4 space–time coordinates, which can be associated with 4×4 matrices $\Lambda(\boldsymbol{v})$. When $\boldsymbol{v}$ is parallel [13] to Ox, relation (III-105) enables us to write:

$$\Lambda(v,0,0)=\begin{pmatrix} \gamma & 0 & 0 & \gamma v \\ 0 & 1 & 0 & 0 \\ 0 & 0 & 1 & 0 \\ \gamma v/c^2 & 0 & 0 & \gamma \end{pmatrix} \tag{III-106}$$

Note that these matrices (associated with pure Lorentz transformations) do *not* form a group. We shall come back to this point later.

(iv) Point reflection (symmetry with respect to a fixed point O);

[13] See for example Jackson's book [46], p. 517 and 541 for an expression of $\Lambda(\boldsymbol{v})$ in the general case.

(v) Time reversal.

Combining these different transformations leads to different groups:

- (i) and (iii): Rotations and pure Lorentz transformations form the "*proper* (sometimes called the *restricted*) Lorentz group". It is associated with a set of 4×4 matrices, each having a positive determinant, describing homogeneous transformations of space–time coordinates, all generated by infinitesimal transformations.

- (i), (ii), and (iii): connected Poincaré group, where all operations are continuous.

- (i), (iii) and (iv): This leads to the "*orthochronous* Lorentz group". Time is never reversed: if two events are separated by a timelike space–time interval, anteriority is preserved in any reference frame. Since parity is a discrete, non-continuous, operation, this group is not connected.

- Adding (v) to the previous operations leads to the "*complete* Lorentz group" (homogeneous transformations associated with 4×4 matrices). The combined action of time and space reflections creates in this group 4 disconnected "mappings".

Until now, we have ignored the transformations (ii), as they are not homogeneous. If we take all the transformations, including the ones generated by (ii), we get the "*inhomogeneous* Lorentz group" or " *Poincaré group*", the set of all the linear space–time transformations that preserve the space–time intervals.

As for the Galilean group, we limit ourselves to the study of transformations [obtained from (i), (ii), and (iii)] that can be generated from infinitesimal transformations. Writing:

$$\begin{pmatrix} x' \\ y' \\ z' \\ t' \\ 1 \end{pmatrix} = M\left(\boldsymbol{a}, \boldsymbol{b}, \boldsymbol{v}, \tau\right) \begin{pmatrix} x \\ y \\ z \\ t \\ 1 \end{pmatrix} \tag{III-107}$$

we deduce the commutation relations between the infinitesimal operators. It is not necessary to completely redo the calculations of the previous § C-1. The only differences arise from the fact that space and time are more intimately mixed in the Lorentz group than in the Galilean one. For all the purely spatial transformations, where time plays no role, the transformation laws and the M

matrices remain unchanged. Consequently, we still have:

$$\left[X_{J_i}, X_{J_j}\right] = \varepsilon_{ijk}\, X_{J_k} \tag{III-108a}$$

$$\left[X_{J_i}, X_{P_j}\right] = \varepsilon_{ijk}\, X_{P_k} \tag{III-108b}$$

$$[X_{\boldsymbol{P}}, X_{\boldsymbol{P}}] = 0 \tag{III-108c}$$

X_H can still be written as in (III-100c), and we have:

$$[X_{\boldsymbol{J}}, X_H] = 0 \tag{III-108d}$$

$$[X_{\boldsymbol{P}}, X_H] = 0 \tag{III-108e}$$

Changes appear when $X_{\boldsymbol{K}}$ come into play. For example, we see on (III-106) that:

$$X_{K_x} = \left(\begin{array}{ccc|cc} 0 & 0 & 0 & 1 & 0 \\ 0 & 0 & 0 & 0 & 0 \\ 0 & 0 & 0 & 0 & 0 \\ \hline 1/c^2 & 0 & 0 & 0 & 0 \\ 0 & 0 & 0 & 0 & 0 \end{array}\right) \tag{III-109}$$

and, exchanging the Ox and Oy axes:

$$X_{K_y} = \left(\begin{array}{ccc|cc} 0 & 0 & 0 & 0 & 0 \\ 0 & 0 & 0 & 1 & 0 \\ 0 & 0 & 0 & 0 & 0 \\ \hline 0 & 1/c^2 & 0 & 0 & 0 \\ 0 & 0 & 0 & 0 & 0 \end{array}\right) \tag{III-110}$$

A similar expression is obtained for X_{K_z}, by shifting to the right the $1/c^2$ element, and to the bottom the element equal to 1. By comparison with (III-100c), we now have a new element equal to $1/c^2$ (which goes to 0 if $c \to \infty$), which is responsible for all the relativistic effects.

An elementary computation yields:

$$\left[X_{J_i}, X_{K_j}\right] = \varepsilon_{ijk}\, X_{K_k} \tag{III-111a}$$

$$\left[X_{K_{x_i}}, X_{P_{x_j}}\right] = -\frac{1}{c^2}\delta_{ij}\, X_H \tag{III-111b}$$

$$\left[X_{K_i}, X_{K_j}\right] = -\frac{1}{c^2}\varepsilon_{ijk}\, X_{J_k} \tag{III-111c}$$

$$[X_{K_i}, X_H] = -X_{P_i} \tag{III-111d}$$

Whereas the first and fourth relations are unchanged, the second and third change. The second expresses the way a pure Lorentz transformation changes momentum[14]: whereas the normal components are unchanged (hence the δ_{ij}), the parallel one follows the transformation law of the four-vector momentum-energy. The third commutation relation confirms that pure Lorentz transformations do not form a group: the product of successive transformations (commutator) yields a *rotation*. This explains the origin of the *Thomas precession.*

Exercise: Write the matrices explicitly, and give a demonstration of the above commutation relations.

Comments:

(i) More symmetric formulas are obtained by replacing time t by ct, so that all four space–time coordinates have the same dimension (length). The matrix of a pure Lorentz transformation (III-106) becomes:

$$\Lambda(v,0,0) = \begin{pmatrix} \gamma & 0 & 0 & \gamma v/c \\ 0 & 1 & 0 & 0 \\ 0 & 0 & 1 & 0 \\ \gamma v/c & 0 & 0 & \gamma \end{pmatrix} \tag{III-112}$$

and the X_{K_x} matrix becomes:

$$X_{K_x} = \left(\begin{array}{ccc|cc} 0 & 0 & 0 & 1/c & 0 \\ 0 & 0 & 0 & 0 & 0 \\ 0 & 0 & 0 & 0 & 0 \\ \hline 1/c & 0 & 0 & 0 & 0 \\ 0 & 0 & 0 & 0 & 0 \end{array}\right) \tag{III-113}$$

As we saw earlier, the X_{K_y} matrix is obtained by shifting the element in the first column to the right, and the element on the first row to the bottom; repeating the operation yields the X_{K_z} matrix. These three matrices are symmetric, whereas the X_J matrices corresponding to the rotation group are anti-symmetric. However, multiplying all those matrices by the relativistic metric tensor $g_{\mu\nu}$, we can ensure that all the matrices of the infinitesimal generators of the proper Lorentz group are anti-symmetric.

(ii) Instead of choosing the velocity $\boldsymbol{v}$ as a parameter in a Lorentz transformation, it is often convenient to use the "rapidity" q defined [13, 32] as:

$$\frac{v}{c} = \tanh q \tag{III-114}$$

[14] A translation and a change of Galilean frame in that order or in the reverse order do not yield the same result because of the Lorentz contraction of length.

As an example, this parameter is a natural choice in the computations of § 1-b of complement A_V on the proper Lorentz group. In the composition of two Lorentz transformations along parallel directions and with parameters q_1 and q_2, this choice has the advantage of yielding another Lorentz transformation with a $(q_1 + q_2)$ parameter(*cf.* complement A_V, § 1-e). In other words, the Φ_2 composition law defined in (III-3) is now simply linear. As $\gamma = \cosh q$, the matrix $\Lambda(v,0,0)$ is written:

$$\Lambda(v,0,0) = \begin{pmatrix} \cosh(q/c) & 0 & 0 & \sinh(q/c) \\ 0 & 1 & 0 & 0 \\ 0 & 0 & 1 & 0 \\ \sinh(q/c) & 0 & 0 & \cosh(q/c) \end{pmatrix} \tag{III-115}$$

Complement A_{III}

Adjoint representation, Killing form, Casimir operator

In this complement, we introduce (§ 1) the concept of "adjoint representation" of a Lie algebra $\mathscr{L}$. Starting from the structure constants C_{ij}^k of the group, we derive a number of useful quantities: the K_{ij} components of a covariant tensor, the so-called Killing form (§ 2); the anti-symmetric and completely covariant structure constants $\overline{C}_{ijk}$ (§ 3); the Casimir operator (§ 4), which is of interest as it commutes with all the X_i operators.

1. Adjoint representation of a Lie algebra

A Lie algebra $\mathscr{L}$ is a real vector space having as a basis the n operators X_i ($i = 1, 2, \ldots, n$). Any X operator of $\mathscr{L}$ can be written as:

$$X = \sum_{i=i}^{n} \lambda^i X_i \tag{1a}$$

where the λ^i are real numbers. When X is considered as a vector of $\mathscr{L}$, it will be noted $|X\rangle\rangle$; relation (1a) is then written:

$$|X\rangle\rangle = \sum_{i=i}^{n} \lambda^i \, |X_i\rangle\rangle \tag{1b}$$

In addition to its vector space structure, which implies the existence of addition as an internal composition law, a Lie algebra is also equipped with a product as a second internal composition law. When the Lie algebra concerns operators, this product is actually the commutator between two operators.

With any operator A of the Lie algebra, we can associate an "adjoint operator" $\mathscr{A}$ acting in the $\mathscr{L}$ vector space, and whose action on an arbitrary vector $|X\rangle\rangle$ is written:

$$\mathscr{A}|X\rangle\rangle = |[A, X]\rangle\rangle \tag{2}$$

Expanding A and X on the X_i basis:

$$\begin{cases} |A\rangle\rangle = \sum\limits_i a^i \, |X_i\rangle\rangle \\ |X\rangle\rangle = \sum\limits_j \lambda^j \, |X_j\rangle\rangle \end{cases} \tag{3}$$

we can write:

$$\mathscr{A}|X\rangle\rangle = \sum_{ij} a^i \, \lambda^j \, |[X_i, X_j]\rangle\rangle = \sum_{ijk} a^i \, \lambda^j \, C_{ij}^k \, |X_k\rangle\rangle \tag{4}$$

which contains the structure constants C_{ij}^k (real numbers) introduced in chapter III:

$$[X_i, X_j] = \sum_{k=1}^{n} C_{ij}^k \, X_k \tag{5}$$

Relation (4) shows that the (mixed) matrix elements of $\mathscr{A}$ in the $|X_k\rangle\rangle$ basis are:

$$(\mathscr{A})_j^k = \sum_i C_{ij}^k \, a^i \tag{6}$$

where j is the column index and k the row index of the matrix, which is real.

Similarly, for an operator $\mathscr{B}$ associated with $|B\rangle\rangle \in \mathscr{L}$:

$$|B\rangle\rangle = \sum_{i'} b^{i'} |X_{i'}\rangle\rangle \tag{7}$$

we can write:

$$(\mathscr{B})_\ell^j = \sum_{i'} C_{i'\ell}^j \, b^{i'} \tag{8}$$

The product:

$$\mathscr{B}\mathscr{A}|X\rangle\rangle = |[B, [A, X]]\rangle\rangle \tag{9}$$

is associated with the $(\mathscr{B}\mathscr{A})$ matrix whose elements are:

$$(\mathscr{B}\mathscr{A})_\ell^k = \sum_j (\mathscr{B})_\ell^j \, (\mathscr{A})_j^k = \sum_{iji'} a^i \, b^{i'} \, C_{ij}^k \, C_{i'\ell}^j \tag{10}$$

The trace of an operator acting in the $\mathscr{L}$ vector space is, by definition, the sum of its diagonal elements. This yields, for the $\mathscr{B}\mathscr{A}$ operator:

$$\text{Tr}\,\{\mathscr{B}\mathscr{A}\} = \sum_k \, (\mathscr{B}\mathscr{A})_k^k = \sum_{kj} (\mathscr{B})_k^j \, (\mathscr{A})_j^k \tag{11}$$

We can define in $\mathscr{L}$ a bilinear scalar product as the trace of the product $\mathscr{A}\mathscr{B}$:

$$\begin{aligned}\langle\langle A|B\rangle\rangle &= \mathrm{Tr}\{\mathscr{B}\mathscr{A}\}\\ &= \sum_{ij} a^i\, K_{ij}\, b^j\end{aligned} \tag{12}$$

where we have used the Killing form K defined as:

$$K_{ij} = \langle\langle X_i|X_j\rangle\rangle \;= \mathrm{Tr}\{\mathcal{X}_i\mathcal{X}_j\} \tag{13}$$

($\mathcal{X}_{i,j}$ is the adjoint operator of $X_{i,j}$). The scalar product $\langle\langle A|B\rangle\rangle$ is real since the components of the vectors are real. It is equal to $\langle\langle B|A\rangle\rangle$ since we can commute two operators in a trace.

The algebra structure of $\mathscr{L}$ allowed us to establish a correspondence between the vectors $|A\rangle\rangle$ of $\mathscr{L}$ and the operators $\mathscr{A}$ acting in $\mathscr{L}$. This is an homomorphic correspondence: with the product of two elements of the $\mathscr{L}$ algebra (i.e. with the commutator $|[A,B]\rangle\rangle \in \mathscr{L}$) is associated the commutator of operators $\mathscr{A}$ and $\mathscr{B}$, as we now show. Expliciting the product $\mathscr{A}\mathscr{B}$:

$$\mathscr{A}\mathscr{B}\,|X\rangle\rangle = |\,[A,[B,X]]\;\rangle\rangle \tag{14}$$

and the product $\mathscr{B}\mathscr{A}$:

$$\mathscr{B}\mathscr{A}\,|X\rangle\rangle = |\,[B,[A,X]]\;\rangle\rangle \tag{15}$$

we easily obtain, using the Jacobi equality:

$$\begin{aligned}[\mathscr{A},\mathscr{B}]\;|X\rangle\rangle &= |\,[A,[B,X]]\;\rangle\rangle - |\,[B,[A,X]]\;\rangle\rangle\\ &= |\,[[A,B]\,,X]\;\rangle\rangle\end{aligned} \tag{16}$$

The commutators of the $(\mathscr{A})_i^j$ matrices are thus associated with the commutators of the corresponding A operators. An $(\mathscr{A})$ matrix is associated with each infinitesimal operator $M(\delta\boldsymbol{a})$ of the group G ; by integration, one can associate $(\mathcal{M})$ matrices to the finite matrices $M(\boldsymbol{a})$ of the group. This correspondence yields a representation of the group, called the "adjoint representation of $\mathscr{L}$"; it has the same dimension as that of $\mathscr{L}$.

2. Killing form ; scalaire product and change of basis in $\mathscr{L}$

Inserting relations (6) and (8) into (11) and (12), we get another expression for the scalar product $\langle\langle A|B\rangle\rangle$:

$$\langle\langle A|B\rangle\rangle = \sum_{kiji'} a^i\, b^{i'}\, C_{ij}^k\, C_{i'k}^j = \sum_{ijkk'} a^i\, b^j\, C_{ik'}^k\, C_{jk}^{k'} \tag{17}$$

Comparing with relation(12) shows that:

$$K_{ij} = \sum_{kk'} C^k_{ik'}\, C^{k'}_{jk} \tag{18}$$

This expression is another definition of the Killing form, based on the structure constants. We obviously have:

$$K_{ij} = K_{ji} \tag{19}$$

which means that the $n \times n$ matrix K, with coefficients K_{ij}, is symmetric and real (hence diagonalizable).

Expression (12) contains a bilinear symmetric form of the two real vectors $|X\rangle\rangle$ and $|Y\rangle\rangle$; it is called the "Killing form". Note that when the structure of the G group is not specified, the eigenvalues of (K) do not necessarily have the same sign. Consequently, the scalar product $\langle\langle X|X\rangle\rangle$ of vector $|X\rangle\rangle$ by itself may be either positive or negative.

We can perform a basis change in $\mathscr{L}$, going from the $|X_i\rangle\rangle$ to the $|\overline{X}_j\rangle\rangle$ using a matrix S:

$$|\overline{X}_j\rangle\rangle = \sum_i S^i_j\, |X_i\rangle\rangle \tag{20}$$

with the inverse relation:

$$|X_i\rangle\rangle = \sum_j (S^{-1})^j_i\, |\overline{X}_j\rangle\rangle \tag{21}$$

We then naturally introduce the group structure constants pertaining to the $|\overline{X}\rangle\rangle$ basis, i.e. the numbers $\overline{C}^k_{ij}$ such that:

$$\left[\overline{X}_i, \overline{X}_j\right] = \sum_k \overline{C}^k_{ij}\, \overline{X}_k \tag{22}$$

It is easy to show that these structure constants are the twice covariant and once contravariant components of a tensor[1] of order 3. This is because:

$$\begin{aligned}\left[\overline{X}_i, \overline{X}_j\right] &= \sum_{\ell m} S^\ell_i\, S^m_j\, \left[X_\ell, X_m\right] = \sum_{\ell m p} S^\ell_i\, S^m_j\, C^p_{\ell m}\, X_p \\ &= \sum_{\ell m p k} S^\ell_i\, S^m_j\, (S^{-1})^k_p\, C^p_{\ell m}\, \overline{X}_k \end{aligned} \tag{23}$$

so that:

$$\overline{C}^k_{ij} = \sum_{\ell m p} S^\ell_i\, S^m_j\, (S^{-1})^k_p\, C^p_{\ell m} \tag{24}$$

[1] *Cf.* complement A_{VIII}: "Short review of classical tensors".

Based on the structure constants $\overline{C}_{ij}^k$, we can introduce the $(\overline{K})_{ij}$ matrix:

$$\overline{K}_{ij} = \sum_{kk'} \overline{C}_{ik}^{k'} \, \overline{C}_{jk'}^{k} \tag{25}$$

(twice covariant tensor obtained by the double contraction of an order 6 tensor). Using definition (12), we compute the scalar product $\langle\langle \overline{X}_i | \overline{X}_j \rangle\rangle$, and show that:

$$\overline{K}_{ij} = \langle\langle \overline{X}_i | \overline{X}_j \rangle\rangle \tag{26}$$

Demonstration: Inserting (24) into (25), yields:

$$\overline{K}_{ij} = \sum_{kk'} \sum_{lmp} S_i^l \, S_k^m \, (S^{-1})_p^{k'} C_{lm}^p \sum_{l'm'p'} S_j^{l'} \, S_{k'}^{m'} \, (S^{-1})_{p'}^k C_{l'm'}^{p'} \tag{27}$$

As the S and S^{-1} matrices are the inverse of each other:

$$\sum_k S_k^m \, (S^{-1})_{p'}^k = \delta_{p'}^m \quad \text{and} \quad \sum_{k'} S_{k'}^{m'} \, (S^{-1})_p^{k'} = \delta_p^{m'} \tag{28}$$

we can write:

$$\overline{K}_{ij} = \sum_{lml'm'} S_i^l \, C_{lm}^{m'} S_j^{l'} \, C_{l'm'}^m = \sum_{ll'} S_i^l \, K_{ll'} \, S_j^{l'} \tag{29}$$

Changing the name of the dummy indices, we obtain the scalar product $\langle\langle \overline{X}_i | \overline{X}_j \rangle\rangle$.

The scalar product defined in $\mathscr{L}$ is independent of the basis. Computations will obviously be simplified when performed in an orthonormal basis of X operators [i.e. in a basis where $(\overline{K})$ is diagonal]. In that case, one can normalize the operators, multiplying them by adequate constants to ensure that all the eigenvalues of (K) are equal to ± 1 or 0:

$$\overline{K}_{ij} = \eta_i \, \delta_{ij} \quad \text{where} \quad \eta_i = \pm 1, 0 \tag{30}$$

3. Completely antisymmetric structure constants

Introducing the K_{ij} allows the definition of new (completely covariant) structure constants $\tilde{C}_{ij\ell}$:

$$\tilde{C}_{ij\ell} = \sum_k K_{\ell k} \, C_{ij}^k \tag{31}$$

As we now show, these constants are completely antisymmetric (they change sign if any two indices are exchanged, whereas this property is only valid for the two covariant indices i and j of the constants C_{ij}^k). Using expression (18) for K, relation (31) is written:

$$\tilde{C}_{ij\ell} = \sum_{kmp} C_{ij}^k \, C_{kp}^m \, C_{\ell m}^p \tag{32}$$

or, using relation (III-43) of chapter III:

$$\tilde{C}_{ij\ell} = -\sum_{kmp} \left[C^k_{jp}\, C^m_{ki} + C^k_{pi}\, C^m_{kj}\right] C^p_{\ell m}$$
$$= \sum_{kmp} \left[C^m_{ik}\, C^k_{jp}\, C^p_{\ell m} + C^k_{pi}\, C^m_{kj}\, C^p_{m\ell}\right] \tag{33}$$

(since $C^m_{ik} = -C^m_{ki}$, $C^p_{\ell m} = -C^p_{m\ell}$). In both terms on the right-hand side, the i, j and ℓ indices can be exchanged by a circular permutation, changing only the names of dummy indices. On the other hand, relation $C^k_{ij} = -C^k_{ji}$ shows that $\tilde{C}_{ij\ell} = -\tilde{C}_{ji\ell}$. Consequently:

$$\tilde{C}_{ij\ell} = \tilde{C}_{j\ell i} = \tilde{C}_{\ell ij} = -\tilde{C}_{ji\ell} = -\tilde{C}_{\ell ji} = -\tilde{C}_{i\ell j} \tag{34}$$

Exercise: Relation (31) amounts to transforming the contravariant k index into a covariant ℓ index. Check that the following relation is true:

$$\tilde{C}_{ij\ell} = \langle\langle X_\ell |\, [X_i, X_j] \rangle\rangle \tag{35}$$

Comment:

Choosing the $|X_i\rangle\rangle$ basis where the (K) matrix is diagonal and its elements are given by (30), relation (31) becomes simply:

$$\tilde{C}_{ij\ell} = \eta_\ell\, C^\ell_{ij} \qquad \text{where} \quad \eta_\ell = +1, 0 \text{ or } -1 \tag{36}$$

4. Casimir operator

We assume that the (K) matrix is invertible[2] (none of its eigenvalues are equal to zero), and introduce the inverse matrix H:

$$(H) = (K)^{-1} \tag{37}$$

According to (31), the $H^{k\ell}$ matrix elements yield the C^k_{ij} as a function of the $\tilde{C}_{ijk}$:

$$C^k_{ij} = \sum_\ell H^{k\ell}\, \tilde{C}_{ij\ell} \tag{38}$$

Like (K), the (H) matrix is symmetric.

[2]It can be shown that (K) is invertible if, and only if, the group under study does not have an invariant abelian Lie subgroup (the group is then said to be semi-simple).

The Casimir operator C is defined as:

$$C = \sum_{mn} H^{mn}\ X_m X_n \tag{39}$$

This operator is a quadratic function of the X_i. Let us calculate its commutator with X_j:

$$\begin{aligned}[C, X_j] &= \sum_{mn} H^{mn}\ \{X_m [X_n, X_j] + [X_m, X_j]\ X_n\} \\ &= \sum_{mnk} H^{mn}\ \left\{C^k_{nj}\ X_m X_k + C^k_{mj}\ X_k X_n\right\}\end{aligned} \tag{40}$$

We can exchange the m and n dummy indices in the second term of the right-hand side. This yields, since (H) is symmetric:

$$\begin{aligned}[C, X_j] &= \sum_{mnk} H^{mn}\ C^k_{nj}\ [X_m X_k + X_k X_m] \\ &= \sum_{mnk\ell} H^{mn}\ H^{k\ell}\ \tilde{C}_{nj\ell}\,[X_m X_k + X_k X_m]\end{aligned} \tag{41}$$

In this expression, the coefficient of the symmetric operator $[X_m X_k + X_k X_m]$ is equal to:

$$\sum_{n\ell} H^{mn}\ H^{k\ell}\ \tilde{C}_{nj\ell} \tag{42}$$

which, according to (34), contains a quantity that changes sign if n and l are exchanged: the summation over these two indices yields zero. Consequently, the right-hand side of (40) is zero for any index j. The Casimir operator commutes with all the elements of the Lie algebra $\mathscr{L}$.

Comment:

Choosing in $\mathscr{L}$ the $\overline{X}_i$ basis where $(\overline{K})$ is diagonal and of the form (30) [$\eta_i = +1$ or -1, the value 0 being excluded since (K) is assumed to be invertible], the Casimir operator is simply the sum (or the difference) of the $\overline{X}_i^2$ operators:

$$C = \sum_i \eta_i\ \overline{X}_i^2 \tag{43}$$

Exercise: Calculate the $(K)$and (H) matrices, as well as the Casimir operator C for the Lie algebra of the SU(2) and SO(3) groups.

Chapter IV

Induced representations in the state space

In this chapter, we use previously introduced concepts (discrete or continuous groups of transformations, regular or projective representations, etc.), in the framework of quantum mechanics. We first determine, in § A, the types of T operators, acting in the state space of an arbitrary system, that correspond to the $\mathcal{T}$ transformations introduced in chapter I (translations, rotations, change of Galilean reference frame, etc.). We then show, in § B, that these T operators can only be unitary or antiunitary, enabling us, in § C, to study the transformations of the system's observables. In § D, we examine the action of not only one but of a set of $\mathcal{T}$ transformations that form a (continuous or discrete) group.

Introduction to Continuous Symmetries: From Space–Time to Quantum Mechanics, First Edition. Franck Laloë.

Finally, § E shows how the arbitrary phase factors of the state vectors in quantum mechanics can lead to more general representations, the so-called "projective representations".

In the case of a continuous group, this study will provide a number of commutation relations that must be obeyed by the G operators associated with the infinitesimal transformations of the group (infinitesimal generators). This is an important step in our approach: starting from general considerations on the transformations that can be applied to a physical system, we will construct the spaces of the possible states of the system, and derive its time evolution in these spaces.

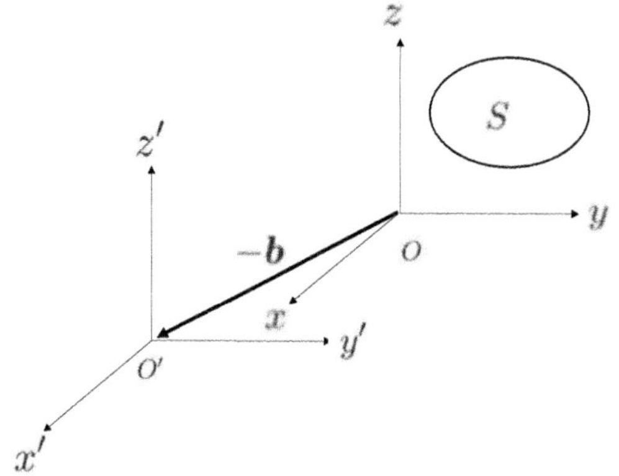

Figure 1: The same physical system S is studied either in a first $Oxyz$ reference frame, or in a second $O'x'y'z'$ reference frame, deduced from the first one by a translation of a vector $-\boldsymbol{b}$. Viewed from the second reference frame, the system seems to have been translated by a vector $+\boldsymbol{b}$.

Before starting this chapter, we recall a few notations and results. We already discussed, at the end of chapter I (§ C-2), the effect of an operation $\mathcal{T}$ on the state vector $|\psi\rangle$ of a physical system. As an example, assume that the operation $\mathcal{T}$ is a translation by a vector $\boldsymbol{b}$. Consider an arbitrary physical system, described in the $Oxyz$ reference frame by the normalized ket $|\psi\rangle \in \mathscr{E}$. The system can also be described in a second reference frame $O'x'y'z'$ (figure 1) deduced from the first one by a vector $-\boldsymbol{b}$ translation. Viewed from this new reference frame, the original physical system has exactly the same properties as when it was viewed from the first reference frame, except for its position in space. This position has been translated by a vector $+\boldsymbol{b}$. Calling $|\psi'\rangle$ the normalized ket[1] of $\mathscr{E}$ describing the physical system in the $O'x'y'z'$ reference frame, we have

[1]The normalized kets $|\psi\rangle$ and $|\psi'\rangle$ are defined up to a phase factor. We are thus looking for

to determine $|\psi'\rangle$ as a function of $|\psi\rangle$. We shall call F the transformation that changes $|\psi\rangle$ into $|\psi'\rangle$:

$$|\psi'\rangle = F\left(|\psi\rangle\right) \tag{IV-1}$$

(we use this type of notations to underline the fact that for the moment we have no proof that the F transformations must be linear; we will come back later to the T notations of the previous chapters).

These considerations can be easily generalized to other $\mathcal{T}$ operations: rotations, symmetry with respect to a point, change of reference frame (pure Galilean or Lorentz transformations), etc. In all these cases, we can use an equality such as (IV-1) to describe the corresponding transformation of the state vector $|\psi\rangle \in \mathscr{E}$. The question then arises: based on physical criteria, what should be imposed on the F transformation? This point will be discussed in § A.

Comment:

We are using here the "passive point of view" (§ A-3 of chapter I) where we consider a unique physical system, described in different reference frames. Remember that, conversely, the "active point of view" considers a single reference frame and introduces a second physical system deduced from the first one (in the example above) by a translation of $+\boldsymbol{b}$: the system is being "moved", not the axes.

A. Conditions imposed on the transformations in the state space

First of all, it is clear that if the F transformation exists, it must be reversible. In the translation example, one can go back from the $O'x'y'z'$ to the $Oxyz$ reference frame. We must be able to write:

$$|\psi\rangle = F^{-1}\left(|\psi'\rangle\right) \tag{IV-2}$$

where the function F^{-1} exists, whatever the vector $|\psi'\rangle$. If this were not true, there would be a fundamental asymmetry between systems described by $|\psi\rangle$ and those described by $|\psi'\rangle$.

Secondly, the measurement postulates of quantum mechanics should not lead to incoherent results. Imagine, for example, that B is an observable of the system (Hermitian operator with a complete spectrum in $\mathscr{E}$) and that $|b_i\rangle$ is a normalized eigenket of B with the eigenvalue b_i:

$$B|b_i\rangle = b_i|b_i\rangle \tag{IV-3}$$

---

a correspondence between classes of kets that have different phases. In the following discussion, we assume that a ket with a given phase has been arbitrarily chosen in each class. This phase ambiguity can be eliminated by reasoning in terms of the projectors $P_\psi = |\psi\rangle\langle\psi|$ and $P_{\psi'} = |\psi'\rangle\langle\psi'|$.

(to keep it simple, we assume that B has a discrete and non-degenerate spectrum). If the physical system is in the state $|b_i\rangle$, the measurement result will certainly yield b_i.

We write:

$$|b'_i\rangle = F(|b_i\rangle) \tag{IV-4}$$

the normalized ket describing the system once the transformation $\mathcal{T}$ has been applied. It is the eigenket of an operator called B', with the same eigenvalue b_i:

$$B'|b'_i\rangle = b_i|b'_i\rangle \qquad \forall\ |b_i\rangle \tag{IV-5}$$

What is the physical meaning of B' ? Before the measurement, it is equivalent to say that the system is described in the initial reference frame by $|b_i\rangle$, and after the action of $\mathcal{T}$, by $|b'_i\rangle$. These two kets yield the same certain result b_i when measurements associated respectively to B and B' are performed[2]. B' describes the same measurement as B (the interaction of the system with the same measurement apparatus), but from the point of view associated with the new reference frame. In a general way, going from B to B' amounts to performing on the measurement apparatus the same operation $\mathcal{T}$ that transforms the ket of the physical system from $|\psi\rangle$ into $|\psi'\rangle$.

If the system is in an arbitrary normalized state $|\psi\rangle$ (linear superposition of the $|b_i\rangle$), the probability of finding the measurement result b_i, calculated in the first reference frame, is:

$$\mathscr{P}(b_i) = |\langle b_i|\psi\rangle|^2 \tag{IV-6}$$

and, in the second:

$$\mathscr{P}(b_i) = |\langle b'_i|\psi'\rangle|^2 \tag{IV-7}$$

As these two expressions must be equal, we must have:

$$|\langle b_i|\psi\rangle| = |\langle b'_i|\psi'\rangle| \tag{IV-8}$$

This result may be generalized since, if $|\varphi\rangle$ is a normalized arbitrary ket of $\mathscr{E}$, it can be considered to be an eigenket associated with one (or several[3]) observables $B_1, B_2, \ldots$ having one (or several) eigenvalues $b_{i_1}, b_{i_2}, \ldots$ The same reasoning leads to[4]:

$$|\langle\varphi'|\psi'\rangle| = |\langle\varphi|\psi\rangle| \tag{IV-9}$$

[2]This is actually the same measurement, described by two different observers. The result is, for example, the position of a pointer on a measurement scale. This result must be the same for the two observers, even though they give different descriptions of the physical system and the measurement apparatus.

[3]For example, a Complete Set of Commuting Observables (CSCO).

[4]If one reasons in terms of projectors P_ψ and $P_{\psi'}$ (*cf.* note 1), condition (IV-9) must be replaced by $\text{Tr}\{P_{\varphi'}P_{\psi'}\} = \text{Tr}\{P_\varphi P_\psi\}$ (these numbers are real and positive).

Transformation F must *preserve the modulus of the scalar product* of the normalized kets. In such cases, F is said to "preserve the physical properties" of the system, and an important theorem considerably limits the set of acceptable transformations.

B. Wigner's theorem

Consider a transformation F that obeys the above conditions, meaning it is bijective and preserves the modulus of the scalar products of normalized kets. We are going to show that F is then necessarily equivalent to either a linear and unitary operation or an antilinear and unitary (i.e. "antiunitary") operation[5]. The argument is as follows:

(i) Let us assume that $\{|u_i\rangle\}$ is an orthonormal basis of $\mathscr{E}(i = 1, 2, \ldots N)$. The set of:

$$|u'_i\rangle = F(|u_i\rangle) \tag{IV-10}$$

is another orthonormal basis of $\mathscr{E}$, as we now show.

– the set of $|u'_i\rangle$ is orthonormal since F preserves the modulus of the scalar product:

$$|\langle u'_i|u'_j\rangle| = |\langle u_i|u_j\rangle| = \delta_{ij} \tag{IV-11}$$

and hence:

$$\langle u'_i|u'_j\rangle = \delta_{ij} \tag{IV-12}$$

(since $\langle u'_i|u'_i\rangle$ is real);

– the set of $|u'_i\rangle$ forms a basis. If this were not the case, we could find a non-zero ket $|\psi'\rangle$ orthogonal to all the $|u'_i\rangle$, and $|\psi\rangle = F^{-1}(|\psi'\rangle)$ would be orthogonal to all the $|u_i\rangle$, which is impossible since the $\{|u_i\rangle\}$ form an orthonormal basis.

(ii) Consider the kets:

$$|\varphi_i\rangle = \frac{1}{\sqrt{2}}[|u_1\rangle + |u_i\rangle] \qquad i = 2, \ldots, N$$
$$|\varphi'_i\rangle = F\left(\frac{1}{\sqrt{2}}[|u_1\rangle + |u_i\rangle]\right) \tag{IV-13}$$

[5] An operator is said to be antilinear if, for any kets $|\varphi_1\rangle$ and $|\varphi_2\rangle$ and for any complex numbers λ_1 and λ_2, we have:

$$A\Big[\lambda_1|\varphi_1\rangle + \lambda_2|\varphi_2\rangle\Big] = \lambda_1^* A|\varphi_1\rangle + \lambda_2^* A|\varphi_2\rangle$$

Properties of antilinear and antiunitary transformations are reviewed in appendix .

The modulus of the scalar product of $|\varphi_i'\rangle$ and $|u_k'\rangle$ must be equal to $1/\sqrt{2}$ if $k = 1$ or i, and zero otherwise. Consequently:

$$|\varphi_i'\rangle = \frac{1}{\sqrt{2}}\left[e^{i\alpha_i}|u_1'\rangle + e^{i\beta_i}|u_i'\rangle\right] \qquad (\alpha_i, \beta_i \text{ real}) \qquad \text{(IV-14)}$$

Let us introduce a new basis of kets $|v_i'\rangle$, proportional to the $|u_i'\rangle$:

$$\begin{cases} |v_1'\rangle = |u_1'\rangle \\ |v_i'\rangle = e^{i(\beta_i-\alpha_i)}|u_i'\rangle \qquad i = 2, \ldots, N \end{cases} \qquad \text{(IV-15)}$$

We can write:

$$F\left(\frac{1}{\sqrt{2}}[|u_1\rangle + |u_i\rangle]\right) = \frac{e^{i\alpha_i}}{\sqrt{2}}\left[|v_1'\rangle + |v_i'\rangle\right] \qquad \text{(IV-16)}$$

where the transform of $|\varphi_i\rangle$ contains, in the new basis, only one global phase factor $e^{i\alpha_i}$.

(iii) Let $|\chi\rangle$ be the normalized ket defined as:

$$|\chi\rangle = \sum_i x_i\,|u_i\rangle \qquad \text{(IV-17)}$$

where the x_i are *real*; $|\chi\rangle$ is said to be a "real ket"[6]. We assume:

$$x_1 \neq 0 \qquad \text{(IV-18)}$$

We can always write:

$$|\chi'\rangle = F(|\chi\rangle) = \sum_i g_i\,|v_i'\rangle \qquad \text{(IV-19)}$$

with:

$$|g_i| = |\langle v_i'|\chi'\rangle| = |\langle u_i|\chi\rangle| = |x_i| = \pm x_i \qquad \text{(IV-20)}$$

Let us take the scalar product of $|\chi\rangle$ and the ket $|\varphi_i\rangle$ introduced above:

$$|\langle\varphi_i|\chi\rangle| = \frac{1}{\sqrt{2}}|x_1 + x_i| = |\langle\varphi_i'|\chi'\rangle| = \frac{1}{\sqrt{2}}\,|g_1 + g_i| \qquad \text{(IV-21)}$$

This leads to:

$$|x_1 + x_i| = |g_1 + g_i| \qquad \text{(IV-22)}$$

Since $|x_i| = |g_i|$, this equality implies that g_i/g_1 is real:

$$\frac{g_i}{g_1} = \frac{x_i}{x_1} \qquad \text{(IV-23)}$$

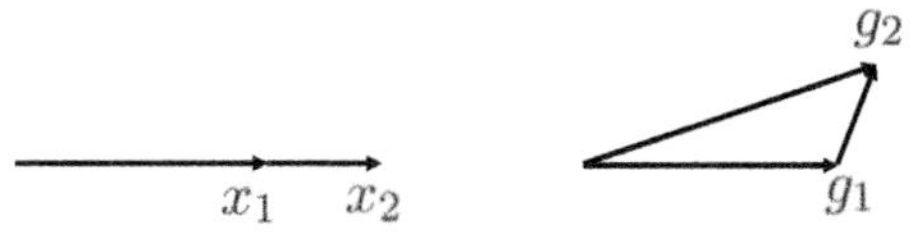

Figure 2: In the complex plane, the vectors representing g_1 and g_2 have the same lengths as those representing the numbers x_1 and x_2. These latter vectors are parallel to each other since x_1 and x_2 are real. Consequently, the length of the vector associated with the sum $g_1 + g_2$ can only be equal to that of the sum $x_1 + x_2$ if the vectors g_1 and g_2 are also parallel to each other.

We have used the fact that, in the complex plane, if the vectors representing g_1 and g_2 were not parallel (or anti-parallel) to each other, as those representing x_1 and x_2, equality (IV-22) could not be satisfied – *cf.* figure 2. This leads to:

$$|\chi'\rangle = \mathrm{e}^{i\beta} \sum_i x_i \, |v_i'\rangle \tag{IV-24}$$

where:

$$\mathrm{e}^{i\beta} = \frac{g_1}{x_1} = \frac{g_2}{x_2} = \ldots \tag{IV-25}$$

is a global phase factor of the ket $|\chi'\rangle$.

Comment:

To establish (IV-24), we used the fact that $|\chi\rangle$ was chosen with $x_1 \neq 0$. This restriction is not essential. If x_1 is equal to zero, we can use the same argument changing the index 1 for the index of the first basis ket on which $|\chi\rangle$ has a non-zero component.

To sum up, with the bases $\{|u_i\rangle\}$ and $\{|v_i'\rangle\}$ we have chosen, the transform of any normalized "real" ket keeps the same components (up to a global phase factor of the ket).

[6]Note that there is nothing absolute in this concept of "real ket", it depends on the chosen basis.

(iv) In the final step of the reasoning, we choose an arbitrary normalized ("complex") ket $|\psi\rangle$, whose transform is $|\psi'\rangle$:

$$|\psi\rangle = \sum_i c_i \, |u_i\rangle$$

$$|\psi'\rangle = F(|\psi\rangle) = \sum_i d_i \, |v_i'\rangle \tag{IV-26}$$

where, necessarily:

$$|d_i| = |c_i| \tag{IV-27}$$

Let us take the scalar product with any of the kets $|\chi\rangle$ considered above:

$$|\langle\chi|\psi\rangle| = |\sum_i x_i \, c_i| = |\langle\chi'|\psi'\rangle| = |\sum_i x_i \, d_i| \tag{IV-28}$$

– Let us first choose $x_1 = x_2 = 1/\sqrt{2}$, $x_3 = x_4 = \ldots = 0$. This leads to:

$$|d_1 + d_2| = |c_1 + c_2| \tag{IV-29}$$

In the complex plane, c_1 and the sum $c_1 + c_2$ can be represented by the points M_1 and M_2 ; c_1' and $c_1' + c_2'$ by the points M_1' and M_2'. The relations written above show that $|OM_1| = |OM_1'|$, $|OM_2| = |OM_2'|$ and $|M_1M_2| = |M_1'M_2'|$. As seen in figure 3 two cases are possible, either:

$$\frac{d_2}{d_1} = \frac{c_2}{c_1} \tag{IV-30a}$$

or:

$$\frac{d_2}{d_1} = \left(\frac{c_2}{c_1}\right)^* \tag{IV-30b}$$

In the first case, the c are transformed into the d though a rotation in the complex plane; in the second case, one must add a symmetry with respect to the axis along d_1.

– Other possible choices for the x_i are $x_3 = 1$ (the other x_i being equal to zero), or $x_2 = x_3 = 1/\sqrt{2}$ (the other being equal to zero), or else $x_1 = x_2 = x_3 = 1/\sqrt{3}$ (the other being equal to zero). We then have:

$$\begin{aligned} |d_3| &= |c_3| \\ |d_2 + d_3| &= |c_2 + c_3| \\ |d_1 + d_2 + d_3| &= |c_1 + c_2 + c_3| \end{aligned} \tag{IV-31}$$

Let us call M_3 the point representing the sum $c_1 + c_2 + c_3$, and M_3' its transform representing $d_1 + d_2 + d_3$. We follow the same argument as above. The three

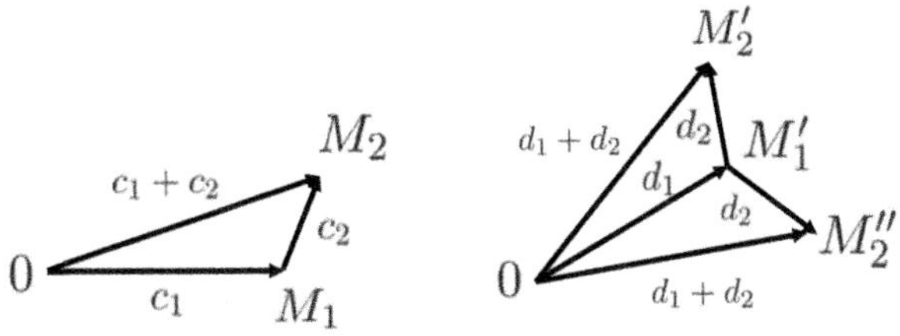

Figure 3: The numbers c_1 and c_2 are represented, in the complex plane, by the vectors OM_1 and M_1M_2. Through a transformation, they become the complex numbers d_1 and d_2, represented by the vectors OM_1' and $M_1'M_2'$. The vectors representing d_1 and d_2 have the same lengths as those of the vectors associated with c_1 and c_2. The length of the vector associated with the sum $d_1 + d_2$ must be the same as that of the vector associated with the sum c_1+c_2. This is only possible in two cases: either the c are transformed into the d through a simple rotation in the complex plane (leading to the point M_2'), or we must add a symmetry with respect to the axis along d_1 (point M_2'').

equalities we just wrote show that the distances of M_3' from the transforms of M_2, M_1, and the origin, must be the same as the distances of M_3 from the corresponding points. A point in a plane, whose distances to three other points in the plane are fixed, can occupy only one position. Consequently, the position of the transform of M_2 determines the position of the transform of M_3.

In the (IV-30a) hypothesis, the transformation of the points amounts to a simple rotation, and we necessarily have:

$$\frac{d_3}{d_1} = \frac{c_3}{c_1} \tag{IV-32a}$$

whereas, in the (IV-30b) hypothesis, the addition of a planar symmetry leads to:

$$\frac{d_3}{d_1} = \left(\frac{c_3}{c_1}\right)^* \tag{IV-32b}$$

– The reasoning can be extended to $d_4, d_5, \ldots$, showing that all the M_i' are the transforms of the M_i either by a rotation around the origin in the complex plane, or by a rotation followed by a symmetry with respect to an axis of the plane. This leads to:

$$|\psi'\rangle = \begin{cases} \text{either} \quad e^{i\delta} \sum_i c_i \, |v_i'\rangle \\ \text{or} \quad e^{i\delta} \sum_i c_i^* \, |v_i'\rangle \end{cases} \tag{IV-33}$$

where δ is a real phase that may depend on $|\psi\rangle$.

We call T the transformation defined as:

$$T|\psi\rangle = \mathrm{e}^{-i\delta}|\psi'\rangle \tag{IV-34}$$

T is the physical equivalent of the initial F transformation, the only difference being a phase factor. Relations (IV-33) show that T is either a linear transformation:

$$T|\psi\rangle = \sum_i c_i\,|v_i'\rangle \tag{IV-35a}$$

or an antilinear one:

$$T|\psi\rangle = \sum_i c_i^*\,|v_i'\rangle \tag{IV-35b}$$

Remember that the $|v_i'\rangle$ form an orthonormal basis. Comparing (IV-35) and (IV-26), we see that, in both cases, the T operator preserves the norm and the norm of the scalar product of the kets.

Conversely, it is easy to show that unitary linear or antilinear transformations preserve the modulus of the scalar product. Wigner's theorem is thus established (for a space with a finite dimension N; we will assume it is also valid for a state space with an infinite dimension).

C. Transformations of observables

Imagine we know the linear (or antilinear) operator T that changes each ket $|\psi\rangle$ of a physical system into the ket $|\psi'\rangle$ of the system, once it has been transformed by $\mathcal{T}$:

$$|\psi'\rangle = T|\psi\rangle \tag{IV-36}$$

One may wonder how the system's observables are transformed in this operation. In the "active point of view", if we call B the (Hermitian) operator describing a measurement performed with a given apparatus (a Stern Gerlach magnet, for example), which operator B' describes the measurement after the apparatus has undergone the transformation $\mathcal{T}$ (rotation of the Stern Gerlach magnet, for instance)? Or, in the "passive point of view", if B is the operator describing a given experimental setup in the first reference frame, which operator B' describes the same measurement in another reference frame that has been transformed by $\mathcal{T}^{-1}$?

C-1. Operator expression

This question has been discussed in § A where we introduced the eigenkets $|b_i\rangle$ of an observable B as well as the transformed kets $|b_i'\rangle$:

$$B|b_i\rangle = b_i|b_i\rangle \qquad\qquad |b_i'\rangle = T|b_i\rangle \tag{IV-37}$$

This led to:

$$B'|b_i'\rangle = b_i|b_i'\rangle \tag{IV-38}$$

(where the eigenvalue remains the same). Consequently[7]:

$$B'T|b_i\rangle = b_i\,T|b_i\rangle = TB|b_i\rangle \tag{IV-39}$$

and:

$$\left(T^{-1}B'T\right)|b_i\rangle = B|b_i\rangle \tag{IV-40}$$

As the $\{|b_i\rangle\}$ form a basis, we have:

$$B = T^{-1}B'T \qquad \text{and, conversely:} \qquad B' = T\,B\,T^{-1} \tag{IV-41}$$

If T is unitary, $T^{-1} = T^\dagger$, and these equalities can be written:

$$B = T^\dagger B'T \qquad \text{and, conversely:} \qquad B' = T\,B\,T^\dagger \tag{IV-42}$$

These are the relations linking B and B'. The demonstration can be generalized to the case where the spectrum of B is degenerate and continuous.

Comments:

(i) To remember where T and $T^\dagger$ appear in (IV-42), we can merely write:

$$B = \sum_i b_i|b_i\rangle\langle b_i| \quad \Rightarrow \quad B' = \sum_i b_i|b_i'\rangle\langle b_i'| = TBT^\dagger \tag{IV-43}$$

(ii) If B commutes with T:

$$[B,T] = 0 \tag{IV-44}$$

we readily see that $B' = B$. The corresponding observable is said to be invariant under the operation $\mathcal{T}$. For example, the observable S_z, the spin component of a particle on the Oz axis, is invariant under any rotation around Oz.

[7] As b_i is real, T commutes with multiplication by b_i, even if T is antilinear [*cf.* note (5) and appendix].

C-2. Physical discussion

Operator B' has been introduced as the operator that, acting on the transformed kets $|\psi'\rangle$, yields the same results as the operator B acting on the initial kets $|\psi\rangle$. Imagine, for example, that the measurement concerns the position of the system S, and that $\mathcal{T}$ describes a translation by a vector $\boldsymbol{b}$. In the active point of view, when S and the measurement apparatus M are both moved by the same amount $\boldsymbol{b}$, their relative position is unchanged. Consequently, the interaction between S and M remains the same, as do the measurement results (the position of a pointer on a measurement apparatus). The results and their probabilities are therefore unchanged. In the passive point of view where neither S nor M is moved, this invariance is even more obvious.

But one can also consider that the transformation is applied to the measurement apparatus only. One should then look for the operator B' describing this new measurement, performed with an unchanged physical system S. Of course, in this case there is no reason why the measurements results should remain identical, and we then need to determine the effect of B' on the initial $|\psi\rangle$.

Imagine that M is translated by a vector $\boldsymbol{b}$, whereas S does not move. The relative position of S with respect to M then changes by the opposite quantity, as does the position measurement result. Noting $\boldsymbol{R}$ the operator that measures the position, and $T(\boldsymbol{b})$ the translation operator in the state space, we must have:

$$\boldsymbol{R}' = T(\boldsymbol{b})\,\boldsymbol{R}\,T^{\dagger}(\boldsymbol{b}) = \boldsymbol{R} - \boldsymbol{b} \tag{IV-45a}$$

and, of course:

$$\boldsymbol{R} = T^{\dagger}(\boldsymbol{b})\,\boldsymbol{R}'\,T(\boldsymbol{b}) = \boldsymbol{R}' + \boldsymbol{b} \tag{IV-45b}$$

The presence of a minus sign in equality (IV-45a) is not surprising, even though we are dealing with a $T(+\boldsymbol{b})$ transformation. In a similar way, when we perform a Galilean transformation that increases the speed(s) of the physical system by a vector $\boldsymbol{v}$, we expect the measurement apparatus M subjected to this transformation to yield results decreased by $\boldsymbol{v}$.

D. Linear representations in the state space

D-1. Action of a group of transformations

Consider a set of transformations $\mathcal{T}$ that form a group $\mathscr{G}$. According to Wigner's theorem, we can associate with each of them a unitary linear (or antilinear) operator T, acting in the state space:

$$|\psi'\rangle = T(\mathcal{T})\ |\psi\rangle$$
$$|\psi\rangle = T^\dagger(\mathcal{T})\ |\psi'\rangle \qquad \text{(IV-46)}$$

Each operator T is, in a way, the image in the state space of a transformation $\mathcal{T}$ in real space. We shall only consider here linear T operators since, in most cases in quantum mechanics, antilinear operators do not come into play. The antilinear operators will be mentioned later in the particular case of time reversal, discussed in appendix .

Comment:

In the case of a continuous group of transformations $\mathcal{T}$, the T operator is necessarily unitary (and not antiunitary) as long as $\mathcal{T}$ remains close to the identity transformation, as we now show. As in § D-2 that follows, we introduce a parameter $\boldsymbol{a}$ to describe the $\mathcal{T}$ transformations , and hence the T operators. Expanding $T(\delta\boldsymbol{a})$ to first-order in $\delta\boldsymbol{a}$ [relations (IV-53) to (IV-55)], we get $T(\delta\boldsymbol{a}) = T(\delta\boldsymbol{a}/2)\ T(\delta\boldsymbol{a}/2)$, an equality that is not compatible with the operator T being antiunitary (the product of two antiunitary operators must be unitary – *cf.* appendix § 2-a). The unitarity of the T operators is not limited to the neighborhood of the identity but can be continuously extended to the entire "map" (connected component) that includes the identity element.

Performing two successive transformations $\mathcal{T}_1$ and $\mathcal{T}_2$, in that order, changes the state vector into:

$$|\psi''\rangle = T(\mathcal{T}_2)\ |\psi'\rangle = T(\mathcal{T}_2)\ T(\mathcal{T}_1)\ |\psi\rangle \qquad \text{(IV-47)}$$

We could also directly perform the product transformation:

$$\mathcal{T}_3 = \mathcal{T}_2\ \mathcal{T}_1 \qquad \text{(IV-48)}$$

and obtain the new state vector:

$$|\tilde{\psi}''\rangle = T(\mathcal{T}_3)\ |\psi\rangle \qquad \text{(IV-49)}$$

Both kets $|\psi''\rangle$ and $|\tilde{\psi}''\rangle$ actually describe the same physical state of the system (in the same reference frame). Consequetly, they must only differ by a phase factor[8]:

$$|\tilde{\psi}''\rangle = e^{i\xi(\mathcal{T}_2,\mathcal{T}_1)}|\psi''\rangle \qquad \text{(IV-50)}$$

[8]We assume that the structure of the state space is such that, whenever two kets $|\psi\rangle$ and $|\psi'\rangle$ are not proportional to each other, they are not physically equivalent (i.e. there exists at least one physical measurement that allows a distinction between them).

where ξ is a real function that depends[9] on $\mathcal{T}_2$ and $\mathcal{T}_1$. According to (IV-47) and (IV-49), we can write (as the ket $|\psi\rangle$ is arbitrary):

$$T(\mathcal{T}_2)\, T(\mathcal{T}_1) = \mathrm{e}^{-i\xi(\mathcal{T}_2,\mathcal{T}_1)}\, T(\mathcal{T}_2\mathcal{T}_1) \tag{IV-51}$$

which shows that the T operators are in general a projective representation (*cf.* chapter II § B-1-c) of the $\mathscr{G}$ group. They can be associated with matrices that are finite or infinite, depending on the dimension of the state space, and whose elements are written:

$$T_{ij}(\mathcal{T}) = \langle u_i|T(\mathcal{T})|u_j\rangle \tag{IV-52}$$

where the $\{|u_i\rangle\}$ form an arbitrary basis of $\mathscr{E}$. If the basis is orthonormal, and since the T are unitary, so are the matrices of the representation.

Assume, without justification for the moment, that we can choose ξ equal to zero for any $\mathcal{T}_1$ and $\mathcal{T}_2$. This corresponds to a representation in the strict sense (i.e. non-projective) of the $\mathscr{G}$ group in the state space $\mathscr{E}$ of the system of dimension N (dimension of $\mathscr{E}$). An important problem we shall address in the coming chapters will be the reducibility of such representations.

D-2. Case of a continuous group, infinitesimal generators

If the group $\mathscr{G}$ is continuous, its elements can be identified by a parameter $\boldsymbol{a}$ (with n real components):

$$\mathcal{T} \equiv \mathcal{T}(\boldsymbol{a}) \tag{IV-53a}$$

The unitary operator T acting in $\mathscr{E}$, and corresponding to the transformation $\mathcal{T}$, is also a function of $\boldsymbol{a}$:

$$T \equiv T(\boldsymbol{a}) \tag{IV-53b}$$

[9] It is useless to assume that ξ may depend on the ket $|\psi\rangle$. If this were the case, one would write:

$$|\tilde{\psi}''\rangle = T(\mathcal{T}_3)\,|\psi\rangle = \mathrm{e}^{i\xi(\mathcal{T}_2,\mathcal{T}_1,\Psi)}\, T(\mathcal{T}_2)\, T(\mathcal{T}_1)\,|\psi\rangle$$

or else, multiplying on the left by $T^{-1}(\mathcal{T}_1)\, T^{-1}(\mathcal{T}_2)$:

$$T^{-1}(\mathcal{T}_1)\, T^{-1}(\mathcal{T}_2)\, T(\mathcal{T}_3)\,|\psi\rangle = \mathrm{e}^{i\xi(\mathcal{T}_2,\mathcal{T}_1,\psi)}|\psi\rangle$$

The linear operator acting on the left-hand side on $|\psi\rangle$ would be simply multiplying an arbitrary ket $|\psi\rangle$ by a number (a phase factor), and would be diagonal in any basis, i.e. would be a scalar operator whose diagonal matrix elements are all equal. This means that ξ must necessarily be independent of $|\psi\rangle$.

As in chapter III, whose notations we use, we shall reason on the $\boldsymbol{a}$ rather than on the elements $\mathcal{T}$ of $\mathscr{G}$. We shall assume that $\boldsymbol{a} = \boldsymbol{0}$ corresponds to the identity operation, and that:

$$T(\boldsymbol{0}) = \mathbb{1} \tag{IV-54}$$

(identity operator acting in $\mathscr{E}$). If the T operators are continuous, differentiable, etc. functions of $\boldsymbol{a}$, and if $\delta\boldsymbol{a}$ is an infinitesimal vector, we can write[10]:

$$T(\delta\boldsymbol{a}) = 1 - \frac{i}{\hbar}\sum_j \delta a^j\, G_j + \ldots = 1 - \frac{i}{\hbar}\delta\boldsymbol{a}\cdot\boldsymbol{G} + \ldots \tag{IV-55}$$

The G_j, included in the notation $\boldsymbol{G}$ (that symbolizes an n-dimensional vector whose components are operators acting in $\mathscr{E}$), are called "*infinitesimal generators*". The facteur $\hbar$ has been introduced arbitrarily, and only changes the definition of the infinitesimal operator; it will be convenient in what follows [11]. The i in the right-hand side of (IV-55) ensures that the G_i are Hermitian operators, as we now show. We can write:

$$T^\dagger(\delta\boldsymbol{a}) = 1 + \frac{i}{\hbar}\sum_j \delta a^j\, G_j^\dagger + \ldots \tag{IV-56}$$

As T is unitary, this leads to:

$$T^\dagger(\delta\boldsymbol{a})\; T(\delta\boldsymbol{a}) = 1 - \frac{i}{\hbar}\sum_j \delta a^j \left(G_j - G_j^\dagger\right) + \ldots \tag{IV-57}$$

which means, as the δa^j are arbitrary:

$$G_j = G_j^\dagger \tag{IV-58}$$

This Hermiticity is essential for assigning a physical meaning to the operators G (it ensures that their eigenvalues are real).

We saw in chapter III the importance of the structure constants of a Lie Group. These constants are given by:

$$C_{ij}^k = f_{ij}^k - f_{ji}^k \tag{IV-59}$$

[10]As is commonly done, we shall simplify the notation $\mathbb{1}$ into 1.

[11]As an example, whereas $X_{\boldsymbol{P}}$ has the dimension of the inverse of a length, its correspondence in the space state is an infinitesimal generator $\boldsymbol{P}$, which now contains the additional factor $\hbar$, as in the quantum relation $\boldsymbol{P} \Rightarrow -i\hbar\boldsymbol{\nabla}$; as for $X_{\boldsymbol{J}}$, it is dimensionless but corresponds to a generator that has the dimension of $\hbar$, etc. At this stage, $\hbar$ can be considered as a free parameter in the theory, useful to identify quantum effects, and whose value must be determined by comparing theoretical predictions and experimental measurements.

where the f_{ij}^k are related to the second derivatives of the function $\boldsymbol{\Phi}_2$ expressing the group law in $\mathscr{G}$. More precisely, relation (III-33) shows that when k varies from 1 to n, the f_{ij}^k are the components of the n^2 vectors $\boldsymbol{F}_{ij}$:

$$\boldsymbol{F}_{ij} = \left. \frac{\partial^2}{\partial a^i \, \partial b^j} \boldsymbol{\Phi}_2(\boldsymbol{a}, \boldsymbol{b}) \right|_{\boldsymbol{a}=\boldsymbol{b}=0} \tag{IV-60}$$

We are going to follow exactly the same reasoning as in § A-2-b of chapter III to find the image, in the state space, of the transformation $\mathcal{T}(\delta \boldsymbol{c})$ (commutator in $\mathscr{G}$) defined as:

$$\mathcal{T}(\delta \boldsymbol{c}) = \mathcal{T}(\delta \boldsymbol{a}) \, \mathcal{T}(\delta \boldsymbol{b}) \, \mathcal{T}^{-1}(\delta \boldsymbol{a}) \, \mathcal{T}^{-1}(\delta \boldsymbol{b}) \tag{IV-61}$$

Remember [relation (III-34) of chapter III] that the kth component of $\delta \boldsymbol{c}$ is given by:

$$\delta c^k = \sum_{ij} C_{ij}^k \, \delta a^i \, \delta b^j \tag{IV-62}$$

In addition, with the transformation (IV-61) is associated in the state space $\mathscr{E}$ the operator (we still assume there is no phase factor in (IV-51)]:

$$T(\delta \boldsymbol{a}) \, T(\delta \boldsymbol{b}) \, T^\dagger(\delta \boldsymbol{a}) \, T^\dagger(\delta \boldsymbol{b}) = T(\delta \boldsymbol{c}) \tag{IV-63}$$

Let us insert (IV-55) in this equality, as well as an equivalent relation where $\delta \boldsymbol{b}$ replaces $\delta \boldsymbol{a}$. We limit the calculation to second-order in δa and δb, as in § A-2-b of chapter III. Relation (IV-63) readily shows that $T(\delta \boldsymbol{c})$ is the identity operator if either $\delta \boldsymbol{a}$ or $\delta \boldsymbol{b}$ is zero. This allows discarding the "square" terms in $\delta \boldsymbol{a}$ and $\delta \boldsymbol{b}$, and justifies using expansion (IV-55) limited to first-order. This leads to:

$$T(\delta \boldsymbol{a}) \, T(\delta \boldsymbol{b}) = 1 - \frac{i}{\hbar} (\delta \boldsymbol{a} \cdot \boldsymbol{G}) - \frac{i}{\hbar} (\delta \boldsymbol{b} \cdot \boldsymbol{G}) - \frac{1}{\hbar^2} (\delta \boldsymbol{a} \cdot \boldsymbol{G}) \, (\delta \boldsymbol{b} \cdot \boldsymbol{G}) + \ldots \tag{IV-64a}$$

and consequently:

$$\begin{aligned} T(\delta \boldsymbol{a}) \, T(\delta \boldsymbol{b}) \, T^\dagger(\delta \boldsymbol{a}) &= 1 - \frac{i}{\hbar} (\delta \boldsymbol{a} \cdot \boldsymbol{G}) - \frac{i}{\hbar} (\delta \boldsymbol{b} \cdot \boldsymbol{G}) - \frac{1}{\hbar^2} (\delta \boldsymbol{a} \cdot \boldsymbol{G}) \, (\delta \boldsymbol{b} \cdot \boldsymbol{G}) \\ &\quad + \frac{i}{\hbar} (\delta \boldsymbol{a} \cdot \boldsymbol{G}) + \frac{1}{\hbar^2} (\delta \boldsymbol{b} \cdot \boldsymbol{G}) \, (\delta \boldsymbol{a} \cdot \boldsymbol{G}) + \ldots \\ &= 1 - \frac{i}{\hbar} (\delta \boldsymbol{b} \cdot \boldsymbol{G}) - \frac{1}{\hbar^2} [\delta \boldsymbol{a} \cdot \boldsymbol{G}, \delta \boldsymbol{b} \cdot \boldsymbol{G}] + \ldots \end{aligned} \tag{IV-64b}$$

where on the right-hand side appears the commutator of $(\delta \boldsymbol{a} \cdot \boldsymbol{G})$ and $(\delta \boldsymbol{b} \cdot \boldsymbol{G})$. We multiply on the right this expression by $T^\dagger(\delta \boldsymbol{b})$, keeping only the linear or cross terms in $\delta \boldsymbol{a}$ and $\delta \boldsymbol{b}$. Using again (IV-56), this amounts to discarding the $(\delta \boldsymbol{b} \cdot \boldsymbol{G})$ term on the right-hand side of (IV-64b). We finally insert this expression on the left-hand side of (IV-63), while replacing $T(\delta \boldsymbol{c})$ by $1 - (i/\hbar)\, \delta \boldsymbol{c} \cdot \boldsymbol{G}$. This leads to:

$$[\delta \boldsymbol{a} \cdot \boldsymbol{G}, \delta \boldsymbol{b} \cdot \boldsymbol{G}] = i\hbar\, \delta \boldsymbol{c} \cdot \boldsymbol{G} \tag{IV-65}$$

Using relation (IV-62) and identifying the $\delta a^i\, \delta b^j$ terms (since $\delta \boldsymbol{a}$ and $\delta \boldsymbol{b}$ have been chosen arbitrarily), we get the important relation:

$$\boxed{[G_i, G_j] = i\hbar \sum_k C_{ij}^k\, G_k} \tag{IV-66}$$

We have obtained, in a general way, commutation relations satisfied by the observables G_k. They are the direct image of the commutation relations (III-36) satisfied by the X operators of the Lie algebra associated with the group of the $\mathcal{T}$ transformations. Note that no hypothesis has been made on the nature of the physical system under study, or on the structure of its state space. The result is based only on the properties of the $\mathscr{G}$ group, via its structure constants.

E. Phase factors and projective representations

Let us go back to the previously neglected phase factor question and examine what should be modified in the previous results.

Written as a function of the parameters $\boldsymbol{a}$ and $\boldsymbol{b}$ associated with $\mathcal{T}_1$ and $\mathcal{T}_2$, relation (IV-51) becomes:

$$T(\boldsymbol{a})\, T(\boldsymbol{b}) = \mathrm{e}^{-i\xi(\boldsymbol{a},\boldsymbol{b})}\, T\left[\boldsymbol{\Phi}_2(\boldsymbol{a},\boldsymbol{b})\right] \tag{IV-67}$$

E-1. Local properties

We first study how the phase factors modify the infinitesimal transformations, and in particular relation (IV-66).

E-1-a. General properties of the phase factors

Choosing $\boldsymbol{a} = 0$ or $\boldsymbol{b} = 0$ in (IV-67), leads to:

$$\xi(\boldsymbol{a}, \boldsymbol{0}) = \xi(\boldsymbol{0}, \boldsymbol{b}) = 0 \qquad \forall\, \boldsymbol{a}, \boldsymbol{b} \tag{IV-68}$$

meaning that in the neighborhood of the origin, the second-order expansion of ξ only includes "cross" terms:

$$\xi(\delta \boldsymbol{a}, \delta \boldsymbol{b}) = \sum_{ij} \gamma_{ij}\, \delta a^i\, \delta b^j + \ldots \tag{IV-69}$$

Furthermore, the ξ function cannot be chosen completely arbitrarily. Consider, for example, three successive $\mathcal{T}$ operations, with respective parameters $\boldsymbol{a}$, $\boldsymbol{b}$, $\boldsymbol{c}$. The parameters associated with the products $\mathcal{T}(\boldsymbol{a})\,\mathcal{T}(\boldsymbol{b})$ and $\mathcal{T}(\boldsymbol{b})\,\mathcal{T}(\boldsymbol{c})$ are:

$$\begin{aligned} \boldsymbol{d}_1 &= \boldsymbol{\Phi}_2(\boldsymbol{a},\boldsymbol{b}) \\ \boldsymbol{d}_2 &= \boldsymbol{\Phi}_2(\boldsymbol{b},\boldsymbol{c}) \end{aligned} \tag{IV-70}$$

and that associated with the product $\mathcal{T}(\boldsymbol{a})\,\mathcal{T}(\boldsymbol{b})\,\mathcal{T}(\boldsymbol{c})$ is:

$$\boldsymbol{f} = \boldsymbol{\Phi}_2(\boldsymbol{d}_1,\boldsymbol{c}) = \boldsymbol{\Phi}_2(\boldsymbol{a},\boldsymbol{d}_2) \tag{IV-71}$$

[product associativity in the $\mathscr{G}$ group, relation (III-44) of chapter III]. This leads to:

$$\begin{cases} T(\boldsymbol{a})\,T(\boldsymbol{b}) = \mathrm{e}^{-i\xi(\boldsymbol{a},\boldsymbol{b})}\;T(\boldsymbol{d}_1) \\ T(\boldsymbol{b})\,T(\boldsymbol{c}) = \mathrm{e}^{-i\xi(\boldsymbol{b},\boldsymbol{c})}\;T(\boldsymbol{d}_2) \end{cases} \tag{IV-72}$$

and hence:

$$\begin{aligned} T(\boldsymbol{a})\,T(\boldsymbol{b})\,T(\boldsymbol{c}) &= \mathrm{e}^{-i\xi(\boldsymbol{a},\boldsymbol{b})}\;\mathrm{e}^{-i\xi(\boldsymbol{d}_1,\boldsymbol{c})}\;T(\boldsymbol{f}) \\ &= \mathrm{e}^{-i\xi(\boldsymbol{a},\boldsymbol{d}_2)}\;\mathrm{e}^{-i\xi(\boldsymbol{b},\boldsymbol{c})}\;T(\boldsymbol{f}) \end{aligned} \tag{IV-73}$$

The product associativity thus leads to:

$$\xi(\boldsymbol{a},\boldsymbol{b}) + \xi(\boldsymbol{d}_1,\boldsymbol{c}) = \xi(\boldsymbol{a},\boldsymbol{d}_2) + \xi(\boldsymbol{b},\boldsymbol{c}) \tag{IV-74}$$

where the vectors $\boldsymbol{d}_{1,2}$ have been defined by relation (IV-70).

E-1-b. Extension of the Lie algebra

We now return to the previous development of § D-2, taking into account the presence of phase factors. The product of the transformations (IV-61) corresponds, in the state space, to the operator:

$$T(\delta\boldsymbol{a})\,T(\delta\boldsymbol{b})\,T(\delta\boldsymbol{a}')\,T(\delta\boldsymbol{b}') \tag{IV-75}$$

where:

$$\begin{aligned} \delta\boldsymbol{a}' &= \boldsymbol{\Phi}_1(\delta\boldsymbol{a}) \\ \delta\boldsymbol{b}' &= \boldsymbol{\Phi}_1(\delta\boldsymbol{b}) \end{aligned} \tag{IV-76}$$

In the space of the $\mathcal{T}$ transformations, the product of the first two transformations of (IV-61) is associated with the parameter $\delta\boldsymbol{d}_1$, the product of the first three with the parameter $\delta \boldsymbol{d}_3$, and finally, the product of the four with the parameter $\delta \boldsymbol{c}$:

$$\begin{aligned}
\delta \boldsymbol{d}_1 &= \boldsymbol{\Phi}_2(\delta \boldsymbol{a}, \delta \boldsymbol{b}) \\
\delta \boldsymbol{d}_3 &= \boldsymbol{\Phi}_2(\delta \boldsymbol{d}_1, \delta \boldsymbol{a}') \\
\delta \boldsymbol{c} &= \boldsymbol{\Phi}_2(\delta \boldsymbol{d}_3, \delta \boldsymbol{b}')
\end{aligned} \tag{IV-77}$$

Following the same reasoning that led to (IV-73), we get:

$$T(\delta \boldsymbol{a})\, T(\delta \boldsymbol{b})\, T(\delta \boldsymbol{a}')\, T(\delta \boldsymbol{b}') = \mathrm{e}^{-i\delta\xi}\; T(\delta \boldsymbol{c}) \tag{IV-78}$$

with:

$$\delta\xi = \xi(\delta \boldsymbol{a}, \delta \boldsymbol{b}) + \xi(\delta \boldsymbol{d}_1, \delta \boldsymbol{a}') + \xi(\delta \boldsymbol{d}_3, \delta \boldsymbol{b}') \tag{IV-79}$$

We now compute the $\delta\xi$ phase, to second-order with respect to the variables δa and δb, using relation (IV-69). As all the ξ phases are of second-order or more, they can be calculated by replacing their variables by their first-order values:

$$\begin{aligned}
\delta \boldsymbol{a}' &= -\delta \boldsymbol{a} + \ldots \\
\delta \boldsymbol{b}' &= -\delta \boldsymbol{b} + \ldots \\
\delta \boldsymbol{d}_1 &= \delta \boldsymbol{a} + \delta \boldsymbol{b} + \ldots \\
\delta \boldsymbol{d}_3 &= \delta \boldsymbol{d}_1 + \delta \boldsymbol{a}' = \delta \boldsymbol{b} + \ldots
\end{aligned} \tag{IV-80}$$

In this approximation, we get:

$$\begin{aligned}
\delta\xi &= \sum_{ij} \gamma_{ij} \left[\delta a^i\, \delta b^j + \left(\delta a^i + \delta b^i\right)\left(-\delta a^j\right) + \left(\delta b^i\right)\left(-\delta b^j\right)\right] + \ldots \\
&= \sum_{ij} \gamma_{ij} \left(\delta a^i\, \delta b^j - \delta b^i\, \delta a^j\right) - \sum_{ij} \gamma_{ij} \left(\delta a^i\, \delta a^j + \delta b^i\, \delta b^j\right)
\end{aligned} \tag{IV-81}$$

We now express $T(\delta \boldsymbol{a}')$ and $T(\delta \boldsymbol{b}')$ as a function of $T(\delta \boldsymbol{a})$ and $T(\delta \boldsymbol{b})$. As we assumed in (IV-54) that the identity transformation is associated with the identity operator in the state space, relation (IV-67) shows that:

$$T(\delta \boldsymbol{a}) T(\delta \boldsymbol{a}') = \mathrm{e}^{-i\zeta(\delta \boldsymbol{a})} \tag{IV-82a}$$

The operator $T(\delta \boldsymbol{a}')$ may thus be replaced in (IV-78) by $\mathrm{e}^{-i\zeta(\delta \boldsymbol{a})} T^\dagger(\delta \boldsymbol{a})$, with:

$$\zeta(\delta \boldsymbol{a}) = \xi(\delta \boldsymbol{a}, -\delta \boldsymbol{a}) = -\sum_{ij} \gamma_{ij} \delta a^i \delta a^j \tag{IV-82b}$$

Once the same transformation has been performed on $T(\delta \boldsymbol{b}')$, we get:

$$T(\delta \boldsymbol{a})\, T(\delta \boldsymbol{b})\, T^\dagger(\delta \boldsymbol{a})\, T^\dagger(\delta \boldsymbol{b}) = \mathrm{e}^{i[\zeta(\delta \boldsymbol{a}) + \zeta(\delta \boldsymbol{b}) - \delta\xi]}\; T(\delta \boldsymbol{c}) \tag{IV-83}$$

The only difference between this relation and the one obtained in § D-2 is the presence on the right-hand side of a phase exponential. Taking into account (IV-81) and (IV-82b), this exponential is written, to second-order:

$$e^{i[\zeta(\delta\boldsymbol{a})+\zeta(\delta\boldsymbol{b})-\delta\xi]} = 1 + i\zeta(\delta\boldsymbol{a}) + i\zeta(\delta\boldsymbol{b}) - i\delta\xi = 1 - i\sum_{ij} \gamma_{ij}\left(\delta a^i\,\delta b^j - \delta b^i\,\delta a^j\right) \tag{IV-84}$$

As second-order terms on each side of (IV-83) must be equal, we have:

$$1 - \frac{1}{\hbar^2}\left[\delta\boldsymbol{a}\cdot\boldsymbol{G},\, \delta\boldsymbol{b}\cdot\boldsymbol{G}\right] = 1 - \frac{i}{\hbar}\delta\boldsymbol{c}\cdot\boldsymbol{G} - i\sum_{ij}\gamma_{ij}\left(\delta a^i\,\delta b^j - \delta b^i\,\delta a^j\right) + \ldots \tag{IV-85}$$

Identifying the $\delta a^i\,\delta b^j$ terms on each side of this equality leads to:

$$[G_i, G_j] = i\hbar\sum_k C_{ij}^k\, G_k + i\hbar^2\left(\gamma_{ij} - \gamma_{ji}\right) \tag{IV-86}$$

which is a generalization of relations (IV-66).

In the case of a projective representation, this equality shows how the commutation relations between infinitesimal operators must be modified. To the structure constants C_{ij}^k terms, we must add antisymmetric constants (proportional to the identity operator):

$$\boxed{[G_i, G_j] = i\hbar\sum_k C_{ij}^k\, G_k + i\hbar^2\beta_{ij}} \tag{IV-87a}$$

with:

$$\boxed{\beta_{ij} = \gamma_{ij} - \gamma_{ji}} \tag{IV-87b}$$

The β_{ij} are called “extension constants”. Such a modification of the Lie algebra by adding the identity operator is called an *extension of the Lie algebra.*

Comment:

A special case occurs when the matrix of the second derivatives yielding the function ξ is antisymmetric:

$$\gamma_{ij} + \gamma_{ji} = 0 \tag{IV-88}$$

This relation implies that the right-hand side of (IV-82b) is zero. This means that two inverse infinitesimal transformations $\mathcal{T}$ are associated with two transformations T that are simply the inverse of each other, without any phase factors.

E-1-c. Relation between the extension constants and the structure constants

Note that the function ξ, and hence the β_{ij} related to its second derivatives, are not arbitrary. We already wrote in (IV-74) a relation necessarily satisfied by ξ. Let us see what it implies concerning the β_{ij} extension constants of the Lie algebra.

Replacing $\boldsymbol{a}$ by $\delta\boldsymbol{a}$, $\boldsymbol{b}$ by $\delta\boldsymbol{b}$, and $\boldsymbol{c}$ by $\delta\boldsymbol{c}$, we write the equality of the expansions of both sides of (IV-74), up to the third-order cross term $\delta a\, \delta b\, \delta c$. We have [*cf.* relation (III-28)]:

$$\begin{aligned}
\delta\boldsymbol{d}_1 &= \boldsymbol{\Phi}_2(\delta\boldsymbol{a},\delta\boldsymbol{b}) = \delta\boldsymbol{a} + \delta\boldsymbol{b} + \sum_{ij} \boldsymbol{F}_{ij}\, \delta a^i\, \delta b^j + \ldots \\
\delta\boldsymbol{d}_2 &= \boldsymbol{\Phi}_2(\delta\boldsymbol{b},\delta\boldsymbol{c}) = \delta\boldsymbol{b} + \delta\boldsymbol{c} + \sum_{jk} \boldsymbol{F}_{jk}\, \delta b^j\, \delta c^k + \ldots
\end{aligned} \tag{IV-89}$$

Since $\xi(\delta\boldsymbol{a},\delta\boldsymbol{b})$ and $\xi(\delta\boldsymbol{b},\delta\boldsymbol{c})$ do not include any $\delta a\, \delta b\, \delta c$ term, relation (IV-74) reduces to:

$$\xi(\delta\boldsymbol{d}_1,\delta\boldsymbol{c}) = \xi(\delta\boldsymbol{a},\delta\boldsymbol{d}_2) + \ldots \tag{IV-90}$$

We then have, taking (IV-69) into account:

$$\sum_{k\ell} \gamma_{\ell k}\Big(\sum_{ij} f_{ij}^{\ell}\, \delta a^i\, \delta b^j\Big)\delta c^k = \sum_{i\ell} \gamma_{i\ell}\, \delta a^i \Big(\sum_{jk} f_{jk}^{\ell}\, \delta b^j\, \delta c^k\Big) \tag{IV-91}$$

Identifying the $\delta a^i\, \delta b^j\, \delta c^k$ terms yields:

$$\sum_{\ell} \gamma_{\ell k}\, f_{ij}^{\ell} = \sum_{\ell} \gamma_{i\ell}\, f_{jk}^{\ell} \qquad \forall\ i,j,k \tag{IV-92}$$

This leads, as we show below:

$$\boxed{\sum_{\ell} \left\{\beta_{\ell i}\, C_{jk}^{\ell} + \beta_{\ell j}\, C_{ki}^{\ell} + \beta_{\ell k}\, C_{ij}^{\ell}\right\} = 0} \tag{IV-93}$$

This equality is true for any i, j, and k, and the three terms on the left-hand side are obtained by a circular permutation.

Demonstration: Let us calculate the double commutator $[[G_i,G_j]\,G_k]$. Taking (IV-86) and (IV-87b) into account, we get:

$$[[G_i,G_j]\,,G_k] = i\hbar \sum_{\ell} \left[C_{ij}^{\ell}\ G_\ell + \hbar\beta_{ij}\ ,\ G_k\right] = -\hbar^2 \sum_{\ell p} C_{ij}^{\ell}\, [C_{\ell k}^{p} G_p + \hbar\beta_{\ell k}] \tag{IV-94}$$

We add to this equality the other two obtained by a circular permutation of the three indices i, j, and k. Using Jacobi relation (III-42), we see that the sum of the three double commutators on the left-hand side is zero. As for the right-hand side, the G_p term is proportional to:

$$\sum_{\ell} \left[C_{ij}^{\ell} C_{\ell k}^{p} + C_{jk}^{\ell} C_{\ell i}^{p} + C_{ki}^{\ell} C_{vj}^{p}\right] \tag{IV-95}$$

which is also equal to zero, according to relation (III-43). We are left with the $\hbar\beta_{\ell k}$ term, whose sum with the terms obtained by the circular permutation of the indices i, j and k must be zero, demonstrating equality (IV-93).

E-2. Finite representations

Let us study the effect of the factor $e^{-i\xi(\boldsymbol{a},\boldsymbol{b})}$ in the simple case where the dimension N of the representation space is finite[12]. The representation matrices, whose elements are given in (IV-52), are $N \times N$ square matrices.

In this case, the phase factors can be readily simplified, as we now show. We call Δ the determinant of the matrices:

$$\Delta(\boldsymbol{a}) = \det \begin{pmatrix} T_{1\,1}(\boldsymbol{a}) \; T_{1\,2}(\boldsymbol{a}) \dots \\ T_{2\,1}(\boldsymbol{a}) \\ \vdots \\ T_{N\,1}(\boldsymbol{a}) \end{pmatrix} \tag{IV-96}$$

As a unitary matrix is diagonalizable and all its eigenvalues have a modulus equal to one, we have:

$$|\Delta(\boldsymbol{a})| = 1 \tag{IV-97}$$

Relation (IV-67) yields (the determinant of a product is the product of the determinants):

$$\Delta(\boldsymbol{a})\,\Delta(\boldsymbol{b}) = e^{-iN\xi(\boldsymbol{a},\boldsymbol{b})}\;\Delta(\boldsymbol{d}) \tag{IV-98}$$

where $\boldsymbol{d}$ is the parameter associated with the product $\mathcal{T}(\boldsymbol{a})\;\mathcal{T}(\boldsymbol{b})$:

$$\boldsymbol{d} = \boldsymbol{\Phi}_2(\boldsymbol{a},\boldsymbol{b}) \tag{IV-99}$$

It is easy to impose the value 1 for the determinants Δ of all the matrices: we simply replace the operators T by the T' defined as[13]:

$$T'(\boldsymbol{a}) = T(\boldsymbol{a}) / \left[\Delta(\boldsymbol{a})\right]^{1/N} \tag{IV-100}$$

[12]We may assume, without N being necessarily finite, that the T representation of the $\mathscr{G}$ group has been decomposed into several representations, and that at least one of them is finite (meaning the T leave globally invariant a finite dimensional subspace $\mathscr{E}'$ of $\mathscr{E}$). It is this finite representation (if it exists) that is studied in § E-2.

[13]$\Delta(\boldsymbol{a})^{1/N}$ represents any one of the Nth complex root of $\Delta(\boldsymbol{a})$.

Relation (IV-98) shows that the phase factors ξ can only take the values 0, $2\pi/N$, $4\pi/N, \ldots, (N-1)2\pi/N$. Therefore, a simple phase change of each T operator reduces to a finite number the possible values of the phase factors. Furthermore, as $\xi(\mathbf{0}, \mathbf{0}) = 0$, the function ξ is either discontinuous or always equal to zero. We show in complement A_{IV} that, if the group is simply connected, all the ξ must be equal to zero, so that the projective representation becomes a representation in the strict sense of the term.

Complement A_{IV}

Unitary projective representations, with finite dimension, of connected Lie groups. Bargmann's theorem.

We examine in more detail the unitary representations (with finite dimension) of a continuous group $\mathscr{G}$, introduced in § E-2 of chapter IV. Consider an arbitrary projective representation, given by the $N \times N$ matrices of the $T(\boldsymbol{a})$ operators:

$$\begin{pmatrix} T_{11}(\boldsymbol{a}) \ldots\ldots & T_{1N}(\boldsymbol{a}) \\ T_{21}(\boldsymbol{a}) & \\ \vdots & \\ T_{N1}(\boldsymbol{a}) \ldots\ldots & T_{NN}(\boldsymbol{a}) \end{pmatrix} \tag{1}$$

The representation of a product in $\mathscr{G}$:

$$\mathcal{T}(\boldsymbol{d}) = \mathcal{T}(\boldsymbol{a})\, \mathcal{T}(\boldsymbol{b}) \tag{2}$$

is written:

$$T(\boldsymbol{d}) = \mathrm{e}^{-i\xi\,(a,b)}\; T(\boldsymbol{a})\; T(\boldsymbol{b}) \tag{3}$$

The determinant of a unitary matrix can be any complex number whose modulus equals one. Nevertheless, we can assume that the determinant of the matrices described in (1) is actually equal to 1, if we proceed as follows. We call $K(\boldsymbol{a})$ one of the Nth complex roots of the determinant $\Delta(\boldsymbol{a})$ of these matrices:

$$[K(\boldsymbol{a})]^N = \Delta(\boldsymbol{a}) \tag{4}$$

Replacing the $T(\boldsymbol{a})$ operators by the new operators:

$$T(\boldsymbol{a})/K(\boldsymbol{a})$$

ensures that the corresponding matrices have a determinant equal to one. We saw in § E-2 of chapter IV that in that case, the phase factor $\mathrm{e}^{i\xi}$ can only take a discrete set of N values, the Nth roots of unity:

$$\mathrm{e}^{i\xi(\boldsymbol{a},\boldsymbol{b})} = 1,\, \mathrm{e}^{2i\pi/N},\, \mathrm{e}^{4i\pi/N},\, \ldots,\, \mathrm{e}^{(N-1)\,2i\pi/N} \tag{5}$$

As $\xi(\mathbf{0},\mathbf{0}) = 0$, the function ξ must be either discontinuous or always equal to zero. We will see in this complement why, depending on the topology of the group, discontinuities are sometimes unavoidable.

The redefinition of the $T(\boldsymbol{a})$ operators is obviously not unique since there are N complex numbers $K(\boldsymbol{a})$ obeying relation (4). A priori, N distinct $T(\boldsymbol{a})$ operators, differing by a phase factor, could be associated with each transformation $\mathcal{T}(\boldsymbol{a})$ of the group $\mathscr{G}$. The aim of this complement is to show how, for a connected group $\mathscr{G}$, this undetermined phase can help simplify the phase factors $e^{-i\xi}$ in (3). Furthermore, if the $\mathscr{G}$ group is simply connected, we are going to show that all the $e^{-i\xi}$ phase factors can be completely omitted.

Comment:

The concept of a group representation can be generalized to that of a multi-valued representation: with each element of the group $\mathscr{G}$ we associate not one but a set of N matrices or operators $T_1(\boldsymbol{a}), T_2(\boldsymbol{a}), \ldots, T_N(\boldsymbol{a})$. Consider two operations $\mathcal{T}(\boldsymbol{a})$ and $\mathcal{T}(\boldsymbol{b})$, whose product is $\mathcal{T}(\boldsymbol{c})$. If all the couples $T_i(\boldsymbol{a})$, $T_j(\boldsymbol{b})$ are equal to one of the $T_k(\boldsymbol{c})$ matrices, the representation is said to be "multi-valued". We will return to this point in § A-3-b of chapter VII.

1. Case where $\mathscr{G}$ is simply connected

Relation (4) offers N possible determinations of the phase of $K(\boldsymbol{a})$, and hence of the $T(\boldsymbol{a})$ operator. If this determination is chosen at random for the different values of $\boldsymbol{a}$, we can make the variations of T as a function of $\boldsymbol{a}$ as complex as we wish. For example, selecting two different determinations for two very close values of $\boldsymbol{a}$ creates a discontinuity for the $T(\boldsymbol{a})$ operator [the $T_{ij}(\boldsymbol{a})$ elements are then discontinuous functions of $\boldsymbol{a}$]. We shall make here the opposite choice, and render the variations of $T(\boldsymbol{a})$ as regular as possible.

The operator $T(\boldsymbol{a} = 0)$ associated with the identity transformation $\mathcal{T}(\boldsymbol{a} = 0)$ is necessarily a scalar diagonal operator, since we assumed that two kets Ψ describing the same physical states are necessarily proportional to each other (note 8 of chapter IV). We shall choose this operator equal to unity. We assume that $T(\boldsymbol{a})$ varies in a continuous way in the vicinity of the origin $\boldsymbol{a} = 0$, so that we can write[1], as in (IV-55):

$$T(\delta\boldsymbol{a}) = 1 - \frac{i}{\hbar}(\delta\boldsymbol{a} \cdot \boldsymbol{G}) + \ldots \tag{6}$$

In a certain domain around the origin, the choice of the phases has been made so as to ensure the continuity of the T operators.

[1]Because the determinant of the matrix of T has been fixed to unity, the trace of all the components of $\boldsymbol{G}$ must vanish.

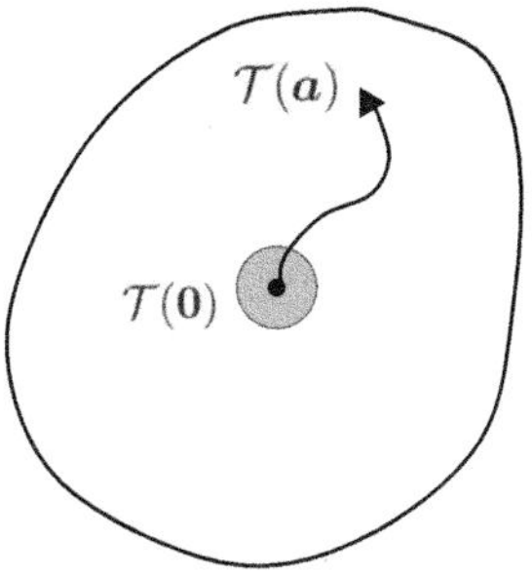

Figure 1: In an infinitesimal domain, displayed in gray, around the identity transformation $\mathcal{T}(\mathbf{0})$, the phase factors of any projective representation can be chosen equal to zero ($\xi = 0$). To reach a finite transformation $\mathcal{T}(\boldsymbol{a})$, we follow a continuous path that starts from the identity transformation.

This means that there exists, around the origin, a certain domain of $\boldsymbol{a}$ and $\boldsymbol{b}$ where the function $\xi(\boldsymbol{a}, \boldsymbol{b})$ is necessarily zero. This can be readily seen on (3), which can be written, for infinitesimal transformations:

$$T^\dagger(\delta\boldsymbol{b})\; T^\dagger(\delta\boldsymbol{a})\; T(\delta\boldsymbol{d}) = \mathrm{e}^{-i\xi(\delta\boldsymbol{a},\delta\boldsymbol{b})} \tag{7}$$

As the left-hand side of this equality is an operator (actually a scalar) very close to 1, we can only choose for $\mathrm{e}^{-i\xi}$, on the right-hand side, the first of the values given in (5). The function $\xi(\delta\boldsymbol{a}, \delta\boldsymbol{b})$ is thus necessarily zero.

When $\boldsymbol{a}$ and $\boldsymbol{b}$ have finite values, we do not know, a priori, the value of $\mathrm{e}^{-i\xi}$. We shall follow the same line of reasoning, choosing the phase of the T operators that renders their dependance on $\boldsymbol{a}$ as regular as possible. Consider a "path" in $\mathscr{G}$ (figure 1), starting from the origin $\mathcal{T}(\mathbf{0})$ and whose endpoint is the $\mathcal{T}(\boldsymbol{a})$ operator. We define a $\mathcal{T}[\boldsymbol{c}(t)]$ operator, function of t ($0 \leq t \leq 1$), such that:

$$\begin{aligned} \mathcal{T}[\boldsymbol{c}(0)] &= \mathcal{T}(\mathbf{0}) \\ \mathcal{T}[\boldsymbol{c}(1)] &= \mathcal{T}(\boldsymbol{a}) \end{aligned} \tag{8}$$

To go from $\mathcal{T}[\boldsymbol{c}(t)]$ to $\mathcal{T}[\boldsymbol{c}(t + \mathrm{d}t)]$, we can use an infinitesimal $\mathcal{T}$ operator (the path is continuous in $\mathscr{G}$), noted $\mathcal{T}[\boldsymbol{v}(t)\,\mathrm{d}t]$:

$$\mathcal{T}\left[\boldsymbol{c}(t + \mathrm{d}t)\right] = \mathcal{T}\left[\boldsymbol{v}(t)\,\mathrm{d}t\right]\; \mathcal{T}\left[\boldsymbol{c}(t)\right] \tag{9}$$

where $\boldsymbol{v}(t)$ is the "velocity" along the path.

This relation can be used to build an "image" of the path we consider in the set of T operators . In the infinitesimal domain around the origin, the T are the operators already chosen to represent the $\mathcal{T}$ (those very close to 1). We then

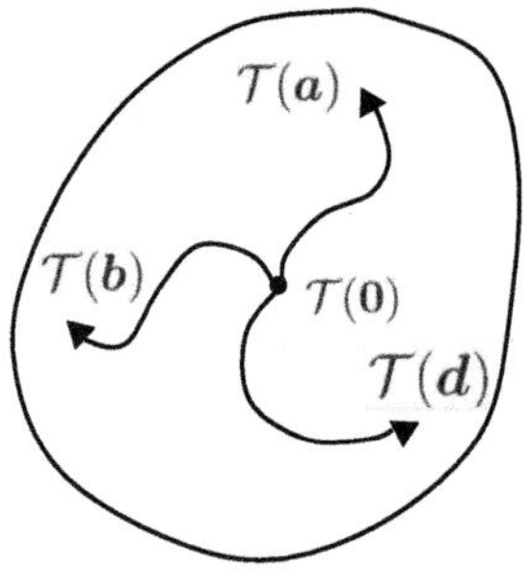

Figure 2: Two transformations $\mathcal{T}(\boldsymbol{a})$ and $\mathcal{T}(\boldsymbol{b})$ are the end points of two paths starting from the identity tranformation. The product transformation follows the path that ends at $\mathcal{T}(\boldsymbol{d})$. Along those three paths, the variations of the representation matrices must be continuous. The phase factors cannot jump, from their initial zero value, to another value: they remain equal to zero.

write the relation image of (9):

$$T\left[\boldsymbol{c}(t+\mathrm{d}t)\right] = \left[1 - \frac{i}{\hbar}\,\boldsymbol{v}(t)\cdot\boldsymbol{G}\,\mathrm{d}t\right]\; T\left[\boldsymbol{c}(t)\right] \tag{10}$$

This means that, to represent $\mathcal{T}\left[\boldsymbol{c}(t+\mathrm{d}t)\right]$, we select one of the a priori N possible T operators, and hence choose the phase of the T operators. Their very construction (continuity of a matrix product) ensures that the T operators vary continuously along the path taken (no phase jump).

We assume that the $\mathcal{G}$ group is connected and that any $\mathcal{T}$ can be reached by a (continuous) path starting from the origin. For each $\mathcal{T}(\boldsymbol{a})$, we will choose an arbitrary path in $\mathcal{G}$, and define by continuity the phase of $T(\boldsymbol{a})$. If the group $\mathcal{G}$ is simply connected, all the paths starting from $\mathcal{T}(\mathbf{0})$ and ending at $\mathcal{T}(\boldsymbol{a})$ are homotopic, i.e. they can be continuously deformed into one another. During such a continuous deformation, the phase of the $T(\boldsymbol{a})$ operator cannot jump by $2\pi/N$, or by any multiple of this quantity. Consequently, all the paths in $\mathcal{G}$ starting from the origin $\mathcal{T}(\mathbf{0})$ and ending at $\mathcal{T}(\boldsymbol{a})$ correspond to the same definition of $T(\boldsymbol{a})$. In other words, the phase of each operator $T(\boldsymbol{a})$ does not depend on the path taken to reach $\mathcal{T}(\boldsymbol{a})$ in the $\mathcal{G}$ group.

Under these conditions, all the ξ phase factors are necessarily zero, as we now show. Consider two paths in $\mathcal{G}$, both starting from $\mathcal{T}(\mathbf{0})$. The first one, described by $\mathcal{T}_a(t)$, ends at $\mathcal{T}(\boldsymbol{a})$, and the second, described by $\mathcal{T}_b(t)$, at $\mathcal{T}(\boldsymbol{b})$ (figure 2). The product transformation of the two values (at the same instant t) of $\mathcal{T}$ on each of these paths is noted $\mathcal{T}_d(t) = \mathcal{T}_a(t)\;\mathcal{T}_b(t)$. It follows a third path, ending at $\mathcal{T}(\boldsymbol{d})$.

The images, in the T set, of these three paths are three continuous paths (where the phases of the T operators do not jump), labeled $T_a(t)$, $T_b(t)$, and

$T_d(t)$. We can, a priori, have:

$$T_d(t) = \mathrm{e}^{-i\xi}\ T_a(t)\ T_b(t) \tag{11}$$

where $\mathrm{e}^{-i\xi}$ takes any of the values (5). However, as the product of two operators is a continuous function of these operators, $\mathrm{e}^{-i\xi}$ cannot make a discontinuous jump, and remains equal to one.

We have demonstrated that, for a simply connected $\mathscr{G}$ group, a redefinition of the phase of the T operators allows transforming any projective representation into a "true" representation. This result is sometimes referred to as "Bargmann's theorem".

2. Case where $\mathscr{G}$ is p-connected

A connected group $\mathscr{G}$ has been defined as p-connected if there exists a maximum of p distinct (non homotopic) closed paths (loops) starting from (and ending at) the same fixed point (note 5 of chapter III). In other words, the homotopy group Π_1 of the $\mathscr{G}$ groups has p distinct elements (*cf.* chapter III, § A-1-b). If the paths considered are not closed, but have fixed end points, one can regroup into the same class all the paths homotopic with each other, and obtain p distinct classes. An important example of a 2-connected group, the $R_{(3)}$ rotation group, will be studied in chapter VI.

The previous reasoning of § 1 remains valid until we begin the discussion on how the choice of path determines the phase of the T operators. As an example, if the group is 2-connected, the phase of T may be fixed in two different ways, depending on the chosen path (for the sake of simplicity, we reason on the $p = 2$ case, but the results may be generalized). Imagine that we have made such an arbitrary choice for each $T(\boldsymbol{a})$ operator associated with $\mathcal{T}(\boldsymbol{a})$. Following the same reasoning as in § 1, schematized on figure 2, two cases may occur:

- either the operator follows a path equivalent to the one that defined $T(\boldsymbol{d})$, and for the same reasons as above:

$$T(\boldsymbol{d}) = T(\boldsymbol{a})\ T(\boldsymbol{b}) \qquad \text{(no phase factor);}$$

- or the path followed belongs to the other class (there are only two classes since we assume $p = 2$) and:

$$T(\boldsymbol{d}) = \mathrm{e}^{-i\xi(\boldsymbol{a},\boldsymbol{b})}\ T(\boldsymbol{a})\ T(\boldsymbol{b}).$$

This means that $\mathrm{e}^{-i\xi}$ is simply the phase difference that appears when choosing the path defining $T(\boldsymbol{d})$ in one or the other class. As we now show, ξ can only take the values:

$$\xi = 0, \frac{2\pi}{p}, \frac{4\pi}{p}, \dots, (p-1)\frac{2\pi}{p} \tag{12}$$

or, as $p = 2$:

$$e^{-i\xi} = \pm 1 \tag{13}$$

Consider two non-equivalent paths in $\mathscr{G}$, both starting from $\mathcal{T}(\mathbf{0})$ and ending in $\mathcal{T}(\boldsymbol{d})$ (right-hand side of figure 3). One can build a loop by following the first path and then the second in the opposite direction. This loop is not homotopic to zero (otherwise the two initial paths would be homotopic with each other). As in (10), we can build a continuous image of the T building this loop. We first follow the path that was used for defining $T(\boldsymbol{d})$, then we follow the other path (but in the opposite direction) connecting by continuity the identity operator to $e^{-i\xi}\, T(\boldsymbol{d})$. Therefore, when the path in $\mathscr{G}$ comes back to the origin, the path of the continuous image in the set of T comes back to the value $e^{i\xi} \times (\mathbb{1})$ [instead of $(\mathbb{1})$]. This continuous image does not follow a loop but an open path. Such a situation only occurs when, by hypothesis, the loop in $\mathscr{G}$ cannot be continuously deformed into a zero path.

Imagine that we follow twice the loop described above: the result is a new loop, not equivalent to the previous one, homotopic to zero if $p = 2$. The image of the end point in the set of T is $e^{2i\xi}$ $(\mathbb{1})$, which must be equal to $(\mathbb{1})$ since the

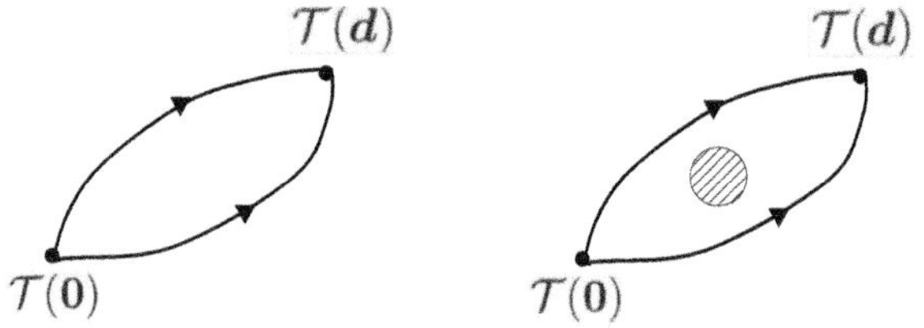

Figure 3: Two different paths go from the identity transformation $\mathcal{T}(\mathbf{0})$ to the transformation $\mathcal{T}(\boldsymbol{d})$. If they are equivalent (left-hand side of the figure), the closed path obtained by following one of them in the direct sense and the other in the opposite sense is a loop equivalent to zero. If they are not equivalent (right-hand side of the figure, where the paths cannot cross the dashed region), the loop is no longer equivalent to zero, and its image in the T space may include a phase factor.

new loop can be continuously deformed into a zero path. This means that:

$$\xi = 0,\ \pi \qquad \text{if} \quad p = 2 \tag{14}$$

Example: the group of rotations, whose parameters $\boldsymbol{a}$ vary within a spherical domain of radius π, as illustrated in figure 14 of chapter III. The matrices representing the rotations can be defined by continuity: one starts from the center O of the sphere, which is associated with the identity matrix, and one follows a continuous path to reach any point inside the sphere; the path is arbitrary, provided it remains inside the sphere. If the matrices vary continuously along the path, this procedure defines the phase of all matrices representing rotations having an angle smaller than π. Now, a difficulty appears for rotations with angle π: they are represented by points at the surface of the sphere, but not in a unique way, since two diametrically opposed points define the same rotation. Nothing then guarantees that the two matrices obtained by the preceding construction are equal. We will see in chapter VI that, for all representations having a half integer value of the quantum number j, they are actually opposite.

Let us now discuss what happens when the parameter describing the path crosses the outside surface of the sphere. If the path corresponds to a continuous variation of the rotations, its parameter reappears at the diametrically opposed point on the surface, so that the rotation vector undergoes a discontinuous jump. As a consequence, each time the parameter $\boldsymbol{d}$ of the product (11) of two rotations crosses the surface of the sphere, the function ξ may undergo a discontinuity equal to π.

Complement B_{IV}

Uhlhorn–Wigner theorem

Wigner's theorem has been stated, in § B of chapter IV, in its most current form where we assumed that symmetry transformations preserve the norm of the scalar product of state vectors. The demonstration of this theorem has been generalized by several authors. Uhlhorn [17] in particular proposed a more general and rigorous form, where the scalar product is supposed to be preserved only when it is equal to zero. Instead of assuming that all the probabilities must be preserved, one simply imposes the preservation of zero probabilities (corresponding to impossible measurement results). Reference [19] provides an overall perspective of the demonstration of Wigner's theorem, as well as of possible generalizations (quaternions). The aim of this complement is to introduce Uhlhorn's version of Wigner's theorem. We shall assume that two orthogonal state vectors are transformed into two other orthogonal vectors, with no hypothesis concerning the "oblique" case where the scalar product is different from zero.

Going back to § B of chapter IV, we consider the transformation of "rays", that is of vectors of the state space defined modulo a multiplicative (complex) factor. Our objective is to show that if a transformation preserves the zero scalar product of rays, it must necessarily be equivalent to either a unitary or an antiunitary transformation in the state space. To present a more intuitive argument we start in § 1 with a transformation that acts in real space, easier to visualize. Generalization to a complex space is presented in § 2.

1. Real space

Consider a real vector space $\mathcal{E}$, with dimension $N > 2$. "Rays" of this space form a "projective space", where each ray is simply identified with a straight line (one-dimensional subspace of $\mathcal{E}$). One can place on this line colinear vectors of $\mathcal{E}$, having different lengths and directions. We assume there exists in this projective space a transformation $\mathscr{W}$ that preserves the orthogonality between rays, hence between the corresponding lines. We are going to show how to build a linear and orthogonal transformation F that acts on the vectors of $\mathcal{E}$ and is equivalent to $\mathscr{W}$ (both yielding the same ray transformation).

We choose in $\mathcal{E}$ an orthonormal basis formed by the vectors $\boldsymbol{e}_1$, $\boldsymbol{e}_2$, ..., $\boldsymbol{e}_N$ defining N orthogonal lines on which these vectors are placed. Applying the $\mathscr{W}$ transformation to these N lines yields, as their image, N other orthogonal lines onto which we can place normalized vectors $\boldsymbol{E}_1$, $\boldsymbol{E}_2$, ..., $\boldsymbol{E}_N$. These vectors, called the image vectors of the $\boldsymbol{e}_i$, form another orthonormal basis of $\mathcal{E}$.

In addition, all the lines of the plane determined by $\boldsymbol{e}_i$ and $\boldsymbol{e}_j$ are perpendicular to all the other vectors $\boldsymbol{e}_k$ of the basis, so that their images through $\mathscr{W}$ are perpendicular to the $\boldsymbol{E}_k$. They must necessarily belong to the plane determined by $\boldsymbol{E}_i$ and $\boldsymbol{E}_j$. This means that the image of the plane determined by two base vectors is another plane determined by the images of these two vectors. In the same way, the image through $\mathscr{W}$ of the 3-dimensional space determined by any 3 vectors $\boldsymbol{e}_i$, $\boldsymbol{e}_j$, and $\boldsymbol{e}_k$ is the 3-dimensional space determined by the three vectors $\boldsymbol{E}_i$, $\boldsymbol{E}_j$, and $\boldsymbol{E}_k$. We can thus reason independently in each of the 3-dimensional subspaces of the $\mathcal{E}$ space.

1-a. Components of a vector and its image

Choosing one of these subspaces, we call $\boldsymbol{e}_x$, $\boldsymbol{e}_y$, and $\boldsymbol{e}_z$ the vectors of a basis, and $\boldsymbol{E}_x$, $\boldsymbol{E}_y$, and $\boldsymbol{E}_z$ their images. The image through $\mathscr{W}$ of the xOy plane, determined by $\boldsymbol{e}_x$ and $\boldsymbol{e}_y$, is the XOY plane determined by $\boldsymbol{E}_x$ and $\boldsymbol{E}_y$. Similarly, the images of the yOz and zOx planes are, respectively, the YOZ and ZOX planes.

In the space spanned by $\boldsymbol{e}_x$, $\boldsymbol{e}_y$, and $\boldsymbol{e}_z$, we choose to identify any line (or ray) by the vector $\boldsymbol{v}$ with components x and y such that:

$$\boldsymbol{v} = x\boldsymbol{e}_x + y\boldsymbol{e}_y + \boldsymbol{e}_z \tag{1}$$

In other words, since $\boldsymbol{v}$ is a priori defined to within a real factor, we assign by convention[1] the value one to its Oz component. We adopt the same convention for the image vectors, so that the image through $\mathscr{W}$ of the ray defined by $\boldsymbol{v}$ is associated with the vector $\boldsymbol{V}$:

$$\boldsymbol{V} = X(x,y)\boldsymbol{E}_x + Y(x,y)\boldsymbol{E}_y + \boldsymbol{E}_z \tag{2}$$

We now introduce a transformation $\boldsymbol{F}(\boldsymbol{v})$ of the $\boldsymbol{v}$ vectors, defined as:

$$\boldsymbol{F}(\boldsymbol{v}) = \boldsymbol{V} = X(x,y)\boldsymbol{E}_x + Y(x,y)\boldsymbol{E}_y + \boldsymbol{E}_z \tag{3}$$

Since it was built using $\mathscr{W}$, which preserves the orthogonality between rays, this transformation preserves the orthogonality between vectors. It follows that the

[1] *cf.* comment (i) below, page 151.

two functions $X(x,y)$ and $Y(x,y)$ cannot take a random value: as we now show, X can only depend on x, and Y can only depend on y.

Consider a vector of the plane yOz:

$$\boldsymbol{v}' = y\boldsymbol{e}_y + \boldsymbol{e}_z \tag{4}$$

We assume $y \neq 0$, so that this vector does not coincide with $\boldsymbol{e}_z$. The vector $\boldsymbol{w}'$, perpendicular to $\boldsymbol{v}'$ in the plane yOz, is written:

$$\boldsymbol{w}' = -\frac{1}{y}\boldsymbol{e}_y + \boldsymbol{e}_z \tag{5}$$

The images through $\boldsymbol{F}$ of these two vectors are in the plane YOZ, and also perpendicular to each other. They are written:

$$\begin{aligned} \boldsymbol{V}' &= Y_1(0,y)\,\boldsymbol{E}_y + \boldsymbol{E}_z \\ \boldsymbol{W}' &= -\frac{1}{Y_1(0,y)}\,\boldsymbol{E}_y + \boldsymbol{E}_z \end{aligned} \tag{6}$$

Now the vector $\boldsymbol{v}$ written in (1) is perpendicular to $\boldsymbol{w}'$, so that its image $\boldsymbol{V}$ must be orthogonal to $\boldsymbol{W}'$. Consequently:

$$Y(x,y) = Y_1(0,y) \tag{7}$$

This means that $Y(x,y)$ does not depend on x, but only on y. In the same way, one can show that $X(x,y)$ does not depend on y, and relation (2) takes the simple form:

$$\boldsymbol{V} = X(x)\boldsymbol{E}_x + Y(y)\boldsymbol{E}_y + \boldsymbol{E}_z \tag{8}$$

1-b. The F transformation is linear and orthogonal

Consider two orthogonal vectors, $\boldsymbol{v}_1$ with x_1 and y_1 components, and $\boldsymbol{v}_2$ with x_2 and y_2 components. As $\boldsymbol{F}$ preserves the zero scalar product, the orthogonality condition for $\boldsymbol{v}_1$ and $\boldsymbol{v}_2$:

$$x_1x_2 + y_1y_2 + 1 = 0 \tag{9}$$

necessarily implies a similar condition for the images of these vectors:

$$X(x_1)X(x_2) + Y(y_1)Y(y_2) + 1 = 0 \tag{10}$$

When the 4 variables x_1, x_2, y_1, and y_2 vary, while obeying relation (9), we can use a Lagrange multiplier λ to take into account the stationarity condition of this scalar product (the functions X and Y are supposed to be continuous and differentiable):

$$\begin{aligned}
&\dot{X}(x_1)X(x_2) + \lambda x_2 = 0 \\
&X(x_1)\dot{X}(x_2) + \lambda x_1 = 0 \\
&\dot{Y}(y_1)Y(y_2) + \lambda y_2 = 0 \\
&Y(y_1)\dot{Y}((y_2) + \lambda y_1 = 0
\end{aligned} \tag{11}$$

where a dot on one of the two functions X or Y indicates a derivation with respect to its variable. Eliminating λ between the first two equations yields:

$$\frac{\dot{X}(x_1)X(x_2)}{x_2} = \frac{X(x_1)\dot{X}(x_2)}{x_1} \tag{12}$$

or:

$$\frac{x_1\dot{X}(x_1)}{X(x_1)} = \frac{x_2\dot{X}(x_2)}{X(x_2)} \tag{13}$$

This means that the function $x\dot{X}(x)/X(x)$ is necessarily independent of x, which leads to the differential equation:

$$\frac{\dot{X}(x)}{X(x)} = \frac{n}{x} \tag{14}$$

where n is a constant coefficient. The solution of this equation is written:

$$X(x) = \mu x^n \tag{15}$$

where μ is another constant. In the same way, we show that:

$$Y(y) = \nu y^p \tag{16}$$

where p and ν are two other constants. For the functions $X(x)$ and $Y(y)$ to be defined for negative values of x and y, the constants n and p must be positive or negative integers.

As equality (9) can be written:

$$y_1 y_2 = -1 - x_1 x_2 \tag{17}$$

relation (10) becomes:

$$\mu^2(x_1x_2)^n + \nu^2(-1 - x_1x_2)^p = -1 \tag{18}$$

This condition is obviously satisfied for any value of the product x_1x_2 if $n = p = 1$ and $\mu^2 = \nu^2$; these values are the only possible ones as we show below.

If the product x_1x_2 goes to infinity, the left-hand side of (18) amounts to $\mu^2(x_1x_2)^n + (-1)^p\nu^2(x_1x_2)^p$, where each term of this sum goes to infinity or zero, depending on whether n or p is positive or negative. Now their sum cannot go to either of these limits, and has to remain constant. The two terms must therefore cancel each other, the squares of μ and ν must be equal and $p = n$ (odd). Writing that the derivative of relation (18) must be zero, we have:

$$n\mu^2\left[(x_1x_2)^{n-1} - \left(1 + x_1x_2\right)^{n-1}\right] = 0 \tag{19}$$

or:

$$\left[\frac{1 + x_1x_2}{(x_1x_2)}\right]^{n-1} = \left[1 + \frac{1}{(x_1x_2)}\right]^{n-1} = 1 \tag{20}$$

Obviously, this expression remains constant only if $n = 1$. Inserting this result in (18) finally shows that $\mu^2 = \nu^2 = 1$.

The functions X and Y have thus simple expressions:

$$\begin{aligned} X(x) &= \pm x \\ Y(y) &= \pm y \end{aligned} \tag{21}$$

Conversely, it is easy to show that these equalities are sufficient to obtain relation (10). The components of the image vector $\boldsymbol{V}$ on the $\boldsymbol{E}_x$, $\boldsymbol{E}_y$ and $\boldsymbol{E}_z$ vectors are equal or opposite to the components of the $\boldsymbol{v}$ vector on the $\boldsymbol{e}_x$, $\boldsymbol{e}_y$ and $\boldsymbol{e}_z$ vectors. This defines a transformation $\boldsymbol{F}$ which is linear. This transformation is also orthogonal since the scalar product of two $\boldsymbol{v}$ and $\boldsymbol{v}'$ vectors, calculated in the orthonormal basis $\boldsymbol{e}_x$, $\boldsymbol{e}_y$, $\boldsymbol{e}_z$, is equal to the scalar product of the two image vectors $\boldsymbol{V}$ and $\boldsymbol{V}'$, calculated in the orthonormal basis $\boldsymbol{E}_x$, $\boldsymbol{E}_y$, $\boldsymbol{E}_z$.

Comments:

(i) Taking the $\boldsymbol{e}_z$ component of the vectors equal to one means that we have excluded from our reasoning vectors of the plane xOy. To include them, we can apply a circular permutation to the axes and follow the same reasoning.

(ii) Our demonstration is limited to a 3-dimensional subspace. It can be repeated in all the subspaces to generalize it to the entire space we are studying.

2. Complex space

If we are dealing with a complex vector space, as is the case for the state space in quantum mechanics, the x and y components, as well as the X and Y functions become complex numbers. The arguments of § 1-a remain valid if we replace (9) by:

$$x_1^*x_2 + y_1^*y_2 + 1 = 0 \tag{22}$$

When the real and imaginary parts of the 4 variables x_1, x_2, y_1, and y_2 vary, we must find the stationary condition of the complex function:

$$G = X^*(x_1)X(x_2) + Y^*(y_1)Y(y_2) + 1 + \lambda\left[x_1^* x_2 + y_1^* y_2 + 1\right] \tag{23}$$

where λ is a complex Lagrange multiplier. Writing this stationarity involves taking the derivative of this expression. Now functions of complex variables are not necessarily differentiable, unless they are analytic and satisfy the Cauchy–Riemann conditions in a certain domain. Two possibilities will be examined.

2-a. X and Y are analytic functions of x and y

We first assume that the complex functions $X(x)$ and $Y(y)$ are analytic functions of the x and y variables, respectively. The variation of function (23) is written:

$$\begin{aligned} \mathrm{d}G = &\left[[\dot{X}(x_1)\mathrm{d}x_1]^* X(x_2) + \lambda x_2 \mathrm{d}x_1^*\right] + \left[[X(x_1)]^* \dot{X}(x_2) + \lambda x_1^*\right] \mathrm{d}x_2 \\ &+ \left[[\dot{Y}(y_1)\mathrm{d}y_1]^* Y(y_2) + \lambda y_2 \mathrm{d}y_1^*\right] + \left[[Y(y_1)]^* \dot{Y}(y_2) + \lambda y_1^*\right] \mathrm{d}y_2 \end{aligned} \tag{24}$$

This variation must be equal to zero when $\mathrm{d}x_1$ and $\mathrm{d}x_2$ are any complex infinitesimal quantities, which leads to relations:

$$\begin{aligned} &[\dot{X}(x_1)]^* X(x_2) + \lambda x_2 = 0 \\ &[X(x_1)]^* \dot{X}(x_2) + \lambda x_1^* = 0 \end{aligned} \tag{25}$$

that replace the first two equations (11). Eliminating λ, yields:

$$\frac{[\dot{X}(x_1)]^* X(x_2)}{x_2} = \frac{[X(x_1)]^* \dot{X}(x_2)}{x_1^*} \tag{26}$$

or:

$$\frac{x_1^*[\dot{X}(x_1)]^*}{[X(x_1)]^*} = \frac{x_2 \dot{X}(x_2)}{X(x_2)} \tag{27}$$

The ratio $x\dot{X}(x)/X(x)$ must be independent of x and real, so that:

$$x\dot{X}(x) = nX(x) \tag{28}$$

where n is a real constant. The function $X(x)$ can be expanded as a Laurent series (where each term contains a positive or negative integer power of x). Using this expansion in (28) shows that only one power of x can satisfy the equation: the Laurent series includes only one term, in x to the power n (which implies that n must be an integer):

$$X(x) = \mu x^n \tag{29}$$

Relation (15) remains valid in the complex domain, and this is obviously also the case for (16). The argument proceeds as before, leading to the same relations (21), extended to the complex x and y variables. In conclusion, the $\mathscr{W}$ transformation is equivalent to a linear and unitary transformation.

2-b. X^* and Y^* are analytic functions of x and y

On the other hand, we can assume that $X^*(x)$ is an analytic function of x, which is equivalent to assuming that $X(x^*)$ is an analytic function of x^* (which is different from the § 2-a hypothesis, since signs are changed in the Cauchy–Riemann relations). We can use the same argument as above, taking the derivative of the function $X^*(x)$ [instead of $X(x)$], but it is easier to make a change of functions, setting $A(x) = X^*(x)$ and $B(x) = Y^*(x)$. Since the orthogonality relation between the images through $\boldsymbol{F}$ of two vectors is expressed in the same way for the two sets of functions, we are facing the same problem as the one solved in § 2-a, once we replace $X(x)$ by $A(x)$ and $Y(x)$ by $B(x)$. We thus have:

$$A(x) = x = X^*(x) \tag{30}$$

as well as a similar relation for $B(y) = Y^*(y)$. The conclusion is:

$$\begin{aligned} X(x) &= x^* \\ Y(y) &= y^* \end{aligned} \tag{31}$$

We find the same results as above, once the x and y variables have been replaced by their complex conjugates. In this case, $\mathscr{W}$ is equivalent to an antiunitary transformation.

To summarize this complement, the Wigner theorem's conclusions can be obtained assuming the preservation of scalar products only when they are equal to zero (as long as the dimension of the state space is larger than 2).

Chapter V

Representations of Galilean and Poincaré groups: mass, spin, and energy

Using the results of chapter IV, we now study the projective representations of the Galilean and Poincaré groups, introduced in § C of chapter III. These fundamental groups define the space–time structure either, for the first, from the Galilean (isochronic) point of view or, for the second, from the Einstein relativistic point of view. Using very general assumptions, we will see that the very structure of these groups shows that a basic physical system has a certain

Introduction to Continuous Symmetries: From Space–Time to Quantum Mechanics, First Edition. Franck Laloë.

number of universal characteristics: a mass, a spin, an energy, as well as other physical quantities whose relations with each other arise from the nature of the relativistic group (Galilean or Poincaré) under study.

This gives an internal coherence to quantum mechanics, linked to that of space–time rather than to more or less precise analogies with classical mechanics (as the analogy between the spin and the rotation of a non-punctual object rotating around itself) or to "quantization rules". The assumptions made are relative to the transformations of observation frames, which are macroscopic. They do not concern the properties of the quantum object under study (one or several particles, a field, a microscopic or macroscopic system, etc.). Nevertheless they lead to important and universal results concerning the structure of its quantum state space.

In § A, we show how to introduce representations in the state space. A priori, these are projective representations, involving extension coefficients of the Lie algebra. We can eliminate most of these extension coefficients for the Galilean group, as shown in (§ B), and even all of them for the Poincaré group, as shown in (§ C). In the first case, the few coefficients that remain play a particularly important role, as they ensure that the canonical commutation relations between position and momentum can be satisfied. In both cases and in a very general way, we shall introduce the concept of a spin operator.

A. Representations in the state space

Imagine that several observers O, O′, O″, etc., each with its own space–time frame, describe the evolution of the same quantum system. They each use their own state vector $|\psi\rangle$, $|\psi'\rangle$, $|\psi''\rangle$, etc. Going from the reference frame of O to that of O′ amounts to a transformation of the space–time coordinates, discussed in § C of chapter III, that depends on the parameters $\boldsymbol{a}$, $\boldsymbol{b}$, $\boldsymbol{v}$, and τ. In the state space, this frame change leads to a transformation:

$$|\psi'\rangle = T(\boldsymbol{a}, \boldsymbol{b}, \boldsymbol{v}, \tau)|\psi\rangle \tag{V-1}$$

where T is a unitary operation, according to the results of chapter IV. To ensure the coherence of this scheme, the transformation $T(\boldsymbol{a}, \boldsymbol{b}, \boldsymbol{v}, \tau)$ must also allow going from the reference frame of O′ to that of O″, or from any reference frame to any other. In other words, the $T(\boldsymbol{a}, \boldsymbol{b}, \boldsymbol{v}, \tau)$ must form a group and be a representation of the group of space–time transformations.

It follows that the infinitesimal generators of the T group must form a Lie algebra obeying the same commutation relations as the space–time transformations. These relations have already been obtained in § C of chapter III where we derived the commutation relations between operators of the Lie algebra of Galilean and Poincaré groups. Consequently, using equalities (IV-66) of chapter IV yields the commutation relations between the infinitesimal generators G acting in the space state.

Note that relation (IV-66) is valid only for a "true" (non-projective) representation. However, projective representations are perfectly acceptable since we know that the physical properties of a system do not depend on the global phase factor of its state vector. We must thus use the more general relation (IV-87a):

$$[G_i, G_j] = i\hbar \sum_k C_{ij}^k \, G_k + i\hbar^2 \beta_{ij} \tag{V-2}$$

where, in addition to the term involving the structure constants C_{ij}^k of the group, β_{ij} coefficients appear. These coefficients extend the Lie algebra of the group as they add the identity operator $\mathbb{1}$. They cannot take arbitrary values but must satisfy relation (IV-93):

$$\sum_\ell \left\{ \beta_{\ell i} \, C_{jk}^\ell + \beta_{\ell j} \, C_{ki}^\ell + \beta_{\ell k} \, C_{ij}^\ell \right\} = 0 \qquad \forall \; i, j, k \tag{V-3}$$

B. Galilean group

B-1. Generalities

Combining commutation relations (III-101), which yield the structure constants C_{ij}^k, with relation (V-2) that shows their extension to a projective representation, we obtain:

$$\begin{cases} [J_i, J_j] = i\hbar \, \varepsilon_{ijk} \, J_k + i\hbar \, A_{ij} & \text{(V-4a)} \\ [J_i, P_j] = i\hbar \, \varepsilon_{ijk} \, P_k + i\hbar \, B_{ij} & \text{(V-4b)} \\ [P_i, P_j] = \qquad\qquad\quad i\hbar \, D_{ij} & \text{(V-4c)} \end{cases}$$

$$\begin{cases} [J_i, K_j] = i\hbar \, \varepsilon_{ijk} \, K_k + i\hbar \, N_{ij} & \text{(V-5a)} \\ [P_i, K_j] = \qquad\qquad\quad i\hbar \, M_{ij} & \text{(V-5b)} \\ [K_i, K_j] = \qquad\qquad\quad i\hbar \, Q_{ij} & \text{(V-5c)} \end{cases}$$

$$\begin{cases} [\boldsymbol{J}, H] = \qquad\quad i\hbar \, \boldsymbol{V} & \text{(V-6a)} \\ [\boldsymbol{P}, H] = \qquad\quad i\hbar \, \boldsymbol{W} & \text{(V-6b)} \\ [\boldsymbol{K}, H] = -i\hbar \, \boldsymbol{P} \quad + i\hbar \, \boldsymbol{Z} & \text{(V-6c)} \end{cases}$$

In these equalities, operators $\boldsymbol{J}$, $\boldsymbol{P}$, $\boldsymbol{K}$, and H correspond to the infinitesimal operators $X_{\boldsymbol{J}}$, $X_{\boldsymbol{P}}$, $X_{\boldsymbol{K}}$, and X_H. For example:

$$X_{\boldsymbol{J}} \Rightarrow -\frac{i}{\hbar} \boldsymbol{J} \tag{V-7}$$

As in chapter III, we have simplified the notations of the operators J_{x_i}, K_{x_i}, etc. into J_i, K_i, etc. According to the usual notation, $\varepsilon_{ijk} = \pm 1$ depending on whether the i, j and k indices are an even or odd permutation of 1,2,3, and $\varepsilon_{ijk} = 0$ if (at least) two of the indices are the same. The writing of relations (V-4) and (V-5) implies, on the right-hand side, a summation of the ε_{ijk} terms over k (summation over the repeated dummy index k); this summation actually includes either one single non-zero term (if $i \neq j$) or none (if $i = j$). As for the numbers A_{ij}, B_{ij}, ... and the vectors $\boldsymbol{V}$, $\boldsymbol{W}$, and $\boldsymbol{Z}$, they come from the β_{ij} coefficients appearing in (IV-87a), showing the presence of operators commuting with all the others, introduced by the projective character of the representation.

B-2. First simplifications

Since $[J_x, J_y] = -[J_y, J_x]$, we must have $A_{12} = -A_{21}$; in a general way, it is easy to show that the A_{ij} coefficients are antisymmetric:

$$A_{ij} = -A_{ji} \tag{V-8}$$

(in particular, the A_{ii} are equal to zero). Defining a vector $\boldsymbol{A}$ as[1]:

$$\begin{aligned} A_x &= A_{23} = -A_{32} \\ A_y &= A_{31} = -A_{13} \\ A_z &= A_{12} = -A_{21} \end{aligned} \tag{V-9}$$

relations (V-4a) can be written:

$$[J_i, J_j] = i\hbar\, \varepsilon_{ijk}\, (J_k + A_k) \tag{V-10}$$

Replacing the $\boldsymbol{J}$ operator by a new operator $\boldsymbol{J}'$ defined as:

$$\boldsymbol{J}' = \boldsymbol{J} + \boldsymbol{A} \tag{V-11}$$

amounts to redefining $\boldsymbol{J}$ by adding a diagonal operator (proportional to the identity operator). Such an operation obviously does not change the commutators in relations (V-4) to (V-6), where $\boldsymbol{J}$ can be simply replaced by $\boldsymbol{J}'$. It does, however, eliminate the constant $\boldsymbol{A}$ from the right-hand side of (V-4a). We shall assume, in what follows, that this redefinition of $\boldsymbol{J}$ has been performed, which avoids introducing a new $\boldsymbol{J}'$ notation.

The same reasoning cannot be used in relation (V-4c) to eliminate the constants D, since P_k does not appear on the right-hand side. One could try eliminating in a similar way the B or N coefficients by a simple redefinition of

[1] In three dimensions, a second-order antisymmetric tensor is equivalent to a vector; *cf.* Complement A$_{\text{VIII}}$.

$\boldsymbol{P}$ or $\boldsymbol{K}$, but the above reasoning does not apply: there is no guarantee that the extension constants are antisymmetric, as in (V-8). We shall use another method, based on condition (V-3) that links the structure and extension constants of the algebra.

B-3. Displacement group

The three relations (V-4) concern the commutation relations in a projective representation of the geometric displacement group. We are free to choose the values of i, j, and k in the (V-3) condition. If we want this condition to include a given coefficient β_{qs}, we must (for example) choose $i = s$, and, for j and k, two indices such that the commutator of the G_j and G_k operators has a component on G_q. In the summation over ℓ, the $\beta_{\ell i}$ coefficient takes the expected value with a non-zero coefficient C_{jk}^{q}.

The following computations are not difficult but sometimes a bit tedious. The reader should not hesitate to skip passages in small print, especially in a first reading.

• To obtain a relation involving $\beta_{P_x P_y}$, we choose:

$$i \rightsquigarrow P_y \tag{V-12a}$$

We want the index ℓ of $\beta_{\ell i}$ to correspond to another component of $\boldsymbol{P}$. For the structure constant C_{jk}^{ℓ} to be different from zero, we must use the only commutation relation where $\boldsymbol{P}$ appears on the right-hand side, i.e. relation (V-4b). We thus set:

$$j \rightsquigarrow P_z \qquad\qquad k \rightsquigarrow J_y \tag{V-12b}$$

which amounts to using the fact that:

$$C_{P_z J_y}^{G_\ell} = \begin{cases} -1 & \text{if } G_\ell = P_x \\ 0 & \text{otherwise} \end{cases} \tag{V-13}$$

Furthermore, only the first of the three terms in (V-3) is different from zero since among the three operators selected in (V-12), only P_z and J_y do not commute; the other structure constants must therefore be equal to zero. We are thus left with:

$$\beta_{P_x P_y} = D_{12} = 0 \tag{V-14}$$

A circular permutation of the x, y, and z indices shows that all the D values are equal to zero (as we already know that the D_{ii} are zero by definition).

• To eliminate the B_{ij} constants, we start by those where $i \neq j$, choosing:

$$i \rightsquigarrow P_y \qquad\qquad j \rightsquigarrow J_y \qquad\qquad k \rightsquigarrow J_z$$

Since:

$$C^{G_\ell}_{J_y\,J_z} = \begin{cases} 1 & \text{if } G_\ell = J_x \\ 0 & \text{otherwise} \end{cases}$$

$$C^{G_\ell}_{J_z\,P_y} = \begin{cases} -1 & \text{if } G_\ell = P_x \\ 0 & \text{otherwise} \end{cases}$$

$$C^{G_\ell}_{P_y\,J_y} = 0$$

we get:

$$\beta_{J_x\,P_y} - \beta_{P_x\,J_y} = 0 \tag{V-15a}$$

meaning:

$$B_{1\,2} = -B_{2\,1} \tag{V-15b}$$

In a similar way, we show that all the non-diagonal elements of B are antisymmetric.

• To show that all the diagonal elements B_{ii} are zero, we choose the correspondence:

$$i \rightsquigarrow J_x \qquad\qquad j \rightsquigarrow J_y \qquad\qquad k \rightsquigarrow P_z$$

Since:

$$C^{G_\ell}_{J_y\,P_z} = \begin{cases} 1 & \text{if } G_\ell = P_x \\ 0 & \text{otherwise} \end{cases}$$

$$C^{G_\ell}_{P_z\,J_x} = \begin{cases} 1 & \text{if } G_\ell = P_y \\ 0 & \text{otherwise} \end{cases}$$

$$C^{G_\ell}_{J_x\,J_y} = \begin{cases} 1 & \text{if } G_\ell = J_z \\ 0 & \text{otherwise} \end{cases}$$

we get:

$$\beta_{P_x\,J_x} + \beta_{P_y\,J_y} + \beta_{J_z\,P_z} = \beta_{P_x\,J_x} + \beta_{P_y\,J_y} - \beta_{P_z\,J_z} = 0 \tag{V-16a}$$

In a similar way, a circular permutation of x, y, and z shows:

$$\beta_{P_y\,J_y} + \beta_{P_z\,J_z} - \beta_{P_x\,J_x} = 0 \tag{V-16b}$$

$$\beta_{P_z\,J_z} + \beta_{P_x\,J_x} - \beta_{P_y\,J_y} = 0 \tag{V-16c}$$

These three linear equations (V-16) force the $\beta_{P_i\,J_i}$ to be equal to zero.

We just saw that all the antisymmetric coefficients B_{ij} can be eliminated by a redefinition of $\boldsymbol{P}$ analogous to that of $\boldsymbol{J}$. Consequently, any projective representation of the geometric displacement group can be changed into a "true" representation: we eventually have to redefine some operators by a simple addition of constants.

B-4. Maximum simplification of the commutation relations

The same sort of calculation can be carried out on the other commutation relations.

• We note that the three equations (V-4a), (V-5a), and (V-5c) have the same structure as the three equations (V-4), with the $\boldsymbol{K}$ operator replacing $\boldsymbol{P}$. The same reasoning can be applied to both cases, eliminating the N and Q extension constants in the commutation relations between $\boldsymbol{J}$ and $\boldsymbol{K}$.

• We now examine relation (V-6a), choosing:

$$i \rightsquigarrow H \qquad j \rightsquigarrow J_x \qquad k \rightsquigarrow J_y$$

The only structure constant corresponding to the commutation of two of these operators is $C^{J_z}_{J_y\,J_x}$, so that relation (V-3) simply becomes:

$$\beta_{J_z\,H} = V_z = 0 \tag{V-17}$$

This means that $\boldsymbol{V}$ is necessarily equal to zero. This can be understood intuitively, since a choice of the arbitrary direction of $\boldsymbol{V}$ would violate the rotational invariance.

• In a similar way, we choose for relation (V-6b):

$$i \rightsquigarrow H \qquad j \rightsquigarrow J_x \qquad k \rightsquigarrow P_y$$

which leads to:

$$\beta_{P_z\,H} = W_z = 0 \tag{V-18}$$

and $\boldsymbol{W}$ is eliminated for the same reasons as $\boldsymbol{V}$.

• As for relation (V-6c), setting:

$$i \rightsquigarrow H \qquad j \rightsquigarrow J_x \qquad k \rightsquigarrow K_y$$

leads to:

$$\beta_{K_z\,H}\, C^{K_z}_{J_x\,K_y} + \beta_{P_y\,J_x}\, C^{P_y}_{K_y\,H} = \beta_{K_z\,H} + \beta_{P_y\,J_x} = 0 \tag{V-19}$$

Since we know that $\beta_{P_y\,J_x}$ is equal to zero, it follows that:

$$\beta_{K_z\,H} = Z_z = 0 \tag{V-20}$$

We saved relation (V-5b) for the end, since it is a particularly interesting case where all the constants M do not disappear. The constants concerning two different indices are equal to zero. For example, choosing:

$$i \rightsquigarrow P_x \qquad j \rightsquigarrow J_x \qquad k \rightsquigarrow K_z$$

leads to:

$$\beta_{K_y\,P_x} = M_{21} = 0 \tag{V-21}$$

But if we try applying the same method, choosing:

$$i \rightsquigarrow P_x \qquad j \rightsquigarrow J_y \qquad k \rightsquigarrow K_z$$

we get (since neither the commutator of X_{J_y} and X_{K_z} nor the one of X_{P_x} and X_{J_y} is equal to zero):

$$\beta_{K_x P_x} + \beta_{P_z K_z} = 0 \qquad \text{or:} \qquad \beta_{K_x P_x} = \beta_{K_z P_z} \tag{V-22}$$

Consequently (using a circular permutation):

$$M_{xx} = M_{yy} = M_{zz} = M \tag{V-23}$$

As the x, y, and z axes are equivalent, we expected these three values to be equal. It is not possible, however, to show that M equals zero in the general case.

Comments:

(i) This impossibility is not surprising: we have already seen [*cf.* relation (I-51), for example, or the comment at the end of § C-1 of chapter III] that $\boldsymbol{K}$ is related to the position of the center of mass of the quantum system under study. We know that, for a spinless particle, we can write:

$$[X_i, P_j] = i\hbar\, \delta_{ij}$$

which is in agreement with $M_{ij} \propto \delta_{ij}$.

(ii) If M is found to be negative, it can be brought back to a positive value by changing the signs of the operators $\boldsymbol{K}$ and H.

B-5. Extended Lie algebra, after simplification; mass, internal energy, and spin

Let us summarize all the commutation relations we have obtained:

$$\begin{cases} [J_i, J_j] = i\hbar\, \varepsilon_{ijk}\, J_k & \text{(V-24a)} \\ [J_i, P_j] = i\hbar\, \varepsilon_{ijk}\, P_k & \text{(V-24b)} \\ [P_i, P_j] = 0 & \text{(V-24c)} \end{cases}$$

$$\begin{cases} [J_i, K_j] = i\hbar\, \varepsilon_{ijk}\, K_k & \text{(V-25a)} \\ [P_i, K_j] = i\hbar\, M\, \delta_{ij} & \text{(V-25b)} \\ [K_i, K_j] = 0 & \text{(V-25c)} \end{cases}$$

$$\begin{cases} [\boldsymbol{J}, H] = 0 & \text{(V-26a)} \\ [\boldsymbol{P}, H] = 0 & \text{(V-26b)} \\ [\boldsymbol{K}, H] = -i\hbar\, \boldsymbol{P} & \text{(V-26c)} \end{cases}$$

They are, after extension (and maximum simplification), the commutation relations of the Lie algebra of the Galilean group. The generator $\boldsymbol{P}$ of space translations is called the "momentum operator", the generator H of time translation is called the "Hamiltonian", the generator $\boldsymbol{J}$ of the rotations is called the "angular momentum"; as for $\boldsymbol{K}$, we will see later that it is related (in the Galileo group only) to the position operator of the system.

We notice that $\boldsymbol{P}$, $\boldsymbol{J}$, and $\boldsymbol{K}$ obey similar commutation relations with $\boldsymbol{J}$. Generally speaking, any set of three operators V_i $(i = x, y, z)$ obeying the following commutation relation with $\boldsymbol{J}$:

$$[J_i\,, V_j] = i\hbar\, \varepsilon_{ijk}\, V_k \qquad \text{(V-27)}$$

is called a "vector operator". The operators $\boldsymbol{P}$, $\boldsymbol{J}$, and $\boldsymbol{K}$ are therefore vector operators.

Operators commuting with the three components of $\boldsymbol{J}$ (rotation invariant operators) are called "scalar operators". According to (V-26a), H is a scalar operator, and (V-26b) indicates that it is also translation invariant. The scalar product $\boldsymbol{V} \cdot \boldsymbol{W}$ of two vector operators $\boldsymbol{V}$ and $\boldsymbol{W}$, and in particular the scalar product $\boldsymbol{V} \cdot \boldsymbol{V} = \boldsymbol{V}^2$ of $\boldsymbol{V}$ by itself (square of $\boldsymbol{V}$), are scalar operators since:

$$[J_x\,, V_x W_x + V_y W_y + V_z W_z] = i\hbar\,(V_z W_y + V_y W_z - V_y W_z - V_z W_y) = 0 \qquad \text{(V-28)}$$

Comments:

(i) The commutation relation (V-27) can be interpreted geometrically. Consider a classical vector $\boldsymbol{v}$, with three components v_x, v_y, and v_z. Relation (I-81) shows that its variation in an infinitesimal rotation with parameter $\delta\boldsymbol{a}$ is:

$$\delta\boldsymbol{v} = \delta\boldsymbol{a} \times \boldsymbol{v} \qquad \text{(V-29)}$$

If the rotation axis is parallel to Ox, we have $\delta\boldsymbol{a} = \delta a\, \boldsymbol{e}_x$, and:

$$\frac{\delta v_z}{\delta a} = v_y \qquad \text{(V-30a)}$$

which we compare to the quantum equation (V-27):

$$[J_{,}, V_z] = -i\hbar V_y \qquad \text{(V-30b)}$$

The value of the commutator therefore corresponds to the infinitesimal variation of the components of a classical vector, multiplied by $-i\hbar$. The properties of vector operators and their relation with classical vectors will be discussed in more detail in § A of chapter VIII.

(ii) As we saw on several occasions, the choice of signs is sometimes confusing when discussing time translations. We have introduced a parameter τ as the time an observer O, in a first reference frame, attributes to an event defining the time origin for an observer O′ in a second reference frame (*cf.* figure 2 of chapter I). We then defined the time-translation operator by the time τ as the operator that transforms the state vector $|\psi(t)\rangle$, which describes the physical system in the first reference frame, into that $|\psi'(t)\rangle$ describing it in the second reference frame. Since $t' = t - \tau$, the equality between $|\psi'(t')\rangle$ and $|\psi(t)\rangle$ implies that $|\psi'(t-\tau)\rangle = |\psi(t)\rangle$, that is, $|\psi'(t)\rangle = |\psi(t+\tau)\rangle$. Now, if one keeps the same reference frame, it is equivalent to say that the τ time-translation operator lets time progress (if $\tau > 0$) in the solution of the Schrödinger equation. This is the way we will define the time translation operator and the Hamiltonian H throughout the rest of this book.

B-5-a. Mass, internal energy

Let us build operators that commute with all the operators $\boldsymbol{J}$, $\boldsymbol{P}$, $\boldsymbol{K}$, and H. First of all, we have the operator $M = m \times \mathbb{1}$, that can be identified with the *total mass* of the system. We then introduce the operator:

$$\boldsymbol{P}^2 = P_x^2 + P_y^2 + P_z^2 \tag{V-31}$$

It commutes with $\boldsymbol{P}$ [*cf.* (V-24c)] and with $\boldsymbol{J}$, since we just saw that the square of a vector operator is a scalar. Its commutator with H is also equal to zero, according to (V-26b). Finally, taking (V-25b) into account:

$$\left[P_x^2, K_x\right] = P_x\left[P_x, K_x\right] + \left[P_x, K_x\right] P_x = 2M\ i\hbar\ P_x \tag{V-32}$$

The H operator has similar properties since it also commutes with $\boldsymbol{J}$, $\boldsymbol{P}$, and of course with itself, but it does not commute with $\boldsymbol{K}$. Comparing (V-26c) with (V-32), we see that to build an operator commuting with all the others, we must set:

$$U_{\text{int}} = H - \frac{\boldsymbol{P}^2}{2M} \tag{V-33a}$$

The physical meaning of the equation:

$$H = \frac{\boldsymbol{P}^2}{2M} + U_{\text{int}} \tag{V-33b}$$

is simple: since the constant M is the total mass of the system, we can identify $\boldsymbol{P}^2/2M$ to the system's global translation kinetic energy, U_{int} to its *internal energy*, and H to its total energy. It is satisfying to note that neither H, nor the kinetic energy $\boldsymbol{P}^2/2M$, is the same in two different Galilean reference frames. Their difference, however, is invariant (as is, for example, in classical mechanics, the rotational energy of a solid around itself). The system's energy is minimum in the reference frame where the momentum $\boldsymbol{P}$ is zero.

B-5-b. Orbital angular momentum

To find another invariant, we introduce the operator $\boldsymbol{L}$ (a three component operator) as:

$$\boldsymbol{L} = (\boldsymbol{P} \times \boldsymbol{K}) / M = -(\boldsymbol{K} \times \boldsymbol{P}) / M \tag{V-34}$$

Since the components of $\boldsymbol{P}$ and $\boldsymbol{K}$ on different axes commute with each other, these two definitions are equivalent. We shall derive in (V-50) a relation between the operator $\boldsymbol{K}$ and the position operator $\boldsymbol{R}$, which shows that this definition coincides with the usual definition of the orbital angular momentum.

• Using (V-25b), (V-24c), and (V-25c), the commutation relations between the components of $\boldsymbol{L}$ are written:

$$\begin{aligned}[L_x, L_y] &= M^{-2}\left[K_y P_z - K_z P_y \,,\, K_z P_x - K_x P_z\right] \\ &= M^{-2}\left[K_y P_z \,,\, K_z P_x\right] + M^{-2}\left[K_z P_y \,,\, K_x P_z\right] \\ &= \frac{i\hbar}{M} K_y P_x - \frac{i\hbar}{M} P_y K_x = i\hbar\, L_z \end{aligned} \tag{V-35}$$

The commutation relations between the components of $\boldsymbol{L}$ have therefore the same properties as the commutation relations between the components of $\boldsymbol{J}$.

• The same type of calculation yields:

$$[L_x, J_x] \propto P_y\left[K_z, J_x\right] + \left[P_y, J_x\right] K_z - P_z\left[K_y, J_x\right] - \left[P_z, J_x\right] K_y = 0 \tag{V-36a}$$

and:

$$\begin{aligned}[L_x, J_y] &= \frac{1}{M} P_y\left[K_z, J_y\right] + \frac{1}{M}\left[P_y, J_y\right] K_z - \frac{1}{M} P_z\left[K_y, J_y\right] - \frac{1}{M}\left[P_z, J_y\right] K_y \\ &= \frac{i\hbar}{M}\left(P_x K_y - P_y K_x\right) \end{aligned} \tag{V-36b}$$

We thus obtain:

$$[J_i, L_j] = i\hbar\, \varepsilon_{ijk}\, L_k \tag{V-36c}$$

which is analogous to (V-24a), and shows that $\boldsymbol{L}$ is a vector operator.

More generally, this calculation indicates that the vector product of two vector operators is also a vector operator (see also § 6 of complement A_{VIII}).

We also have:

$$[L_x, K_x] = \frac{1}{M}[P_y K_z - P_z K_y, K_x] = 0$$
$$[L_x, K_y] = \frac{1}{M}[P_y K_z, K_y] = i\hbar\, K_z \tag{V-37}$$

so that:

$$[L_i, K_j] = i\hbar\, \varepsilon_{ijk}\, K_k \tag{V-38}$$

which is the analog of (V-25a). In the same way, we can write:

$$[L_i, P_j] = i\hbar\, \varepsilon_{ijk}\, P_k \tag{V-39}$$

which is similar to (V-24b). Finally:

$$\begin{aligned}[L_x, H] &= \frac{1}{M}[P_y, H]\, K_z + \frac{1}{M}P_y\, [K_z, H] - \frac{1}{M}[P_z, H]\, K_y - \frac{1}{M}P_z\, [K_y, H] \\ &= \frac{i\hbar}{M}(P_z P_y - P_y P_z) = 0\end{aligned} \tag{V-40}$$

Operators $\boldsymbol{L}$ and $\boldsymbol{J}$ have therefore a similar nature: neither do their components commute with each other, nor with $\boldsymbol{P}$ or $\boldsymbol{K}$, but they commute with H.

B-5-c. Spin

It is easy to build from $\boldsymbol{L}$ and $\boldsymbol{J}$ an operator $\boldsymbol{S}$ that commutes with $\boldsymbol{P}$, $\boldsymbol{K}$, and H. Using the fact that the right-hand sides of relations (V-24b) and (V-39) are identical, and that the same is true for relations (V-25a) and (V-38), we set:

$$\boldsymbol{S} = \boldsymbol{J} - \boldsymbol{L} \tag{V-41}$$

so that:

$$[\boldsymbol{S}, \boldsymbol{P}] = [\boldsymbol{S}, \boldsymbol{K}] = 0 \tag{V-42}$$

The operator $\boldsymbol{S}$ is thus invariant under a translation or a velocity change of the Galilean reference frame. In addition:

$$\begin{aligned}[S_i, J_j] &= [J_i - L_i, J_j] = i\hbar\, \varepsilon_{ijk}\,(J_k - L_k) \\ &= i\hbar\, \varepsilon_{ijk}\, S_k\end{aligned} \tag{V-43}$$

and, similarly:

$$\begin{aligned}[S_i, S_j] &= [J_i - L_i, J_j - L_j] = i\hbar\, \varepsilon_{ijk}\,(J_k - L_k - L_k + L_k) \\ &= i\hbar\, \varepsilon_{ijk}\, S_k\end{aligned} \tag{V-44}$$

These commutation rules show that operator $\boldsymbol{S}$ has the same nature as $\boldsymbol{J}$ or $\boldsymbol{L}$ (angular momentum).

Finally, as $\boldsymbol{L} \propto \boldsymbol{P} \times \boldsymbol{K}$ and since $\boldsymbol{S}$ commutes with $\boldsymbol{K}$ and $\boldsymbol{P}$, we have:

$$[\boldsymbol{S}, \boldsymbol{L}] = 0 \tag{V-45}$$

and since H commutes with $\boldsymbol{J}$ and $\boldsymbol{L}$:

$$[\boldsymbol{S}, H] = 0 \tag{V-46}$$

Except for $\boldsymbol{J}$, operator $\boldsymbol{S}$ commutes with all the other operators of the Lie algebra, but the different components of $\boldsymbol{S}$ do not commute with each other. The system's angular momentum $\boldsymbol{J}$ is thus the sum $\boldsymbol{L} + \boldsymbol{S}$ of two contributions, the second being affected neither by a translation nor by a change of Galilean reference frame (pure Galilean transformation). Operator $\boldsymbol{S}$ is identified with the internal *angular momentum*, or spin, of the physical system. Operator $\boldsymbol{L}$ corresponds to the *orbital angular momentum* associated with the motion of the center of mass.

The square $\boldsymbol{S}^2$ of the spin angular momentum operator is defined as:

$$\boldsymbol{S}^2 = S_x^2 + S_y^2 + S_z^2 \tag{V-47}$$

It is easy to show that $\boldsymbol{S}^2$ commutes with $\boldsymbol{J}$ and $\boldsymbol{S}$ and thus with all the operators introduced until now ($\boldsymbol{P}$, $\boldsymbol{J}$, $\boldsymbol{K}$, $\boldsymbol{L}$, $\boldsymbol{S}$, and H). Consequently, it is a Casimir operator.

B-5-d. Irreducible representations of elementary physical systems

Altogether, we have introduced three Casimir operators:

$$M, \quad U_{\text{int}}\,, \quad \boldsymbol{S}^2 \tag{V-48}$$

In any irreducible (projective) representation of the Galilean group, these operators are diagonal with identical matrix elements. Otherwise, the representation space would be the direct sum of subspaces associated with different eigenvalues, whereas no transformations of the Galilean group would have matrix elements between these distinct subspaces; the representation would then be reducible.

It is natural to postulate (Wigner) that, for an elementary system, the structure of the state subspaces is as simple as possible; it therefore constitutes an irreducible representation of the symmetry group under study – otherwise, the state space would be the direct sum of simpler subspaces. This shows that the mass, the internal energy, and the spin of an elementary particle are fundamental properties of the particle (they are independent of the observation reference frame). In the state space, the three components of $\boldsymbol{P}$, the energy H, and one component S_z of $\boldsymbol{S}$ form a set of commuting operators, thus having a common basis of eigenvectors.

These conclusions were reached through very general arguments. We never started from a classical description of the system (point-like particle, classical Lagrangian or Hamiltonian, etc.) and then used a particular quantization method. Furthermore, a classical approach would probably not attribute a spin to an elementary particle. In classical mechanics, an internal angular momentum only exists for an extended system including several elementary particles, and we know that the half-integer spin of the electron has no classical equivalent. In addition to general hypotheses on the existence of a vector space state (superposition principle, etc.), we only made assumptions about space–time properties. The group structure of the transformations we introduced does not correspond to properties of the system (supposed to be "passive" in the change of reference frame), but to the way we think of macroscopic changes in space and time reference frames.

B-5-e. Position operator

We have not yet introduced a position operator $\boldsymbol{R}$ for the physical system. We note, however, that relations (V-25b) are very similar to the canonical commutation relations:

$$[R_i, P_j] = i\hbar\, \delta_{ij} \tag{V-49}$$

As will be seen in more detail in § A-1-a of chapter VI, relations (V-25b) are also fundamental as they express the way an object's position is displaced under a translation, which is closely linked to the very definition of a translation. It is natural to make these two fundamental relations identical, setting:

$$\boldsymbol{R} = -\frac{\boldsymbol{K}}{M} \tag{V-50}$$

(The origin of this relation will be discussed more precisely in § A-1-a-α of chapter VI). This equality constitutes the definition of the position operator in the framework of the Galilean group.

Comment:

To look for non-projective representations of the Galilean group, we can use the preceding arguments and set $M = 0$ in all the results. Under these conditions, the $\boldsymbol{R}$ operator of (V-50) is no longer defined, and the canonical relations (V-49) can no longer be written.

Furthermore, if $M = 0$, we have $[\boldsymbol{P}, \boldsymbol{K}] = 0$, while the relations $[\boldsymbol{P}, H] = 0$ and $[\boldsymbol{K}, H] = i\hbar\boldsymbol{P}$ remain unchanged. Consider an eigenvector common to $\boldsymbol{P}$ and H, with eigenvalues $\boldsymbol{p}$ and E:

$$\boldsymbol{P}|\varphi\rangle = \boldsymbol{p}|\varphi\rangle \qquad\qquad H|\varphi\rangle = E|\varphi\rangle$$

The ket $|\chi\rangle = \left(1 - \frac{i}{\hbar}\,\delta\boldsymbol{v}_0 \cdot \boldsymbol{K}\right)|\varphi\rangle$ is an eigenvector of $\boldsymbol{P}$ with the same eigenvalue $\boldsymbol{p}$, but an eigenvector of H with the eigenvalue $E - \delta\boldsymbol{v}_0 \cdot \boldsymbol{p}$. This means that for the

$M = 0$ representations of the Galilean group, a change of reference frame that modifies the velocity in any given way has no effect on the momentum. Furthermore, there is no upper or lower boundary on the spectrum of H, which is not physically acceptable.

It is thus crucial, from a physical point of view, to not eliminate the projective representations of the Galilean group.

B-5-f. No other invariants

We just saw that a unitary, irreducible (projective) representation of the Galilean group is associated with given values of M, U, and S, interpreted as fundamental properties of an elementary particle. We may wonder whether this list is complete: are M, U, and S sufficient to characterize an irreducible (projective) representations of the Galilean group, or should we add other operators that also commute with all the elements of its Lie algebra? Let us show that knowing M, U, and S is indeed sufficient.

We call $\mathscr{E}$ the representation's vector space (state space of the particle). As the components of the $\boldsymbol{P}$ operator commute with each other, we introduce the subspaces $\mathscr{E}_{\boldsymbol{p}}$, spanned by the eigenvectors common to the three components of $\boldsymbol{P}$ with eigenvalues p_x, p_y, and p_z (components of $\boldsymbol{p}$). One of these subspaces, $\mathscr{E}_{\boldsymbol{p}=0}$, is associated with the eigenvalue $\boldsymbol{p} = 0$.

Relations (V-34) and (V-41) lead to:

$$\boldsymbol{J} = -\frac{1}{M}\boldsymbol{K} \times \boldsymbol{P} + \boldsymbol{S}$$

so that, in the subspace $\mathscr{E}_{\boldsymbol{p}=\mathbf{0}}$, operator $\boldsymbol{J}$ reduces to $\boldsymbol{S}$. As this operator commutes with $\boldsymbol{P}$, it is clear that $\mathscr{E}_{\boldsymbol{p}=\mathbf{0}}$ remains stable under the action of any component of $\boldsymbol{J}$. It follows that, in this subspace, the action of the $\boldsymbol{J}$ or $\boldsymbol{S}$ operator yields a representation (eventually projective), of the rotation group $R_{(3)}$. It is not, a priori, obvious that this representation is irreducible, as $R_{(3)}$ is only a subgroup of the Galilean group. We only know that $\mathscr{E}_{\boldsymbol{p}=\mathbf{0}}$ can be decomposed into a direct sum of subspaces, each associated with an irreducible representation:

$$\mathscr{E}_{\boldsymbol{p}=\mathbf{0}} = \mathscr{E}^{\alpha}_{\boldsymbol{p}=0} \oplus \mathscr{E}^{\beta}_{\boldsymbol{p}=0} \oplus \ldots \oplus \mathscr{E}^{\nu}_{\boldsymbol{p}=0} \oplus \ldots \tag{V-51}$$

We are going to show that, if the representation of the Galilean group is irreducible, this sum is limited to a single term.

Before going any further, let us recall a few properties of the $\mathscr{E}^{\nu}_{\boldsymbol{p}=0}$, demonstrated in chapter VII. Operator $\boldsymbol{S}$ has all the properties of an angular momentum, and its square is a Casimir operator. As an irreducible representation, each subspace $\mathscr{E}^{\nu}_{\boldsymbol{p}=0}$ can only contain a single eigenvalue of this square operator, so that its dimension must be equal to $(2S + 1)$, where S is an integer or half-integer number. This subspace can be spanned by the common eigenvectors of $\boldsymbol{S}^2$ (with eigenvalue $S(S+1)\hbar^2$) and of S_z (with eigenvalues $m_S\,\hbar$):

$$|\nu, \boldsymbol{p} = 0, m_S\rangle \qquad m_S = S,\, S-1,\, S-2, \ldots, -S+1,\, -S$$

We introduce the unitary operator:

$$S(\boldsymbol{\lambda}) = \mathrm{e}^{-i\boldsymbol{\lambda}\cdot\boldsymbol{K}/M\hbar}$$

where the notation $\boldsymbol{\lambda}$ symbolizes a set of three parameters, the components of the vector $\boldsymbol{\lambda}$. It is easy to show (a detailed demonstration is given in Complement $\mathrm{E_{II}}$ in [9]) that:

$$[\boldsymbol{P}, S(\boldsymbol{\lambda})] = \boldsymbol{\lambda}\; S(\boldsymbol{\lambda})$$

This relation means that the action of $S(\boldsymbol{\lambda})$ on any eigenvector of $\boldsymbol{P}$ with eigenvalue $\boldsymbol{p}$ yields another eigenvector with eigenvalue $\boldsymbol{p} + \boldsymbol{\lambda}$.

Consider first the set of vectors $|\alpha, \boldsymbol{p}, m_S\rangle$ defined as:

$$|\alpha, \boldsymbol{p}, m_S\rangle = S\,(\boldsymbol{\lambda} = \boldsymbol{p})\; |\alpha, \boldsymbol{p} = 0, m_S\rangle$$

The ket $|\alpha, \boldsymbol{p}, m_S\rangle$ is a common eigenvector of $\boldsymbol{P}$ and S_z, with respective eigenvalues $\boldsymbol{p}$ and $m_S\hbar$. The set of vectors $|\alpha, \boldsymbol{p}, m_S\rangle$ span a subspace $\mathscr{E}^\alpha$ of $\mathscr{E}$, which, as we are going to show, is none other than $\mathscr{E}$ itself, if the representation of the Galilean group is irreducible.

First of all, we see that $\mathscr{E}^\alpha$ is invariant under the action of $\boldsymbol{P}$, with matrix elements that do not depend on the α index:

$$\langle\alpha, \boldsymbol{p}, m_S|\boldsymbol{P}|\alpha, \boldsymbol{p}', m_S'\rangle = \delta_{m_S m_S'}\,\delta\,(\boldsymbol{p} - \boldsymbol{p}')\,\boldsymbol{p}$$

Using again the results of Complement $\mathrm{E_{II}}$ of [9], we see that any ket $|\psi\rangle$ of $\mathscr{E}^\alpha$, written as:

$$|\psi\rangle = \int d^3p \sum_{m_S} \overline{\psi}_{m_S}\,(\boldsymbol{p})|\alpha, \boldsymbol{p}, m_S\rangle \tag{V-52}$$

becomes, under the action of $\boldsymbol{K}$:

$$\boldsymbol{K}\,|\psi\rangle = \int d^3p \sum_{m_S} \left[\frac{\hbar}{i}\,\boldsymbol{\nabla}_{\boldsymbol{p}}\overline{\psi}_{m_S}\,(\boldsymbol{p})\right] |\alpha, \boldsymbol{p}, m_S\rangle \tag{V-53}$$

The index α does not play any role in the action of the operator $\boldsymbol{K}$, which leaves $\mathscr{E}^\alpha$ globally invariant. Consequently, the subspace $\mathscr{E}^\alpha$ is also invariant under the action of $\boldsymbol{L} = -(\boldsymbol{K} \times \boldsymbol{P})/M$. As for $\boldsymbol{S}$, it commutes with $\boldsymbol{P}$, and its action on the $|\alpha, \boldsymbol{p}, m_S\rangle$ depends only on m_S. The subspace $\mathscr{E}^\alpha$ is thus globally invariant under the action of $\boldsymbol{J} = \boldsymbol{L} + \boldsymbol{S}$.

The last infinitesimal generator we have not considered is H. As $H = U_{\text{int}} + \boldsymbol{P}^2/2M$ (where the internal energy U_{int} is a constant), it is easy to show that $\mathscr{E}^\alpha$ is stable under the action of all the infinitesimal generators.

Similarly, the action of the unitary operators $S(\boldsymbol{\lambda})$ on the various subspaces of zero momentum $\mathscr{E}^\beta_{(\boldsymbol{p}=0)}, \ldots, \mathscr{E}^\mu_{(\boldsymbol{p}=0)}$ leads to subspaces $\mathscr{E}^\beta, \ldots, \mathscr{E}^\mu$, globally invariant under the action of all the operators associated with the Galilean group. We thus obtain several times the same representation. The representation in the entire space will be irreducible (and the particle really elementary) only if ν takes a single value, α. Furthermore, since the index α does not play any role in the action of the $\boldsymbol{P}$, $\boldsymbol{J}$, $\boldsymbol{K}$, and H operators, it can be completely omitted. As announced, the representation is entirely defined by the values of M, S, and U.

C. Poincaré group

To study the projective representations of the Poincaré group (group of space–time transformations in Einstein's relativity), we start from relations (III-108) and (III-111), the commutation relations between operators X_G. Relation (IV-87a) yields their image in the extended algebra (projective representation) of the infinitesimal operators acting in the state space:

$$\begin{cases} [J_i, J_j] = i\hbar\, \varepsilon_{ijk}\, J_k + i\hbar\, A_{ij} & \text{(V-54a)} \\ [J_i, P_j] = i\hbar\, \varepsilon_{ijk}\, P_k + i\hbar\, B_{ij} & \text{(V-54b)} \\ [P_i, P_j] = \qquad\qquad i\hbar\, D_{ij} & \text{(V-54c)} \end{cases}$$

$$\begin{cases} [J_i, K_j] = i\hbar\, \varepsilon_{ijk}\, K_k + i\hbar\, N_{ij} & \text{(V-55a)} \\ [P_i, K_j] = \dfrac{i\hbar}{c^2}\, \delta_{ij}\, H + i\hbar\, M_{ij} & \text{(V-55b)} \\ [K_i, K_j] = -\dfrac{i\hbar}{c^2}\, \varepsilon_{ijk}\, J_k + i\hbar\, Q_{ij} & \text{(V-55c)} \end{cases}$$

$$\begin{cases} [\boldsymbol{J}, H] = \qquad i\hbar\, \boldsymbol{V} & \text{(V-56a)} \\ [\boldsymbol{P}, H] = \qquad i\hbar\, \boldsymbol{W} & \text{(V-56b)} \\ [\boldsymbol{K}, H] = -i\hbar\, \boldsymbol{P} + i\hbar\, \boldsymbol{Z} & \text{(V-56c)} \end{cases}$$

These relations are identical to those written in (V-4), (V-5), and (V-6) for the Galilean group, except for (V-55b) and (V-55c), which concern the commutator of $\boldsymbol{K}$ with $\boldsymbol{P}$ or of $\boldsymbol{K}$ with itself. The new terms are easily recognizable as they include a $1/c^2$ coefficient (they tend toward zero if $c \to \infty$). As an example and as noted in relation (III-111c) of chapter III, the J_k term on the right-hand side of (V-55c) explains the origin of the Thomas precession. The term in H on the right-hand side of (V-55b) is explained as follows: if, in a first reference frame, one moves the position of an event, and then changes the reference frame with a Lorentz transformation in the same direction, the new time attributed to the event depends on the initial change of position.

C-1. Elimination of the diagonal operators

As for the Galilean group, we try and eliminate most of the diagonal operators on the right-hand sides of (V-54), (V-55), and (V-56), so as to render the representation "as little projective as possible". Actually, for the Poincaré group, we are going to show that all the diagonal operators can be eliminated, with no exception.

As relations (V-54), concerning the Euclidean displacement group, are identical to (V-4), we can eliminate the A, B, and D in the same way as in § B. Actually, each time the choice of operators associated with i, j, and k in § B does not involve either two K operators, or one K operator and one P operator, the reasoning yields the same result for the Poincaré and the Galilean groups. This shows that $\boldsymbol{V}$, $\boldsymbol{W}$, $\boldsymbol{Z}$, and the N_{ij} are necessarily zero.

For the commutation relations between components of $\boldsymbol{K}$, we choose:

$$i \rightsquigarrow K_x \qquad j \rightsquigarrow J_x \qquad k \rightsquigarrow K_z$$

This yields, since X_{K_x} and X_{K_z} no longer commute:

$$\beta_{K_y\,K_x}\,C^{K_y}_{J_x\,K_z} + \beta_{J_y\,J_x}\,C^{J_y}_{K_z\,K_x} = -\beta_{K_y\,K_x} - \frac{1}{c^2}\beta_{J_y\,J_x} = 0 \tag{V-57}$$

However, since we eliminated $\beta_{J_y\,J_x}$ via a redefinition of $\boldsymbol{J}$, we also have:

$$\beta_{K_y\,K_x} = Q_{21} = 0 \tag{V-58}$$

showing that the coefficients Q are zero.

We are left with the commutation relations between $\boldsymbol{K}$ and $\boldsymbol{P}$. The reasoning that led to (V-21) and (V-22) is still valid. We had chosen, for the operators associated with i and k, operator P_x and operator K_z. They are two different components, associated with structure constants $C^{G}_{K_z\,P_x}$ equal to zero. The reasoning of the Galilean group remains valid.

At this stage, all the extension operators have been eliminated, except[2]:

$$M_{xx} = M_{yy} = M_{zz} = \tilde{M}$$

This is where a major difference appears between the Galilean and Poincaré groups: the presence of the H operator on the right-hand side of (V-55b) allows eliminating operator $\tilde{M}$ by setting:

$$H' = H + \tilde{M}c^2 \tag{V-59}$$

Thanks to this redefinition of the energy operator associated with time translations, the projective representation has been reduced to a "true" representation, with no phase factor. Relation (V-59) expresses the fundamental link existing between the concept of mass in Galilean relativity and the concept of energy in Einstein's relativity. In what follows, we shall no longer distinguish H' and H.

[2]We use the notation $\tilde{M}$ to avoid any confusion with the mass operator M, which will be defined below, *cf.* relation (V-67).

C-2. Lie algebra

Finally, the commutation relations can be written:

$$\begin{cases} [J_i, J_j] = i\hbar\, \varepsilon_{ijk}\, J_k \\ [J_i, P_j] = i\hbar\, \varepsilon_{ijk}\, P_k \\ [P_i, P_j] = 0 \end{cases} \tag{V-60}$$

$$\begin{cases} [J_i, K_j] = i\hbar\, \varepsilon_{ijk}\, K_k \\ [P_i, K_j] = \dfrac{i\hbar}{c^2}\, \delta_{ij}\, H \\ [K_i, K_j] = -\dfrac{i\hbar}{c^2}\, \varepsilon_{ijk}\, J_k \end{cases} \tag{V-61}$$

$$\begin{cases} [\boldsymbol{J}, H] = [\boldsymbol{P}, H] = 0 \\ [\boldsymbol{K}, H] = -i\hbar\, \boldsymbol{P} \end{cases} \tag{V-62}$$

As above, the terms in $1/c^2$ are the new terms appearing when one goes from Galileo's to Einstein's relativity.

Comment:

The definition (V-27) of a vector or scalar operator with respect to rotations is still valid, since the rotations are a common sub-group of the Galileo and Poincaré groups. We can then generalize the concept of vector operator to a four-vector operator, by taking into account the transformations of its four components in pure Lorentz transformations. Consider, for instance, a Lorentz transformation along Ox, whose infinitesimal parameter is $\delta\boldsymbol{v} = \delta v_x \boldsymbol{e}_x$; relation (III-113) of chapter III shows that the compoments of a four-vector $(ct, \boldsymbol{r})$, or more generally $(w_0, \boldsymbol{w})$, are transformed according to:

$$\delta w_0 = \frac{\delta v_x}{c} w_x \qquad \delta w_x = \frac{\delta v_x}{c} w_0 \tag{V-63a}$$

where x, y, and z are the components of vector $\boldsymbol{r}$. For a Lorentz transformation with a parameter $\delta\boldsymbol{v}$ in any direction, these relations become:

$$\delta w_0 = \frac{\boldsymbol{w} \cdot \delta\boldsymbol{v}}{c} \qquad \delta\boldsymbol{w} = \frac{w_0}{c}\, \delta\boldsymbol{v} \tag{V-63b}$$

Let us then introduce a 4-component operator $(W_0, \boldsymbol{W})$, whose time component W_0 is a scalar under rotation, and whose spatial components $\boldsymbol{W}$ is a vector operator. We have seen in (V-30) that the infinitesimal variations of a classical vector correspond to the quantum commutators, muiltiplied by $-i\hbar$. We will therefore define a four-vector operator $(W_0, \boldsymbol{W})$ as an operator obeying the following commutation relations:

$$
\begin{aligned}
&[K_i\,,\,W_0] = -\frac{i\hbar}{c}W_i \\
&[K_i\,,\,W_j] = -\frac{i\hbar}{c}\delta_{ij}W_0 \qquad i,j = x,y,z
\end{aligned}
\tag{V-64}
$$

For instance, relations (V-61) and (V-62) show that $(H, c\boldsymbol{P})$ is a four-vector operator. We will see another example in § C-3-d.

The Minkowski scalar product $W_0 Z_0 - \boldsymbol{W}\cdot\boldsymbol{Q}$ of two four-vector operators W and Q is a Lorentz invariant. This is the case, for example, of the Minkowski norm $W^2 = (W_0)^2 - \boldsymbol{W}\cdot\boldsymbol{W}$ of a four-vector operator. For the rotations, the invariance is a direct consequence of the properties of scalar and vector operators. For the Lorentz transformations, we check that the commutator vanishes:

$$
\begin{aligned}
&[K_x\,,\,W_0Q_0 - W_xQ_x - W_yQ_y - W_zQ_z] \\
&= [K_x\,,\,W_0]\,Q_0 + W_0\,[K_x\,,\,Q_0] - [K_x\,,\,W_x]\,Q_x - W_x\,[K_x\,,\,Q_x] \\
&= -\frac{i\hbar}{c}(W_xQ_0 + W_0Q_x - W_0Q_x - W_xQ_0) = 0
\end{aligned}
\tag{V-65}
$$

with the same result for the commutators containing K_y and K_z.

C-3. Mass, energy, and spin

In the Poincaré group, mass, spin, and energy appear in a very different way from what we saw in the Galilean group.

C-3-a. Mass operator M

No operator M appears on the right-hand side of the commutation relations (V-60), (V-61), and (V-62). Moreover, we cannot introduce an internal energy operator as we did for the Galileo group: for the Poincaré group, the commutation relation (V-32) between P_x^2 and K_x no longer holds, and we find:

$$
\left[P_x^2, K_x\right] = P_x\,[P_x, K_x] + [P_x, K_x]\,P_x = \frac{2i\hbar}{c^2}H\,P_x
\tag{V-66}
$$

which shows that the operator U_{int} introduced in (V-33a) does not commute with $\boldsymbol{K}$. But we should not conclude that going from the Galilean to the Poincaré group eliminates the concept of mass!

The mass operator arises from the fact that $(H, c\boldsymbol{P})$ is a four-vector operator. Its squared Minkowski norm $H^2 - c^2\boldsymbol{P}^2 = H^2 - P_x^2 - P_y^2 - P_z^2$ is therefore a Lorentz invariant, which ensures that it commutes with $\boldsymbol{J}$ and $\boldsymbol{K}$. It actually commutes with the 10 generators of the Poincaré group, since the operators H and $c\boldsymbol{P}$ commute. We therefore otbain a Casimir operator of the group by writing:

$$\boxed{M^2 = \left(H^2 - c^2\boldsymbol{P}^2\right)/c^4} \qquad \text{(V-67)}$$

The diagonal operator M defined in this way is called the mass of the system[3].

The equality:

$$H^2 = \boldsymbol{P}^2c^2 + M^2c^4 \qquad \text{(V-68)}$$

is the analog of the well-known classical relation $E^2 = \boldsymbol{p}^2c^2 + m^2c^4$. The mass therefore arises in a very different way in the Galileo group (where it appears in the commutation relations of the Lie algebra only after it is extended) and the Poincaré group. The expression of H now contains the square root of an operator, thereby creating difficulties when one tries to generalize a Galilean theory into a Lorentz invariant theory; we will return to this subject in § B of chapter VI.

C-3-b. Operator L

As in (V-34), we first introduce operator[4]:

$$\boldsymbol{L} = (\boldsymbol{P} \times \boldsymbol{K})/M \qquad \text{(V-69)}$$

and compute its commutation relations with $\boldsymbol{J}$, $\boldsymbol{P}$, H, and $\boldsymbol{K}$.

- The computation leading to (V-36c) remains valid ($\boldsymbol{L}$ is a vector):

$$[J_i, L_j] = i\hbar\, \varepsilon_{ijk}\, L_k \qquad \text{(V-70)}$$

However, since the commutation relations between $\boldsymbol{P}$ and $\boldsymbol{K}$ are different from those of the Galilean group, relation (V-39) is no longer satisfied. We now have:

$$[L_x, P_x] = \frac{1}{M}\left[P_yK_z - P_zK_y\,, P_x\right] = 0 \qquad \text{(V-71a)}$$

$$\begin{aligned} [L_x, P_y] &= \frac{1}{M}\left[P_yK_z - P_zK_y\,, P_y\right] = -\frac{1}{M}P_z\left[K_y, P_y\right] \\ &= \frac{i\hbar}{Mc^2}P_z\, H \end{aligned} \qquad \text{(V-71b)}$$

[3]Defining this operator by its square M^2, implicitely assumes that it is positive or zero. If it were negative, it would lead us to assume a purely imaginary mass for the particle, with a velocity always greater than c, etc. A theory of such physical systems ("tachyonic particles") has been proposed (see [31], for example), but will not be discussed here.

[4]We assume here that $M \neq 0$; the case $M = 0$ is discussed in § C-4.

that is:

$$[L_i, P_j] = \frac{i\hbar}{Mc^2}\, \varepsilon_{ijk}\, P_k\, H \tag{V-71c}$$

• We also have:

$$[L_x, H] = \frac{1}{M}\,[P_y K_z - P_z K_y, H] = \frac{i\hbar}{M}\,(-P_y P_z + P_z P_y) = 0 \tag{V-72a}$$

meaning:

$$[\boldsymbol{L}, H] = 0 \tag{V-72b}$$

• We now have to compute the commutator of the $\boldsymbol{L}$ components with those of $\boldsymbol{K}$:

$$\begin{aligned}[L_i, K_j] &= \frac{1}{M}\sum_{k\ell} \varepsilon_{ik\ell}\,[P_k K_\ell, K_j] \\ &= \frac{i\hbar}{M}\sum_{k\ell} \varepsilon_{ik\ell}\left\{-\frac{1}{c^2}\sum_m \varepsilon_{\ell jm}\, P_k J_{x_m} + \frac{1}{c^2}\,\delta_{jk}\, H\, K_\ell\right\}\end{aligned} \tag{V-73}$$

Using relation:

$$\sum_\ell \varepsilon_{ik\ell}\,\varepsilon_{\ell jm} = \sum_\ell \varepsilon_{ik\ell}\,\varepsilon_{jm\ell} = \delta_{ij}\,\delta_{km} - \delta_{im}\,\delta_{jk} \tag{V-74}$$

we get:

$$\begin{aligned}[L_i, K_j] &= \frac{i\hbar}{Mc^2}\left\{-\delta_{ij}\sum_k P_k J_k + P_j J_i + \sum_\ell \varepsilon_{ij\ell}\, H\, K_\ell\right\} \\ &= \frac{i\hbar}{Mc^2}\left\{P_j J_i - \delta_{ij}\,(\boldsymbol{P}\cdot\boldsymbol{J}) + \sum_k \varepsilon_{ijk}\, H\, K_k\right\}\end{aligned} \tag{V-75}$$

C-3-c. Spin operator S

How should we define the spin $\boldsymbol{S}$? If we take it equal to $\boldsymbol{J} - \boldsymbol{L}$ as in the Galilean group, the presence of H on the right-hand side of (V-71b) would mean that $\boldsymbol{S}$ and $\boldsymbol{P}$ do not commute, which is physically questionable (we expect the spin, an internal variable, to be invariant under a translation of the physical system). On the other hand, if we set[5]:

$$\boxed{\boldsymbol{S} = \frac{H}{Mc^2}\,\boldsymbol{J} - \boldsymbol{L}} \tag{V-76}$$

[5] One can find in the literature several definitions of the spin operator as functions of the generators of the Poincaré algebra (see, for instance, refs [33, 34]). Nevertheless, none fulfills all the conditions that are desirable for a spin operator: being a Hermitian vector operator, having the autocommutation relations of an angular momentum, commuting with $\boldsymbol{P}$ and H. The definition we choose here does not obey the autocommutation relations (*cf.* § 1 of complement B_V).

then relation:

$$\left[\frac{H}{Mc^2}J_x, P_y\right] = \frac{H}{Mc^2}[J_x, P_y] = i\hbar\frac{H}{Mc^2}P_z \tag{V-77}$$

combined with (V-71b), leads to $[S_x, P_y] = 0$, and more generally to:

$$[\boldsymbol{S}, \boldsymbol{P}] = 0 \tag{V-78}$$

The spin $\boldsymbol{S}$ is then translation invariant, as we expect for the intrinsinc angular momentum of a system. In the non-relativistic limit (momentum smaller than Mc, so that the energy is close to Mc^2), it becomes identical to the Galilean definition (V-41) of the spin.

It is easy to compute the commutator:

$$[HJ_i, J_j] = i\hbar\, \varepsilon_{ijk}\, HJ_k$$

so that definition (V-76) and relation (V-70) show that $\boldsymbol{S}$ is a vector operator:

$$[J_i, S_j] = i\hbar\, \varepsilon_{ijk}\, S_k \tag{V-79}$$

Furthermore, relation:

$$[\boldsymbol{S}, H] = 0 \tag{V-80}$$

is easy to prove, using (V-72b) and (V-62).

Let us examine the commutation relations between $\boldsymbol{S}$ and the components of $\boldsymbol{K}$. Using (V-75) and relations (V-61), we have:

$$\begin{aligned}[S_i, K_j] &= \frac{i\hbar}{Mc^2}\left(P_jJ_i + \sum_k \varepsilon_{ijk}\, HK_k - P_jJ_i + \delta_{ij}\, \boldsymbol{P}\cdot\boldsymbol{J} - \sum_k \varepsilon_{ijk}\, HK_k\right) \\ &= \frac{i\hbar}{Mc^2}\delta_{ij}\, \boldsymbol{P}\cdot\boldsymbol{J} \end{aligned} \tag{V-81}$$

This relation shows that, if:

$$\boldsymbol{S}^2 = S_x^2 + S_y^2 + S_z^2$$

we have:

$$[\boldsymbol{S}^2, K_x] = \frac{i\hbar}{Mc^2}\{S_x(\boldsymbol{P}\cdot\boldsymbol{J}) + (\boldsymbol{P}\cdot\boldsymbol{J})S_x\} \tag{V-82}$$

Operator $\boldsymbol{S}^2$ does not commute with $\boldsymbol{K}$, and operator $(\boldsymbol{P}\cdot\boldsymbol{J})$ comes into play. It will be useful, in what follows, to compute the commutator of that operator with $\boldsymbol{K}$:

$$\begin{aligned}[(\boldsymbol{P}\cdot\boldsymbol{J}), K_x] &= [P_xJ_x + P_yJ_y + P_zJ_z, K_x] \\ &= [P_x, K_x]J_x + P_y[J_y, K_x] + P_z[J_z, K_x] \\ &= \frac{i\hbar}{c^2}HJ_x - i\hbar\, P_yK_z + i\hbar\, P_zK_y \\ &= i\hbar M\, S_x \end{aligned} \tag{V-83}$$

and:

$$[(\boldsymbol{P}\cdot\boldsymbol{J}), P_x] = -i\hbar P_y P_z + i\hbar P_z P_y = 0 \quad \text{that is:} \quad [(\boldsymbol{P}\cdot\boldsymbol{J}), \boldsymbol{P}] = 0 \tag{V-84}$$

The commutation relations between the components of the $\boldsymbol{S}$ operator are calculated in § 1 of complement B_V.

C-3-d. Four-vector associated with the spin, invariant operator W^2

Operators $(\boldsymbol{P}\cdot\boldsymbol{J})$ and $\boldsymbol{S}$ commute with H and $\boldsymbol{P}$; for $(\boldsymbol{P}\cdot\boldsymbol{J})$, this is a consequence of relations (V-62) and (V-84), for $\boldsymbol{S}$ of (V-78) and (V-80). But this does not ensure a commutation with $\boldsymbol{J}$ and $\boldsymbol{K}$. We nevertheless notice that relations (V-81) and (V-83) may be written:

$$\begin{aligned} \left[K_i\,, \frac{\boldsymbol{P}\cdot\boldsymbol{J}}{Mc}\right] &= -\frac{i\hbar}{c}\,S_i \\ [K_i\,, S_j] &= -\frac{i\hbar}{c}\,\delta_{ij}\frac{\boldsymbol{P}\cdot\boldsymbol{J}}{Mc} \end{aligned} \tag{V-85}$$

According to (V-64), these relations indicate that the pair $(\boldsymbol{P}\cdot\boldsymbol{J}/Mc,\, \boldsymbol{S})$ is a four-vector operator[6]. The square of its norm is therefore a Lorentz invariant, which suggests the definition:

$$\boxed{W^2 = \boldsymbol{S}^2 - \frac{(\boldsymbol{P}\cdot\boldsymbol{J})^2}{M^2c^2} = \boldsymbol{S}^2 - \frac{(\boldsymbol{J}\cdot\boldsymbol{P})^2}{M^2c^2}} \tag{V-86}$$

($\boldsymbol{P}\cdot\boldsymbol{J} = \boldsymbol{J}\cdot\boldsymbol{P}$ since each $\boldsymbol{J}$ component commutes with the same $\boldsymbol{P}$ component). We can then show that:

$$\begin{aligned} \left[W^2, K_x\right] = \frac{i\hbar}{Mc^2}\left\{S_x(\boldsymbol{P}\cdot\boldsymbol{J}) + (\boldsymbol{P}\cdot\boldsymbol{J})S_x\right\} \\ - \frac{i\hbar}{Mc^2}\left\{S_x(\boldsymbol{P}\cdot\boldsymbol{J}) + (\boldsymbol{P}\cdot\boldsymbol{J})S_x\right\} = 0 \end{aligned} \tag{V-87a}$$

that is:

$$\left[W^2, \boldsymbol{K}\right] = 0 \tag{V-87b}$$

In addition, it is easy to show that $\boldsymbol{S}^2$ and $(\boldsymbol{P}\cdot\boldsymbol{J})$ commute with $\boldsymbol{J}$ (the scalar product of two vector operators is a scalar), which leads to:

$$\left[W^2, \boldsymbol{J}\right] = 0 \tag{V-88}$$

[6]The Pauli–Lubanski four-vector is studied in more detail in § 2 of complement B_V.

We saw that $\boldsymbol{S}$ commutes with H and $\boldsymbol{P}$; the same is true for its square $\boldsymbol{S}^2$. Since:

$$\begin{aligned}
[(\boldsymbol{P}\cdot\boldsymbol{J}),H] &= 0\\
[(\boldsymbol{P}\cdot\boldsymbol{J})\ ,\ P_x] &= [P_xJ_x+P_yJ_y+P_zJ_z\ ,\ P_x] = P_y\,[J_y,P_x]+P_z\,[J_z,P_x]\\
&= P_y\,(-i\hbar\,P_z)+P_z\,(i\hbar\,P_y)=0
\end{aligned} \tag{V-89}$$

it is clear that:

$$\left[W^2,H\right]=\left[W^2,\boldsymbol{P}\right]=0 \tag{V-90}$$

In conclusion, operator W^2 commutes with all the generators of the Poincaré group. It plays a role analog to that of $\boldsymbol{S}^2$ for the Galilean group. We are going to show that its eigenvalues are also of the form $S(S+1)\hbar^2$ where S is an integer or a half-integer.

Since operator W^2 commutes with all the generators of the Poincaré group, it can only have one single eigenvalue if the state space under study (one elementary particle) is an irreducible representation of the group. If there were two distinct eigenvalues, the representation space would be the direct sum of two orthogonal subspaces associated with these different eigenvalues, subspaces between which none of the infinitesimal generators could have a non-zero matrix element. The representation would then be reducible.

C-3-e. Eigenvalues of S^2 and W^2

• Operator $\boldsymbol{S}$ was built so as to commute with the momentum operator $\boldsymbol{P}$; its square $\boldsymbol{S}^2$ must have the same property. The eigensubspaces of the momentum are thus stable under the action of $\boldsymbol{S}^2$, and we can look for the eigenvalues of this operator in the eigensubspace of $\boldsymbol{P}$ with a zero eigenvalue ($\boldsymbol{p}=0$). We assume that this eigenvalue exists; as we shall see later, this is always the case if the particle's mass M is different from zero (the case where $M=0$ is discussed in the following § C-4).

We have set, in (V-76):

$$\boldsymbol{S}=\boldsymbol{J}\frac{H}{Mc^2}+\frac{1}{M}\boldsymbol{K}\times\boldsymbol{P} \tag{V-91}$$

(since $\boldsymbol{J}$ and H commute, as well as the different components of $\boldsymbol{K}$ and $\boldsymbol{P}$ appearing in their cross product). Taking into account (V-68), we see that operator $\boldsymbol{S}_0$, which represents the action of $\boldsymbol{S}$ inside the subspace with eigenvalue $\boldsymbol{p}=0$, reduces to[7]:

$$\boldsymbol{S}_0=\boldsymbol{J} \tag{V-92}$$

[7]Since $\boldsymbol{J}$ commutes with $\boldsymbol{P}^2$, this subspace is indeed stable under the action of $\boldsymbol{J}$.

The equality (V-92) shows that the commutation relations in this subspace are those of an angular momentum. The eigenvalues of $\boldsymbol{S}_0^2$ in this $\boldsymbol{p} = 0$ subspace can only be equal to $S(S+1)\hbar^2$, where S is an integer or half-integer number.

• We now look for the eigenvalues of operator W^2. As opposed to $\boldsymbol{S}^2$, it commutes with all the generators of the Poincaré group, so that its eigenvalues are the same in all the eigensubspaces of the momentum. We note on (V-86) (where $\boldsymbol{P} \cdot \boldsymbol{J}$ is equal to $\boldsymbol{J} \cdot \boldsymbol{P}$) that, if W_0^2 is the restriction of W^2 to the $\boldsymbol{p} = 0$ subspace, we have:

$$W_0^2 = \boldsymbol{S}_0^2 \qquad \text{(V-93)}$$

The (sole) eigenvalue of W^2 is the eigenvalue of $\boldsymbol{S}_0^2$ in the $\boldsymbol{p} = 0$ subspace. Consequently:

$$W^2 = S(S+1)\hbar^2 \qquad \text{(V-94)}$$

Since in Einstein's relativity $\boldsymbol{S}^2$ is not invariant under a change of reference frame, stating that the physical system has a spin S does not mean that the $\boldsymbol{S}^2$ operator takes a constant value $S(S+1)\hbar^2$. It is operator W^2 that takes this sole value, regardless of the quantum state of the system.

In complement B_V, we generalize the reasoning that led us to (V-94) via another method that does not require using a reference frame where the particle's momentum is zero.

Comments:

(i) In the Lie algebra of the Galilean group, we introduced in § B a position operator:

$$\boldsymbol{R} = -\boldsymbol{K}/M$$

that satisfies the canonical commutation relations (V-49). These relations are actually a constraint that must be imposed to any position operator. They imply that:

$$[\boldsymbol{R}, \delta\boldsymbol{b} \cdot \boldsymbol{P}] = i\hbar\, \delta\boldsymbol{b}$$

which leads to:

$$\boldsymbol{R}' = \mathrm{e}^{-i\delta\boldsymbol{b}\cdot\boldsymbol{P}/\hbar}\, \boldsymbol{R}\, \mathrm{e}^{i\delta\boldsymbol{b}\cdot\boldsymbol{P}/\hbar} \simeq \boldsymbol{R} - \frac{i}{\hbar}[\delta\boldsymbol{b} \cdot \boldsymbol{P}, \boldsymbol{R}] = \boldsymbol{R} - \delta\boldsymbol{b}$$

We already obtained this equality in (IV-45a) of chapter IV, which had to be expected from the behavior of any position operator during a displacement. We would like to find such an operator, in the Lie algebra of the Poincaré group, satisfying the canonical commutation relations with $\boldsymbol{P}$. Such an operator cannot be found: the second of equalities (V-61) yields these canonical relations only if we replace H by Mc^2 ("non-relativistic limit"). As expected, the general definition of a position operator is not an easy problem to solve in Einstein's relativity.

(ii) Let us examine if the $\boldsymbol{p} = 0$ eigenvalue of $\boldsymbol{P}$ necessarily exists, or in other words, if there is always a reference frame where the particle is at rest. Consider an eigenvector $|E, \boldsymbol{p}, \alpha\rangle$ common to H and the three components of $\boldsymbol{P}$ (it does exist since H and $\boldsymbol{P}$ commute with each other), with eigenvalues E, p_x, p_y, p_z (components of $\boldsymbol{P}$); the quantum number α stands for the eigenvalues of the other operators eventually necessary to form, with H and $\boldsymbol{p}$, a CSCO (Complete Set of Commuting Observables) in the state space. Relation (V-68) shows that E and p are related by $E^2 = \boldsymbol{p}^2c^2 + m^2c^4$.

Consider first an infinitesimal rotation associated with operator:

$$\mathbb{1} - \frac{i}{\hbar}\,\delta\boldsymbol{a}\cdot\boldsymbol{J}$$

According to the second of relations (V-60), we have:

$$\begin{aligned}
&\boldsymbol{P}\left(\mathbb{1} - \frac{i}{\hbar}\,\delta\boldsymbol{a}\cdot\boldsymbol{J}\right)|E, \boldsymbol{p}, \alpha\rangle \\
&\quad = \left\{\left(\mathbb{1} - \frac{i}{\hbar}\,\delta\boldsymbol{a}\cdot\boldsymbol{J}\right)\boldsymbol{P} + \delta\boldsymbol{a}\times\boldsymbol{P}\right\}|E, \boldsymbol{p}, \alpha\rangle \\
&\quad = (\boldsymbol{p} + \delta\boldsymbol{a}\times\boldsymbol{p})\left(\mathbb{1} - \frac{i}{\hbar}\,\delta\boldsymbol{a}\cdot\boldsymbol{J}\right)|E, \boldsymbol{p}, \alpha\rangle + 0\left(\delta a^2\right)
\end{aligned} \tag{V-95}$$

since, for example:

$$[P_x\,,\,\delta a_x\,J_x + \delta a_y\,J_y + \delta a_z\,J_z] = i\hbar\;(\delta a_y\,P_z - \delta a_z\,P_y) = i\hbar(\delta\boldsymbol{a}\times\boldsymbol{P})_x \tag{V-96}$$

Consequently, the ket:

$$\left(\mathbb{1} - \frac{i}{\hbar}\,\delta\boldsymbol{a}\cdot\boldsymbol{J}\right)|E, \boldsymbol{p}, \alpha\rangle$$

is another eigenvector of $\boldsymbol{P}$, with eigenvalue $\boldsymbol{p} + \delta\boldsymbol{a}\times\boldsymbol{p}$. We find again that the rotation operators change the components of $\boldsymbol{P}$ in the same way as those of a vector in ordinary space. On the other hand, it is easy to see that the energy E is unchanged ($\boldsymbol{P}$ is a vector, H a scalar). This argument is valid both in the Galilean and in the Poincaré groups.

Consider now a change of the Lorentz reference frame. We could proceed with the same computation, starting with $\boldsymbol{K}$ commutators given by the second

(V-61) and third (V-62) relations. One would establish that the ket:

$$\left(\mathbb{1} - \frac{i}{\hbar}\,\delta\boldsymbol{v}\cdot\boldsymbol{K}\right)|E,\boldsymbol{p},\alpha\rangle$$

is another eigenvector common to H and $\boldsymbol{P}$, with respective eigenvalues:

$$\boldsymbol{p}' = \boldsymbol{p} + \frac{E}{c^2}\,\delta\boldsymbol{v} \qquad\qquad E' = E + \delta\boldsymbol{v}\cdot\boldsymbol{p} \tag{V-97}$$

As this result is established in a more general way in § 2-b of complement B_{VI}, we shall not carry out this calculation here.

Relations (V-97) yield, to first-order in $\delta\boldsymbol{v}$, the formulas for the Lorentz transformations (III-105) (where the factor γ equals 1 to first-order) for the four-vector[8] momentum-energy $\{\boldsymbol{p}, E/c^2\}$. Integrating equalities (V-97) we get, for a finite value of $\boldsymbol{v}$:

$$\boldsymbol{p}' = \gamma\left(\boldsymbol{p} + \frac{E}{c^2}\boldsymbol{v}\right) \qquad\qquad E' = \gamma\,(E + \boldsymbol{v}\cdot\boldsymbol{p}) \tag{V-98}$$

with $\gamma = \left[1 - v^2/c^2\right]^{-1/2}$. These relations show that the couple $\{\boldsymbol{p}, E/c\}$ is transformed as a relativistic four-vector. We thus know under which condition it can be reduced, by a change of Lorentz reference frame, to having its space components equal to zero (here $\boldsymbol{p} = 0$): it must be a time-like four-vector, with a positive norm. This means that:

$$c^2\left(\frac{E}{c^2}\right)^2 - \boldsymbol{p}^2 = \frac{1}{c^2}\left(E^2 - c^2\,\boldsymbol{p}^2\right) = M^2\,c^2 \tag{V-99}$$

must be positive. Consequently, if the particle's mass is different from zero, the eigenvalue $\boldsymbol{p} = 0$ exists and the above reasoning establishing (V-94) is valid. On the other hand, if $M = 0$, there is no reference frame where the particle is at rest, and the argument does not hold.

C-4. Massless particle

When working with the Galilean group, we saw the importance of giving a finite (non-zero) mass to the particle; otherwise, the representation is radically changed (it becomes true instead of projective), and the H spectrum no longer has a lower boundary (*cf.* comment on page B-5-e). The situation is quite different for the Poincaré group, and there is no particular reason to impose a non-zero M. Moreover, we know that zero mass particles, such as photons, do exist. We must go back to the previous reasoning since, in the very definition of $\boldsymbol{S}$, we assumed $M \neq 0$.

[8] Note that $E^2 - c^2\,\boldsymbol{p}^2$ is an invariant since, to first-order in $\delta\boldsymbol{v}$:
$E'^2 - c^2\,\boldsymbol{p}'^2 = E^2 + 2E\,\delta\boldsymbol{v}\cdot\boldsymbol{p} - c^2\,\boldsymbol{p}^2 - 2c^2(E/c^2)\,\delta\boldsymbol{v}\cdot\boldsymbol{p} = E^2 - c^2\,\boldsymbol{p}^2$.

Instead of $\boldsymbol{S}$, we introduce an operator $\boldsymbol{\Sigma}$, which is simply equal to $M\boldsymbol{S}$ if M is different from zero, but is still defined when M equals zero:

$$\boldsymbol{\Sigma} = \frac{H}{c^2}\boldsymbol{J} + \boldsymbol{K} \times \boldsymbol{P} \tag{V-100}$$

The above reasoning on $\boldsymbol{S}$ when $M \neq 0$ is easily transposed to $\boldsymbol{\Sigma}$, and we get [*cf.* relations (V-78) in (V-82)]:

$$[\boldsymbol{\Sigma}, \boldsymbol{P}] = [\boldsymbol{\Sigma}, H] = 0 \tag{V-101a}$$

$$[\Sigma_i, J_j] = i\hbar\, \varepsilon_{ijk}\, \Sigma_k \tag{V-101b}$$

$$\left[\boldsymbol{\Sigma}^2, \boldsymbol{J}\right] = 0 \tag{V-101c}$$

This shows that the quantity:

$$Z^2 = \boldsymbol{\Sigma}^2 - \frac{1}{c^2}\left(\boldsymbol{P} \cdot \boldsymbol{J}\right)^2 \tag{V-102}$$

commutes with all the generators of the Poincaré group: if $M \neq 0$ we have $Z^2 = M^2 W^2$, and we simply have to repeat the reasoning of § C-3-d (multiplying by M^2 all the equalities annulling the commutators) to obtain this result. As for the commutation relations between the spin components (complement $\mathrm{B_V}$, § 1), they become, after inserting the M factors:

$$[\Sigma_i, \Sigma_j] = i\hbar \left\{\frac{H}{c^2}\, \varepsilon_{ijk}\, \Sigma_k - \frac{1}{c^2}\left(\boldsymbol{J} \cdot \boldsymbol{P}\right) \varepsilon_{ijk}\, P_k\right\} \tag{V-103}$$

Definition (V-100) for $\boldsymbol{\Sigma}$ leads to:

$$\boldsymbol{\Sigma} \cdot \boldsymbol{P} = \frac{H}{c^2}\boldsymbol{J} \cdot \boldsymbol{P} \tag{V-104}$$

which contains the $\boldsymbol{J} \cdot \boldsymbol{P}$ operator, hence the projection of the angular momentum on the linear momentum. The component of $\boldsymbol{J}$ parallel to $\boldsymbol{P}$ (the scalar product $\boldsymbol{J} \cdot \boldsymbol{P}$ divided by the modulus of $\boldsymbol{P}$) is called "helicity".

C-4-a. Operators restricted to an eigensubspace of H and P

Consider the eigensubspace common to H and $\boldsymbol{P}$, with respective eigenvalues E and $\boldsymbol{p} = p\boldsymbol{e}_z$, where the eigenvalues p_x and p_y of P_x and P_y are zero. In this subspace, $\boldsymbol{\Sigma}$ becomes a restricted operator $\boldsymbol{\Sigma}^p$ and, using (V-104) to compute $\boldsymbol{J} \cdot \boldsymbol{P}$, relations (V-103) become:

$$\left[\Sigma_x^p, \Sigma_y^p\right] = i\hbar \left\{\frac{E}{c^2}\Sigma_z^p - \frac{p}{c^2}\boldsymbol{J} \cdot \boldsymbol{p}\right\} = i\hbar \left\{\frac{E}{c^2}\Sigma_z^p - \frac{p^2}{E}\Sigma_z^p\right\} = 0 \tag{V-105}$$

(we used the fact that, if $M = 0$, we simply have $E = pc$). As for the other commutators between the $\boldsymbol{\Sigma}$ components, the $\boldsymbol{J} \cdot \boldsymbol{P}$ term on the right-hand side of (V-103) is zero; this is because only the Oz component of $\boldsymbol{P}$ contributes to the scalar product, which annuls the ε_{ijk} coefficient. The commutation relations between the $\boldsymbol{\Sigma}^p$ components are written:

$$\left[\Sigma_x^p , \Sigma_y^p\right] = 0$$
$$\left[\Sigma_y^p , \Sigma_z^p\right] = i\hbar \, \frac{E}{c^2} \, \Sigma_x^p$$
$$\left[\Sigma_z^p , \Sigma_x^p\right] = i\hbar \, \frac{E}{c^2} \, \Sigma_y^p \tag{V-106}$$

Since an additional structure constant is zero, the three components of $\boldsymbol{\Sigma}$ form a Lie algebra that is degenerated compared to that of $\boldsymbol{S}$ when $M \neq 0$, and which generated the algebra of SO(3) or SU(2). As for the Casimir operator Z^2, using the fact that $E = pc$, it becomes in that subspace:

$$Z^2 = (\boldsymbol{\Sigma}^p)^2 - (\boldsymbol{\Sigma}_z^p)^2 = (\boldsymbol{\Sigma}_x^p)^2 + (\boldsymbol{\Sigma}_y^p)^2 \tag{V-107}$$

Introducing operators:

$$\Sigma_\pm^p = \Sigma_x^p \pm i \, \Sigma_y^p \tag{V-108}$$

we get:

$$\left[\Sigma_+^p , \Sigma_-^p\right] = 0 \tag{V-109a}$$
$$\Sigma_+^p \, \Sigma_-^p = \Sigma_-^p \, \Sigma_+^p = (\Sigma_x^p)^2 + (\Sigma_y^p)^2 = (\boldsymbol{\Sigma}^p)^2 - (\Sigma_z^p)^2 \tag{V-109b}$$
$$\left[\Sigma_z^p , \Sigma_\pm^p\right] = \pm\hbar \, \frac{E}{c^2} \, \Sigma_\pm^p \tag{V-109c}$$

These relations are similar to that of an angular momentum, see equalities (VII-5).

In the eigensubspace common to operators H and $\boldsymbol{P}$, consider a basis $|E, \boldsymbol{p}, \sigma, \nu, \alpha\rangle$ of eigenvectors common to $\boldsymbol{\Sigma}^2$ and Σ_z (remember that the Oz axis is parallel to $\boldsymbol{p}$); we label the respective eigenvalues of these two operators as:

$$\sigma^2 \hbar^2 \, \frac{E^2}{c^4} \qquad \text{and} \qquad \nu\hbar \, \frac{E}{c^2}$$

The Casimir constant Z^2 becomes:

$$Z^2 = (\sigma^2 - \nu^2) \, \frac{\hbar^2 E^2}{c^4} \tag{V-110}$$

Since the norm of the ket $\Sigma_+^p |E, \boldsymbol{p}, \sigma, \nu, \alpha\rangle$ must be positive, and taking (V-109b) into account, we have:

$$\sigma^2 - \nu^2 \geq 0 \tag{V-111a}$$

Assuming $\sigma > 0$, leads to:

$$-\sigma \leq \nu \leq +\sigma \tag{V-111b}$$

Operators $\Sigma^p_\pm$ change the values of the quantum numbers σ and ν. Relation (V-109c) shows that their action on $|E, \boldsymbol{p}, \sigma, \nu, \alpha\rangle$ yields a ket with a $\nu \pm 1$ quantum number. Relation (V-109b) then shows that the squared norm of this ket is multiplied by $(\sigma^2 - \nu^2)\hbar^2 E^2/c^4$. Furthermore, since operator Z^2 is proportional to the identity operator, relation (V-107) shows that this new ket is still an eigenket of $(\boldsymbol{\Sigma}^p)^2$, with a new value of σ^2:

$$\sigma^2 \Rightarrow \sigma^2 + 1 \pm 2\nu \tag{V-112}$$

that balances the change in ν to keep $\sigma^2 - \nu^2$ constant. A direct calculation of the commutator between $(\boldsymbol{\Sigma}^p)^2$ and $\Sigma^p_\pm$ confirms this result.

Two cases may occur, depending whether the constant Z^2 is zero or not.

C-4-b. Zero Z^2 constant

If the Z^2 constant is zero, relation (V-110) shows that ν can only take the two extreme values of (V-111a):

$$\nu = \pm\sigma \tag{V-113}$$

The action of $\Sigma_\pm$ on the $|E, \boldsymbol{p}, \sigma, \pm\sigma, \alpha\rangle$ kets always yields a zero-norm ket, so that:

$$\Sigma^p_\pm = 0 \qquad \text{and thus:} \qquad \Sigma^p_x = \Sigma^p_y = 0 \tag{V-114}$$

The $\boldsymbol{\Sigma}$ component on an axis parallel to $\boldsymbol{p}$ can, a priori, take two values, one for each helicity:

$$\Sigma^p_z = \pm\sigma\hbar\frac{E}{c^2} \tag{V-115}$$

To establish that σ must be an integer or a half-integer, we use relation (V-104) showing that the eigenvalues $\pm\sigma\hbar Ep/c^2$ of $\boldsymbol{\Sigma}\cdot\boldsymbol{P}$ are also those of $H\,(\boldsymbol{J}\cdot\boldsymbol{P})/c^2$. In a basis of eigenvectors common to H, P_z, and J_z (with eigenvalue m_J, where m_J is an integer or half-integer), these eigenvalues are of the form $m_J\hbar Ep/c^2$. Hence:

$$m_J = \pm\sigma \tag{V-116}$$

so that σ is necessarily either an integer or a half-integer. This number is called the particle's spin; for the photon, $\sigma = 1$.

C-4-c. Non-zero Z^2 constant

If the Z^2 constant is not zero, the representation gets more complicated. We saw above that the action of $\Sigma^p_\pm$ is no longer the cancellation of the $|E, \boldsymbol{p}, \sigma, \nu, \alpha\rangle$ ket; it yields another ket whose squared norm is multiplied by Z^2, and which corresponds to a new quantum number σ^2, according to (V-112). This leads to a representation of the Poincaré group where the spin can have numerous values.

Massless particles with spins that can take a large number of values are unknown in particle physics. Representations where the Z^2 constant is different from zero are not taken into account.

Comments:

(i) For a massless particle, the condition $Z^2 = 0$ is naturally obtained by considering a particle with a non-zero mass, for which $Z^2 = S(S+1)\hbar^2 M^2$, and letting M go to zero. The representations where Z^2 is not zero are thus singular limits of the $M \neq 0$ case when $M \to 0$.

(ii) The ν value is necessarily unique if we consider an irreducible representation of the connected part of the Poincaré group, continuously related to the identity (Poincaré group without space inversion or time reversal). Indeed, ν cannot vary in a discontinuous way when applying infinitesimal Lorentz transformations. A single value of ν is thus sufficient to obtain a representation of that part of the group.

To include in the state space of the particle the space parity operation, we must remember that $\boldsymbol{\Sigma}$ (or $\boldsymbol{S}$) are, as well as $\boldsymbol{J}$, axial vectors (parity invariant), as opposed to $\boldsymbol{P}$, which is polar (its sign flips under parity). The quantity $\boldsymbol{\Sigma} \cdot \boldsymbol{P}$ is thus an operator whose sign flips under space symmetry (it is a pseudoscalar). To include parity in the state space, there must be two opposite values of ν in this space. As an example, the photon has two helicity states, $\nu = \pm 1$. They correspond to right or left circular polarizations.

Currently, space symmetry with respect to a point is not considered as a fundamental symmetry of nature. It is possible to assume that, for certain particles, only one ν value exists, and not its opposite value (two-components neutrinos-antineutrinos theory), which can lead to parity violations.

C-5. Finite transformations

Until now, we have focused on infinitesimal transformations and their associated generators in the Lie algebra. These elements will be particularly useful for building space states or wave equations. By exponentiation, we can build finite transformations. In complement C_V we consider, as an example, the displacement group (common to the Galilean and Poincaré groups). In complement B_{VI}, we study in a more general way the finite transformations associated with the Poincaré group.

Let us write, for example, the product of unitary operators:

$$U(\boldsymbol{a}, \boldsymbol{b}, \boldsymbol{v}, \tau) = \mathrm{e}^{-iH\tau/\hbar}\, \mathrm{e}^{-i\boldsymbol{b}\cdot\boldsymbol{P}/\hbar}\, \mathrm{e}^{-i\boldsymbol{a}\cdot\boldsymbol{J}/\hbar}\, \mathrm{e}^{-ic\boldsymbol{q}\cdot\boldsymbol{K}/\hbar} \tag{V-117}$$

It expresses a change of Galilean reference frame with rapidity $\boldsymbol{q}$ (defined in the comment of page 106), followed by a rotation of vector $\boldsymbol{a}$, then a translation of vector $\boldsymbol{b}$, and finally a time translation of a quantity τ. All these operations do not commute with each others: even in the relatively simple Galilean group, it is not equivalent to perform a change in the Galilean reference frame followed by a time translation, or to perform these operations in the reverse order – *cf.* commutation relations (V-26c) of chapter V. In the Poincaré group, to this same non-commutativity we must add for example the commutation relations between the $\boldsymbol{K}$ components, which introduce rotations (Thomas precession). Combinations of the different group transformations are complex. Details are given in §§ 19.1 and 19.2 of reference [13] as well as in chapter 4 of [21] where these calculations are explicitly addressed.

Complement A_V

Proper Lorentz group and SL(2C) group

The so-called "proper" Lorentz group is formed by the four-vector transformations that are products of rotations and proper Lorentz transformations (Lorentz boosts). A space–time four-vector is characterized by its 4 time and space coordinates ct, x, y, z, the last 3 being the coordinates of a vector $\boldsymbol{r}$ in 3-dimensional ordinary space. The transformations of the proper Lorentz group preserve the four-vector norm $c^2t^2 - x^2 - y^2 - z^2$ and are all connected to the identity transformation by a continuous path (their group is connected).

1. Link to the SL(2, C) group

The SL(2, C) group is formed by the 2×2 matrices with complex elements and a determinant equal to 1. It contains the SU(2) subgroup of the unitary 2×2 matrices that will be examined in more detail when studying rotations (complément A_{VII}). The aim of § 1 is to show that there exists a correspondence between any matrix of SL(2, C) and a proper Lorentz transformation, such that, to the product of two matrices associated with two transformations corresponds the product of the transformations. This is not a one-to-one correspondence; it is therefore a homomorphism rather than an isomorphism.

We will return to this issue in complement A_{VII}, in the particular case of the correspondence between rotations and the SU(2) matrices. It will be shown that two opposite SU(2) matrices correspond to the same SO(3) rotation, showing that the rotations representation is double-valued (§ A-3-b of chapter VII). This is due to the fact that SU(2) is simply connected, whereas the rotation group is not. SU(2) forms the universal mapping of the rotation group, and SL(2, c) that of the proper Lorentz group.

1-a. Four-vector transformation

With any four-vector $(ct, \boldsymbol{r})$, having ct, x, y, and z components, we associate the Hermitian matrix (Q) defined as:

$$(Q) = (ct)\mathbb{1} + x\sigma_x + y\sigma_y + z\sigma_z = \begin{pmatrix} ct+z & x-iy \\ x+iy & ct-z \end{pmatrix} \tag{1}$$

The four-vector's norm is simply equal to the determinant of (Q):

$$\det(Q) = c^2t^2 - x^2 - y^2 - z^2 \tag{2}$$

We introduce a complex 2×2 matrix (L) subject only to the constraint of having a determinant equal to 1 (it is supposed to be unimodular, but not necessarily Hermitian):

$$\det(L) = 1 \tag{3}$$

As we did for (Q), we can expand it on the Pauli matrices:

$$(L) = \lambda_0\mathbb{1} + \lambda_1\sigma_x + \lambda_2\sigma_y + \lambda_3\sigma_z = \begin{pmatrix} \lambda_0+\lambda_3 & \lambda_1 - i\lambda_2 \\ \lambda_1 + i\lambda_2 & \lambda_0 - \lambda_3 \end{pmatrix} \tag{4}$$

The λ_0, λ_1, λ_2, and λ_3 components are now complex and must obey the condition:

$$\det(L) = (\lambda_0)^2 - (\lambda_1)^2 - (\lambda_2)^2 - (\lambda_3)^2 = 1 \tag{5}$$

which is equivalent to two condtions in terms of real numbers. Therefore, the (L) matrix depends on 6 real parameters.

We now calculate the (Q') matrix defined as:

$$(Q') = (L)(Q)(L)^\dagger = (ct')\mathbb{1} + x'\sigma_x + y'\sigma_y + z'\sigma_z \tag{6}$$

where $(L)^\dagger$ is the adjoint of matrix (L); its determinant is the complex conjugate of that of (L), hence also equal to 1. Since the determinant of a product of square matrices is equal to the product of their determinants, we have:

$$\det(Q') = c^2t'^2 - x'^2 - y'^2 - z'^2 = \det(Q) = c^2t^2 - x^2 - y^2 - z^2 \tag{7}$$

Going from (Q) to (Q') preserves the four-vector's norm, as does a Lorentz transformation. L can be expected to actually define such a transformation.

Comments:

(i) A change of sign of matrix (L) does not modify either its determinant nor the (Q') matrix: two opposite matrix (L) and $-(L)$ are thus associated with the same transformation of the four-vectors (double-valued representation of these transformations).

(ii) It is the adjoint of L and not its inverse that appears in relation (6). This is not equivalent since, in general, (L) is not a unitary matrix. We have thus defined a "congruence relation" and not an equivalence relation between the matrices (Q) and (Q').

1-b. Pure Lorentz transformations

To see the correspondence between the (L) matrices and the Lorentz transformations, we first assume that (L) has components only on the identity matrix and on σ_z:

$$(L) = \lambda_0 \mathbb{1} + \lambda_3 \sigma_z = \begin{pmatrix} \lambda_0 + \lambda_3 & 0 \\ 0 & \lambda_0 - \lambda_3 \end{pmatrix} \tag{8}$$

A simple calculation leads to:

$$(Q') = \begin{pmatrix} (\lambda_0 + \lambda_3)(\lambda_0^* + \lambda_3^*)(ct + z) & (\lambda_0 + \lambda_3)(\lambda_0^* - \lambda_3^*)(x - iy) \\ (\lambda_0 - \lambda_3)(\lambda_0^* + \lambda_3^*)(x + iy) & (\lambda_0 - \lambda_3)(\lambda_0^* - \lambda_3^*)(ct - z) \end{pmatrix} \tag{9}$$

that is:

$$\begin{aligned} ct' &= (|\lambda_0|^2 + |\lambda_3|^2)(ct) + (\lambda_0\lambda_3^* + \lambda_3\lambda_0^*)z \\ z' &= (\lambda_0\lambda_3^* + \lambda_3\lambda_0^*)(ct) + (|\lambda_0|^2 + |\lambda_3|^2)z \\ x' + iy' &= (\lambda_0 - \lambda_3)(\lambda_0^* + \lambda_3^*)(x + iy) \end{aligned} \tag{10}$$

The standard relations of a pure Lorentz transformation of velocity v along the Oz axis are given in relation (III-105) of chapter III:

$$\begin{aligned} ct' &= \gamma\,[ct + \beta z] \\ z' &= \gamma\,[\beta(ct) + z] \\ x' + iy' &= x + iy \end{aligned} \tag{11}$$

with:

$$\beta = \frac{v}{c} \qquad \gamma = \frac{1}{\sqrt{1-\beta^2}} \tag{12}$$

Assume that matrix (L) is real. Its two diagonal elements must have the same sign since its determinant equals 1. They can be chosen positive since the sign of (L) is arbitrary. We can thus set:

$$\lambda_0 + \lambda_3 = e^{q/2c} \qquad \text{and:} \qquad \lambda_0 - \lambda_3 = e^{-q/2c} \tag{13}$$

where q is a real number, having the dimension of a velocity. We then have:

$$\lambda_0 = \cosh(\frac{q}{2c}) \qquad \text{and:} \qquad \lambda_3 = \sinh(\frac{q}{2c}) \tag{14}$$

Relations (10) are identical to (11) if:

$$\begin{aligned} \gamma &= \cosh q/c \\ \beta &= \tanh q/c \end{aligned} \tag{15}$$

The parameter q is thus equal to:

$$q = c\,\text{artanh}(\beta) = c \ln \sqrt{\frac{1+\beta}{1-\beta}} \tag{16}$$

which is nothing other than the value of the "rapidity". Defining q by this relation, and then λ_0 and λ_3 by (14), we can insert these (real) values in (8), to show that relation (6) does define a pure Lorentz transformation parallel to the Oz axis. The corresponding (L) matrix is written:

$$(L) = \cosh(\frac{q}{2c})\, \mathbb{1} + \sinh(\frac{q}{2c})\, \sigma_z \tag{17}$$

As the three space axes play a symmetrical role, the previous computation can be performed for a Lorentz transformation in the Ox or the Oy directions. We simply have to replace in (17) the σ_z component by σ_x or σ_y. The same is true for a Lorentz transformation in an arbitrary direction $\boldsymbol{u}$, which introduces the $\sigma_{\boldsymbol{u}}$ component of $\boldsymbol{\sigma}$ on the $\boldsymbol{u}$ direction. A pure Lorentz transformation with rapidity q along the direction $\boldsymbol{u}$ is described by the matrix:

$$(L) = \cosh(\frac{q}{2c})\, \mathbb{1} + \sinh(\frac{q}{2c})\, \boldsymbol{u} \cdot \boldsymbol{\sigma} \tag{18}$$

1-c. Pure rotations

We now change q/c into $-i\theta$, with θ being real. Relations (14) become:

$$\begin{aligned} \lambda_0 &= \cos\theta/2 \\ \lambda_3 &= -i\sin\theta/2 \end{aligned} \tag{19}$$

and relations (10) simplify to:

$$\begin{aligned} ct' &= ct \\ z' &= z \\ x' + iy' &= \mathrm{e}^{i\theta}(x + iy) \end{aligned} \tag{20}$$

that is:

$$x' = x\cos\theta - y\sin\theta$$
$$y' = x\sin\theta + y\cos\theta \tag{21}$$

The (L) matrix generates a rotation of the vector $\boldsymbol{r}$ by an angle θ around Oz, and has no effect on the time (pure spatial rotation). The explicit expression of (L) is the unitary matrix:

$$(L) = \begin{pmatrix} \mathrm{e}^{-i\theta/2} & 0 \\ 0 & \mathrm{e}^{i\theta/2} \end{pmatrix} = \mathrm{e}^{-i\sigma_z\theta/2} = \cos\frac{\theta}{2} - i\sin\frac{\theta}{2}\sigma_z \tag{22a}$$

A similar computation shows that a rotation around an arbitrary unit vector $\boldsymbol{u}$ by an angle θ is obtained by replacing in that equality σ_z by the $\boldsymbol{u}$ component σ_u of $\boldsymbol{\sigma}$ on $\boldsymbol{u}$:

$$(L) = \mathrm{e}^{-i\sigma_u\theta/2} = \cos(\frac{\theta}{2})\mathbb{1} - i\sin(\frac{\theta}{2})\,\boldsymbol{u}\cdot\boldsymbol{\sigma} \tag{22b}$$

All these (L) matrices belong to the unitary subgroup SU(2) of the SL$(2,C)$ group. In complement $\mathrm{A_{VII}}$, the matrices of this subgroup are interpreted as those describing the rotations of a spin 1/2 and of its density matrix. In addition and as mentioned above, we further discuss the homomorphism relation between the SU(2) group and the SO(3) rotation group; the first is simply connected and constitutes the universal mapping group of the second, which is not simply connected.

1-d. General Lorentz transformations

Combining pure Lorentz transformations with space rotations, we can get any transformation (pure or not pure) of the proper Lorentz group. Relation (6) shows that the (L) matrices associated with the product of the transformations are the products of the (L) matrices of each transformation. They are also 2×2 matrices with determinants equal to 1, and belong to SL$(2,C)$; this matrix group therefore constitutes a representation of the entire Lorentz group. As expected, this transformation depends on 6 real parameters: 3 for the pure Lorentz transformations, 3 for the pure rotations.

• In the case of infinitesimal rotations, we replace in (4) $\lambda_{1,2,3}$ by $\delta\lambda_{1,2,3}$. The real parts of these coefficients determine an infinitesimal pure Lorentz transformation, the imaginary parts an infinitesimal rotation. For finite transformations, a Hermitian (L) matrix is associated with a pure Lorentz transformation, a unitary (L) matrix with a pure rotation. Since any (L) matrix can be written[1] as the product $(B)(A)$ of a Hermitian matrix (A) and a unitary matrix (B), any transformation of the Lorentz group can be seen as the product of two pure transformations.

The $(L)^{-1}$ matrix associated with the inverse transformation of $L = \mathbb{1} + \boldsymbol{\lambda} \cdot \boldsymbol{\sigma}$ is simply obtained by changing the sign of $\boldsymbol{\lambda}$ as we now show. Using the multiplication and anticommutation properties of the Pauli matrices, we have:

$$(\lambda_0 \mathbb{1} + \boldsymbol{\lambda} \cdot \boldsymbol{\sigma})\,(\lambda_0 \mathbb{1} - \boldsymbol{\lambda} \cdot \boldsymbol{\sigma}) = (\lambda_0)^2\,\mathbb{1} - (\boldsymbol{\lambda} \cdot \boldsymbol{\sigma})(\boldsymbol{\lambda} \cdot \boldsymbol{\sigma}) = (\lambda_0)^2 - \boldsymbol{\lambda}^2 \tag{23}$$

where the right-hand side is the determinant of (L), hence equal to 1.

• We finally show how a given (L) matrix of SL$(2, C)$ transforms the space–time coordinates. We use the usual covariant notation to label the 4 components of a four-vector:

$$x^0 = ct \quad ; \quad x^1 = x \quad ; \quad x^2 = y \quad ; \quad x^3 = z \tag{24}$$

Similarly, we use the same notation for the three Pauli matrices and the identity matrix (σ_ν with $\nu = 0, 1, 2, 3$) setting:

$$\sigma_0 = (\mathbb{1}) \quad ; \quad \sigma_1 = \sigma_x \quad ; \quad \sigma_2 = \sigma_y \quad ; \quad \sigma_3 = \sigma_z \tag{25}$$

We multiply relation (6) by σ_ν and take the trace; using the properties of the Pauli matrices we can write:

$$\begin{aligned} x'^\mu &= \frac{1}{2}\mathrm{Tr}\Big\{(Q')\,\sigma_\mu\Big\} = \frac{1}{2}\mathrm{Tr}\Big\{(L)(Q)(L)^\dagger \sigma_\mu\Big\} \\ &= \frac{1}{2}\mathrm{Tr}\Big\{(L)(x^\nu \sigma_\nu)(L)^\dagger \sigma_\mu\Big\} \end{aligned} \tag{26}$$

where we have used the Einstein summation notation (on the second line, the repeated index ν is summed from 0 to 3). It follows that the $\Lambda^\mu_{\ \nu}$ coefficients of the Lorentz transformation of the four-vector coordinates:

$$x'^\mu = \Lambda^\mu_{\ \nu}\, x^\nu \tag{27}$$

are given by:

$$\Lambda^\mu_{\ \nu} = \frac{1}{2}\mathrm{Tr}\Big\{(L)\sigma_\nu (L)^\dagger \sigma_\mu\Big\} = \frac{1}{2}\mathrm{Tr}\Big\{(L)^\dagger \sigma_\mu (L) \sigma_\nu\Big\} \tag{28}$$

[1]Matrix $(L)^\dagger(L)$ is Hermitian, hence diagonalizable. Its determinant equals 1, meaning none of its eigenvalues equals zero. These eigenvalues are positive since $(L)^\dagger(L)$, taken between a bra and a ket, yields the norm of a ket. We can introduce an invertible Hermitian matrix (A) such that $(A)^2 = (L)^\dagger(L)$ and define a matrix (B) as $(B) = (L)(A)^{-1}$; matrix $(A)^{-1}$ is Hermitian. Since $(B)^\dagger(B) = (A)^{-1}(L)^\dagger(L)(A)^{-1} = (A)^{-1}(A)^2(A)^{-1} = (\mathbb{1})$, the (B) matrix is unitary, and its definition proves that $(L) = (B)(A)$.

For a pure Lorentz transformation, (L) is Hermitian and the Lorentz matrix $(\Lambda^{\mu}_{\ \nu})$ is symmetric.

• Matrices $L^{\dagger}(\Lambda^{-1})$ are another representation of the Lorentz group. The matrix associated with the product $\Lambda_b\Lambda_a$ of two transformations Λ_a and Λ_b is written:

$$\begin{aligned} L^{\dagger}(\Lambda_b^{-1})L^{\dagger}(\Lambda_a^{-1}) &= \left[L(\Lambda_a^{-1})L(\Lambda_b^{-1})\right]^{\dagger} = \left[L^{-1}(\Lambda_a)L^{-1}(\Lambda_b)\right]^{\dagger} \\ &= \left[L^{-1}(\Lambda_b\Lambda_a)\right]^{\dagger} = L^{\dagger}\Big((\Lambda_b\Lambda_a)^{-1}\Big) \end{aligned} \tag{29}$$

which shows that the product of these matrices is indeed the matrix associated with the product in this representation. This representation will be used, for example, in § B-3-b-δ.

1-e. Thomas precession

As an example of the above discussion, let us compute the Thomas precession. Consider the successive application of two pure Lorentz transformations, associated respectively with the matrices (L) and (L'):

$$\begin{aligned} (L) &= \cosh(\frac{q}{2c})\ \mathbb{1} + \sinh(\frac{q}{2c})\ \boldsymbol{u}\cdot\boldsymbol{\sigma} \\ (L') &= \cosh(\frac{q'}{2c})\ \mathbb{1} + \sinh(\frac{q'}{2c})\ \boldsymbol{u}'\cdot\boldsymbol{\sigma} \end{aligned} \tag{30}$$

The SL$(2, C)$ matrix associated with the product of these two transformations is:

$$\begin{aligned} (L)(L') = {} & \cosh(\frac{q}{2c})\cosh(\frac{q'}{2c})\mathbb{1} \\ & + \left[\cosh(\frac{q}{2c})\sinh(\frac{q'}{2c})\boldsymbol{u}' + \cosh(\frac{q'}{2c})\sinh(\frac{q}{2c})\boldsymbol{u}\right]\cdot\boldsymbol{\sigma} \\ & + \sinh(\frac{q}{2c})\sinh(\frac{q'}{2c})(\boldsymbol{u}\cdot\boldsymbol{\sigma})(\boldsymbol{u}'\cdot\boldsymbol{\sigma}) \end{aligned} \tag{31a}$$

that is, taking into account the usual multiplication relations of the Pauli matrices:

$$\begin{aligned} (L)(L') = {} & \left[\cosh(\frac{q}{2c})\cosh(\frac{q'}{2c}) + \sinh(\frac{q}{2c})\sinh(\frac{q'}{2c})\ \boldsymbol{u}\cdot\boldsymbol{u}'\right]\mathbb{1} \\ & + \left[\cosh(\frac{q}{2c})\sinh(\frac{q'}{2c})\boldsymbol{u}' + \cosh(\frac{q'}{2c})\sinh(\frac{q}{2c})\boldsymbol{u}\right]\cdot\boldsymbol{\sigma} \\ & + i\sinh(\frac{q}{2c})\sinh(\frac{q'}{2c})(\boldsymbol{u}\times\boldsymbol{u}')\cdot\boldsymbol{\sigma} \end{aligned} \tag{31b}$$

The (real) term on the second line of this relation is the parameter of the pure Lorentz transformation obtained in the product transformation. The term on the third line is purely imaginary; it shows that, in addition, the product transformation contains a rotation around the direction defined by the vector product $\boldsymbol{u} \times \boldsymbol{u}'$ (unless, of course, $\boldsymbol{u}$ and $\boldsymbol{u}'$ are parallel). At the origin of this rotation, called Thomas precession, is the commutation relation (V-61) between the various components of vector $\boldsymbol{K}$.

Comment: In the particular case where $\boldsymbol{u} = \boldsymbol{u}'$ (Lorentz transformations along parallel directions), relation (31b) simplifies to:

$$(L)(L') = \cosh(\frac{q+q'}{2c})\mathbb{1} + \sinh(\frac{q+q'}{2c})\,\boldsymbol{u}\cdot\boldsymbol{\sigma} \tag{31c}$$

whose parameter $q+q'$ is simply the sum of the parameters of the individual transformations. This property was already mentioned in the last comment of chapter III: when using rapidity as the parameter of pure Lorentz transformations, the parameter of the product of two transformations along the same direction is simply the sum of each rapidity.

2. Little group associated with a four-vector

Consider a four-vector with given coordinates ct, x, y, z. The set of Lorentz transformations that preserve this four-vector form a subgroup, the so-called "little group", of the proper Lorentz group. It is obvious that all the rotations around an axis parallel to the vector $\boldsymbol{r}$ (x, y, z) are part of this little group, which also contains certain combinations of pure Lorentz transformations and rotations, as we now show.

Proper Lorentz transformations depend on 6 real parameters, 3 for the rotations, 3 for the Lorentz boosters defined by a velocity. A priori, the invariance of a four-vector imposes 4 conditions, that are reduced to 3 in our case since Lorentz transformations preserve the four-vector norm: preserving the modulus of $\boldsymbol{r}$ automatically preserves the modulus of time, and also its sign since the transformations are orthochronous (no time reversal). We thus expect the transformations of the little group to depend on 3 real parameters.

2-a. Conditions to be obeyed by an infinitesimal transformation

We construct an infinitesimal transformation of the little group, defined by the (real) parameters $\delta\boldsymbol{v}$ and $\delta\boldsymbol{a}$, by adding the first-order variations of (L) obtained from relations (18) and (22b):

$$(L) = \mathbb{1} + \frac{\delta\boldsymbol{v}}{2c}\cdot\boldsymbol{\sigma} - i\frac{\delta\boldsymbol{a}}{2}\cdot\boldsymbol{\sigma} \tag{32}$$

This yields:

$$(Q') = (L)(Q)(L)^\dagger = \left[\mathbb{1} + \frac{\delta \boldsymbol{v}}{2c} \cdot \boldsymbol{\sigma} - i\frac{\delta \boldsymbol{a}}{2} \cdot \boldsymbol{\sigma}\right]$$
$$\left[(ct)\mathbb{1} + x\sigma_x + y\sigma_y + z\sigma_z\right] \left[\mathbb{1} + \frac{\delta \boldsymbol{v}}{2c} \cdot \boldsymbol{\sigma} + i\frac{\delta \boldsymbol{a}}{2} \cdot \boldsymbol{\sigma}\right] \tag{33}$$

To first-order, the $\delta \boldsymbol{v}$ terms introduce in the equation anticommutators of Pauli matrices, and the $\delta \boldsymbol{a}$ terms, commutators of Pauli matrices. Taking into account the anticommutation and commutation properties of these matrices, and the fact that their square equals the identity matrix, we can write to first-order:

$$(Q') = (Q) + (ct)\frac{\delta \boldsymbol{v}}{c} \cdot \boldsymbol{\sigma} + (ct)\frac{\delta \boldsymbol{v}}{c} \cdot \boldsymbol{r}\ (\mathbb{1}) + (\delta \boldsymbol{a} \times \boldsymbol{r}) \cdot \boldsymbol{\sigma} \tag{34}$$

Writing that the first-order terms must be equal to zero, we get:

$$\begin{cases} \delta \boldsymbol{v} \cdot \boldsymbol{r} = 0 & \text{(35a)} \\ (ct)\dfrac{\delta \boldsymbol{v}}{c} = -\delta \boldsymbol{a} \times \boldsymbol{r} & \text{(35b)} \end{cases}$$

Equality (35a) shows that $\delta \boldsymbol{v}$ is perpendicular to $\boldsymbol{r}$; it was to be expected since a Lorentz boost does not change the four-vector space components that are orthogonal to the boost velocity. Relation (35b) also implies this orthogonality, and thus includes (35a) that can be ignored. Only the δa_x and δa_y components of $\delta \boldsymbol{a}$ perpendicular to $\boldsymbol{r}$ come into play; the third δa_z components remains free, which is understandable since a rotation around $\boldsymbol{r}$ does not change the four-vector. Therefore, as expected, we have three real parameters, the first two a_x and a_y determine the Lorentz boost via (35b), and the third one a_z has no effect on $\boldsymbol{r}$.

2-b. Transformation matrix

Inserting (35b) in the (L) variation written in (32) we get:

$$\delta(L) = -\frac{1}{2ct}(\delta \boldsymbol{a} \times \boldsymbol{r}) \cdot \boldsymbol{\sigma} - \frac{i}{2}\delta \boldsymbol{a} \cdot \boldsymbol{\sigma} \tag{36}$$

Setting:

$$\delta \boldsymbol{a}' = -\frac{\delta \boldsymbol{a}}{2ct} \tag{37}$$

we can write:

$$\delta(L) = \left[\delta \boldsymbol{a}' \times \boldsymbol{r} + i(ct)\delta \boldsymbol{a}'\right] \cdot \boldsymbol{\sigma} + i\delta\varepsilon\ \boldsymbol{r} \cdot \boldsymbol{\sigma} \tag{38}$$

In this expression, we have arbitrarily added a $\delta\varepsilon$ term, which represents a rotation around the vector $\boldsymbol{r}$ but leaves it unchanged. Introducing this parameter will prove useful in the computations discussed in § 3.

If the four-vector invariant under the transformations of the little group is an energy-momentum four-vector $(E, c\boldsymbol{p})$, the variation $\delta(L)$ is written[2]:

$$\begin{aligned}\delta(L) &= \left[\delta\boldsymbol{a}' \times c\boldsymbol{p} + iE\,\delta\boldsymbol{a}'\right] \cdot \boldsymbol{\sigma} + i\delta\varepsilon\; c\boldsymbol{p}\cdot\boldsymbol{\sigma} \\ &= \delta\boldsymbol{a}' \cdot \left[c\boldsymbol{p} \times \boldsymbol{\sigma} + iE\,\boldsymbol{\sigma}\right] + i\delta\varepsilon\; c\boldsymbol{p}\cdot\boldsymbol{\sigma}\end{aligned} \tag{39}$$

As we already saw, the Hermitian parts of this operator are associated with pure Lorentz transformations, the anti-Hermitian parts with pure spatial rotations. The above expression for $\delta(L)$ shows that any transformation that preserves the four-vector $(E, c\boldsymbol{p})$ includes an infinitesimal Lorentz transformation along any direction perpendicular to $\boldsymbol{p}$, a rotation around a direction perpendicular to the previous one (by an infinitesimal angle proportional to the Lorentz transformation parameter, with a factor E/cp), and finally, an arbitrary infinitesimal rotation around the $\boldsymbol{p}$ direction.

2-c. Transformation of an energy-momentum four-vector with a zero spatial component

Another example involving the SL$(2, C)$ group is the study of the way in which the proper Lorentz transformations modify the four-vector of a particle at rest (the spatial component of its momentum is zero). The (Q) matrix introduced in (1) simply reduces to the identity matrix multiplied by E, so that the (Q') matrix is written $(Q') = E(L)(L)^\dagger$, that is:

$$\begin{aligned}(Q') &= E\,(\lambda_0\mathbb{1} + \boldsymbol{\lambda}\cdot\boldsymbol{\sigma})(\lambda_0^\star\mathbb{1} + \boldsymbol{\lambda}^\star\cdot\boldsymbol{\sigma}) \\ &= E\left[(|\lambda_0|^2 + \boldsymbol{\lambda}\cdot\boldsymbol{\lambda}^\star)\mathbb{1} + (\lambda_o\boldsymbol{\lambda}^\star + \lambda_0^\star\boldsymbol{\lambda})\cdot\boldsymbol{\sigma} + i(\boldsymbol{\lambda}\times\boldsymbol{\lambda}^\star)\cdot\boldsymbol{\sigma}\right]\end{aligned} \tag{40}$$

We note λ_0^R and λ_0^I the real and imaginary parts of the constant λ_0, as well as:

$$\boldsymbol{\lambda} = \boldsymbol{b} - i\frac{\boldsymbol{a}}{2} \tag{41}$$

where $\boldsymbol{b}$ and $\boldsymbol{a}$ are real. The three-dimensional momentum $\boldsymbol{p}'$, after the Lorentz transformation L, is simply the vector coefficient of $\boldsymbol{\sigma}$ of this matrix:

$$\boldsymbol{p}' = \frac{E}{c}\,(2\lambda_0^R\,\boldsymbol{b} + \lambda_0^I\,\boldsymbol{a} + \boldsymbol{a}\times\boldsymbol{b}) \tag{42}$$

The parameters λ_0, $\boldsymbol{b}$, and $\boldsymbol{a}$ define a Lorentz transformation on the condition that the determinant of (L) equals 1:

$$(\lambda_0)^2 - \boldsymbol{\lambda}\cdot\boldsymbol{\lambda} = (\lambda_0^R)^2 - (\lambda_0^I)^2 - (\boldsymbol{b})^2 + (\boldsymbol{a})^2 + i(2\lambda_0^R\lambda_0^I - \boldsymbol{b}\cdot\boldsymbol{a}) = 1 \tag{43}$$

[2]With this new four-vector, definition (37) becomes $\delta\boldsymbol{a}' = -\delta\boldsymbol{a}/2E$.

We now try to find which Lorentz transformations yield a given value of $\boldsymbol{p}'$. We are looking for values of the λ_0, $\boldsymbol{b}$, and $\boldsymbol{a}$ parameters that satisfy both equations above, with a given value of $\boldsymbol{p}'$. An obvious solution is to choose a pure Lorentz transformation, with a zero $\boldsymbol{a}$ parameter. The imaginary part of relation (43) shows that either λ_0^R or λ_0^I must be zero. We exclude the $\lambda_0^R = 0$ solution as it would imply, according to (42), that $\boldsymbol{p}'$ equals zero. Relation (43) reduces to $(\lambda_0^R)^2 = 1 + (\boldsymbol{b})^2$, which always has a solution[3] for λ_0^R. Equation (42) simplifies to:

$$\boldsymbol{p}' = \frac{2E}{c}\sqrt{1+(\boldsymbol{b})^2}\,\boldsymbol{b} \tag{44}$$

This relation determines the value of the $\boldsymbol{b}$ parameter that allows reaching any target momentum $\boldsymbol{p}'$, starting from a particle at rest. This is the trivial solution of a pure Lorentz transformation, in a direction parallel to the target value of the momentum.

We now discuss adding an imaginary part to the vector $\boldsymbol{\lambda}$, i.e. a rotation, without changing the value of the target momentum $\boldsymbol{p}'$. We could again use relation (42) and compute the coupled variations of $\boldsymbol{a}$, $\boldsymbol{b}$, and λ_0 that preserve the $\boldsymbol{p}'$ value. Imposing in this way 3 conditions in a group depending on 6 parameters would lead to solutions depending on only 3 real parameters.

To avoid these calculations, we can simply note that any rotation acting on the initial four-vector, having a zero spatial component, preserves that four-vector. This means that any matrice (L) written as:

$$(L) = \left(\lambda_0^R \mathbb{1} + \boldsymbol{b}\cdot\boldsymbol{\sigma}\right)\left(\cos\frac{\theta}{2}\mathbb{1} - i\sin\frac{\theta}{2}\boldsymbol{u}\cdot\boldsymbol{\sigma}\right) \tag{45}$$

will lead to the momentum target value, whatever the values of the unit vector $\boldsymbol{u}$ and angle θ, provided $\boldsymbol{b}$ takes the value determined by (44). Using the standard properties of the Pauli matrices, we get:

$$(L) = \left(\cos\frac{\theta}{2}\lambda_0^R - i\sin\frac{\theta}{2}\,\boldsymbol{u}\cdot\boldsymbol{b}\right)\mathbb{1} + \left(\cos\frac{\theta}{2}\,\boldsymbol{b} - i\lambda_0^R\sin\frac{\theta}{2}\boldsymbol{u}\right)\cdot\boldsymbol{\sigma} - \sin\frac{\theta}{2}(\boldsymbol{u}\times\boldsymbol{b})\cdot\boldsymbol{\sigma} \tag{46}$$

As expected, this transformation depends on three real parameters, two for the rotation axis $\boldsymbol{u}$, and one for the rotation angle θ.

Comments:

(i) We multiply, on the right, the above expression for (L) by $(\lambda_0^R\mathbb{1} - \boldsymbol{b}\cdot\boldsymbol{\sigma})$. This product yields a transformation that, starting from an initial spatial momentum $\boldsymbol{p}'$, brings it to zero, performs a rotation, and finally brings it back to its initial value. We have found by another method the transformations of the little group studied in §§ 2-a and 2-b above.

(ii) The spatial component of a particle's momentum can only be zero if the particle's mass m is different from zero, which was our hypothesis. In that case, one can not only annul but also reverse a spatial momentum. If the particle has a spin, this operation changes the sign of its helicity – *cf.* relation (V-104) of chapter V; this would be impossible for a massless particle.

[3]The solution can be positive or negative, since two opposite (L) matrices are associated with the same Lorentz transformation.

3. W^2 operator

Going back to quantum mechanics, we transpose the above results in terms of operators acting in the state space. We focus on subspaces that are eigenvectors common to the Hamiltonian H and the momentum $\boldsymbol{P}$. Relation (39) is transformed into an operator relation yielding a δT operator acting in the state space. The general form of an infinitesimal (L) transformation is given by relation (32). We identify its coefficients with those of the Hermitian and anti-Hermitian terms of (39). This yields the following values for the $\delta \boldsymbol{v}''$ and $\delta \boldsymbol{a}''$ parameters:

$$\begin{aligned} \delta \boldsymbol{v}'' &= 2c\,(\delta \boldsymbol{a}' \times c\boldsymbol{p}) \\ \delta \boldsymbol{a}'' &= -2\,(E\,\delta \boldsymbol{a}' + \delta\varepsilon\, c\boldsymbol{p}) \end{aligned} \tag{47}$$

Using the general correspondence between infinitesimal operators and generators of the symmetry algebra [relation (IV-55) of chapter IV] yields the following operator acting in the state space:

$$T = \mathbb{1} - \frac{2ic}{\hbar}(\delta \boldsymbol{a}' \times c\boldsymbol{p}) \cdot \boldsymbol{K} + \frac{2i}{\hbar}(E\,\delta \boldsymbol{a}' + \delta\varepsilon\, c\boldsymbol{p}) \cdot \boldsymbol{J} \tag{48}$$

We now replace the $E, c\boldsymbol{p}$ components of the four-vector by operators whose eigenvalues change in each eigensubspace that is common to H and $\boldsymbol{P}$. These subspaces are invariant under the action of this operator, whose infinitesimal variation δT is written:

$$\begin{aligned} \delta T &= \frac{2}{i\hbar}\Big[\delta \boldsymbol{a}' \cdot (c^2 \boldsymbol{P} \times \boldsymbol{K} - H\boldsymbol{J}) - \delta\varepsilon\; c\boldsymbol{P} \cdot \boldsymbol{J}\Big] \\ &= -\frac{2}{i\hbar}\Big[Mc^2\;\delta \boldsymbol{a}' \cdot \boldsymbol{S} + \delta\varepsilon\; c\boldsymbol{P} \cdot \boldsymbol{J}\Big] \end{aligned} \tag{49}$$

This expression includes the spin operator $\boldsymbol{S}$ defined by relation (V-76) of chapter V:

$$\boldsymbol{S} = \frac{H}{Mc^2}\boldsymbol{J} - \frac{1}{M}\boldsymbol{P} \times \boldsymbol{K} \tag{50}$$

By construction, the eigensubspaces common to H and $\boldsymbol{P}$ are invariant under the action of the operator on the right-hand side of (49), whatever the values of parameters $\delta \boldsymbol{a}'$ and $\delta\varepsilon$. This operator commutes with these two elements of the Lie algebra of the Poincaré group. There is no guarantee at the moment that it also commutes with $\boldsymbol{J}$ and $\boldsymbol{K}$. As far as $\boldsymbol{J}$ is concerned, the commutation with the scalar operator $\boldsymbol{P} \cdot \boldsymbol{J}$, product of two vector operators, is ensured. As for the vector part $\boldsymbol{S}$, it does not commute with $\boldsymbol{J}$, but one can square the vector to obtain a scalar operator.

In order to ensure the commutation with $\boldsymbol{K}$, we introduice the following linear combination of two scalar operators:

$$W^2(\lambda) = \boldsymbol{S}^2 - \frac{\lambda}{M^2c^2} \left(\boldsymbol{P}\cdot\boldsymbol{J}\right)^2 \tag{51}$$

Relations (V-82) and (V-83) of chapter V show that the commutation with $\boldsymbol{K}$ is obtained for $\lambda = 1$, so that the operator that commutes with all the generators of the Poincaré group is:

$$W^2 = \boldsymbol{S}^2 - \frac{\left(\boldsymbol{P}\cdot\boldsymbol{J}\right)^2}{M^2c^2} \tag{52}$$

This expression, already given in relation (V-86) of chapter V, was derived by a new method relating expression (50) for the spin operator to the stability of the eigensubspaces common to the components of the energy-momentum operator.

Comment:

We show in § 2-b of complement $\mathrm{B_V}$ that the set of the four operators $(\boldsymbol{P}\cdot\boldsymbol{J}/Mc)$, S_x, S_y, and S_z, which are the generators of the little group in the space of states, transforms as the components of a relativistic four-vector. Since the four-vector norm is a relativistic invariant, it is normal that operator W^2 has this property, which explains why $\lambda = 1$.

Complement B_V

Commutation relations of the spin components, Pauli–Lubanski four-vector

In this complement, we first calculate the commutation relations of the components of the spin operator $\boldsymbol{S}$, which have not been written in chapter V. We then introduce the Pauli–Lubanski operator, which has 4 components transforming as a relativistic four-vector; this helps clarifying the origin of the Casimir operator W^2 of the Poincaré group. We then generalize the calculation of § C-3-e, without assuming the existence of an eigensubspace of the momentum with zero eigenvalue.

1. Operator S

In the framework of the Poincaré group, we now examine the commutation relations between the components of $\boldsymbol{S}$. Equality (V-44), which attributes to $\boldsymbol{S}$ all the characteristics of an angular momentum in the framework of the Galilean group, becomes in the Poincaré group:

$$\begin{aligned}[S_i, S_j] &= \left[\frac{H}{Mc^2}J_i - \frac{\varepsilon_{inp}}{M}P_n K_p \;,\; \frac{H}{Mc^2}J_j - \frac{\varepsilon_{j\ell m}}{M}P_\ell K_m\right] \\ &= \alpha + \beta + \gamma + \delta \end{aligned} \tag{1a}$$

(we use again the common convention according to which any repeated index is a summed dummy index) with:

$$\begin{aligned} \alpha &= \left(\frac{H}{Mc^2}\right)^2 i\hbar\, \varepsilon_{ijk}\, J_k \\ \beta &= -\frac{1}{M}\, \varepsilon_{inp} \left[P_n K_p \;,\; \frac{H}{Mc^2} J_j\right] \\ \delta &= \frac{1}{M^2}\, \varepsilon_{inp}\, \varepsilon_{j\ell m} \left[P_n K_p \;,\; P_\ell K_m\right] \end{aligned} \tag{1b}$$

(γ is obtained by permuting i and j in β, and changing the sign). Relation:

$$[AB, CD] = A[B,C]D + AC[B,D] + [A,C]DB + C[A,D]B \tag{2}$$

leads to:

$$-M\beta = i\hbar\, \varepsilon_{inp} \left\{ P_n \left(-\frac{1}{Mc^2} P_p \right) J_j + P_n \frac{H}{Mc^2}\, \varepsilon_{pjq}\, K_q + \frac{H}{Mc^2}\, \varepsilon_{njq} P_q K_p \right\} \tag{3}$$

On the right-hand side, the $\varepsilon_{inp}\, P_n\, P_p\, J_j$ term, odd when two n and p indices are exchanged, disappears by summation over these indices. To compute the other terms, we use equalities:

$$\sum_p \varepsilon_{inp}\, \varepsilon_{jqp} = \delta_{ij}\, \delta_{nq} - \delta_{iq}\, \delta_{nj}$$

$$\sum_n \varepsilon_{inp}\, \varepsilon_{qnj} = \delta_{iq}\, \delta_{pj} - \delta_{ij}\, \delta_{pq} \tag{4}$$

and we get:

$$-M\beta = \frac{i\hbar}{Mc^2} \left\{ \delta_{ij}\, H\, (\boldsymbol{P} \cdot \boldsymbol{K}) - P_j\, H\, K_i + H\, P_i K_j - \delta_{ij}\, H\, (\boldsymbol{P} \cdot \boldsymbol{K}) \right\} \tag{5a}$$

or:

$$\beta = i\hbar \frac{H}{M^2c^2} \left\{ P_j K_i - P_i K_j \right\} = -i\hbar \frac{H}{Mc^2} \varepsilon_{ijk} L_k \tag{5b}$$

Following the rule stated above for computing γ yields $\beta = \gamma$, and we simply have to calculate δ. Applying several times the identities (4), we get:

$$\begin{aligned} M^2\delta &= \varepsilon_{inp}\, \varepsilon_{j\ell m} \left(-\frac{i\hbar}{c^2}\right) \left\{ \delta_{p\ell}\, P_n\, H K_m + \varepsilon_{pmk}\, P_n P_\ell J_k - \delta_{nm} P_\ell H K_p \right\} \\ &= \left(-\frac{i\hbar}{c^2}\right) \left\{ P_j\, H\, K_i - P_i\, H\, K_j - \delta_{ij} H (\boldsymbol{P} \cdot \boldsymbol{K}) + \delta_{ij} H (\boldsymbol{P} \cdot \boldsymbol{K}) \right. \\ &\qquad \left. + \varepsilon_{ij\ell}\, P_\ell\, (\boldsymbol{P} \cdot \boldsymbol{J}) - \varepsilon_{j\ell n} P_n P_\ell P_j \right\} \end{aligned} \tag{6a}$$

The last term on the right-hand side, odd under an ℓ and n exchange, disappears in the summation. We then get:

$$\begin{aligned} \delta &= \frac{i\hbar}{M^2c^2} \left\{ H(P_i K_j - P_j K_i) - \varepsilon_{ij\ell}\, P_\ell\, (\boldsymbol{P} \cdot \boldsymbol{J}) \right\} \\ &= \frac{i\hbar}{Mc^2}\, \varepsilon_{ijk} \left\{ H L_k - \frac{1}{M} P_k\, (\boldsymbol{P} \cdot \boldsymbol{J}) \right\} \end{aligned} \tag{6b}$$

Inserting in (1a) the values of α, β, γ, and δ, and since relation (V-89) shows that $\boldsymbol{P} \cdot \boldsymbol{J} = \boldsymbol{J} \cdot \boldsymbol{P}$ commutes with $\boldsymbol{P}$, we get:

$$\begin{aligned} [S_i, S_j] &= i\hbar \left\{ \frac{H^2}{M^2c^4}\, \varepsilon_{ijk}\, J_k - \frac{H}{Mc^2}\, \varepsilon_{ijk}\, L_k - \frac{\boldsymbol{J} \cdot \boldsymbol{P}}{M^2c^2}\, \varepsilon_{ijk}\, P_k \right\} \\ &= i\hbar\, \varepsilon_{ijk} \left\{ \frac{H}{Mc^2}\, S_k - \frac{\boldsymbol{J} \cdot \boldsymbol{P}}{M^2c^2}\, P_k \right\} \end{aligned} \tag{7}$$

These relations show that, in an eigensubspace common to H and $\boldsymbol{P}$ (the spin commutes with these observables) and where the momentum's eigenvalue is zero, operator $\boldsymbol{S}$ obeys the usual commutation relations of angular momentum. The eigenvalues of $\boldsymbol{S}^2$ in this subspace are of the form $S(S+1)\hbar^2$.

2. Pauli–Lubanski pseudovector

In chapter V we wrote in (V-85) 4 operators that are the components of a four-vector operator, whose squared norm W^2 is a Casimir operator of the Poincaré group. This set of operators is called the Pauli–Lubanski vector. It is actually a pseudo-vector (or axial vector), hence invariant under parity (*cf.* complement D_V, § 2), as opposed to a polar vector. We are going to further discuss this vector and in particular its link with the invariant subspaces of momentum-energy.

2-a. Antisymmetric tensor J-K

The 3 components of $\boldsymbol{J}$ and the 3 components of $\boldsymbol{K}$ form a basis of the Lie algebra of the proper (connected) Lorentz group. It is useful to regroup these 6 components into the components of a completely antisymmetric second-order tensor N, written in a basis whose α and β indices take the 4 values $0, 1, 2, 3$:

$$\overline{\overline{N}} = (N^{\alpha\beta}) = \begin{pmatrix} 0 & cK_x & cK_y & cK_z \\ -cK_x & 0 & J_z & -J_y \\ -cK_y & -J_z & 0 & J_x \\ -cK_z & J_y & -J_x & 0 \end{pmatrix} \tag{8}$$

It will be called the J-K tensor; its form is similar to that of the second-order tensor that, in classical electromagnetism, contains the fields $\boldsymbol{E}$ (instead of operators K) and $\boldsymbol{B}$ (instead of operators J).

Similarly, we introduce a tensor $\delta\overline{\overline{\Omega}}$ containing $\delta\boldsymbol{a}$ and the velocity $\delta\boldsymbol{v}$ defining an infinitesimal Lorentz transformation:

$$\delta\overline{\overline{\Omega}} = (\delta\Omega_{\alpha\beta}) = \frac{1}{2}\begin{pmatrix} 0 & \delta v_x/c & \delta v_y/c & \delta v_z/c \\ -\delta v_x/c & 0 & \delta a_z & -\delta a_y \\ -\delta v_y/c & -\delta a_z & 0 & \delta a_x \\ -\delta v_z/c & \delta a_y & -\delta a_x & 0 \end{pmatrix} \tag{9}$$

The associated transformation $T(\delta\boldsymbol{a}, \delta\boldsymbol{v})$ in the state space is written:

$$T(\delta\boldsymbol{a}, \delta\boldsymbol{v}) = 1 - \frac{i}{\hbar}[\delta\boldsymbol{a}\cdot\boldsymbol{J} + \delta\boldsymbol{v}\cdot\boldsymbol{K}] = 1 - \frac{i}{\hbar}\,\delta\Omega_{\alpha\beta}\,N^{\alpha\beta} = 1 - \frac{i}{\hbar}\,\delta\overline{\overline{\Omega}}\cdot\overline{\overline{N}} \tag{10}$$

We now show that the J-K tensor is transformed as a relativistic tensor[1]. The condition such a tensor must obey is that, in a Lorentz transformation Λ, its components

[1]See complement A_{VIII} for a short review of tensors.

$N^{\alpha\beta}$ vary according to:

$$(N')^{\alpha\beta} = T(\boldsymbol{a},\boldsymbol{v})\, N^{\alpha\beta}\, T^\dagger(\boldsymbol{a},\boldsymbol{v}) = (\Lambda^{-1})^\alpha_{\ \gamma}\, (\Lambda^{-1})^\beta_{\ \mu}\, N^{\gamma\mu} \tag{11}$$

which is the equivalent of relation (15) of complement A_{VIII} for $n=2$ and for Lorentz transformations not restricted to rotations; see also relations (VIII-17) and (VIII-18). For an infinitesimal transformation with parameters $\delta\boldsymbol{a}$ and $\delta\boldsymbol{v}$, the variation of the left-hand side of this equality is given by the commutator:

$$\delta(N^{\alpha\beta}) = -\frac{i}{\hbar}\left[\delta\boldsymbol{a}\cdot\boldsymbol{J} + \delta\boldsymbol{v}\cdot\boldsymbol{K}\,,\, N^{\alpha\beta}\right] \tag{12}$$

whereas the right-hand side is written:

$$\left(\mathbb{1} - \delta\boldsymbol{a}\cdot\boldsymbol{X}_J - \delta\boldsymbol{v}\cdot\boldsymbol{X}_K\right)^\alpha_{\ \gamma}\left(\mathbb{1} - \delta\boldsymbol{a}\cdot\boldsymbol{X}_J - \delta\boldsymbol{v}\cdot\boldsymbol{X}_K\right)^\beta_{\ \mu} N^{\gamma\mu} \tag{13}$$

where the three components of $\boldsymbol{X}_J$ are the matrices $X_{J_{x,y,z,}}$ defined in (III-100a), and the three components of $\boldsymbol{X}_K$, the matrices $X_{K_{x,y,z,}}$ defined in (III-113), with a permutation of the x, y, and z indices.

We first write the equality of the $\delta\boldsymbol{v}$ terms, or more precisely of the δv_x terms. Taking into account the commutation relations (V-61), we get for the left-hand side:

$$\delta\overline{\overline{N}} = \delta v_x \begin{pmatrix} 0 & 0 & -J_z/c & J_y/c \\ 0 & 0 & -K_y & -K_z \\ J_z/c & K_y & 0 & 0 \\ -J_y/c & K_z & 0 & 0 \end{pmatrix} \tag{14}$$

For the right-hand side, relation (13) shows that, to first-order in δv_x:

$$\delta N^{\alpha\beta} = -\frac{\delta v_x}{c}\left[(X_{K_x})^\beta_{\ \mu}\, N^{\alpha\mu} + (X_{K_x})^\alpha_{\ \gamma}\, N^{\gamma\beta}\right] \tag{15}$$

However, expression (III-113) of matrix (X_{K_x}) shows that only two elements are different from zero and equal to $1/c$: the one in the time row and x column, i.e. $(X_{K_x})^0_{\ 1}$, and the one in the x row and time column, i.e. $(X_{K_x})^1_{\ 0}$. As a result:

$$\delta N^{\alpha\beta} = -\frac{\delta v_x}{c}\left[\delta_{\beta 0} N^{\alpha 1} + \delta_{\beta 1} N^{\alpha 0} + \delta_{\alpha 0} N^{1\beta} + \delta_{\alpha 1} N^{0\beta}\right] \tag{16}$$

The first term in the right-hand bracket shifts the second column of the matrix, with index 1, toward the first column, with index 0. The second terms produces the opposite displacement, from the first column toward the second. The third term shifts the second row of the matrix, with index 1, toward the first row. Finally, the fourth term moves the first row toward the second. The diagonal terms cancel each other, and we get:

$$\delta\overline{\overline{N}} = -\frac{\delta v_x}{c} \begin{pmatrix} 0 & 0 & J_z & -J_y \\ 0 & 0 & cK_y & cK_z \\ -J_z & -cK_y & 0 & 0 \\ J_y & -cK_z & 0 & 0 \end{pmatrix} \tag{17}$$

This expression is identical to (14), which shows that relation (11) is verified for infinitesimal Lorentz transformations along the Ox axis; by integration, the equality is extended to finite transformations. As the three axes Ox, Oy and Oz play an equivalent role, the equality is also valid for any pure Lorentz transformation.

The same type of reasoning can be applied for the rotation terms in δa. However, this computation may be avoided by noticing that it only involves spatial rotations that do not change time. Now the terms in the 1, 2, and 3 rows and columns of the matrix (8) are given by relation:

$$(N^{ij}) = \varepsilon_{ijk} J_k \tag{18}$$

This shows that the second-order tensor (in three dimensions) is the result of the contraction of a completely antisymmetric third-order tensor ε_{ijk} (*cf.* § 3-c and § 6 in complement A_{VIII}) with a vector, which confirms its tensorial nature.

We have thus established the tensorial character of expression (8) for all the transformations of the Lorentz group.

Comment:

Relation (10) can be generalized to any infinitesimal transformation of the Poincaré group, and is written:

$$\begin{aligned} T(\delta\boldsymbol{a}, \delta\boldsymbol{b}, \delta\boldsymbol{v}, \tau) &= 1 - \frac{i}{\hbar}\left[\delta\overline{\overline{\Omega}} \cdot \overline{\overline{N}} + \delta\boldsymbol{b} \cdot \boldsymbol{P} + \delta\tau\, H\right] \\ &= 1 - \frac{i}{\hbar}\left[\delta\Omega_{\alpha\beta}\, N^{\alpha\beta} + \delta b_\alpha\, P^\alpha\right] \end{aligned} \tag{19}$$

where the δb_α are the components of the four-vector $(\tau, \boldsymbol{b}/c)$ and the P^α those of the four-vector operator $(H, c\boldsymbol{P})$.

2-b. Construction of the Pauli–Lubanski four-vector

The Pauli–Lubanski four-vector, with components W_μ^{PL}, is defined as:

$$W_\mu^{\mathrm{PL}} = \frac{1}{2}\varepsilon_{\mu\nu\rho\sigma}\, N^{\nu\rho} P^\sigma \tag{20}$$

where the P^σ are the components of the energy-momentum four-vector $(H, c\boldsymbol{P})$ and $\varepsilon_{\mu\nu\rho\sigma}$ the totally antisymmetric 4-dimensional Levi-Civita tensor (equal to $+1$ when μ, ν, ρ, σ are an even permutation of the $0, 1, 2, 3$ indices, equal to -1 when they are an odd permutation, and equal to 0 if at least two of the 4 indices are the same). The 4 components of W^{PL} are the result of the contraction of a fourth-order tensor with a second-order tensor and a first-order tensor. They are thus the components of a four-vector. Explicitly, the components of W^{PL} are:

$$\begin{aligned} W_0^{\mathrm{PL}} = (W^{\mathrm{PL}})^0 &= \frac{1}{2}\,\varepsilon_{ijk} N^{ij} P^k \\ &= \frac{c}{2}\left[(\varepsilon_{123} - \varepsilon_{213}) J_z P_z - (\varepsilon_{132} - \varepsilon_{312}) J_y P_y + (\varepsilon_{231} - \varepsilon_{321}) J_x P_x\right] \\ &= c\,\boldsymbol{J} \cdot \boldsymbol{P} \end{aligned} \tag{21}$$

and:

$$W_x^{\mathrm{PL}} = -(W^{\mathrm{PL}})^x = \frac{1}{2}\left[\varepsilon_{1ij0}N^{ij}H + c\varepsilon_{1i0j}N^{i0}P^j + c\varepsilon_{10ij}N^{0i}P^j\right] \tag{22}$$

This yields (remember that ε_{ijkl} changes sign under a circular permutation of its indices):

$$\boldsymbol{W}^{\mathrm{PL}} = H\boldsymbol{J} + c^2\boldsymbol{K} \times \boldsymbol{P} = Mc^2\,\boldsymbol{S} \tag{23}$$

Since $\boldsymbol{J}$ is an axial vector (pseudovector) and the cross product of two polar vectors yields an axial vector, $\boldsymbol{W}$ is also a pseudovector.

The Minkowski scalar product of the four-vector W^{PL} and the four-vector $(H, c\boldsymbol{P})$ is zero since:

$$H\,W_0^{\mathrm{PL}} - c\boldsymbol{P}\cdot\boldsymbol{W}^{\mathrm{PL}} = cH(\boldsymbol{J}\cdot\boldsymbol{P}) - cH(\boldsymbol{P}\cdot\boldsymbol{J}) = 0 \tag{24}$$

In addition, the three spatial components of W^{PL}, as well as its time component, commute with H and $\boldsymbol{P}$ – *cf.* relations (V-78), (V-80) and (V-84) of chapter V, or § 3 and relation (49) of complement A$_V$. Therefore any eigensubspace common to H and $\boldsymbol{P}$ remains invariant under the action of the 4 components of the Pauli–Lubanski four-vector. They generate the corresponding "little group".

However, they do not commute with the generators $\boldsymbol{J}$ and $\boldsymbol{K}$ of the Lorentz group. To look for an operator that commutes with $\boldsymbol{J}$ and $\boldsymbol{K}$, we take the square $(W^{\mathrm{PL}})^2$ of the relativistic norm of this four-vector, which is a Lorentz invariant. The relation between $(W^{\mathrm{PL}})^2$ and the square of the Casimir operator W^2 of the Poincaré group defined in (V-86) of chapter V is simply written:

$$(W^{\mathrm{PL}})^2 = Mc^2 W = \boldsymbol{S}^2 - \frac{(\boldsymbol{J}\cdot\boldsymbol{P})^2}{M^2c^2} \tag{25}$$

The Pauli–Lubanski construction yields this invariant in a more systematic way than the method used in § C-3-d of chapter V.

3. Energy-momentum eigensubspace with any eigenvalues.

For the sake of simplicity, in § C-3-e of chapter V we used a Galilean reference frame where the momentum was zero. This made it easy to prove relation (V-92) and find the eigenvalues of $\boldsymbol{S}^2$, which we made equal to those of W^2. It is interesting to discuss what becomes of this reasoning in the more general case of an eigenspace common to H and any $\boldsymbol{P}$, when the eigenvalue of $\boldsymbol{P}$ is not necessarily equal to zero.

Operator $\boldsymbol{S}$ commutes with H and $\boldsymbol{P}$. Furthermore, it commutes with $\boldsymbol{J}$, but not with $\boldsymbol{K}$ [*cf.* (V-82)]. Equalities (7) show that the commutation relations between the components of $\boldsymbol{S}$ are not the "classical" (Galilean) relations since we can no longer replace H by Mc^2 and $\boldsymbol{P}$ by $\mathbf{0}$. However, relation (V-81) shows that it is only the $\boldsymbol{S}$ and $\boldsymbol{K}$ components on the same axes that do not commute. In other words, a Lorentz transformation with velocity $\delta\boldsymbol{v}$ does not change the $\boldsymbol{S}$ components perpendicular to $\delta\boldsymbol{v}$.

Consequently, as long as $\delta\boldsymbol{v}$ remains parallel to Oz, operators:

$$S_{\pm} = S_x \pm iS_y \tag{26}$$

are invariant. This suggests how to deal with an eigenspace of $\boldsymbol{P}$ with a non-zero eigenvalue $\boldsymbol{p}$: we must choose the $Oxyz$ axes such that Oz is parallel to $\boldsymbol{p}$. Once the change of axes has been performed, we call $\boldsymbol{S}^p$ the operator that describes the action of $\boldsymbol{S}$ inside the eigensubspace common to H and $\boldsymbol{P}$, with eigenvalues E and $\boldsymbol{p} = p\,\boldsymbol{e}_z$ (where $\boldsymbol{e}_z$ is the unit vector parallel to Oz). Relation (7) yields:

$$\left[S_x^p\,, S_y^p\right] = i\hbar\left\{\frac{E}{Mc^2}\,S_z^p - \frac{p^2}{M^2c^2}J_z\right\} \tag{27}$$

In addition, since the $\boldsymbol{K}$ and $\boldsymbol{P}$ components on orthogonal axes commute with each others, we have:

$$\boldsymbol{L} = \frac{1}{M}(\boldsymbol{P}\times\boldsymbol{K}) = -\frac{1}{M}(\boldsymbol{K}\times\boldsymbol{P}) \tag{28}$$

In an eigensubspace where the P_x and P_y momentum components are zero, the action of operator L_z^p is thus a simple multiplication by zero. Relation (V-76) that defines $\boldsymbol{S}$ shows that the action of J_z in the subspace $(E, \boldsymbol{p} = p\boldsymbol{e}_z)$ reduces to that of S_z^p, followed by a multiplication by the constant Mc^2/E. The right-hand side of (27) is thus written:

$$i\hbar\left\{\frac{E}{Mc^2}\,S_z^p - \frac{p^2}{ME}\,S_z^p\right\} = i\hbar\frac{E^2 - p^2c^2}{Mc^2E}\,S_z^p = i\hbar\frac{Mc^2}{E}\,S_z^p \tag{29}$$

We finally obtain, using both (27) and relations directly obtained from (7), and the fact that $p_x = p_y = 0$:

$$\begin{cases} \left[S_x^p\,, S_y^p\right] = i\hbar\,\dfrac{Mc^2}{E}\,S_z^p \\ \left[S_y^p\,, S_z^p\right] = i\hbar\,\dfrac{E}{Mc^2}\,S_x^p \\ \left[S_z^p\,, S_x^p\right] = i\hbar\,\dfrac{E}{Mc^2}\,S_y^p \end{cases} \tag{30}$$

We now introduce operator $\boldsymbol{T}^p$ as:

$$\begin{cases} T_x^p = S_x^p \\ T_y^p = S_y^p \\ T_z^p = \dfrac{Mc^2}{H}\,S_z^p \end{cases} \tag{31}$$

This leads to simpler relations:

$$\begin{cases} \left[T_x^p, T_y^p\right] = i\hbar\, T_z^p \\ \left[T_y^p, T_z^p\right] = i\hbar\, T_x^p \\ \left[T_z^p, T_x^p\right] = i\hbar\, T_y^p \end{cases} \tag{32}$$

which are exactly the commutation relations of an angular momentum. The same reasoning as above shows that operator $(\boldsymbol{T}^p)^2$ has eigenvalues of the form:

$$S(S+1)\hbar^2$$

where S is an integer or half-integer number.

Finally, equalities (31) yield:

$$(\boldsymbol{T}^p)^2 = (\boldsymbol{S}^p)^2 - (S_z^p)^2 \left[1 - \left(\frac{Mc^2}{H}\right)^2\right] = (\boldsymbol{S}^p)^2 - (S_z^p)^2 \frac{p^2c^2}{E^2} \tag{33}$$

We thus have:

$$(\boldsymbol{S}^p)^2 - \frac{c^2}{E^2}\left(\boldsymbol{S}^p \cdot \boldsymbol{p}\right)^2 = S(S+1)\hbar^2 \tag{34}$$

The square of the spin is the sum of its squared components in any orthonormal reference frame, so that $(\boldsymbol{S}^p)^2 = (\boldsymbol{S})^2$.

Furthermore, since we saw that the actions of both operators J_z and $(Mc^2/E)S_z$ are equivalent, we have:

$$(\boldsymbol{S})^2 - \frac{1}{M^2c^2}\left(\boldsymbol{J} \cdot \boldsymbol{P}\right)^2 = W^2 = S(S+1)\hbar^2 \tag{35}$$

We find again that operator W^2, defined in (V-86), has eigenvalues of the form $S(S+1)\hbar^2$, but this time without assuming that the particle could be studied in an eigensubspace of $\boldsymbol{P}$ where it is at rest.

Complement C_V

Group of geometric displacements

Commutation relations (V-24) to (V-26) of the Galilean group are identical to those, (V-60) to (V-62), of the Poincaré group: the subgroup of geometric displacements is common to the Galilean and Poincaré groups. The differences between these two groups only concern transformations where time comes into play, which is not the case for displacements. This complement is devoted to a more detailed study of operators associated with the set of space displacements, from a more geometrical point of view than the one of chapter V. We shall also demonstrate the additivity property of linear and angular momenta of two physical subsystems.

The Euclidian *isometry* group $\mathscr{G}$ is the set of transformations having the following properties. In the three-dimensional Euclidean point space, to any point N corresponds a point N' (image of N) so that:

- the vectors lengths are preserved; if N_1' and N_2' are the images of N_1 and N_2, we always have:

$$|\overrightarrow{N_1' N_2'}| = |\overrightarrow{N_1 N_2}| \tag{1}$$

- the scalar product and hence the (non-oriented) angles are preserved; if N_3' and N_4' are the images of N_3 and N_4:

$$\overrightarrow{N_1' N_2'} \cdot \overrightarrow{N_3' N_4'} = \overrightarrow{N_1 N_2} \cdot \overrightarrow{N_3 N_4} \tag{2}$$

This set of transformations includes rotations, translations, and space reflections (symmetry with respect to any given point), as well as products of these operations[1].

We can build several subgroups of $\mathscr{G}$. For example:

α . Group $T_{(3)}$ of 3-dimensional translations;

β . Group $R_{(3)}$ of 3-dimensional rotations around any given point;

γ . Symmetry with respect to any given point (parity) and identity operation.

The Euclidean group $\mathscr{E}_{(3)}$ of displacements is a combination of α and β. One can also combine β and γ to form the group $O_{(3)}$. Other subgroups are the crystallographic "point" groups, which will not be discussed here. A brief study of the parity operator is given in complement D_V.

1. Brief review: classical properties of displacements

1-a. Translations

To a point N, represented in a given reference frame by a position vector $\boldsymbol{r}$, we associate a point N' with a position vector $\boldsymbol{r}' = \boldsymbol{r} + \boldsymbol{b}$. The translations $\mathcal{T}$ thus depend on the 3 components of vector $\boldsymbol{b}$:

$$\mathcal{T}(\boldsymbol{b})\,\boldsymbol{r} = \boldsymbol{r} + \boldsymbol{b} \tag{3}$$

We shall write the translations either as $\mathcal{T}(\boldsymbol{b})$, or as $\mathcal{T}_{\boldsymbol{u}}(b)$, where $\boldsymbol{u}$ is the unit vector parallel to $\boldsymbol{b}$. The group law is simply:

$$\mathcal{T}(\boldsymbol{b}')\,\mathcal{T}(\boldsymbol{b}) = \mathcal{T}(\boldsymbol{b})\,\mathcal{T}(\boldsymbol{b}') = \mathcal{T}(\boldsymbol{b}' + \boldsymbol{b}) \tag{4}$$

This equality shows that the translation group is abelian: the product translation corresponds to a vector that is the sum of two vectors, and vector addition is commutative. This group is not compact (the domain accessible to $\boldsymbol{b}$ has an infinite volume).

We already wrote in § C-1 of chapter III matrices describing the effects of a translation on the coordinates. We recall these results, slightly simplified since time no longer comes into play, with 4×4 instead of 5×5 matrices. The x', y',

[1] When the transformation under study leaves unchanged a point O, it is easy to show that the coordinates undergo a linear transformation, to which we associate a 3×3 matrix. If the determinant of this matrix equals $+1$, the transformation is a rotation; if it equals -1, it is the product of a rotation and a symmetry with respect to point O. For an arbitrary transformation, where O' is the image of O, a translation by a vector $\overrightarrow{O'O}$ takes us back to the previous case.

and z' coordinates of $\boldsymbol{r}'$ are given as a function of the x, y, and z coordinates of $\boldsymbol{r}$:

$$\begin{pmatrix} x' \\ y' \\ z' \\ 1 \end{pmatrix} = M(\boldsymbol{b}) \begin{pmatrix} x \\ y \\ z \\ 1 \end{pmatrix} = \begin{pmatrix} 1 & 0 & 0 & b_x \\ 0 & 1 & 0 & b_y \\ 0 & 0 & 1 & b_z \\ 0 & 0 & 0 & 1 \end{pmatrix} \begin{pmatrix} x \\ y \\ z \\ 1 \end{pmatrix} \tag{5}$$

This expression shows that the infinitesimal operation $M(\delta\boldsymbol{b})$ is written:

$$\begin{aligned} M(\delta\boldsymbol{b}) &= (1) + \delta b_x\,\Pi_x + \delta b_y\,\Pi_y + \delta b_z\,\Pi_z \\ &= (1) + \delta\boldsymbol{b} \cdot \boldsymbol{\Pi} \end{aligned} \tag{6}$$

where the three components of vector $\boldsymbol{\Pi}$ are the matrices:

$$\Pi_x = \begin{pmatrix} 0 & 0 & 0 & 1 \\ 0 & 0 & 0 & 0 \\ 0 & 0 & 0 & 0 \\ 0 & 0 & 0 & 0 \end{pmatrix} \qquad \Pi_y = \begin{pmatrix} 0 & 0 & 0 & 0 \\ 0 & 0 & 0 & 1 \\ 0 & 0 & 0 & 0 \\ 0 & 0 & 0 & 0 \end{pmatrix} \qquad \Pi_z = \begin{pmatrix} 0 & 0 & 0 & 0 \\ 0 & 0 & 0 & 0 \\ 0 & 0 & 0 & 1 \\ 0 & 0 & 0 & 0 \end{pmatrix} \tag{7}$$

It is easy to show that:

$$\Pi_i\,\Pi_j = 0 \qquad \forall\; i, j \tag{8}$$

Consequently, all the Π matrices commute with each others:

$$[\Pi_i\,, \Pi_j] = 0 \qquad \forall\; i, j \tag{9a}$$

or else:

$$[\boldsymbol{\Pi}\,, \boldsymbol{\Pi}] = 0 \tag{9b}$$

As in § C of chapter III, we see that the structure constants of the translation group are all equal to zero (abelian group).

In addition, using relation (8), we see that:

$$\mathrm{e}^{(\boldsymbol{b}\cdot\boldsymbol{\Pi})} = 1 + (\boldsymbol{b}\cdot\boldsymbol{\Pi}) + \frac{1}{2}(\boldsymbol{b}\cdot\boldsymbol{\Pi})^2 + \ldots = 1 + \boldsymbol{b}\cdot\boldsymbol{\Pi} \tag{10a}$$

which is equal to the matrix written in (5). We thus have:

$$M(\boldsymbol{b}) = \mathrm{e}^{\boldsymbol{b}\cdot\boldsymbol{\Pi}} \tag{10b}$$

This relation shows that any translation matrix can be expressed as a function of the matrix components of $\boldsymbol{\Pi}$.

1-b. Rotations, topology of the group

A rotation can be defined as a one-to-one correspondence between the real 3-dimensional space and itself; it is a transformation $\mathscr{R}$ that preserves a point in space (chosen as the origin O), the distances (hence the angles), and the orientation of the reference frame (which eliminates symmetries with respect to O or to a plane including O).

α. *Group parameters*

Rotations $\mathscr{R}$ depend on 3 parameters (the dimension of the group is $n = 3$). We already introduced in chapter III a vector parameter to characterize the rotation group:

$$\boldsymbol{a} = \varphi \boldsymbol{u} \tag{11}$$

where $\boldsymbol{u}$ is the unit vector of the rotation axis around which the rotation angle φ takes a positive value.

Another way to associate 3 parameters with a rotation is to use its three "Euler angles" α, β, and γ, represented in figure 1. We call $Oxyz$ the initial frame. Under a rotation $\mathscr{R}$, the points on the initial axes $Oxyz$ now coincide with those of a new frame $OXYZ$; α and β are, by definition, the azimuthal and polar angles of OZ with respect to the initial $Oxyz$ frame.

The three rotation angles α, β, and γ are defined as follows. We call Ou the intersection of the plane containing Oz and OZ with the xOy plane, and Ov the axis that completes the orthogonal $Ouvz$ frame (Ov is thus the intersection of the plane orthogonal to OZ with the xOy plane). To bring $Oxyz$ onto $OXYZ$, we must successively:

- perform a rotation around Oz by an angle α (this rotation brings Ox onto Ou and Oy onto Ov);

- perform a rotation around Ov by an angle β (this brings Oz onto its final position OZ, while Oy remains onto Ov);

- finally perform a rotation around OZ by an angle γ, which brings Ov onto OY. Two of the three axes of the rotated orthogonal frame coincide with OY and OZ, so that the third one is also in its final position.

The 3 angles α, β, γ are signed angles, positive or negative (on figure 1, α is negative). We thus have:

$$\mathscr{R}(\alpha, \beta, \gamma) = \mathscr{R}_Z(\gamma)\, \mathscr{R}_v(\beta)\, \mathscr{R}_z(\alpha) \tag{12}$$

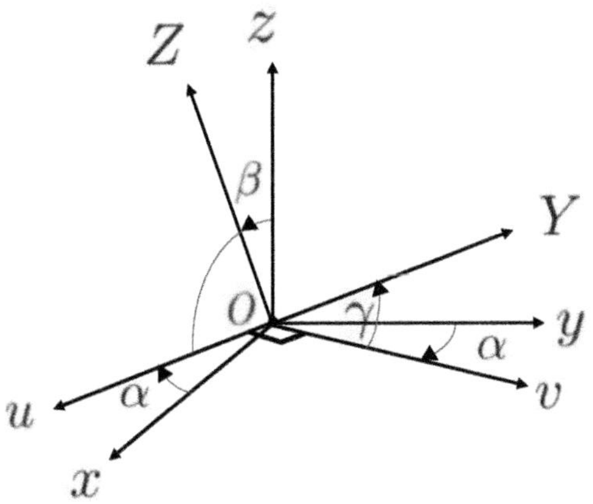

Figure 1: Euler angles in a rotation. A first rotation around the Oz axis by an angle α brings the Ox axis onto Ou in the zOZ plane, as well as the Oy axis onto the Ov axis perpendicular to OZ. A second rotation around Ov by an angle β brings the Oz axis onto its final position OZ. Finally, a rotation around OZ by an angle γ brings the axes perpendicular to OZ in their final positions OX (not shown on the figure) and OY.

This expression involves rotations around OZ and Ov, axes which depend on the rotation under study. This rotation can be changed so as to involve only rotations around coordinate axes, since, in (12), we can replace $\mathscr{R}_v(\beta)$ by:

$$\mathscr{R}_v(\beta) = \mathscr{R}_z(\alpha)\,\mathscr{R}_y(\beta)\,\mathscr{R}_z(-\alpha) \tag{13}$$

To understand this last equality, consider an object (a set of points) on which we apply the succession of the 3 rotations written on the right-hand side. We first bring the object axis, assumed to coincide with Ov, onto Oy, perform around Oy a rotation of the object by an angle β, then bring back the object axis onto Ov. It is clear that this axis is invariant in the global operation.

Other point of view: formula (13), written in a matrix form, can be considered to be the transformation of a matrix associated with an operator under a change of coordinate axes. In such a change of axes, defined by a matrix S (whose columns contain the components of the new basis vectors on the old ones), we know that a matrix M, representing any linear operator, becomes $M' = S^{-1}\,M\,S$. In this point of view, the right-hand side of (13) is interpreted as the operation, which, in coordinate axes that have turned by an angle $-\alpha$ around Oz, describes a rotation by β around Oy. The same matrix, in the $Oxyz$ axes, will describe a rotation by an angle β around an axis of the plane xOy obtained from Oy by a rotation by $+\alpha$ around Oz, i.e. around the axis Ov:

it is indeed $\mathscr{R}_v(\beta)$.

In the same way, one can show that:

$$\begin{aligned}\mathscr{R}_Z(\gamma) &= \mathscr{R}_{\boldsymbol{v}}(\beta)\,\mathscr{R}_z(\gamma)\,\mathscr{R}_{\boldsymbol{v}}(-\beta)\\ &= \mathscr{R}_z(\alpha)\,\mathscr{R}_y(\beta)\,\mathscr{R}_z(-\alpha)\,\mathscr{R}_z(\gamma)\,\mathscr{R}_z(\alpha)\,\mathscr{R}_y(-\beta)\,\mathscr{R}_z(-\alpha)\\ &= \mathscr{R}_z(\alpha)\,\mathscr{R}_y(\beta)\,\mathscr{R}_z(\gamma)\,\mathscr{R}_y(-\beta)\,\mathscr{R}_z(-\alpha)\end{aligned} \tag{14}$$

(in a product of rotations along the same axis, the angles add up). We thus have:

$$\begin{aligned}\mathscr{R}_Z(\gamma)\,\mathscr{R}_v(\beta) &= \mathscr{R}_z(\alpha)\,\mathscr{R}_y(\beta)\,\mathscr{R}_z(\gamma)\,\mathscr{R}_y(-\beta)\,\mathscr{R}_z(-\alpha)\mathscr{R}_z(\alpha)\,\mathscr{R}_y(\beta)\,\mathscr{R}_z(-\alpha)\\ &= \mathscr{R}_z(\alpha)\,\mathscr{R}_y(\beta)\,\mathscr{R}_z(\gamma-\alpha)\end{aligned} \tag{15}$$

Inserting this result in (12), we finally have:

$$\mathscr{R}(\alpha,\beta,\gamma) = \mathscr{R}_z(\alpha)\,\mathscr{R}_y(\beta)\,\mathscr{R}_z(\gamma) \tag{16}$$

β. *Rotation matrices*

We call $\boldsymbol{e}_1$, $\boldsymbol{e}_2$, and $\boldsymbol{e}_3$ the unit vectors of the three coordinate axes, and:

$$\boldsymbol{e}'_i = \mathscr{R}\,\boldsymbol{e}_i \qquad i = 1,2,3 \tag{17}$$

their transforms through $\mathscr{R}$. Matrix $(\mathscr{R})$, with elements $\mathscr{R}_{ji}$, is defined as:

$$(\boldsymbol{e}'_1,\boldsymbol{e}'_2,\boldsymbol{e}'_3) = (\boldsymbol{e}_1,\boldsymbol{e}_2,\boldsymbol{e}_3)\begin{pmatrix}\mathscr{R}_{11} & \mathscr{R}_{12} & \mathscr{R}_{13}\\ \mathscr{R}_{21} & \mathscr{R}_{22} & \mathscr{R}_{23}\\ \mathscr{R}_{31} & \mathscr{R}_{32} & \mathscr{R}_{33}\end{pmatrix} \tag{18}$$

The columns of the $(\mathscr{R})$ matrix contain the coordinates of $\boldsymbol{e}'_i$, transforms of the base vectors. We have:

$$\boldsymbol{e}'_i = \sum_j \mathscr{R}_{ji}\,\boldsymbol{e}_j \qquad i,j = 1,2,3 \tag{19}$$

Writing $\overrightarrow{ON}$ as:

$$\overrightarrow{ON} = \sum_{i=1}^{3} x_i\,\boldsymbol{e}_i \tag{20}$$

the transform N' of N through the rotation is given by:

$$\overrightarrow{ON}' = \sum_i x_i\,\boldsymbol{e}'_i = \sum_{ij} x_i\,\mathscr{R}_{ji}\,\boldsymbol{e}_j \tag{21}$$

meaning:

$$x'_j = \sum_i \mathscr{R}_{ji}\, x_i \qquad \text{or:} \qquad \begin{pmatrix} x' \\ y' \\ z' \end{pmatrix} = (\mathscr{R}) \begin{pmatrix} x \\ y \\ z \end{pmatrix} \tag{22}$$

The 3×3 $(\mathscr{R})$ matrices are orthogonal (real and unitary), unimodular (with a determinant equal to 1), i.e. they are the matrices of the SO(3) group. We have:

$$\begin{cases} (\mathscr{R})^{-1} = (\mathscr{R})^t \\ \det(\mathscr{R}) = 1 \\ \mathrm{Tr}(\mathscr{R}) = 1 + 2\cos\varphi \end{cases} \tag{23}$$

where $(\mathscr{R})^t$ is the transpose of matrix $(\mathscr{R})$, and φ the rotation angle.

Whether we use parameter $\boldsymbol{a}$, or rather α, β, and γ, equality (16) can be used to get an explicit expression for $(\mathscr{R})$. For example:

$$\mathscr{R}(\alpha,\beta,\gamma) = \begin{pmatrix} \cos\alpha\,\cos\beta\,\cos\gamma - \sin\alpha\,\sin\gamma & -\cos\alpha\,\cos\beta\,\sin\gamma - \sin\alpha\,\cos\gamma & \cos\alpha\,\sin\beta \\ \sin\alpha\,\cos\beta\,\cos\gamma + \cos\alpha\,\sin\gamma & -\sin\alpha\,\cos\beta\,\sin\gamma + \cos\alpha\,\cos\gamma & \sin\alpha\,\sin\beta \\ -\sin\beta\,\cos\gamma & \sin\beta\,\sin\gamma & \cos\beta \end{pmatrix} \tag{24}$$

The identity rotation is only obtained for the zero value of parameter $\boldsymbol{a}$, whereas it is associated with an infinite number of values for the parameters α, β and γ (all the values such that $\beta = 0$, $\alpha + \gamma = 0$); this redundancy is a disadvantage of Euler angles.

γ. *Infinitesimal rotations*

We already studied infinitesimal rotations and their commutation relations in § B-4 of chapter III. We show here their geometric link with the vector product. In an infinitesimal rotation of vector:

$$\delta\boldsymbol{a} = \boldsymbol{u}\,\delta\varphi \tag{25}$$

the transform N' of N (figure 2) is such that, to first-order in $\delta\varphi$:

$$\overrightarrow{NN'} = \delta\varphi\; \boldsymbol{u} \times \overrightarrow{ON} + \dots \tag{26}$$

which means that:

$$\overrightarrow{ON'} = \overrightarrow{ON} + \delta\varphi\; \boldsymbol{u} \times \overrightarrow{ON} + \dots = \overrightarrow{ON} + \delta\boldsymbol{a} \times \overrightarrow{ON} + \dots \tag{27}$$

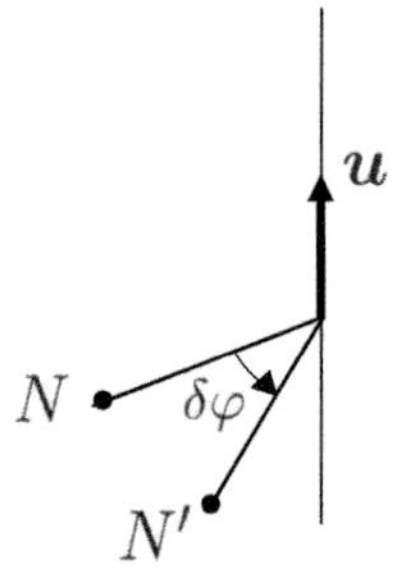

Figure 2: In an infinitesimal rotation by an angle $\delta\varphi$ around $\boldsymbol{u}$, point N is displaced to N' according to the vector product relation (26).

We can write:

$$\mathscr{R}(\boldsymbol{u}\delta\varphi) = 1 + \delta\varphi\; \boldsymbol{u} \times \quad + \ldots \tag{28}$$

Consequently, the matrix associated with $\mathscr{R}$ is:

$$(\mathscr{R}(\boldsymbol{u}\delta\varphi)) = 1 + \delta\varphi\; \mathscr{M}_u + \ldots \tag{29}$$

where $\mathscr{M}_u$ is given as a function of the u_x, u_y, and u_z components of $\boldsymbol{u}$ by:

$$\mathscr{M}_u = u_x \mathscr{M}_x + u_y \mathscr{M}_y + u_z \mathscr{M}_z \tag{30}$$

This yields again relations (III-88) and (III-90) of chapter III:

$$\mathscr{M}_x = \begin{pmatrix} 0 & 0 & 0 \\ 0 & 0 & -1 \\ 0 & 1 & 0 \end{pmatrix} \qquad \mathscr{M}_y = \begin{pmatrix} 0 & 0 & 1 \\ 0 & 0 & 0 \\ -1 & 0 & 0 \end{pmatrix} \qquad \mathscr{M}_z = \begin{pmatrix} 0 & -1 & 0 \\ 1 & 0 & 0 \\ 0 & 0 & 0 \end{pmatrix} \tag{31}$$

The $\mathscr{M}_i$ ($i = x$, y, or z) form the Lie algebra of the group of rotation matrices $(\mathscr{R})$. This easily leads again to the commutation relations (III-92):

$$[\mathscr{M}_i, \mathscr{M}_j] = \varepsilon_{ijk}\, \mathscr{M}_k \tag{32}$$

These relations show that $R_{(3)}$ is not an abelian group, and they yield the structure constants:

$$C_{ij}^k = \varepsilon_{ijk} \tag{33}$$

Let us calculate, to second-order in $\delta\varphi$ and $\delta\varphi'$, the product rotation [commutator between $\mathscr{R}_{\boldsymbol{u}}(\delta\varphi)$ and $\mathscr{R}_{\boldsymbol{v}}(\delta\varphi')$]:

$$\mathscr{R}_{\boldsymbol{u}}(\delta\varphi)\,\mathscr{R}_{\boldsymbol{v}}(\delta\varphi')\,\mathscr{R}_{\boldsymbol{u}}(-\delta\varphi)\,\mathscr{R}_{\boldsymbol{v}}(-\delta\varphi') \tag{34}$$

The general reasoning of § A-2-b of chapter III can be applied: the $\delta\varphi^2$ or $\delta\varphi'^2$ squared terms are zero (the product rotation being the identity if $\delta\varphi$ or $\delta\varphi'$ is zero), which makes it easy to invert (28). This yields:

$$\begin{aligned}
\boldsymbol{r}_1 &= \mathscr{R}_{\boldsymbol{v}}(-\delta\varphi')\boldsymbol{r} = \boldsymbol{r} - \delta\varphi'\ \boldsymbol{v}\times\boldsymbol{r} + \ldots \\
\boldsymbol{r}_2 &= \mathscr{R}_{\boldsymbol{u}}(-\delta\varphi)\boldsymbol{r}_1 = \boldsymbol{r}_1 - \delta\varphi\ \boldsymbol{u}\times\boldsymbol{r}_1 + \ldots \\
&= \boldsymbol{r} - \delta\varphi'\ \boldsymbol{v}\times\boldsymbol{r} - \delta\varphi\ \boldsymbol{u}\times\left[\boldsymbol{r} - \delta\varphi'\ \boldsymbol{v}\times\boldsymbol{r}\right] + \ldots \\
\boldsymbol{r}_3 &= \mathscr{R}_{\boldsymbol{v}}(\delta\varphi')\boldsymbol{r}_2 = \boldsymbol{r}_2 + \delta\varphi'\ \boldsymbol{v}\times\boldsymbol{r}_2 \\
&= \boldsymbol{r} - \delta\varphi\ \boldsymbol{u}\times\left[\boldsymbol{r} - \delta\varphi'\ \boldsymbol{v}\times\boldsymbol{r}\right] - \delta\varphi\delta\varphi'\ \boldsymbol{v}\times\left[\boldsymbol{u}\times\boldsymbol{r}\right] + \ldots \\
\boldsymbol{r}_4 &= \mathscr{R}_{\boldsymbol{u}}(\delta\varphi)\,\boldsymbol{r}_3 = \boldsymbol{r}_3 + \delta\varphi\ \boldsymbol{u}\times\boldsymbol{r}_3 \\
&= \boldsymbol{r} + \delta\varphi\delta\varphi'\left[\boldsymbol{u}\times(\boldsymbol{v}\times\boldsymbol{r}) - \boldsymbol{v}\times(\boldsymbol{u}\times\boldsymbol{r})\right] \\
&= \boldsymbol{r} + \delta\varphi\delta\varphi'\,(\boldsymbol{u}\times\boldsymbol{v})\times\boldsymbol{r}
\end{aligned} \tag{35}$$

Finally, to second-order in $\delta\varphi$ and $\delta\varphi'$:

$$\mathscr{R}_{\boldsymbol{u}}(\delta\varphi)\,\mathscr{R}_{\boldsymbol{v}}(\delta\varphi')\,\mathscr{R}_{\boldsymbol{u}}(-\delta\varphi)\,\mathscr{R}_{\boldsymbol{v}}(-\delta\varphi') = \mathscr{R}_{\boldsymbol{u}\times\boldsymbol{v}}(\delta\varphi\delta\varphi') + \ldots \tag{36a}$$

Using the notations of relation (III-19), we thus have:

$$\delta\boldsymbol{c} = \left[\boldsymbol{u}\times\boldsymbol{v}\right]\delta\varphi\,\delta\varphi' \tag{36b}$$

δ. Topology of the rotation group

We discussed the particular topology of the rotation group in § B-4 of chapter III. We saw that even though it is a compact and connected group, it is not simply connected. For example, we do not get a path equivalent to zero in the rotation group by taking a path that starts from the identity and is followed by rotations whose angle goes (once) through the value π, then continues from the value $-\pi$ toward zero (figure 15 of chapter III). The same is true if the angle goes through the value $\pm\pi$ an odd number of times. On the other hand, if it goes twice, or an even number of times, through the value π (figure 16), the path is equivalent to zero.

1-c. Arbitrary displacement

An arbitrary displacement can be the result of a rotation, with parameter $\boldsymbol{a}$, and a translation by a vector $\boldsymbol{b}$, in that order (but we could have chosen the reverse order). The xyz coordinates of a point are transformed according to:

$$\begin{pmatrix} x' \\ y' \\ z' \\ 1 \end{pmatrix} = \mathscr{D}(\boldsymbol{a}, \boldsymbol{b}) \begin{pmatrix} x \\ y \\ z \\ 1 \end{pmatrix} \tag{37a}$$

where $\mathscr{D}$ is a 4×4 matrix, which can be written as a function of the a_x, a_y, a_z components of $\boldsymbol{a}$ and the b_x, b_y, b_z components of $\boldsymbol{b}$:

$$\mathscr{D}(\boldsymbol{a}, \boldsymbol{b}) = \left(\begin{array}{c|c} & b_x \\ \mathscr{R}(\boldsymbol{a}) & b_y \\ & b_z \\ \hline 0\ 0\ 0 & 1 \end{array}\right) \tag{37b}$$

In this expression, $\mathscr{R}(\boldsymbol{a})$ is one of the 3×3 rotation matrix written in § 1-b above. Equalities (37) are equivalent to:

$$\boldsymbol{r}' = \mathscr{R}(\boldsymbol{a})\, \boldsymbol{r} + \boldsymbol{b} \tag{38a}$$

or, conversely:

$$\boldsymbol{r} = \mathscr{R}^{-1}(\boldsymbol{a})\ \left[\boldsymbol{r}' - \boldsymbol{b}\right] \tag{38b}$$

As in chapter III, derivation of (37b) yields the matrices of the Lie algebra. We thus get the three Π_x, Π_y, Π_z matrices already written in (7), plus the 3 matrices obtained by adding both a row and a column of zeros to the $\mathscr{M}$ written in (31). We shall not go through the calculation of the commutation relations between these 6 matrices, as it has been performed in § C of chapter III (in the case of 5×5 matrices, but the result is the same).

We rather do an elementary vector analysis calculation, which will give us again the commutation relations between infinitesimal operators of translation and rotation. The most general infinitesimal displacement operation transforms $\boldsymbol{r}$ into $\boldsymbol{r}'$, as given by [*cf.* (3) and (27)]:

$$\boldsymbol{r}' = \boldsymbol{r} + (\delta \boldsymbol{a} \times \boldsymbol{r}) + \delta \boldsymbol{b} \tag{39}$$

We compute, to second-order in $\delta\varphi$ and δb, the effect of the commutator:

$$\mathscr{R}_{\boldsymbol{u}}(\delta\varphi)\, \mathcal{T}_{\boldsymbol{v}}(\delta b)\, \mathscr{R}_{\boldsymbol{u}}(-\delta\varphi)\, \mathcal{T}_{\boldsymbol{v}}(-\delta b) \tag{40}$$

Following the same reasoning as in § 1-b-γ:

$$\begin{aligned}
\boldsymbol{r}_1 &= \mathcal{T}_{\boldsymbol{v}}(-\delta b)\,\boldsymbol{r} = \boldsymbol{r} - \boldsymbol{v}\delta b \\
\boldsymbol{r}_2 &= \mathscr{R}_{\boldsymbol{u}}(-\delta\varphi)\,\boldsymbol{r}_1 = \boldsymbol{r}_1 - \delta\varphi\;\boldsymbol{u}\times\boldsymbol{r}_1 + \ldots \\
&= \boldsymbol{r} - \boldsymbol{v}\delta b - \delta\varphi\;\boldsymbol{u}\times[\boldsymbol{r} - \boldsymbol{v}\delta b] \\
\boldsymbol{r}_3 &= \mathcal{T}_{\boldsymbol{v}}(\delta b)\,\boldsymbol{r}_2 = \boldsymbol{r}_2 + \boldsymbol{v}\delta b \\
&= \boldsymbol{r} - \delta\varphi\;\boldsymbol{u}\times\boldsymbol{r} + \delta\varphi\,\delta b\;\boldsymbol{u}\times\boldsymbol{v} + \ldots \\
\boldsymbol{r}_4 &= \mathscr{R}_{\boldsymbol{u}}(\delta\varphi)\,\boldsymbol{r}_3 = \boldsymbol{r}_3 + \delta\varphi\;\boldsymbol{u}\times\boldsymbol{r}_3 + \ldots \\
&= \boldsymbol{r} + \delta\varphi\,\delta b\;\boldsymbol{u}\times\boldsymbol{v} + \ldots
\end{aligned} \tag{41}$$

We thus have, to second-order in $\delta\varphi$ and δb:

$$\mathscr{R}_{\boldsymbol{u}}(\delta\varphi)\,\mathcal{T}_{\boldsymbol{v}}(\delta b)\,\mathscr{R}_{\boldsymbol{u}}(-\delta\varphi)\,\mathcal{T}_{\boldsymbol{v}}(-\delta b) = \mathcal{T}_{\boldsymbol{u}\times\boldsymbol{v}}(\delta\varphi\,\delta b) \tag{42}$$

This relation shows that the infinitesimal commutator is a translation by a vector $\delta\varphi\,\delta b(\boldsymbol{u}\times\boldsymbol{v})$.

We shall use the following correspondence:

$$\begin{array}{ccccccc} i = & 1 & 2 & 3 & 4 & 5 & 6 \\ & \mathscr{M}_x & \mathscr{M}_y & \mathscr{M}_z & \Pi_x & \Pi_y & \Pi_z \end{array} \tag{43}$$

If we choose $i, j, k \leq 3$:

$$C_{ij}^{k} = \varepsilon_{ijk} \qquad C_{ij}^{k+3} = 0 \tag{44a}$$

(rotations do not commute with each others), we can write:

$$\begin{cases} C_{i+3,\,j+3}^{k+3} = 0 \\ C_{i+3,\,j+3}^{k} = 0 \end{cases} \tag{44b}$$

(translations commute with each others) and:

$$\begin{cases} C_{i,\,j+3}^{k} = 0 \\ C_{i,\,j+3}^{k+3} = \varepsilon_{ijk} \end{cases} \tag{44c}$$

[translations do not commute with rotations, as written in (42)]. These equalities are equivalent to relations (III-101a), (III-101b), and (III-101c) of chapter III.

2. Associated operators in the state space

Following the method of § D of chapter IV, we now look for an image, in the state space of the system, of the displacement group $\mathscr{E}_{(3)}$.

2-a. Infinitesimal generators. Commutation relations

We call $|\psi\rangle$ the state vector of the initial system, $|\psi'\rangle$ the system's state vector after a displacement with parameters $\boldsymbol{a}$ and $\boldsymbol{b}$. The unitary operator T transforms $|\psi\rangle$ into $|\psi'\rangle$:

$$|\psi'\rangle = T(\boldsymbol{a}, \boldsymbol{b})\, |\psi\rangle \tag{45}$$

For infinitesimal values of $\boldsymbol{a}$ and $\boldsymbol{b}$ we have:

$$\boxed{T(\delta\boldsymbol{a}, \delta\boldsymbol{b}) = 1 - \frac{i}{\hbar}(\delta\boldsymbol{a} \cdot \boldsymbol{J} + \delta\boldsymbol{b} \cdot \boldsymbol{P}) + \ldots} \tag{46}$$

This equality must be considered as the *definition* of operators $\boldsymbol{J}$ (i.e. of its 3 components J_x, J_y, and J_z, each being an operator) and of $\boldsymbol{P}$ (which also symbolizes 3 operators P_x, P_y, P_z). Writing as in (IV-56) and (IV-57) that $T^\dagger T = 1$ (T is unitary), we get:

$$\begin{cases} P_i = P_i{}^\dagger & i = x, y, z \\ J_i = J_i{}^\dagger & \quad'' \quad '' \end{cases} \tag{47}$$

Operators $\boldsymbol{P}$ and $\boldsymbol{J}$ are thus Hermitian. The first is, by definition, the total (linear) momentum of the physical system under study; the second, its total angular momentum.

We now use the general reasoning of § D-2 of chapter IV to find the commutation relations between the components of $\boldsymbol{P}$ and $\boldsymbol{J}$. We have the following correspondence, first between operators $\boldsymbol{\Pi}$ and $\boldsymbol{\mathscr{M}}$ acting in the ordinary $R^{(3)}$ space, and then between operators $\boldsymbol{P}$ and $\boldsymbol{J}$ acting in the state space $\mathscr{E}$:

$$\boldsymbol{\Pi} \Leftrightarrow -\frac{i}{\hbar}\boldsymbol{P} \qquad\qquad \boldsymbol{\mathscr{M}} \Leftrightarrow -\frac{i}{\hbar}\boldsymbol{J} \tag{48}$$

We saw in chapter V that, for the infinitesimal transformations in the displacement group, the eventual projective character of the representation could be eliminated by a proper redefinition of the operators. Writing relations of the type (IV-66) with the C_{ij}^k constants given in (44), we get:

$$\begin{cases} [P_i, P_j] = 0 \\ [J_i, J_j] = i\hbar\,\varepsilon_{ijk}\, J_k \\ [J_i, P_j] = i\hbar\,\varepsilon_{ijk}\, P_k \end{cases} \tag{49}$$

The basic origin of these relations is purely geometric, and in no way does it depend on the physical system under study.

2-b. Finite displacement operators

While phase factors can be eliminated for infinitesimal transformations, it is not obvious that this holds for all finite transformations. We are going to show that it is actually the case for translations, but not always for rotations.

α. *Translations*

We are looking for an operator $T(0,\boldsymbol{b})$ associated with a pure translation (without rotation), in a given direction $\boldsymbol{u}$: $\boldsymbol{b} = b\boldsymbol{u}$. We know the infinitesimal operator $T(0,\delta b\,\boldsymbol{u})$ where $\delta b \to 0$. According to (46):

$$T\left(0,\delta b\,\boldsymbol{u}\right) = 1 - \frac{i}{\hbar}\delta b\;\boldsymbol{u}\cdot\boldsymbol{P} + \ldots = 1 - \frac{i}{\hbar}\delta b\,P_u + \ldots \tag{50}$$

where:

$$P_u = \boldsymbol{u}\cdot\boldsymbol{P} \tag{51}$$

We now compute the operator $T(0,b\boldsymbol{u})$ associated with a finite translation $b\boldsymbol{u}$. A translation by $(b+\delta b)\boldsymbol{u}$ is a translation by $b\boldsymbol{u}$ followed by another by $\delta b\boldsymbol{u}$, so that:

$$T\Big(0,(b+\delta b)\boldsymbol{u}\Big) = \mathrm{e}^{-i\xi(b,\delta b)}\,T(0,\delta b\boldsymbol{u})\,T(0,b\boldsymbol{u}) \tag{52}$$

We introduced the phase factor $\mathrm{e}^{-i\xi(b,\delta b)}$, but will show below that it is not needed, when dealing with translations; it will be ignored for the moment.

Under these conditions we obtain:

$$T\Big(0,(b+\delta b)\boldsymbol{u}\Big) - T\left(0,b\boldsymbol{u}\right) = \left[-\frac{i}{\hbar}\delta b\,P_u\right]\;T\left(0,b\boldsymbol{u}\right) \tag{53}$$

or:

$$\frac{\mathrm{d}}{\mathrm{d}b}T\left(0,b\boldsymbol{u}\right) = -\frac{i}{\hbar}\,P_u\;T\left(0,b\boldsymbol{u}\right) \tag{54}$$

This first-order differential equation can be solved, taking into account the initial condition:

$$T(0,b=0) = \mathbb{1} \tag{55}$$

and we get:

$$T\left(0,b\boldsymbol{u}\right) = T\left(0,\boldsymbol{b}\right) = \exp\left\{-\frac{i}{\hbar}b\,P_u\right\} \tag{56}$$

Deriving this equality, which is easy to do since it involves a single operator (P_u), we check that we again get (54).

We obtained a value for the translation operator "going in a straight line" from the origin to the final value $\boldsymbol{b}$ of the parameter. It is written in the form:

$$\boxed{T(0, \boldsymbol{b}) = \exp\left\{-\frac{i}{\hbar}\boldsymbol{b} \cdot \boldsymbol{P}\right\}} \tag{57}$$

We may wonder whether the value of operator T, obtained when using one path (the most direct) from the origin, changes if we choose another path[2]. For example, imagine that, instead of going in a straight line, we make a detour, as shown on figure 3. The integral yielding the final value of $T(0, \boldsymbol{b})$ is made out of 5 segments. In 3 of these, the integration variable varies along the direction $\boldsymbol{u}$ of the final value of $\boldsymbol{b}$, and in the other two, it varies in the perpendicular direction $\boldsymbol{v}$. We call b_1, b_2, b_3, b_2, and b_4 the length of the paths associated with each segment. They each contribute to the transformation operator by a factor of the form (57), so that the final value at the end of the path is:

$$\begin{aligned} T(0, b\boldsymbol{u}) = \exp\left\{-\frac{i}{\hbar}b_1 P_u\right\} \exp\left\{-\frac{i}{\hbar}b_2 P_v\right\} \exp\left\{-\frac{i}{\hbar}b_3 P_u\right\} \\ \times \exp\left\{+\frac{i}{\hbar}b_2 P_v\right\} \exp\left\{-\frac{i}{\hbar}b_4 P_u\right\} \end{aligned} \tag{58}$$

Since the P_u and P_v operators commute, we can change the exponentials' order. The two exponentials containing P_v cancel each other, and regrouping the three containing P_u, we again obtain (56). The two different paths used to build operator $T(0, b\boldsymbol{u})$ thus yield the same value. This reasoning can be generalized to a path containing an increasing number of segments in directions perpendicular to $\boldsymbol{u}$, approaching more and more any curved path. Any path describing the same translation $\mathcal{T}(\boldsymbol{b})$ will yield the same operateur T, with no phase factor depending on the path.

Expression (57) serves as a definition of translation operators. Since all the components of P commute with each others, we see using (57) that:

$$T(0, \boldsymbol{b})\ T(0, \boldsymbol{b}') = T(0, \boldsymbol{b}')\ T(0, \boldsymbol{b}) = T(0, \boldsymbol{b} + \boldsymbol{b}') \tag{59}$$

Operators T yield a representation of the translation group with no phase factors. We will see that the situation is not always that simple for the rotation group.

Another demonstration of the absence of a phase factor. If we include a phase factor in (52), we must write:

$$T(0, \boldsymbol{b} + \mathrm{d}\boldsymbol{b}) = \mathrm{e}^{-i\xi(\boldsymbol{b}, \mathrm{d}\boldsymbol{b})}\ T(0, \mathrm{d}\boldsymbol{b})\, T(0, \boldsymbol{b}) \tag{60}$$

[2]This question has already been discussed in complement A_{IV} for representations of finite dimension. Since the group of translations is non-compact, we do not assume here that the representation is of finite dimension, but we use the fact that the group is simply connected.

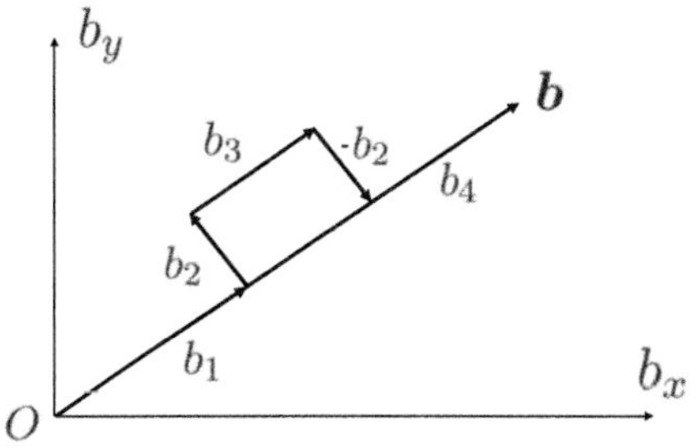

Figure 3: In the space of parameters **b** *defining the translation operators, we follow two paths starting from the origin (identity transformation) and ending at the same value of* **b**. *Integrations along these two paths lead to the same final value $T(0, \boldsymbol{b})$. For the sake of simplicity, the figure only shows the two b_x and b_y components, ignoring the third one b_z.*

Since, by definition, the function $\xi(\boldsymbol{b}, 0)$ is zero, we get to first-order:

$$\xi(\boldsymbol{b}, \mathrm{d}\boldsymbol{b}) = \mathrm{d}\boldsymbol{b} \cdot \boldsymbol{\nabla}_2 \xi(\boldsymbol{b}) \tag{61}$$

where the index 2 in the gradient means that it is taken with respect to the second variable of ξ. This leads to:

$$T(0, \boldsymbol{b} + \mathrm{d}\boldsymbol{b}) - T(0, \boldsymbol{b}) = \left[-\frac{i}{\hbar} \mathrm{d}\boldsymbol{b} \cdot \boldsymbol{P} - i \mathrm{d}\boldsymbol{b} \cdot \boldsymbol{\nabla}_2 \xi(\boldsymbol{b}) \right] T(0, \boldsymbol{b}) \tag{62}$$

Integrating this equation to go from a $\boldsymbol{b} = 0$ translation to a finite $\boldsymbol{b}$ translation, the $\boldsymbol{\nabla}_2 \xi$ term introduces an additional phase factor:

$$\exp \left[-i \int \mathrm{d}\boldsymbol{b} \cdot \boldsymbol{\nabla}_2 \xi(\boldsymbol{b}) \right] \tag{63}$$

This factor contains a line integral in the space of the translation parameters, leading from the origin $\boldsymbol{b} = 0$ to the final value $\boldsymbol{b}$. Since the translation group is simply connected, the value of this integral over a gradient is independent of the path followed. Each translation operator acquires its own additional phase factor, which consequently has no physical consequence and can be ignored. Note that relation (62) shows that suppressing this phase factor amounts to subtracting from the $\boldsymbol{P}$ operator a term proportional to $\boldsymbol{\nabla}_2 \xi$. This is equivalent to redefining $\boldsymbol{P}$ by subtracting a diagonal operator, which enabled us, in § B-3 of chapter V, to simplify the commutation relations of the Galilean group.

β. *Rotations*

We follow, for the rotations, the same type of argument as for the translations. An infinitesimal rotation of vector $\delta \boldsymbol{a}$ corresponds, according to (46), to the operator:

$$T(\delta\boldsymbol{a},0) = 1 - \frac{i}{\hbar}\delta\boldsymbol{a}\cdot\boldsymbol{J} + \ldots = 1 - \frac{i}{\hbar}\delta a\, J_u + \ldots \tag{64}$$

where:

$$\boldsymbol{u} = \frac{\delta\boldsymbol{a}}{|\delta\boldsymbol{a}|} \qquad\qquad J_u = \boldsymbol{u}\cdot\boldsymbol{J} \tag{65}$$

Equations (50) to (56) can be rewritten, replacing $T(0, b\boldsymbol{u})$ by $T(a\boldsymbol{u}, 0)$ and P_u by J_u. We get:

$$\boxed{T(\boldsymbol{a},0) = \exp\left\{-\frac{i}{\hbar}\,\boldsymbol{a}\cdot\boldsymbol{J}\right\}} \tag{66}$$

an expression that will serve as the definition of a rotation operator.

An important difference with the translations occurs at this stage: the $\boldsymbol{J}$ components do not commute with each others. Furthermore, we saw in § B-4 that the rotation group is not simply connected, meaning that several classes of non-equivalent paths can lead from one point to another. The unicity of the operator obtained by integration from the origin is no longer guaranteed. Depending on the structure of the state space $\mathscr{E}$ of the system, the representation is either a true representation or a projective representation such that[3]:

$$T(\mathscr{R}_2)\; T(\mathscr{R}_1) = \pm T(\mathscr{R}_2\mathscr{R}_1) \tag{67}$$

We shall also discuss the link between double-valued representations and the 2-connected structure of the rotation group.

γ. *Arbitrary displacement*

For a displacement defined as in (37) or (39) by the parameters $\boldsymbol{a}$ and $\boldsymbol{b}$, the associated operator T acting in the state space $\mathscr{E}$ is by definition:

$$\boxed{\begin{aligned} T(\boldsymbol{a},\boldsymbol{b}) &= T(\boldsymbol{0},\boldsymbol{b})\; T(\boldsymbol{a},\boldsymbol{0}) \\ &= \exp\left\{-\tfrac{i}{\hbar}\,\boldsymbol{b}\cdot\boldsymbol{P}\right\}\; \exp\left\{-\tfrac{i}{\hbar}\,\boldsymbol{a}\cdot\boldsymbol{J}\right\} \end{aligned}} \tag{68}$$

As is the case for rotations, the representation thus obtained is not always a true (non-projective) representation of the displacement group $\mathscr{E}_{(3)}$.

[3]In complement A_{IV} we show that, since the rotation group is 2-connected, two values of the phase are possibles.

2-c. Union of two subsystems, additivity of linear and angular momenta

Consider a physical system obtained by the union of several subsystems. It can be, for example, a system composed of several particles, or of particles and fields, etc. For the sake of simplicity, we shall take the number of subsystems equal to two. The space state $\mathscr{E}$ of the global physical system is the tensor product:

$$\mathscr{E} = \mathscr{E}_1 \otimes \mathscr{E}_2 \tag{69}$$

of the state spaces $\mathscr{E}_1$ and $\mathscr{E}_2$ associated respectively with subsystems (1) and (2).

We can define in each of these subspaces[4] momentum operators $\boldsymbol{P}_1$ and $\boldsymbol{P}_2$ as well as angular momentum operators $\boldsymbol{J}_1$ and $\boldsymbol{J}_2$. Similarly, we can introduce in $\mathscr{E}_1$, as well as in $\mathscr{E}_2$, displacement operators. As an example, for a translation by a vector $\boldsymbol{b}$:

$$\begin{aligned} T_1(\boldsymbol{0}, \boldsymbol{b}) &= \exp\left\{-\frac{i}{\hbar}\, \boldsymbol{b} \cdot \boldsymbol{P}_1\right\} \\ T_2(\boldsymbol{0}, \boldsymbol{b}) &= \exp\left\{-\frac{i}{\hbar}\, \boldsymbol{b} \cdot \boldsymbol{P}_2\right\} \end{aligned} \tag{70}$$

We can associate with any operator A_1 acting in $\mathscr{E}_1$ an "extended" operator acting in $\mathscr{E} = \mathscr{E}_1 \otimes \mathscr{E}_2$. If $|\varphi_1\rangle$ is a ket of $\mathscr{E}_1$ and $|\varphi_2\rangle$ a ket of $\mathscr{E}_2$, we set, by definition:

$$A_1 \left(|\varphi_1\rangle \otimes |\varphi_2\rangle\right) = \left(A_1|\varphi_1\rangle\right) \otimes |\varphi_2\rangle \tag{71}$$

The same holds true for an operator acting initially in $\mathscr{E}_2$. It is easy to show that any operator labeled with index 1 commutes with any operator labeled with index 2. For example:

$$\begin{aligned} &[\boldsymbol{J}_1, \boldsymbol{J}_2] = 0 \\ &[\boldsymbol{J}_1, \boldsymbol{P}_2] = 0 \\ &[T_1, T_2] = 0 \qquad \text{etc.} \end{aligned} \tag{72}$$

Consider a state of the global system of the form:

$$|\psi\rangle = |\varphi_1\rangle \otimes |\varphi_2\rangle \tag{73}$$

It is clear that to perform a translation of this system, one must perform the same translation in each of the two subsystems. This leads to the final state $|\psi'\rangle$:

$$|\psi'\rangle = |\varphi_1'\rangle \otimes |\varphi_2'\rangle \tag{74}$$

[4]In this complement, the 1 and 2 indices do not label the x and y components of the operators, but the physical subsystems on which the operators act.

with:

$$
\begin{aligned}
|\varphi_1'\rangle &= T_1(\mathbf{0}, \boldsymbol{b})\, |\varphi_1\rangle \\
|\varphi_2'\rangle &= T_2(\mathbf{0}, \boldsymbol{b})\, |\varphi_2\rangle
\end{aligned} \tag{75}
$$

This means that $|\psi'\rangle$ can be expressed as a function of $|\psi\rangle$:

$$
|\psi'\rangle = T_1(\mathbf{0}, \boldsymbol{b})\, T_2(\mathbf{0}, \boldsymbol{b})\, |\psi\rangle \tag{76}
$$

If $|\psi\rangle$ is an arbitrary ket of $\mathscr{E}$ (not necessarily a tensor product), it can be linearly decomposed as a sum of tensor products of the form $|\varphi_1\rangle \otimes |\varphi_2\rangle$, and it is easy to see that formula (76) remains valid. This shows that translation (or more generally displacement) operators are simply obtained as products of operators yielding the same displacement in each individual state space $\mathscr{E}_1$ and $\mathscr{E}_2$. The generalization to a larger number of subspaces is straightforward.

We now show that, because of this property, the total momentum $\boldsymbol{P}$ of the system and its total angular momentum $\boldsymbol{J}$ are simply the sums:

$$
\boxed{\begin{cases} \boldsymbol{P} = \boldsymbol{P}_1 + \boldsymbol{P}_2 \\ \boldsymbol{J} = \boldsymbol{J}_1 + \boldsymbol{J}_2 \end{cases}} \tag{77}
$$

Consider a translation by an infinitesimal vector $\mathrm{d}\boldsymbol{b}$. We have:

$$
\begin{aligned}
T(\mathbf{0}, \mathrm{d}\boldsymbol{b}) &= T_1(\mathbf{0}, \mathrm{d}\boldsymbol{b})\, T_2(\mathbf{0}, \mathrm{d}\boldsymbol{b}) \\
&= \left[1 - \frac{i}{\hbar}\, \mathrm{d}\boldsymbol{b} \cdot \boldsymbol{P}_1 + \ldots\right] \left[1 - \frac{i}{\hbar}\, \mathrm{d}\boldsymbol{b} \cdot \boldsymbol{P}_2 + \ldots\right] \\
&= 1 - \frac{i}{\hbar}\, \mathrm{d}\boldsymbol{b} \cdot (\boldsymbol{P}_1 + \boldsymbol{P}_2) + \ldots
\end{aligned} \tag{78}
$$

The first of equalities (77) comes from the definition of the total momentum of the system (Hermitian operator generating the infinitesimal translations). For the rotations, the reasoning is the same as for the translations. This proves equalities (77): the total momentum and the total angular momentum of the system are simply obtained by adding the momenta and angular momenta of the subsystems.

In addition, the commutation relations between the $\boldsymbol{P}$ or $\boldsymbol{J}$ components are the same for the global system as for each of the subsystems; this immediately follows from (72) and (77).

Comments:

(i) Equalities (77) remain valid even if the subsystems interact.

(ii) Spaces $\mathscr{E}_1$ and $\mathscr{E}_2$ can pertain to different variables of the same system (orbital and spin variables of the same particle, for example).

(iii) If system (1) is a particle and (2) a field, the structures of the state spaces $\mathscr{E}_1$ and $\mathscr{E}_2$ are quite different. A field has an infinite number of degrees of liberty and a "much larger" state space than that of a single particle. Nevertheless, equalities (77) remain valid in this case.

2-d. Translation or rotation invariance

In a translation of vector $\boldsymbol{b}$, an observable B becomes (*cf.* chapter IV, § C):

$$\begin{aligned} B' &= T(0,\boldsymbol{b})\ B\ T^{\dagger}(0,\boldsymbol{b}) \\ &= \exp\left\{-\frac{i}{\hbar}\,\boldsymbol{b}\cdot\boldsymbol{P}\right\}\ B\ \exp\left\{\frac{i}{\hbar}\,\boldsymbol{b}\cdot\boldsymbol{P}\right\} \end{aligned} \tag{79}$$

and in a rotation of vector $\boldsymbol{a}$:

$$\begin{aligned} B'' &= T(\boldsymbol{a},\boldsymbol{0})\ B\ T^{\dagger}(\boldsymbol{a},\boldsymbol{0}) \\ &= \exp\left\{-\frac{i}{\hbar}\,\boldsymbol{a}\cdot\boldsymbol{J}\right\}\ B\ \exp\left\{\frac{i}{\hbar}\,\boldsymbol{a}\cdot\boldsymbol{J}\right\} \end{aligned} \tag{80}$$

An observable B is said to be translation invariant if $B' = B$, and rotation invariant if $B'' = B$. Equality $B' = B$ is equivalent to:

$$[B, T(0,\boldsymbol{b})] = 0 \tag{81}$$

or:

$$[B, \boldsymbol{P}] = 0 \tag{82}$$

This last equality can serve as a definition: a translation invariant observable is an observable that commutes with the total momentum $\boldsymbol{P}$ of the system.

Example: We saw in chapter I that for a system with a time-independent Hamiltonian H (conservative system), the translation operations are symmetry transformations provided that:

$$[H, T(\boldsymbol{0},\boldsymbol{b})] = 0 \qquad\qquad \forall\ \boldsymbol{b} \tag{83}$$

meaning:

$$[H, \boldsymbol{P}] = 0 \tag{84}$$

The Hamiltonian is thus translation invariant, which amounts to saying that the momentum $\boldsymbol{P}$ is a constant of motion. We can then seek the eigenstates of H among the eigenstates of momentum (plane waves).

Comment:

For a system composed of several interacting subsystems, it is important to distinguish commutation relations with individual momenta $\boldsymbol{P}_1, \boldsymbol{P}_2, \ldots$ of the subsystems from those with the total momentum:

$$\boldsymbol{P} = \sum_i \boldsymbol{P}_i \tag{85}$$

If the global system is free (no external potential), the Hamiltonian H necessarily commutes with $\boldsymbol{P}$:

$$[H, \boldsymbol{P}] = 0 \tag{86}$$

but the commutators $[H, \boldsymbol{P}_i]$ are not necessarily zero (although their sum is).

All the above can be easily transposed to rotation invariance, simply replacing $\boldsymbol{P}$ by $\boldsymbol{J}$. Observable B is rotation invariant if:

$$[B, \boldsymbol{J}] = 0 \tag{87}$$

For a system with a time-independent Hamiltonian H, any rotation around a given point O is a symmetry operation provided the angular moment $\boldsymbol{J}$ about this point is a constant of motion:

$$[H, \boldsymbol{J}] = 0 \tag{88}$$

Example: Consider a free particle (or physical system) in space. The angular moment $\boldsymbol{J}$ of this system with respect to any given point is a constant of motion.

Complement D_V

Space reflection (parity)

This complement discusses purely geometrical operations, like the displacements studied in complement C_V, but of a different nature: symmetries (or reflections) with respect to a point, or parity[1]. In general, an arbitrary object cannot be superposed by a displacement onto that same object obtained by symmetry.

1. Action in real space

A given point in space is chosen as the origin O of the coordinate axes $Oxyz$. A symmetry operation with respect to O associates with point M defined by:

$$\overrightarrow{OM} = \boldsymbol{r} \tag{1}$$

the point M' such that (figure 1):

$$\overrightarrow{OM'} = \boldsymbol{r}' = -\boldsymbol{r} \tag{2}$$

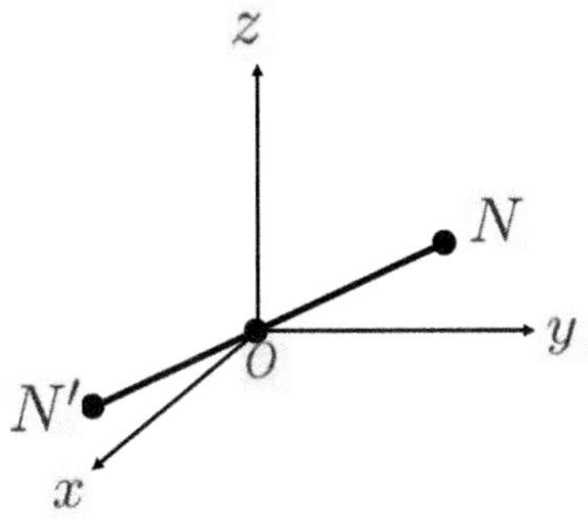

Figure 1: In a parity operation, each point N is tranformed into a point N', symmetric to N with respect to the origin O.

[1]One often uses the name "space inversion", or simply "inversion" (hence the notation $\mathcal{I}_O$, which we choose instead of P to avoid any confusion with the momentum operator).

This is a linear operation $\mathcal{I}_O$, associated with the matrix:

$$(\mathcal{I}_O) = \begin{pmatrix} -1 & 0 & 0 \\ 0 & -1 & 0 \\ 0 & 0 & -1 \end{pmatrix} \tag{3}$$

The corresponding operator is linear and orthogonal (unitary and real), as for a rotation. Its matrix, however, is different from a rotation matrix since:

$$\det(\mathcal{I}_O) = -1 \qquad \text{(insead of } +1 \text{ for a rotation)} \tag{4}$$

A right-hand orthogonal reference frame is transformed into a left-hand orthogonal frame. It is easy to show that:

$$(\mathcal{I}_O)^2 = (1) \tag{5a}$$

$$(\mathcal{I}_O)^{-1} = (\mathcal{I}_O) \tag{5b}$$

The set formed by the identity operator and the symmetry with respect to a fixed point O, is a two-element group, isomorphic to the Z_2 cyclic group. It is a discrete and non-continuous group (there is no operation of infinitesimal symmetry with respect to a point).

Since matrix $(\mathcal{I}_O)$ is simply the opposite of the identity matrix (scalar matrix), it commutes with any 3×3 matrix, and in particular with the rotation matrices. Therefore:

$$[\mathcal{I}_O, \mathscr{R}] = 0 \tag{6}$$

Adding matrix $(\mathcal{I}_O)$ to the $(\mathscr{R})$ matrices, we get the orthogonal group 0(3) (all the 3×3 orthogonal matrices, with determinant ± 1), whose elements are, either the rotations $\mathscr{R}$, or their products with $\mathcal{I}_O$.

To obtain the commutation relations between $\mathcal{I}_O$ and the translations $\mathcal{T}$ we can, as in § C of chapter V, introduce 5×5 matrices to describe the operations. We must add to the matrix written in (3) two rows and two columns of zeros, except for the diagonal elements, which are equal to 1. We can also use definition (2) and, with a geometric reasoning, show that:

$$\begin{aligned} \mathcal{I}_O \mathcal{T}_b\, \boldsymbol{r} &= \mathcal{I}_O \left[\boldsymbol{r} + \boldsymbol{b}\right] = -\boldsymbol{r} - \boldsymbol{b} \\ \mathcal{T}_b \mathcal{I}_O\, \boldsymbol{r} &= \mathcal{T}_b \left[-\boldsymbol{r}\right] = -\boldsymbol{r} + \boldsymbol{b} \end{aligned} \tag{7}$$

and thus:

$$[\mathcal{I}_O, \mathcal{T}_b] = \mathcal{T}_{(-2b)} - \mathbb{1} \tag{8}$$

where $\mathbb{1}$ is the identity operator.

Comments:

(i) The $\mathcal{I}$ symmetry operation can also be defined with respect to an arbitrary point A. It is then noted $\mathcal{I}_A$. This leads to:

$$\mathcal{I}_A\, \boldsymbol{r} = 2\,\boldsymbol{r}_A - \boldsymbol{r} \tag{9}$$

where $\boldsymbol{r}_A$ is the position vector of point A.
The set of operations $\mathcal{I}_A$, plus the identity operation, do not form a group, since:

$$\begin{aligned} \mathcal{I}_A \mathcal{I}_B\, \boldsymbol{r} &= 2\,\boldsymbol{r}_A - 2\,\boldsymbol{r}_B + \boldsymbol{r} \\ &= \mathcal{T}_{[2(\boldsymbol{r}_A - \boldsymbol{r}_B)]}\, \boldsymbol{r} \end{aligned} \tag{10}$$

(ii) It follows from this equation that if the equations of motion of a physical system show that $\mathcal{I}_A$ and the set of translations are symmetry operations, the same is true for the $\mathcal{I}_B$ symmetries with respect to an arbitrary point B.

2. Associated operator in the state space

Consider first a single spinless particle. Its quantum state can be described either by a state vector $|\psi\rangle$, or by the wave function:

$$\psi(\boldsymbol{r}) = \langle \boldsymbol{r}|\psi\rangle \tag{11}$$

For the displacements, we define the action of the parity operator I_O as the image, in the state space, of the symmetry operation ($\mathcal{I}_O$) with respect to a point[2]:

$$I_O|\boldsymbol{r}\rangle = |-\boldsymbol{r}\rangle \tag{12}$$

It transforms an orthonormal basis (in the sense of continuous bases) into another orthonormal basis, it is a unitary operator. Its square equals the identity operator:

$$[I_O]^2 = 1 \tag{13}$$

and thus:

$$I_O = I_O^\dagger = I_O^{-1} \tag{14}$$

[2]Do not confuse $|-\boldsymbol{r}\rangle$ and the opposite $-|\boldsymbol{r}\rangle$ of the ket $|\boldsymbol{r}\rangle$! The kets $|-\boldsymbol{r}\rangle$ and $|\boldsymbol{r}\rangle$ are orthogonal (unless $\boldsymbol{r} = 0$) and not colinear.

(see, for example, complement F_{II} of [9] for a more detailed discussion). Consequently:

$$\langle \boldsymbol{r}|I_O = \langle -\boldsymbol{r}| \tag{15}$$

and the wave function $\psi'(\boldsymbol{r})$ associated with the ket $I_O|\psi\rangle$ is given by:

$$\begin{aligned}\psi'(r) &= \langle \boldsymbol{r}|I_O|\psi\rangle \\ &= \langle -\boldsymbol{r}|\psi\rangle = \psi(-\boldsymbol{r})\end{aligned} \tag{16}$$

This equality describes the action of I_O in the $\{|\boldsymbol{r}\rangle\}$ representation. It shows, for example, that a plane wave $e^{i\boldsymbol{k}\cdot\boldsymbol{r}}$ with momentum $\hbar\boldsymbol{k}$ is transformed into another plane wave $e^{-i\boldsymbol{k}\cdot\boldsymbol{r}}$ with an opposite momentum $-\hbar\boldsymbol{k}$, so that:

$$I_O|\boldsymbol{p}\rangle = |-\boldsymbol{p}\rangle \tag{17}$$

Therefore:

$$\overline{\psi'}(\boldsymbol{p}) = \langle \boldsymbol{p}|I_O|\psi\rangle = \langle -\boldsymbol{p}|\psi\rangle = \overline{\psi}(-\boldsymbol{p}) \tag{18}$$

Using (12) and (17) it is easy to show that:

$$\begin{cases} I_O\, \boldsymbol{R}\, I_O = -\boldsymbol{R} \\ I_O\, \boldsymbol{P}\, I_O = -\boldsymbol{P} \end{cases} \tag{19}$$

$\boldsymbol{R}$ and $\boldsymbol{P}$ are therefore observables that are not invariant under parity, which is physically understandable. On the other hand:

$$I_O\ (\boldsymbol{R}\times\boldsymbol{P})\ I_O = I_O\, \boldsymbol{L}\, I_O = \boldsymbol{L} \tag{20}$$

meaning that $\boldsymbol{L}$ is parity invariant (figure 2).

$\boldsymbol{R}$ and $\boldsymbol{P}$ correspond to polar vectors ("true vectors"), $\boldsymbol{L}$ is an axial vector (or "pseudovector"). In electromagnetism, the electric field $\boldsymbol{E}$ is a polar vector, the magnetic field $\boldsymbol{B}$ an axial vector.

If the system is composed of several particles 1, 2, 3, ..., the parity operator I_O is simply the product of the parity operators associated with each particle:

$$I_O = I_O(1)\ I_O(2)\ I_O(3)\ldots \tag{21}$$

If the particle(s) has (have) a spin $\boldsymbol{S}$, based on (20) showing that the orbital angular momentum $\boldsymbol{L}$ is parity invariant, we set:

$$I_O\, \boldsymbol{S}\, I_O = \boldsymbol{S} \tag{22}$$

meaning that the parity operator does not act in the spin state space (as is the case for the translation operators).

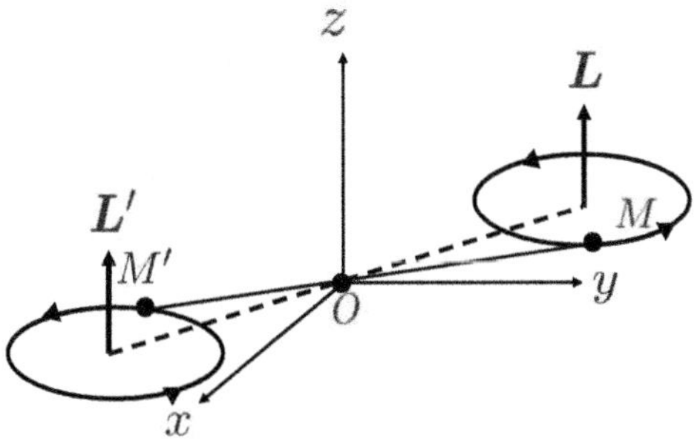

Figure 2: When mobile M rotates according to the right-hand rule on a circular orbit, its parity transform M' rotates in the same direction, and $\boldsymbol{L} = \boldsymbol{L}'$. Contrary to $\boldsymbol{R}$ and $\boldsymbol{P}$ whose signs change under parity, the orbital angular momentum operator $\boldsymbol{L} = \boldsymbol{R} \times \boldsymbol{P}$ is parity invariant.

3. Parity conservation

If the Hamiltonian of the system under study is time-independent, parity will be a symmetry operation if:

$$[H, I_O] = 0 \tag{23}$$

I_O is then a constant of motion. The eigenstates of H can be classified on the basis of the eigenvalues of I_O, which, according to (13), are equal to ± 1. In the first case, the eigenstates are said to be even, and in the other, odd. If the initial state vector $|\psi(0)\rangle$ of the system is even (or odd), it will keep this property over time: $|\psi(t)\rangle$ will remain even (or odd), and one says that parity is conserved.

If the initial state vector is neither even nor odd, condition (23) shows that $I_O|\psi(t)\rangle$ describes another possible motion of the system (with the same Hamiltonian): the parity transform (point symmetry) of a possible motion is another possible motion. When parity is conserved, one sometimes says in a more illustrative way that looking through a mirror at the motion of a system, one observes another possible motion. This statement implicitly assumes that rotations are symmetry transformations of the moving system. It uses the fact that the product of a symmetry with respect to a plane (that of the mirror) and a rotation by π around an axis perpendicular to that plane and crossing it at some point O, is simply the symmetry with respect to point O (figure 3).

It is clear that the right and left notions are inverted in a mirror symmetry (or in a symmetry with respect to a point). This is why parity conservation is sometime referred to as Left-Right symmetry.

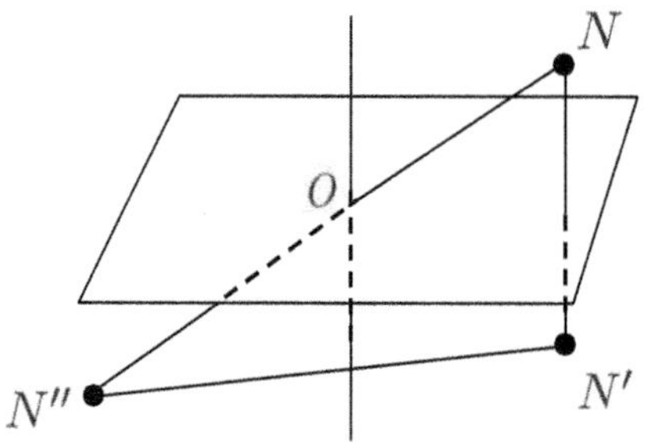

Figure 3: The product of a planar symmetry (mirror symmetry that transforms point N into N') and a rotation by an angle π around a perpendicular axis (that transforms N' into N'') yields a symmetry with respect to the point of intersection O of the axis and the plane.

For a long time and more or less implicitly, physicists have considered parity symmetry as straightforward. Looking through a mirror does not provide a spectacle that goes against our day to day experience or our physical intuition. For example, a movie can be projected after inverting the right and left of the reel of film without showing anything immediately recognizable as physically absurd.

It was not until the mid-1950's that this parity conservation started to be questioned. The physical phenomenon that led to this questioning was the weak interaction disintegration of a particle we now call the K^+ meson into two or three pions. These two types of disintegration lead to final states with respective parities $+1$ and -1. If $\mathcal{I}_O$ is a symmetry of the interactions responsible for the disintegration, the parities of the initial states before disintegration must necessarily be different, meaning that we started from two different particles. This is why it was considered for a long time that there existed two distinct particles, the θ^+ and τ^+ mesons, that disintegrated, respectively, into two or three pions.

It was noted, however, that the θ^+ and the τ^+ were identical in every respect (same mass, same lifetime, etc.). This led Lee and Yang to propose that the θ^+ and τ^+ particles were actually one and the same particle (the K^+ meson), but that weak interactions violated parity symmetry.

This hypothesis was brilliantly confirmed experimentally in 1957 by Wu and coworkers [41]. Their experiment studied the nuclear disintegration:

$$^{60}\text{Co} \rightarrow {}^{60}\text{Ni} + \text{e}^- + \overline{\nu}_\text{e} + 2\gamma \tag{24}$$

(due to weak interaction), starting from polarized cobalt nuclei; e^- is an electron, $\overline{\nu}_\text{e}$ an electronic antineutrino, and γ a gamma photon. The cobalt nuclei have a

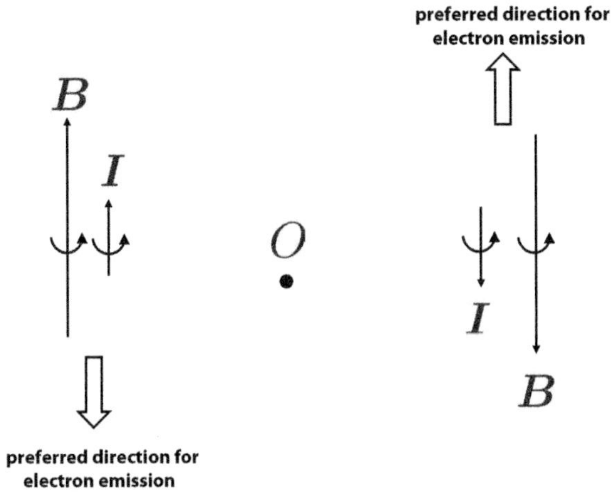

*Figure 4: If parity were conserved, two experiments symmetric with respect to a point O, hence with the same directions of **I** and **B**, would yield two opposite directions for the preferred direction of β^- emission. As the same causes cannot produce opposite effects, the spin direction (axial vector) cannot determine the emission direction (polar vector), unless parity is violated.*

spin $I = 5$, a lifetime $\tau = 5.26$ years (they undergo β^- decay) and a magnetic moment $\mu = +3{,}8\ \mu_n$ (the moment $\boldsymbol{\mu}$ is parallel to $\boldsymbol{I}$). Placing ^{60}Co nuclei in a strong magnetic field $\boldsymbol{B}$ and at very low temperature, thermodynamic equilibrium favors the number of nuclei with spin parallel to $\boldsymbol{B}$ compared to those with spin antiparallel to $\boldsymbol{B}$.

The experiment proceeds as follows. Once the field $\boldsymbol{B}$ is applied, one measures the number of electrons emitted per unit time first in the direction of $\boldsymbol{B}$, then in the opposite direction. The result shows that more electrons are emitted in the direction opposite to $\boldsymbol{B}$ (hence to $\boldsymbol{I}$) than in the same direction. The observed asymmetry is high (ratio of 1,2 / 0,8 between the emission intensities in both directions). As a control, one can invert the field $\boldsymbol{B}$ and the spins $\boldsymbol{I}$, and check that the emission intensities in both directions are also inverted.

The effect of the direction of an axial vector ($\boldsymbol{B}$ or $\boldsymbol{I}$) on a polar vector (preferred direction for electron emission) is a violation of parity conservation. If parity were to be conserved, two experiments symmetric with respect to a point, hence with the same directions of $\boldsymbol{I}$ and $\boldsymbol{B}$, would yield two opposite directions for the preferred directions of β^- emission (*cf.* figure 4).

A similar result is obtained in a mirror symmetry with respect to a plane parallel to $\boldsymbol{B}$. In this case, the directions of $\boldsymbol{I}$ and $\boldsymbol{B}$ are inverted, but the preferred emission direction would remain the same (*cf.* figure 5) if parity was conserved. Experimental results show that the parity conservation is violated.

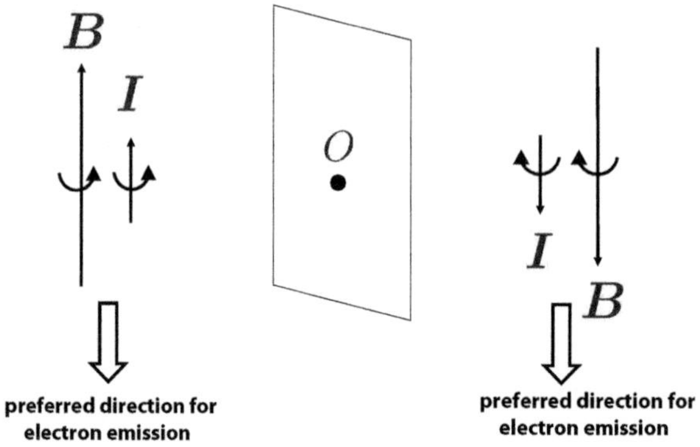

Figure 5: Comparison of two symmetric experiments, not with respect to a point as in figure 4, but with respect to a plane (mirror symmetry). If parity were conserved, the preferred direction of the β^- emission should remain the same although the spin direction is inverted.

At the present time, the idea of parity conservation in weak interactions has been abandoned. Note that it is not a small effect but rather a 100 % effect (neutrino theory with two, and not four, components). On the other hand, electromagnetic and strong interactions strictly conserve parity.

Chapter VI

Construction of state spaces and wave equations

We now apply the general results of the previous chapter to the construction of state spaces $\mathscr{E}$ for the action of operators that have a Galileo or Einstein Lie algebra structure. One may expect that many quantum systems should have such state spaces, with rather complex structures if, for example, they contain an very large number of particles. For clarity, we choose to focus on spaces $\mathscr{E}$ with the simplest possible structure, knowing that they can later be combined by tensor products to obtain more complex spaces. We shall look for "the smallest" spaces, those corresponding to an irreducible representation of the symmetry group under study. We will determine which wave equations can describe, in these spaces, the time evolution of quantum systems. In the case of the Galileo group symmetries, the physical system under study will be a single particle, with or without spin, free or interacting with an external electromagnetic field, and described by the Schrödinger equation. In the case of the Poincaré group, the situation is more complex because of the relativistic possibility of converting energy into particles or vice versa. This limits the scope of one-particle wave equations. Nevertheless,

Introduction to Continuous Symmetries: From Space–Time to Quantum Mechanics, First Edition. Franck Laloë.

we shall discuss three relativistic equations: the Klein–Gordon equation for a spinless particle, the Dirac equation for spin 1/2 particles with a non-zero mass such as the electron, and finally, the Weyl equation for massless particles with a non-zero spin.

A. Galilean group, the Schrödinger equation

For the Galilean group, the commutation relations of interest are relations (V-24) to (V-26) of chapter V.

A-1. Spinless free particle

To discuss, as we did before, the system's displacement we must be able to define a position operator $\boldsymbol{R}$ for the system (i.e. 3 Hermitian operators X, Y, and Z, its components). To build the "minimal" state space $\mathscr{E}$, we assume that these three components X, Y, and Z form a CSCO (Complete Set of Commuting Observables). Consequently, there exists an orthonormal basis whose vectors are eigenkets common to X, Y, and Z, and that can be specified solely (within a phase factor) by the corresponding eigenvalues x, y, and z. Note that if X, Y, and Z did not form a CSCO, additional quantum numbers would be necessary to define a basis in $\mathscr{E}$, hence "enlarging" that space. We shall limit ourselves to this " minimal hypothesis", and call "spinless particle" the corresponding physical system.

To any state $|\psi\rangle$ of the particle corresponds the wave function:

$$\psi(\boldsymbol{r}) = \langle \boldsymbol{r}|\psi\rangle \tag{VI-1}$$

where $\boldsymbol{r}$ is a condensed notation for the 3 numbers x, y, and z:

$$|\boldsymbol{r}\rangle = |x, y, z\rangle \tag{VI-2}$$

It is equivalent to define $|\psi\rangle$ belonging to $\mathscr{E}$ or a complex wave function $\psi(\boldsymbol{r})$ with an integrable squared modulus.

A-1-a. Transformations of the position operator

α. *Translations*

We now examine how the position operator $\boldsymbol{R}$ is transformed when the system undergoes a translation by a vector $\boldsymbol{b}$. As in complement C_V, we note $T(\boldsymbol{a}, \boldsymbol{b})$ the displacement operator that includes a rotation with parameter $\boldsymbol{a}$ and a translation with parameter $\boldsymbol{b}$. We studied in § C of chapter IV the transformation of an observable when a T operator acts in the state space. Under a translation, described by the unitary operator $T(0, \boldsymbol{b})$, operator $\boldsymbol{R}$ becomes operator $\boldsymbol{R}'$ given by:

$$\boldsymbol{R}' = T(0, \boldsymbol{b})\ \boldsymbol{R}\, T^{-1}(0, \boldsymbol{b}) \tag{VI-3}$$

As we saw in § C-2 of chapter IV, when the apparatus measuring the position undergoes a translation by vector $+\boldsymbol{b}$, the operator corresponding to this measurement varies by the quantity $-\boldsymbol{b}$, which led us to equality (IV-45a):

$$\boldsymbol{R}' = \boldsymbol{R} - \boldsymbol{b} \tag{VI-4}$$

When $\boldsymbol{b}$ takes an infinitesimal value $\delta\boldsymbol{b}$, we saw that:

$$T\left(0, \delta\boldsymbol{b}\right) = 1 - \frac{i}{\hbar}\delta\boldsymbol{b} \cdot \boldsymbol{P} + \ldots \tag{VI-5}$$

and relations (VI-3) and (VI-4) led to:

$$\boldsymbol{R} - \frac{i}{\hbar}\left[\delta\boldsymbol{b} \cdot \boldsymbol{P}\,,\ \boldsymbol{R}\right] = \boldsymbol{R} - \delta\boldsymbol{b} + \ldots \tag{VI-6}$$

We identify the terms containing the three components of $\delta\boldsymbol{b}$ on both sides of this equality. Taking $\delta\boldsymbol{b}$ parallel to the Ox axis, we get:

$$\begin{aligned}
[X, P_x] &= i\hbar \\
[Y, P_x] &= 0 \\
[Z, P_x] &= 0
\end{aligned} \tag{VI-7}$$

By circular permutation of x, y, and z, we get the general relations:

$$\boxed{[R_i, P_j] = i\hbar\ \delta_{ij}} \qquad i, j = x, y, z \tag{VI-8}$$

which are the basic commutation relations of position and momentum in quantum mechanics. Instead of assuming quantization rules, we obtained them in a purely "geometrical" way.

β. Change of Galilean reference frame

We now study the transformations of the position operator under a change of Galilean reference frame, also supposed to be infinitesimal. The unitary operation T describing such a transformation with the velocity parameter $\delta\boldsymbol{v}$, is:

$$T = 1 - \frac{i}{\hbar}\delta\boldsymbol{v}\cdot\boldsymbol{K} + \ldots \tag{VI-9}$$

We can follow the same argument as for a translation, but with a significant difference: a change of reference frame performed at $t = 0$ (when both spatial frames coincide) does not change the positions. Imposing the same rule to the quantum position operator $\boldsymbol{R}$, we must have:

$$[\boldsymbol{R}, \boldsymbol{K}] = 0 \tag{VI-10}$$

It follows that operator $\boldsymbol{K}$ must be diagonal in the position representation. Since we assumed that the components of $\boldsymbol{R}$ form a CSCO, $\boldsymbol{K}$ must be a function of operator $\boldsymbol{R}$:

$$\boldsymbol{K} = \boldsymbol{K}(\boldsymbol{R}) \tag{VI-11}$$

Commutation relation (VI-8) then shows:

$$[K_i(\boldsymbol{R}), P_j] = i\hbar \frac{\partial}{\partial R_j} K_i(\boldsymbol{R}) \tag{VI-12}$$

We obtained in chapter V the commutation relations of $\boldsymbol{K}$ and the momentum operator [equation (V-25b)]:

$$[K_i, P_j] = -i\hbar\, \delta_{ij} M \tag{VI-13}$$

Equality (VI-12) then shows that the derivative of the function $\boldsymbol{K}(\boldsymbol{R})$ is necessarily constant, so that:

$$\boldsymbol{K}(\boldsymbol{R}) = -M\ \boldsymbol{R} + \boldsymbol{C} \tag{VI-14}$$

The constant $\boldsymbol{C}$ can be readily eliminated by a redefinition of operator $\boldsymbol{K}$ into $\boldsymbol{K}' = \boldsymbol{K} - \boldsymbol{C}$, which does not change any of the commutation relations of the Lie algebra. From now on we will assume that this redefinition has been performed, and we shall abandon the prime in the notation $\boldsymbol{K}'$. This yields a simple relation between the position operator $\boldsymbol{K}$ and the position operator $\boldsymbol{R}$:

$$\boxed{\boldsymbol{R} = -\frac{\boldsymbol{K}}{M}} \tag{VI-15}$$

that we already anticipated at the end of § C-1 of chapter III.

A-1-b. Velocity operators and Hamiltonian

Taking (VI-15) into account, relation (V-26c) of chapter V yields the commutation relation between operators $\boldsymbol{R}$ and H:

$$[H, \boldsymbol{R}] = -\frac{i\hbar}{M}\ \boldsymbol{P} \tag{VI-16}$$

Since expression $(i/\hbar)\,[H, \boldsymbol{R}]$ directly determines the time evolution of the position operator in the Heisenberg point of view, it can be associated with the velocity operator $\boldsymbol{V}$, and we set:

$$\boldsymbol{V} = \frac{\boldsymbol{P}}{M} \tag{VI-17}$$

This well known relation between velocity and momentum comforts the identification we made between $\boldsymbol{P}$, the infinitesimal generator of translations, and the usual momentum of a physical system.

In addition, relation (V-26b) shows that H commutes with operator $\boldsymbol{P}$. Just as the commutation of $\boldsymbol{K}$ and $\boldsymbol{R}$ indicated above that $\boldsymbol{K}$ was a function of $\boldsymbol{R}$, this commutation indicates that:

$$H = H(\boldsymbol{P}) \tag{VI-18}$$

Relation (VI-16) then yields:

$$[H(\boldsymbol{P}), \boldsymbol{R}] = -i\hbar \boldsymbol{\nabla}_{\boldsymbol{P}} H(\boldsymbol{P}) = -i\hbar \frac{\boldsymbol{P}}{M} \tag{VI-19}$$

so that:

$$H = \frac{\boldsymbol{P}^2}{2M} + U_{\text{int}} = \frac{1}{2} M \boldsymbol{V}^2 + U_{\text{int}} \tag{VI-20}$$

where U_{int} is a constant playing the role of an internal energy that can be easily eliminated by a redefinition of the energy origin. This yields again expression (V-33b) for the kinetic energy of a free particle.

A-1-c. Position representation, the Schrödinger equation

In the position $\{|\boldsymbol{r}\rangle\}$ representation , the wave function $\psi(\boldsymbol{r})$ is defined in (VI-1). Operator $\boldsymbol{R}$ corresponds to the multiplication by $\boldsymbol{r}$ of this wave function:

$$\langle \boldsymbol{r} |\, \boldsymbol{R}\, | \psi \rangle = \boldsymbol{r}\ \psi(\boldsymbol{r}) \tag{VI-21}$$

Applying a translation operator to the ket $|\psi\rangle$ yields the wave function $\psi'(\boldsymbol{r})$:

$$\psi'(\boldsymbol{r}) = \langle \boldsymbol{r} |\, \mathrm{e}^{-i\boldsymbol{b}\cdot\boldsymbol{P}/\hbar}\, | \psi \rangle = \langle \tilde{\boldsymbol{r}}\ | \psi \rangle \tag{VI-22}$$

with:

$$|\tilde{\boldsymbol{r}}\rangle = \mathrm{e}^{+i\boldsymbol{b}\cdot\boldsymbol{P}/\hbar}\, |\boldsymbol{r}\rangle \tag{VI-23}$$

The ket $|\tilde{\boldsymbol{r}}\rangle$ is simply the transform under a translation by vector $-\boldsymbol{b}$ of the position eigenket $|\boldsymbol{r}\rangle$. It is proportional to $|\boldsymbol{r} - \boldsymbol{b}\rangle$ and, since the operator $\mathrm{e}^{+i\boldsymbol{b}\cdot\boldsymbol{P}}$

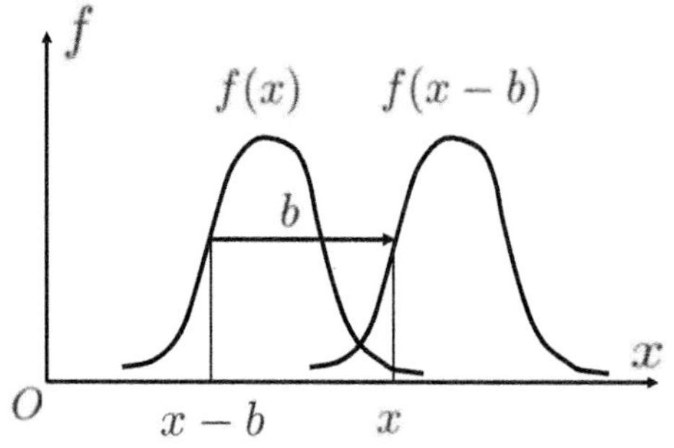

Figure 1: In one dimension, a function $f(x)$ is translated by $+b$ when subtracting the quantity b from its variable x.

preserves the norm, it only differs from it by an a priori arbitrary phase factor $\mathrm{e}^{i\theta}$. Choosing the relative phase (unspecified until now) of all the position eigenvectors $|\boldsymbol{r}\rangle$ as:

$$|\boldsymbol{r}\rangle = \mathrm{e}^{-i\boldsymbol{r}\cdot\boldsymbol{P}/\hbar}\,|\boldsymbol{r}=\boldsymbol{0}\rangle \tag{VI-24}$$

eliminates this phase factor, and we have:

$$|\tilde{\boldsymbol{r}}\rangle = \mathrm{e}^{i\boldsymbol{b}\cdot\boldsymbol{P}/\hbar}\mathrm{e}^{-i\boldsymbol{r}\cdot\boldsymbol{P}/\hbar}\,|\boldsymbol{r}=\boldsymbol{0}\rangle = \mathrm{e}^{i(\boldsymbol{b}-\boldsymbol{r})\cdot\boldsymbol{P}/\hbar}\,|\boldsymbol{r}=\boldsymbol{0}\rangle = |\boldsymbol{r}-\boldsymbol{b}\rangle \tag{VI-25}$$

This yields:

$$\psi'(\boldsymbol{r}) = \langle\tilde{\boldsymbol{r}}\,|\psi\rangle = \psi(\boldsymbol{r}-\boldsymbol{b}) \tag{VI-26}$$

Figure 1 illustrates in one dimension how, subtracting a quantity b from the variable, translates the function by a quantity $+b$.

Considering now an infinitesimal translation by a vector $\mathrm{d}\mathbf{b}$, we get:

$$\begin{aligned}\psi'(\mathbf{r}) &= \psi(\boldsymbol{r}) - \mathrm{d}\mathbf{b}\cdot\boldsymbol{\nabla}\psi(\mathbf{r}) = \langle\boldsymbol{r}|\,1 - \frac{i}{\hbar}\,\mathrm{d}\boldsymbol{b}\cdot\boldsymbol{P}\,|\psi\rangle \\ &= \psi(\boldsymbol{r}) - \frac{i}{\hbar}\,\mathrm{d}\mathbf{b}\cdot\langle\boldsymbol{r}|\,\boldsymbol{P}\,|\psi\rangle\end{aligned} \tag{VI-27}$$

or:

$$\langle\boldsymbol{r}|\,\boldsymbol{P}\,|\psi\rangle = \frac{\hbar}{i}\boldsymbol{\nabla}\psi(\boldsymbol{r}) \tag{VI-28}$$

We find again that the action of the momentum operator in the position representation is that of the gradient operator multiplied by $\hbar/i$.

Finally, the time evolution of the wave function $\psi(\mathbf{r})$ is given by:

$$|\psi(t+\mathrm{d}t)\rangle - |\psi(t)\rangle = -\frac{i}{\hbar} H \mathrm{d}t \, |\psi(t)\rangle \tag{VI-29}$$

Projecting on the $\{|\boldsymbol{r}\rangle\}$ basis, we get:

$$i\hbar \frac{\partial}{\partial t}\psi(\boldsymbol{r},t) = i\hbar\frac{\partial}{\partial t}\langle \boldsymbol{r}\,|\psi(t)\rangle = \langle \boldsymbol{r}|\,H\,|\psi(t)\rangle \tag{VI-30}$$

Taking into account (VI-20) and (VI-28) yields:

$$i\hbar \frac{\partial}{\partial t}\psi(\boldsymbol{r},t) = \left[-\frac{\hbar^2}{2M}\Delta + U_{\text{int}}\right]\psi(\boldsymbol{r},t) \tag{VI-31}$$

which is the well-known form of the Schrödinger equation for a free particle (the constant internal energy U_{int} only changes the global phase of the wave function at each instant, with no physical consequence).

Comment:

The above reasoning explains why operator H corresponds to $+i\hbar\partial/\partial t$, whereas $\boldsymbol{P}$ corresponds to $-i\hbar\boldsymbol{\nabla}$. This is the quantum analog of the classical situation discussed in the comment of page 8.

The quantum description of position is an operator associated with eigenkets $|\boldsymbol{r}\rangle$. We know that when going from a bra to a ket there is a sign change between ket and bra. This is because $\mathrm{e}^{-i\boldsymbol{b}\cdot\boldsymbol{P}/\hbar}\,|\boldsymbol{r}\rangle$ is equal to $|\boldsymbol{r}+\boldsymbol{b}\rangle$, whereas $\langle\boldsymbol{r}|\mathrm{e}^{-i\boldsymbol{b}\cdot\boldsymbol{P}/\hbar}$ is equal to $\langle\boldsymbol{r}-\boldsymbol{b}|$ – *cf.* for example (VI-22) and (VI-23). As a result, the translation of a wave function $\psi(\boldsymbol{r})$ by a vector $+\boldsymbol{b}$ yields a wave function $\psi(\boldsymbol{r}-\boldsymbol{b})$. However, time t is associated neither with an operator nor with a representation in the state space; such a sign change between ket and bra cannot happen.

A-1-d. Rotations

We now examine the action of the rotation operators and of the components of $\boldsymbol{J}$. When dealing with translations, we eliminated in (VI-24) the phase factors by a suitable choice of the relative phases of the kets $|\boldsymbol{r}\rangle$. It is not obvious, a priori, that such an elimination will work for the rotations though, in a first step, we shall assume it does. Later on, we will show that the results are unchanged, should we have to take into account the possible existence of phase factors.

The action of a rotation operator $T(\boldsymbol{a},\boldsymbol{0})$ on an eigenket $|\boldsymbol{r}\rangle$ of the position operator yields another eigenket of this operator, with an eigenvalue $\boldsymbol{r}'$ obtained from $\boldsymbol{r}$ by a rotation $\mathscr{R}(\boldsymbol{a})$ in ordinary space. We assume for the moment that no phase factor comes into play (this will be demonstrated later). We can write:

$$T(\boldsymbol{a},\boldsymbol{0})\,|\boldsymbol{r}\rangle = |\mathscr{R}(\boldsymbol{a})\,\boldsymbol{r}\rangle \tag{VI-32}$$

Since T is unitary, we have:

$$T^{\dagger}(\boldsymbol{a},\boldsymbol{0})\,|\boldsymbol{r}\rangle = |\mathscr{R}^{-1}(\boldsymbol{a})\,\boldsymbol{r}\rangle \qquad \text{that is:} \qquad \langle\boldsymbol{r}|T(\boldsymbol{a},\boldsymbol{0}) = \langle\mathscr{R}^{-1}(\boldsymbol{a})\boldsymbol{r}| \tag{VI-33}$$

where $\mathscr{R}^{-1}(\boldsymbol{a})$ is the inverse of rotation $\mathscr{R}(\boldsymbol{a})$. The transform $\psi'(\boldsymbol{r})$ of a wave function $\psi(\boldsymbol{r})$ is written:

$$\psi'(\boldsymbol{r}) = \langle\boldsymbol{r}|T(\boldsymbol{a},\boldsymbol{0})\,|\psi\rangle = \psi\left(\mathscr{R}^{-1}(\boldsymbol{a})\,\boldsymbol{r}\right) \tag{VI-34a}$$

or as:

$$\psi'(\boldsymbol{r}') = \psi(\boldsymbol{r}) \qquad \text{with: } \boldsymbol{r}' = \mathscr{R}\boldsymbol{r} \tag{VI-34b}$$

For an infinitesimal rotation, we get [*cf.* relation (28) of complement C_V]:

$$\psi'(\boldsymbol{r}) = \psi(\boldsymbol{r}) - (\delta\boldsymbol{a}\times\boldsymbol{r})\cdot\boldsymbol{\nabla}\psi = \psi(\boldsymbol{r}) - \frac{i}{\hbar}\langle\boldsymbol{r}|\delta\boldsymbol{a}\cdot\boldsymbol{J}|\psi\rangle \tag{VI-35}$$

that is:

$$\langle\boldsymbol{r}|\delta\boldsymbol{a}\cdot\boldsymbol{J}|\psi\rangle = \frac{\hbar}{i}\,(\delta\boldsymbol{a}\times\boldsymbol{r})\cdot\boldsymbol{\nabla}\psi = \frac{\hbar}{i}\,\delta\boldsymbol{a}\cdot(\boldsymbol{r}\times\boldsymbol{\nabla}\psi) \tag{VI-36}$$

Therefore:

$$\langle\boldsymbol{r}|\boldsymbol{J}|\psi\rangle = \frac{\hbar}{i}\boldsymbol{r}\times\boldsymbol{\nabla}\psi(\boldsymbol{r}) \tag{VI-37}$$

so that the action of $\boldsymbol{J}$ in the $\{|\boldsymbol{r}\rangle\}$ representation is given by:

$$\boldsymbol{J} \Rightarrow \frac{\hbar}{i}\boldsymbol{r}\times\boldsymbol{\nabla} \tag{VI-38}$$

Taking (VI-28) into account leads us to set:

$$\boldsymbol{J} = \boldsymbol{L} = \boldsymbol{R}\times\boldsymbol{P} \tag{VI-39}$$

which is the well-known expression for the orbital angular momentum of a particle. Relation (V-41) of chapter V then shows that its spin is zero:

$$\boldsymbol{S} = 0 \tag{VI-40}$$

which is a direct consequence of our assumption that the three components of operator $\boldsymbol{R}$ form a CSCO.

Effect of phase factors: We now take into account the possible existence of phase factors $e^{i\theta}$ in (VI-32), which becomes:

$$T(\boldsymbol{a},\boldsymbol{0})\,|\boldsymbol{r}\rangle = e^{i\theta(\boldsymbol{a},\boldsymbol{r})}|\mathscr{R}(\boldsymbol{a})\,\boldsymbol{r}\rangle \tag{VI-41}$$

Relation (VI-34a) must be replaced by:

$$\psi'(\boldsymbol{r}) = \mathrm{e}^{-i\theta(\boldsymbol{a},\boldsymbol{r})}\,\psi\left(\mathscr{R}^{-1}(\boldsymbol{a})\,\boldsymbol{r}\right) \tag{VI-42}$$

The phase factor $\theta(\boldsymbol{a},\boldsymbol{r})$ is identically zero if $\boldsymbol{a}=0$. When $\boldsymbol{a}$ is infinitesimal, expression (VI-36) becomes:

$$\langle\boldsymbol{r}|\delta\boldsymbol{a}\cdot\boldsymbol{J}|\psi\rangle = \frac{\hbar}{i}\,\delta\boldsymbol{a}\cdot(\boldsymbol{r}\times\boldsymbol{\nabla}\psi) + \hbar\,\delta\boldsymbol{a}\cdot\boldsymbol{\nabla}_{\boldsymbol{a}}\theta(\boldsymbol{a},\boldsymbol{r}) + \ldots \tag{VI-43}$$

which includes the gradient of phase θ with respect to $\boldsymbol{a}$, taken at $\boldsymbol{a}=0$. Correspondence (VI-38) must be modified, as the action of $\boldsymbol{J}$ in the $\{|\boldsymbol{r}\rangle\}$ representation is now written:

$$\frac{\hbar}{i}\boldsymbol{r}\times\boldsymbol{\nabla} + \hbar\boldsymbol{\nabla}_{\boldsymbol{a}}\theta(\boldsymbol{a},\boldsymbol{r}) \tag{VI-44}$$

and the right-hand side of (VI-39) becomes:

$$\boldsymbol{R}\times\boldsymbol{P} + \hbar\boldsymbol{\nabla}_{\boldsymbol{a}}\theta(\boldsymbol{a},\boldsymbol{r}) \tag{VI-45}$$

This leads to a change in the commutation relations between $\boldsymbol{J}$ and $\boldsymbol{P}$. We must add to the angular momentum operator (VI-39) a function of the position operator. This introduces the operator $\boldsymbol{J}''$ given by:

$$\boldsymbol{J}'' = \boldsymbol{R}\times\boldsymbol{P} + \boldsymbol{\chi}(\boldsymbol{R}) \qquad \text{with:} \quad \boldsymbol{\chi}(\boldsymbol{R}) = \hbar\boldsymbol{\nabla}_{\boldsymbol{a}}\theta(\boldsymbol{R}) \tag{VI-46}$$

Calculating the commutation relations of this operator with the three components of $\boldsymbol{P}$ we get:

$$[J''_x, P_x] = [J_x, P_x] + [\chi_x(\boldsymbol{R}), P_x] = i\hbar\frac{\partial}{\partial x}\,\chi_x(\boldsymbol{R})$$

Similarly the commutation relation between the x and y becomes:

$$[J''_x, P_y] = i\hbar\frac{\partial}{\partial y}\,\chi_x(\boldsymbol{R}) + i\hbar\,P_z$$

An analogous result is obtained for $[J''_x, P_z]$. Under these conditions, to preserve the commutation relations (V-24b) of chapter V, the function χ_x must necessarily be a constant independent of x, y, and z (by symmetry, the same conclusion is valid for the 3 components of $\boldsymbol{\chi}$, which must all be constants). As a result, we would get, for example, relations such as:

$$\left[J''_x, J''_y\right] = i\hbar\,(J''_z - \chi_z)$$

which would lead to modified commutation relations for the components of the angular momentum. We saw in § B of chapter V that such constants in the commutation relations can be eliminated. This amounts to setting $\chi\equiv 0$ and shows that the phase factors do not play any role in our problem.

For the simple case we are studying (single spinless particle), we have obtained representations in the strict sense of the term (non-projective representations) for the rotation and translation groups.

A-2. Interacting particle

Imagine now that the particle is no longer free, but placed in an external potential. The origin of this potential could be, for example, the interactions with one or several other particles, supposed to be infinitely heavy and hence immobile in the reference frame of a given Galilean observer O_{ref}. It follows that all the Galilean observers are no longer equivalent, which invalidates the reasoning of §§ A-1-b and A-1-c, more precisely those parts that concern time evolution and the value of H, which is expected to include terms other than the kinetic energy. The Hamiltonian H may, a priori, be different for all the possible O observers.

Nevertheless, we shall try to extend the approach of § A-1, which will prove to be possible under a few reasonable assumptions. Among all the commutation relations of the Lie algebra of the Galilean group described at the beginning of § B-5 of chapter V, it is clear that we must abandon those containing H, since all the Galilean reference frames are not equivalent as far as time evolution of the system is concerned. We shall focus on the other commutation relations.

At a given instant, we can still study the physical system in a different displaced reference frame. Mathematically, this operation is the same as for a free particle, even though it may no longer be a symmetry operation since the particle's subsequent evolution may be changed. Nevertheless, the conclusions of § A-1-a-α remain valid.

We can also consider transformations involving a change of Galilean frame, provided we only study their instantaneous effect at time $t = 0$ (a wave function is multiplied at this instant by exponential $\mathrm{e}^{i\boldsymbol{k}\cdot\boldsymbol{r}}$). Classically, these transformations change the velocities without changing the positions, which are only modified in the subsequent evolution controlled by the Hamiltonian. The reasoning of § A-1-a-β remains valid, including the relation (VI-15) between operators $\boldsymbol{K}$ and $\boldsymbol{R}$.

We now study how the momentum $\boldsymbol{P}$ is transformed under a change of Galilean reference frame. For an infinitesimal transformation with parameter $\delta\boldsymbol{v}$, the momentum $\boldsymbol{P}$ becomes:

$$\begin{aligned}\boldsymbol{P}' &= \left[1 - \frac{i}{\hbar}\delta\boldsymbol{v}\cdot\boldsymbol{K}\right]\boldsymbol{P}\left[1 + \frac{i}{\hbar}\delta\boldsymbol{v}\cdot\boldsymbol{K}\right] + \ldots = \boldsymbol{P} - \frac{i}{\hbar}\left[\delta\boldsymbol{v}\cdot\boldsymbol{K}\;,\;\boldsymbol{P}\right] \\ &= \boldsymbol{P} - M\delta\boldsymbol{v}\end{aligned} \qquad \text{(VI-47)}$$

where the second line comes from the commutation relation (V-25b) of $\boldsymbol{K}$ and $\boldsymbol{P}$. As in relation (VI-4), the minus sign comes from the apparent motion of the measurement apparatus upon a change of reference frame (when an apparatus, moving at a velocity $\delta\boldsymbol{v}$ in a reference frame, measures the velocity of a particle moving at a velocity $\boldsymbol{v}_0$ in the same reference frame, the measurement result will be $\boldsymbol{v}_0 - \delta\boldsymbol{v}$).

We keep the definition of a velocity operator $\boldsymbol{V}$ as the rate of variation of the position $\boldsymbol{R}$:

$$\boldsymbol{V} = \frac{1}{i\hbar}\left[\boldsymbol{R}\ ,\ H\right] \tag{VI-48}$$

In the previous change of reference frame, we expect $\boldsymbol{V}$ to vary by $-\delta\boldsymbol{v}$. Comparison with (VI-47) shows that $\boldsymbol{P} - M\boldsymbol{V}$ must be invariant under this operation, meaning that it commutes with the infinitesimal generator $\boldsymbol{K}$. Since this operator is proportional to the $\boldsymbol{R}$ operator, $\boldsymbol{P} - M\boldsymbol{V}$ must be diagonal in the position representation. Two cases are then possible:

• For a spinless particle, the three components of the position operator form a CSCO (an eigenket is completely determined, up to a constant, by the three eigenvalues of the position components), and operator $\boldsymbol{P} - M\boldsymbol{V}$ is simply expressed as a function[1] of $\boldsymbol{R}$:

$$\boldsymbol{P} - M\boldsymbol{V} = q\boldsymbol{A}(\boldsymbol{R}) \tag{VI-49}$$

Using (VI-48), we then have:

$$M\boldsymbol{V} = \boldsymbol{P} - q\boldsymbol{A}(\boldsymbol{R}) = \frac{M}{i\hbar}\left[\boldsymbol{R}\ ,H\ \right] \tag{VI-50}$$

For the commutator of H and $\boldsymbol{R}$ to have the value given by this equality, we must have:

$$\boxed{H = \frac{1}{2M}\left[\boldsymbol{P} - q\boldsymbol{A}(\boldsymbol{R})\right]^2 + qU(\boldsymbol{R}) = \frac{1}{2}M\boldsymbol{V}^2 + qU(\boldsymbol{R})} \tag{VI-51}$$

where $U(\boldsymbol{R})$ is a function of $\boldsymbol{R}$ that we multiply, as above, by the arbitrary constant q.

• If the particle has a non-zero spin, we can keep relations (VI-49) and (VI-50), but must modify expression (VI-51) for the Hamiltonian. Let us set:

$$H = \frac{1}{2M}\left[\boldsymbol{P} - q\boldsymbol{A}(\boldsymbol{R})\right]^2 + X \tag{VI-52}$$

where X is an operator to be determined. As before, relation (VI-48) demands that this operator be diagonal in the position representation. If the particle's spin is $1/2$, this operator is represented by a 2×2 matrix in each eigen subspace of the position operator with eigenvalue $\boldsymbol{r}$. Each matrix can be decomposed on the Pauli matrices $\boldsymbol{\sigma}$ and the identity matrix. The restriction of X to this subspace can be written as:

$$(X)_{\boldsymbol{r}} = qU(\boldsymbol{r}) - \mu\,\boldsymbol{B}(\boldsymbol{r}) \cdot \boldsymbol{\sigma} \tag{VI-53}$$

[1]By convention, instead of writing this function as $\boldsymbol{A}(\boldsymbol{R})$, we multiply it by a constant q, whose value is not specified at this stage. This leads for the Hamiltonian to expression (VI-51), identical to the standard expression that includes a charge q.

where $U(\boldsymbol{r})$ is a scalar function, and $\boldsymbol{B}(\boldsymbol{r})$ a vector function, both a priori arbitrary; μ is a constant, also arbitrary. Coming back to an operator formalism valid in the entire state space, we can write:

$$\boxed{H = \frac{1}{2M}\left[\boldsymbol{P} - q\boldsymbol{A}(\boldsymbol{R})\right]^2 + qU(\boldsymbol{R}) - \mu\,\boldsymbol{B}(\boldsymbol{R})\cdot\boldsymbol{\sigma}} \tag{VI-54}$$

Based only on symmetry arguments we found the Hamiltonian of a particle placed in a vector potential $\boldsymbol{A}$ and a scalar potential U. The μ term is interpreted as the coupling between the magnetic moment of the particle's spin and the magnetic field. Note, however, that we have not demonstrated that $\boldsymbol{B}(\boldsymbol{r})$ is proportional to the curl $\boldsymbol{\nabla}\times\boldsymbol{A}(\boldsymbol{r})$; for a more complete demonstration including the $g = 2$ factor for the particle's spin, see reference [29].

Comments:

(i) Note that we have remained in the framework of Galilean relativity, whereas electromagnetism is compatible with Einstein relativity. We should not expect finding the exact Hamiltonian of a charged particle placed in an electromagnetic field. For example, the function $\boldsymbol{A}(\boldsymbol{R})$ appearing in the Hamiltonian is invariant under a change of Galilean reference frame at $t = 0$, whereas Maxwell's electromagnetism predicts that the vector and scalar potentials in the Lorenz gauge should vary as a relativistic four-vector[2].

(ii) Taking the reasoning further, we compare the Hamiltonians H and H' used by two different Galilean observers O and O', which will establish how the potential operator $U(\boldsymbol{R}, t)$ is transformed. From now on, we explicitly take into account a possible time dependence of U and $\boldsymbol{A}$. We introduce the $\boldsymbol{\Gamma}$ operator associated with acceleration. It is defined as the time variation of operator (VI-50) in the Heisenberg point of view, which is the sum of a Hamiltonian variation of the velocity operator and a variation coming from the explicit time dependence of $\boldsymbol{A}(\boldsymbol{r}, t)$:

$$\boldsymbol{\Gamma} = \frac{1}{i\hbar}\left[\boldsymbol{V}, H\right] - \frac{q}{M}\frac{\partial}{\partial t}\boldsymbol{A}(\boldsymbol{r}, t) = \frac{i}{\hbar}\left[\frac{M\boldsymbol{V}^2}{2} + qU(\boldsymbol{R}, t)\,,\boldsymbol{V}\right] - \frac{q}{M}\frac{\partial}{\partial t}\boldsymbol{A}(\boldsymbol{r}, t) \tag{VI-55}$$

A simple calculation yields[3]:

$$\left[\frac{M\boldsymbol{V}^2}{2}\,,\boldsymbol{V}\right] = \frac{\hbar q}{2iM}\left[\boldsymbol{V}\times\boldsymbol{B}(\boldsymbol{R}, t) - \boldsymbol{B}(\boldsymbol{R}, t)\times\boldsymbol{V}\right] \tag{VI-56a}$$

where:

$$\boldsymbol{B}(\boldsymbol{R}, t) = \boldsymbol{\nabla}\times\boldsymbol{A}(\boldsymbol{R}, t) \tag{VI-56b}$$

[2]To first-order in $\delta\boldsymbol{v}$, their variations are given by $U' = U + \delta\boldsymbol{v}\cdot\boldsymbol{A} + \ldots$ and $\boldsymbol{A}' = \boldsymbol{A} + (\delta\boldsymbol{v}/c^2)U + \ldots$.

[3]The explicit calculation is given in complement $\mathrm{E_{VI}}$ of reference [9], equation (45).

is the operator corresponding to the magnetic field derived from the vector potential $\boldsymbol{A}$. This leads to the analog of the classical law for the Laplace force. Since:

$$[U(\boldsymbol{R},t),\boldsymbol{V}] = \frac{1}{M}\,[U(\boldsymbol{R},t),\boldsymbol{P}] = \frac{i\hbar}{M}\,\boldsymbol{\nabla} U(\boldsymbol{R},t) \tag{VI-57a}$$

we have:

$$M\boldsymbol{\Gamma} = \frac{q}{2}\,[\boldsymbol{V}\times\boldsymbol{B}(\boldsymbol{R},t) - \boldsymbol{B}(\boldsymbol{R},t)\times\boldsymbol{V}] - q\left[\boldsymbol{\nabla} U(\boldsymbol{R},t) + \frac{\partial}{\partial t}\,\boldsymbol{A}(\boldsymbol{R},t)\right] \tag{VI-57b}$$

This equality yields, in addition to the magnetic force, the electric force associated with the field $\boldsymbol{E}$:

$$\boldsymbol{E}(\boldsymbol{R},t) = -\boldsymbol{\nabla} U(\boldsymbol{R},t) - \frac{\partial}{\partial t}\,\boldsymbol{A}(\boldsymbol{R},t) \tag{VI-58}$$

It is clear that the calculations of the acceleration by two different Galilean observers O and O' must lead to the same result. Since $\boldsymbol{A}$ is an invariant quantity under a change of Galilean reference frame, we can write:

$$\boldsymbol{A}(\boldsymbol{r},t) \equiv \boldsymbol{A}'(\boldsymbol{r}',t) \tag{VI-59a}$$

if:

$$\boldsymbol{r}' = \boldsymbol{r} + \boldsymbol{v}\,t \tag{VI-59b}$$

We thus have:

$$\mathrm{d}A_x = \boldsymbol{dr}\cdot\boldsymbol{\nabla} A_x + \mathrm{d}t\,\frac{\partial}{\partial t}\,A_x = \boldsymbol{dr}'\cdot\boldsymbol{\nabla}' A_x + \mathrm{d}t'\,\frac{\partial}{\partial t'}\,A_x \tag{VI-59c}$$

or:

$$\frac{\partial}{\partial t}\,\boldsymbol{A} = \frac{\partial}{\partial t'}\,\boldsymbol{A} + (\boldsymbol{v}\cdot\boldsymbol{\nabla})\,\boldsymbol{A} \tag{VI-59d}$$

The accelerations in the two reference frames O and O′ will be the same if:

$$\begin{aligned}\frac{1}{2}\,[\boldsymbol{V}\times\boldsymbol{B}(\boldsymbol{R},t) - \boldsymbol{B}(\boldsymbol{R},t)\times\boldsymbol{V}] - \boldsymbol{\nabla} U(\boldsymbol{R},t) - \frac{\partial}{\partial t}\,\boldsymbol{A}(\boldsymbol{R},t)\\ = \frac{1}{2}\,([\boldsymbol{V}+\boldsymbol{v}]\times\boldsymbol{B}(\boldsymbol{R},t) - \boldsymbol{B}(\boldsymbol{R},t)\times[\boldsymbol{V}+\boldsymbol{v}])\\ - \boldsymbol{\nabla} U'(\boldsymbol{R},t) - \frac{\partial}{\partial t}\,\boldsymbol{A}(\boldsymbol{R},t) + (\boldsymbol{v}\cdot\boldsymbol{\nabla})\,\boldsymbol{A}(\boldsymbol{R},t)\end{aligned} \tag{VI-60a}$$

meaning if:

$$\boldsymbol{\nabla}\,[U'(\boldsymbol{R},t) - U(\boldsymbol{R},t)] = \boldsymbol{v}\times\boldsymbol{B}(\boldsymbol{R},t) + (\boldsymbol{v}\cdot\boldsymbol{\nabla})\boldsymbol{A}(\boldsymbol{R},t) \tag{VI-60b}$$

Since:

$$\boldsymbol{\nabla}\,[\boldsymbol{v}\cdot\boldsymbol{A}] = \boldsymbol{v}\times(\boldsymbol{\nabla}\times\boldsymbol{A}) + (\boldsymbol{v}\cdot\boldsymbol{\nabla})\,\boldsymbol{A} \tag{VI-61}$$

we simply have:

$$U'(\boldsymbol{R},t) - U(\boldsymbol{R},t) = \boldsymbol{v}\cdot\boldsymbol{A}(\boldsymbol{R},t) + U_0(t) \qquad \text{(VI-62)}$$

where the diagonal operator U_0 can be chosen equal to zero. As opposed to $\boldsymbol{A}$, the scalar potential is not invariant under a Galilean transformation[4]. Consequently, going from Hamiltonian H in a first reference frame to Hamiltonian H' in a second one is not done by simply applying to H the same unitary transformation as for the state vector (such a transformation would change H into $H - M\,\boldsymbol{v}\cdot\boldsymbol{V}$, which is different from the value of H').

B. Poincaré group, Klein–Gordon, Dirac, and Weyl equations

We return to the previous approach, but in the framework of the Poincaré group and the commutation relations (V-60) to (V-62) of § C of chapter V. We obtained two operators that commute with all those of the group's Lie algebra: the mass M and the operator W^2, which characterizes the value of the particle's spin. In §§ B-1 and B-2, we will assume that the mass is different from zero and will consider successively two different values of the spin; the case of a massless particle is studied in § B-3.

B-1. Klein–Gordon equation

In the Lie algebra of the Poincaré group, there is no position operator corresponding to position. We saw that, in the Galilean group, the translation transformation of the position operator led to the canonical commutation relations (VI-8). In the Lie algebra we are now studying, there is no operator whose commutator with $\boldsymbol{P}$ yields a constant. This is why, instead of assuming that the three components of the position operator form a CSCO, we now assume it is the three components of $\boldsymbol{P}$ that form a CSCO. As before, this hypothesis implies that the spin is equal to zero, and we introduce[5] the wave function $\tilde{\psi}(\boldsymbol{p})$:

$$\tilde{\psi}(\boldsymbol{p}) = \langle \boldsymbol{p}|\psi\rangle \qquad \text{(VI-63)}$$

B-1-a. Writing the equation

To get the time evolution of this function, we use relation (V-68), which yields the square of the Hamiltonian H as a function of $\boldsymbol{P}$ and mass M:

$$H^2 = \boldsymbol{P}^2c^2 + M^2c^4 \qquad \text{(VI-64)}$$

[4]This is the so-called "magnetic Galilean limit" of electromagnetism – *cf.* equation (2.18) of [49]; see also note 2 where the variation of $\boldsymbol{A}$ is proportional to $1/c^2$.

[5]We use the notation $\tilde{\psi}$ for this wave function in the momentum representation, instead of the usual notation $\overline{\psi}$, because $\overline{\psi}$ has traditionally a different signification in the context of the Dirac equation – *cf.* relation (29b) of complement C_{VI}.

Since $-iH/\hbar$ is associated with the infinitesimal time variation of $|\psi\rangle$, hence to its time derivative, we get[6]:

$$-\hbar^2 \frac{\partial^2}{\partial t^2}\tilde{\psi}(\boldsymbol{p},t) = p^2c^2\ \tilde{\psi}(\boldsymbol{p},t) + M^2c^4\ \tilde{\psi}(\boldsymbol{p},t) \tag{VI-65a}$$

Introducing the Fourier transform $\psi(\boldsymbol{r})$ of $\tilde{\psi}(\boldsymbol{p})$, we take the Fourier transform of this equation:

$$-\hbar^2 \frac{\partial^2}{\partial t^2}\psi(\boldsymbol{r},t) = -\hbar^2c^2\ \Delta\psi(\boldsymbol{r},t) + M^2c^4\ \psi(\boldsymbol{r},t) \tag{VI-65b}$$

These two relations are the two forms of the Klein–Gordon equation, respectively in the moment and position representations.

B-1-b. Problems with the local conservation law

Like the Schrödinger equation, the Klein–Gordon equation obeys a local conservation law:

$$\frac{\partial}{\partial t}\rho(\boldsymbol{r},t) + \boldsymbol{\nabla}\cdot\boldsymbol{j}(\boldsymbol{r},t) = 0 \tag{VI-66}$$

where the probability current $\boldsymbol{j}(\boldsymbol{r},t)$ has the usual expression:

$$\boldsymbol{j}(\boldsymbol{r},t) = \frac{\hbar}{2Mi}\left[\psi^\star(\boldsymbol{r},t)\,\boldsymbol{\nabla}\psi(\boldsymbol{r},t) - \text{c.c.}\right] \tag{VI-67}$$

(c.c. means "complex conjugate") but where $\rho(\boldsymbol{r},t)$ is now written:

$$\rho(\boldsymbol{r},t) = \frac{i\hbar}{2Mc^2}\left[\psi^\star(\boldsymbol{r},t)\,\frac{\partial}{\partial t}\psi(\boldsymbol{r},t) - \text{c.c.}\right] \tag{VI-68}$$

Using these two expressions, it is easy to show[7] that the conservation equation (VI-66) is a consequence of the Klein–Gordon equation (VI-65b). However a difficulty arises when trying to interpret $\rho(\boldsymbol{r},t)$ as a probability density: the right-hand side of (VI-68) is not always positive, contrary to the modulus squared of ψ, which gives the value of ρ for the Schrödinger equation.

Such difficulties had to be expected. As we did not find a position operator in the Lie algebra of the group, we simply used a mathematical definition using a Fourier transform, without further justification. In addition, we are reaching

[6]It may be tempting to take the square root of expression (VI-64) to build a first-order time differential equation. One would then have to deal with the square root of a sum of operators, source of many difficulties.

[7]In complement C_{VI}, § 3-c, we obtain this conservation relation by a more general method (Noether theorem).

the limit of trying to build a simple physical system (single particle) in a relativistic framework valid at all energies. As energy and mass are equivalent in such a framework, with sufficient energy new particles can certainly be created. Imposing uniqueness of the particle is certainly too strict a condition.

A more powerful tool for dealing with such a situation is field theory, where $\psi(\boldsymbol{r})$ is no longer a wave function but an operator acting in a "Fock space" where the particle number can vary (*cf.*, for example, chapter XV of [9]). Position is no longer an operator, but a parameter that plays the role of an index labeling the operators acting in the state pace. In this context, the Klein–Gordon equation is no longer seen as a theoretical base in itself, but rather as an intermediary step towards a more general field theory.

B-1-c. Relativistic transformations

Under a space or time translation, the Klein–Gordon wave function is transformed in the same way as the Schrödinger wave function in (VI-26), hence according to (*cf.* comment page 247):

$$\psi'(\boldsymbol{r}',t') = \psi'(\boldsymbol{r}+\boldsymbol{b},t-\tau) = \psi(\boldsymbol{r},t) \tag{VI-69a}$$

or:

$$\psi'(\boldsymbol{r},t) = \psi(\boldsymbol{r}-\boldsymbol{b},t+\tau) \tag{VI-69b}$$

Similarly, under a rotation the transformation of the wave function is:

$$\psi'(\boldsymbol{r}',t) = \psi'(\mathscr{R}\boldsymbol{r},t) = \psi(\boldsymbol{r},t) \qquad \text{or:} \qquad \psi'(\boldsymbol{r},t) = \psi(\mathscr{R}^{-1}\boldsymbol{r},t) \tag{VI-70}$$

Consider now a general Lorentz transformation Λ (that can include a rotation), which, applied to the four-vector $(ct,\boldsymbol{r})$ with components x^μ, yields the four-vector $(ct',\boldsymbol{r}')$ with components x'^μ. It is associated with a 4×4 matrix, noted Λ, which transforms[8] these components as:

$$x'^\nu = \Lambda^\nu_{\ \mu}\, x^\mu \tag{VI-71}$$

(with the Einstein convention where the repeated index μ is summed). We then have:

$$\psi(\boldsymbol{r}',t') = \psi'(\Lambda(\boldsymbol{r},t)) = \psi(\boldsymbol{r},t) \qquad \text{or:} \qquad \psi'(\boldsymbol{r},t) = \psi(\Lambda^{-1}(\boldsymbol{r},t)) \tag{VI-72}$$

[8]The coordinates of a space–time event undergo the Λ transformation, which amounts to saying that the physical system is seen in the new reference frame as having undergone this same transformation. It is also equivalent to say that the axes of the reference frame have undergone the inverse transformation Λ^{-1}. For example, if Λ is a rotation $\mathscr{R}$ in ordinary space as in (VI-34), one can say that either the system has undergone an $\mathscr{R}$ rotation , or the axes have undergone a $\mathscr{R}^{-1}$ rotation.

In general, the values of $\psi'(\boldsymbol{r},t))$ depend on the values of $\psi(\boldsymbol{r},t)$ at different instants. Consequently, the Lorentz transformation only makes sense when applied to the entire motion – *cf.* comment on page 2.

The Klein–Gordon equation reads:

$$-\hbar^2 \Box\psi(\boldsymbol{r},t) = M^2c^2\psi(\boldsymbol{r},t) \tag{VI-73}$$

where $\Box$ is the d'Alembertian operator $\Box \equiv \partial^2/c^2\partial t^2 - \Delta$. This operator is Lorentz invariant: its values $\Box'$ and $\Box$ are identical when the derivatives are taken, either with respect to the variables t' and $\boldsymbol{r}'$, or with respect to the variables t and $\boldsymbol{r}$. Since according to (VI-72) $\psi(\boldsymbol{r},t)$ is also invariant, it follows that the Klein–Gordon equation keeps the same form in all Lorentz reference frames; it is thus a relativistic invariant.

B-1-d. Minimal coupling

As in § A-2, we may consider generalizing the previous results to the case of a particle that is no longer free and bears a charge q. As before, the procedure is to replace the momentum $\boldsymbol{P}$ by the expression:

$$\boldsymbol{P} \Rightarrow \boldsymbol{P} - q\boldsymbol{A}(\boldsymbol{R}) \tag{VI-74}$$

and do the same for the first component of the energy-momentum four-vector[9], which yields the time derivative of the wave function:

$$H \Rightarrow H - qU(\boldsymbol{R}) \tag{VI-75}$$

For historical reasons this operation is called "minimal coupling". It yields the following equation:

$$\left[i\hbar\frac{\partial}{\partial t} - qU(\boldsymbol{r})\right]^2\psi(\boldsymbol{r},t) - c^2\left[\frac{\hbar}{i}\boldsymbol{\nabla} - q\boldsymbol{A}(\boldsymbol{r})\right]^2\psi(\boldsymbol{r},t) = M^2c^4\,\psi(\boldsymbol{r},t) \tag{VI-76}$$

Taking a Coulomb scalar potential $U(\boldsymbol{r})$ and a zero vector potential $\boldsymbol{A} = 0$, we get a relativistic model for the hydrogen atom. This model yields accurate expressions for certain relativistic corrections on the position of the energy levels (fine structure corrections), but not for the corrections linked to the electron spin, which require the Dirac equation.

[9]This amounts to simply adding two four-vectors of electromagnetism, the $\{H/c, \boldsymbol{P}\}$ four-vector and the $\{U/c, \boldsymbol{A}\}$ four-vector (in the Lorenz gauge) multiplied by the opposite of the charge q.

B-2. Dirac equation

It is clearly the second-order time derivative, and in particular the non-positivity of the probability density appearing in the conservation law (VI-66) that makes it hard to interpret the Klein–Gordon equation. This second-order time derivative comes from the fact that the relativistic expression for the Hamiltonian yields the square of H instead of H itself. Dirac had the idea in 1928 of solving that problem by introducing a first-order time equation that can lead to a second-order equation.

B-2-a. Etablishing the equation

We assume the mass M is different from zero and that the wave function $\tilde{\psi}(\boldsymbol{p},t)$ in the $\{|\boldsymbol{p}\rangle\}$ representation follows the time evolution:

$$i\hbar\frac{\partial}{\partial t}\tilde{\psi}(\boldsymbol{p},t) = \left[\alpha^0\, Mc^2 + c\,\boldsymbol{\alpha}\cdot\boldsymbol{p}\right]\tilde{\psi}(\boldsymbol{p},t) \tag{VI-77}$$

where α^0 and the three components $\alpha^{x,y,z}$ of $\boldsymbol{\alpha}$, are constants (independent of $\boldsymbol{p}$ and t). Applying twice the operator inside the brackets, we get:

$$-\hbar^2\frac{\partial^2}{\partial t^2}\tilde{\psi}(\boldsymbol{p},t) = \left[\alpha^0\, Mc^2 + c\,\boldsymbol{\alpha}\cdot\boldsymbol{p}\right]^2\tilde{\psi}(\boldsymbol{p},t) = \Big\{(\alpha^0)^2\, M^2c^4 + c^2\sum_{i=1}^{3}(\alpha^i)^2p_i^2 + \sum_{i\neq j,1}^{3}\left[\alpha^i\alpha^j + \alpha^j\alpha^i\right]c^2p_ip_j + \sum_{i=1}^{3}\left[\alpha^0\alpha^i + \alpha^i\alpha^0\right]Mc^3p_i\Big\}\tilde{\psi}(\boldsymbol{p},t) \tag{VI-78}$$

where $i,j = 1,2,3$ label the x,y,z components. This yields an expression compatible with relation (VI-64) for H^2 provided the squares of the α^i components are each equal to the identity:

$$(\alpha^\mu)^2 = 1 \qquad \text{for } \mu = 0,1,2,3 \tag{VI-79a}$$

and the different components α^μ and α^ν anticommute. These two conditions are summarized in relation:

$$\alpha^\mu\alpha^\nu + \alpha^\nu\alpha^\mu = 2\delta_{\mu\nu} \qquad \text{for } \mu,\ \nu = 0,1,2,3 \tag{VI-79b}$$

Relations (VI-79b) have obviously no solution if α^0 and the $\boldsymbol{\alpha}$ components are ordinary numbers, which cannot anticommute. However, this is not impossible for matrices. Well-known examples are the Pauli matrices, which are the components of a spin 1/2: not only do these matrices anticommute but their squares are equal to the identity, and hence satisfy relations (VI-79a). Could an elegant solution be to simply assume that the particle has a spin 1/2? Unfortunately, there are only three Pauli matrices σ_i, whereas we need four matrices obeying the above conditions.

• Let us try and build four 2×2 matrices, obeying the required conditions. We expand these μ_λ matrices on the identity matrix and the Pauli matrices:

$$\mu_\lambda = x_\lambda^0 \,\mathbb{1} + \sum_{i=1}^{3} x_\lambda^i \,\sigma_i \tag{VI-80}$$

Since the Pauli matrices anticommute with each other and their square equals unity, we have:

$$(\mu_\lambda)^2 = \left[(x_\lambda^0)^2 + \sum_i (x_\lambda^i)^2\right]\mathbb{1} + 2x_\lambda^0 \sum_i x_\lambda^i \sigma_i \tag{VI-81}$$

This square is equal to unity in two cases only :

– either all the x_λ^i for $i = 1, 2, 3$ are zero, and $x_\lambda^0 = 1$, which yields the trivial case where the matrix is the identity matrix. This case is of no interest for us since it is not compatible with the anticommutation of μ_λ with three other matrices;

– or $x_\lambda^0 = 0$, and the sum of the squares of the three other components is equal to one. We shall examine only this possibility.

We calculate the anticommutator of two matrices μ_λ and μ_ν that have a zero component on the identity matrix. We get:

$$\mu_\lambda \,\mu_\nu + \mu_\nu \,\mu_\lambda = 2\sum_{i=1}^{3} x_\lambda^i x_\nu^i \;\mathbb{1} = 2\,\boldsymbol{x}_\lambda \cdot \boldsymbol{x}_\nu \tag{VI-82}$$

The anticommutation is obeyed if the two vectors $\boldsymbol{x}_\lambda$, with x_λ^i components, and $\boldsymbol{x}_\nu$, with x_ν^i components, are orthogonal. However, in a 3-dimensional space, we cannot find more than 3 unit vectors all orthogonal with each other. The four matrices we are looking for do not exist if we limit ourselves to 2×2 matrices.

• Nor is it possible to find 3×3 matrices satisfying relations (VI-79b) if these matrices are Hermitian (which is necessary to ensure the unitarity of the evolution of $\hat{\psi}$). Since these matrices are diagonalizable, and their square equals unity, their eigenvalues are all equal to ± 1. Now the anticommutation relation $\alpha^0 \alpha^i = -\alpha^i \alpha^0$, and the fact that the square of α^0 equals unity imply that:

$$\alpha^i = -\alpha^0 \alpha^i \alpha^0 \tag{VI-83}$$

Taking the trace of this relation shows that:

$$\mathrm{Tr}\{\alpha^i\} = -\mathrm{Tr}\{\alpha^0 \alpha^i \alpha^0\} = -\mathrm{Tr}\{(\alpha^0)^2 \alpha^i\} = -\mathrm{Tr}\{\alpha^i\} \tag{VI-84}$$

(we have again used the fact that the square of the α^0 matrix equals unity). The trace of each α^i is thus equal to zero. Since this trace is the sum of all the eigenvalues, which are all equal to ± 1, there must be an even number of these eigenvalues, meaning that the space the matrices act in cannot have an odd dimension. It is thus pointless to look for 3×3 matrices satisfying relations (VI-79b).

Considering now 4 × 4 matrices, we can set:

$$(\alpha^0) = \left(\begin{array}{c|c} (\mathbb{1}) & 0 \\ \hline 0 & -(\mathbb{1}) \end{array}\right) \qquad (\alpha^i) = \left(\begin{array}{c|c} 0 & \sigma_i \\ \hline \sigma_i & 0 \end{array}\right) \tag{VI-85}$$

where, as above, $(\mathbb{1})$ is the 2×2 identity matrix and σ_i a Pauli matrix σ_x, σ_y, or σ_z; the zeros symbolize 2×2 matrices whose elements are all equal to zero. It is easy to show that these Hermitian matrices satisfy relations (VI-79). The Dirac equation is then written:

$$\boxed{i\hbar \frac{\partial}{\partial t} \tilde{\psi}_D(\boldsymbol{p}, t) = \left[c\, \boldsymbol{\alpha} \cdot \boldsymbol{p} + \alpha^0\, Mc^2\right] \tilde{\psi}_D(\boldsymbol{p}, t)} \tag{VI-86}$$

where $\tilde{\psi}_D(\boldsymbol{p}, t)$ is the four-component Dirac spinor, which can be represented by a column matrix:

$$\tilde{\psi}_D(\boldsymbol{p}, t) = \begin{pmatrix} \tilde{\psi}_1(\boldsymbol{p}, t) \\ \tilde{\psi}_2(\boldsymbol{p}, t) \\ \tilde{\psi}_3(\boldsymbol{p}, t) \\ \tilde{\psi}_4(\boldsymbol{p}, t) \end{pmatrix} \tag{VI-87}$$

Taking a spatial Fourier transform, this equation becomes:

$$\boxed{i\hbar \frac{\partial}{\partial t} \psi_D(\boldsymbol{r}, t) = \left[-i\hbar c\, \boldsymbol{\alpha} \cdot \boldsymbol{\nabla} + \alpha^0\, Mc^2\right] \psi_D(\boldsymbol{r}, t)} \tag{VI-88}$$

where $\psi_D(\boldsymbol{r}, t)$ is another four-component spinor, similar to (VI-87), but where the four functions ψ_i depend on position and time. This equation is interpreted as describing both a spin 1/2 particle and its antiparticle with the same spin, each contributing two units to the dimension of the state space. Thanks to his equation, Dirac predicted in 1931 that the electron had an antiparticle with the same spin. This particle was observed a year later by Anderson while studying cosmic rays, and is now called the "positron".

B-2-b. A simple solution: the plane wave

A simple case, the plane wave, furnishes a first example of the role of the four states of the Dirac equation. We assume an $e^{i\boldsymbol{k}\cdot\boldsymbol{r}}$ spatial dependence of the four functions $\psi_i(\boldsymbol{r}, t)$. The 4×4 matrix corresponding to the operator inside the brackets has only one contribution to the Oz component of $\boldsymbol{\alpha}$, and is written:

$$\begin{pmatrix} Mc^2 & 0 & \hbar ck & 0 \\ 0 & Mc^2 & 0 & -\hbar ck \\ \hbar ck & 0 & -Mc^2 & 0 \\ 0 & -\hbar ck & 0 & -Mc^2 \end{pmatrix} \tag{VI-89}$$

This matrix couples only the first vector to the third one, and the second to the fourth. It is actualy a tensor product of two 2×2 matrices, which are both written:

$$\begin{pmatrix} Mc^2 & \pm\hbar ck \\ \pm\hbar ck & -Mc^2 \end{pmatrix} = Mc^2\,\sigma_z \pm \hbar ck\,\sigma_x \tag{VI-90}$$

These two matrices have the same eigenvalues:

$$E = \pm\sqrt{M^2c^4 + \hbar^2c^2k^2} \tag{VI-91}$$

Since $p = \hbar k$, the two eigenvalues squared yield again the relativistic relation (VI-64) between energy and momentum.

The positive eigenvalue E corresponds to our objective in establishing the Dirac equation. However, the negative value is more difficult to interpret as it has no lower boundary when $k \to \infty$. What prevents an electron with positive energy from constantly relaxing into this infinity of negative energy states? To answer this question, Dirac suggested that all the negative energy states are occupied, which, because of the Pauli principle, forbids an electron with positive energy to enter these states. The ensemble of these occupied negative energy states is often called the "Dirac sea". He proposed, in addition, that a "hole" could be found in this Dirac sea. This unoccupied state increased the energy and could thus be describing a new particle. This particle would have a charge opposite to that of the electron, but be completely similar otherwise: a positron. Here again, quantum field theory in which the particle number is not constant, allows us to go further while avoiding the difficulties related to the Dirac sea.

B-2-c. Relativistic notation

It is possible to write an explicitly relativistic form of the Dirac equation. We multiply both sides of equation (VI-88) by the matrix α^0 and set:

$$\gamma^0 = \alpha^0 \quad \text{and:} \quad \gamma^i = \alpha^0\alpha^i = \left(\begin{array}{c|c} 0 & +\sigma_i \\ \hline -\sigma_i & 0 \end{array}\right) \quad \text{with } i = x, y, z \tag{VI-92}$$

As opposed to the α matrices, the γ^i are not Hermitian. Since the square of the α^0 matrix is equal to unity, we get:

$$\left[Mc^2 - i\hbar\gamma^0\frac{\partial}{\partial t} - i\hbar c\boldsymbol{\gamma}\cdot\boldsymbol{\nabla}\right]\psi_D(\boldsymbol{r},t) = 0 \tag{VI-93}$$

or, dividing by c:

$$\left[Mc - i\hbar\gamma^\mu\partial_\mu\right]\psi_D(\boldsymbol{r},t) = 0 \tag{VI-94}$$

We use here the standard relativistic notation, where the four-vector ∂_μ has a first component equal to $\partial/c\partial t$, followed by the three components of the spatial gradient $\boldsymbol{\nabla}_{\boldsymbol{r}}$. Following Einstein's convention, there is an implicit summation over the 4 values from 0 to 3 of any index appearing twice, once in the upper position and once in the lower.

The relativistic invariance of the Dirac equation is established in complement A_{VI}, that of its Lagrangian in § 4 of complement C_{VI}.

B-2-d. Conservation laws

The Hermitian conjugate $\psi_D^\dagger(\boldsymbol{r},t)$ of the Dirac spinor $\psi_D(\boldsymbol{r},t)$ is associated with the row matrix:

$$\psi_D^\dagger(\boldsymbol{r},t) = \begin{pmatrix}\psi_1^*(\boldsymbol{r},t) & \psi_2^*(\boldsymbol{r},t) & \psi_3^*(\boldsymbol{r},t) & \psi_4^*(\boldsymbol{r},t)\end{pmatrix} \tag{VI-95}$$

The product of two matrices 1×4 by 4×1 is a number. We get:

$$\psi_D^\dagger(\boldsymbol{r},t)\psi_D(\boldsymbol{r},t) = |\psi_1(\boldsymbol{r},t)|^2 + |\psi_2(\boldsymbol{r},t)|^2 + |\psi_3(\boldsymbol{r},t)|^2 + |\psi_4(\boldsymbol{r},t)|^2 \geq 0 \tag{VI-96}$$

Expressions (VI-85) for the α matrices are Hermitian, so that the differential operator on the right-hand side of (VI-88) is Hermitian. Since the evolution under a Hermitian Hamiltonian preserves the norm, we get a conservation law for the integral over all space of $\psi_D^\dagger(\boldsymbol{r},t)\psi_D(\boldsymbol{r},t)$:

$$\frac{\mathrm{d}}{\mathrm{d}t}\int \mathrm{d}^3r\ \psi_D^\dagger(\boldsymbol{r},t)\psi_D(\boldsymbol{r},t) = 0 \tag{VI-97}$$

This conservation law also has a local form, hence more precise, which involves a four-vector current j^μ defined as:

$$j^\mu = \psi_D^\dagger\gamma^0\gamma^\mu\psi_D \tag{VI-98}$$

Since $(\gamma^0)^2 = (\alpha^0)^2 = 1$, we readily get:

$$j^{\mu=0} = \psi_D^\dagger(\boldsymbol{r},t)\psi_D(\boldsymbol{r},t) \geq 0 \quad ; \quad j^i = \psi_D^\dagger(\boldsymbol{r},t)\alpha^i\psi_D(\boldsymbol{r},t) \tag{VI-99}$$

where $i = x, y, z$. The $\mu = 0$ component of the current is indeed the positive function (VI-96) involved in the global conservation law (VI-97).

We now compute the sum $\partial_\mu j^\mu$. The first term, associated with the time derivative, is obtained from (VI-88):

$$\frac{1}{c}\frac{\partial}{\partial t}\psi_D^\dagger\psi_D = \frac{1}{i\hbar}\psi_D^\dagger[\alpha^0 Mc - i\hbar\boldsymbol{\alpha}\cdot\boldsymbol{\nabla}]\psi_D - \frac{1}{i\hbar}\psi_D^\dagger[\alpha^0 Mc + i\hbar\overleftarrow{\boldsymbol{\nabla}}\cdot\boldsymbol{\alpha}]\psi_D \tag{VI-100}$$

where, in the second term of the second bracket, the $\overleftarrow{\boldsymbol{\nabla}}$ gradient means it is acting on the $\psi_D^\dagger$ on its left. The Mc^2 terms cancel out and, since $i\boldsymbol{\nabla}$ and the $\boldsymbol{\alpha}$ components are Hermitian and commute, we get:

$$\frac{1}{c}\frac{\partial}{\partial t}\psi_D^\dagger\psi_D = -[\psi_D^\dagger(\boldsymbol{\alpha}\cdot\boldsymbol{\nabla}\psi_D) + (\boldsymbol{\nabla}\psi_D^\dagger)\boldsymbol{\alpha}\cdot\psi_D] = -\boldsymbol{\nabla}\cdot\psi_D^\dagger\boldsymbol{\alpha}\psi_D \tag{VI-101}$$

The next three terms of the sum $\partial_\mu j^\mu$ add up to:

$$\boldsymbol{\nabla}\cdot\psi_D^\dagger\gamma^0\boldsymbol{\gamma}\,\psi_D = \boldsymbol{\nabla}\cdot\psi_D^\dagger\boldsymbol{\alpha}\psi_D \qquad \text{(VI-102)}$$

Adding (VI-101) and (VI-102), we obtain (with the standard notation $\partial_\mu = \partial/\partial x^\mu$):

$$\partial_\mu j^\mu = 0 \qquad \text{or:} \qquad \frac{\partial}{\partial t}\psi_D^\dagger\psi_D + c\,\boldsymbol{\nabla}\cdot\psi_D^\dagger\boldsymbol{\alpha}\psi_D \qquad \text{(VI-103a)}$$

If we multiply by c by the space components of j^μ, and if we set:

$$\rho(\boldsymbol{r},t) = \psi_D^\dagger\psi_D \geq 0 \quad ; \quad \boldsymbol{j}(\boldsymbol{r},t) = c\,\psi_D^\dagger\,\boldsymbol{\alpha}\,\psi_D \qquad \text{(VI-103b)}$$

we recover the usual form $\partial\rho/\partial t + \boldsymbol{\nabla}\cdot\boldsymbol{j} = 0$ of a local conservation law for the density (the $\boldsymbol{r}$ and t dependence of the spinors is left implicit). Note in the expression of $\boldsymbol{j}$ that the product $c\boldsymbol{\alpha}$ now plays the role of a velocity operator.

B-2-e. Coupling to an electromagnetic field

Like for the Klein–Gordon equation, we can apply the minimal coupling procedure and include in the Dirac equation (VI-88) the action of an electromagnetic field. This yields:

$$\left[i\hbar\frac{\partial}{\partial t} - qU(\boldsymbol{r},t)\right]\psi_D(\boldsymbol{r},t) = \left[\alpha^0\,Mc^2 - c\boldsymbol{\alpha}\cdot\left[i\hbar\boldsymbol{\nabla} + q\boldsymbol{A}(\boldsymbol{r},t)\right]\right]\psi_D(\boldsymbol{r},t) \quad \text{(VI-104)}$$

where $\boldsymbol{A}(\boldsymbol{r},t)$ and $U(\boldsymbol{r},t)$ are respectively the vector and scalar potentials of the electromagnetic field.

If U is a time-independent Coulomb potential and $\boldsymbol{A} = 0$, this equation enables calculating the fine structure energy levels of hydrogen (*cf.* § XX-27 in [6] and §§ 19.3.4 and 5 in [13]). In a non-zero magnetic field ($\boldsymbol{A} \neq 0$) it also allows computing the value of the "gyromagnetic ratio g of the electron", which is the ratio between its magnetic moment and its spin angular momentum. The value found, $g = 2$, is not a priori obvious since it is equal to 1 for orbital variables. Furthermore, the value 2 is not strictly accurate, since it does not take into account the corrections due to the coupling between the electron and the quantized electromagnetic field. Although very small (roughly one per thousand), these corrections have been measured with a truly remarkable precision, on the order of 10^{-10} in relative value.

B-2-f. Conclusion: status of the Dirac equation

The Dirac equation has successfully treated a certain number of subjects (taking the spin into account, fine structure of the hydrogen atom, the g factor of the electron, the first introduction of the antiparticle concept). Quantum electrodynamics is a theory where the coupling between field and charged matter (obeying the Dirac equation) can be computed, in a perturbative way, up to successive orders. This theory also leads to dead ends, principally linked to the infinity of negative energy states (the Dirac sea). These fundamental difficulties are related to the relativistic possibility of creating or annihilating particles, which invalidates any single particle relativistic theory. One can interpret the Dirac equation as a single charge, and not single particle, theory, to include the possibility of creation of electron–positron pairs, but this does not eliminate all the difficulties. In addition to its historical interest, the Dirac equation is now seen as an intermediate step towards a more powerful theory, the field theory that allows treating many more physical phenomenon. The reader interested in modern field theory may consult references quoted in the bibliography, in particular [11] and [14].

B-3. Zero mass, the Weyl equation

In our discussion of the Dirac equation, we assumed a non-zero mass. This made it necessary to include a mass term in the evolution equation, which led us in § B-2-a to look for 4 anti-commuting matrices. We showed that a minimal dimension of 4 was required for these four matrices, whereas a dimension of 2 would have been enough for 3 matrices. We now discuss the zero mass case, and under which conditions we can use 2-dimensional matrices.

B-3-a. Construction of the evolution equation

When the mass equals zero, the evolution equation of the state vector must reflect relation $H^2 = c^2\boldsymbol{P}^2$, a consequence of (V-68) when $M = 0$.

• Searching for the evolution equation of a one-component (hence without spin) wave function leads to a particular case of the Klein–Gordon equation. Each $\boldsymbol{p}$ component of the wave function $\overline{\psi}(\boldsymbol{p}, t)$ evolves as a superposition of two exponentials $\mathrm{e}^{\pm icpt}$. Relation (VI-65b) becomes:

$$\Box\psi(\boldsymbol{r}, t) = 0 \tag{VI-105}$$

where $\Box$ is the d'Alembert operator. This is similar to a standard propagation equation in classical electromagnetism, such as the propagation equation of the electric potential in vacuum. The fact that the differential equation is of second-order in time leads to the difficulties we already discussed in the framework of the Klein–Gordon equation. Furthermore, the local probability density (VI-68) and its current (VI-67) are not defined for $M = 0$, which is to be expected for a quantum system with zero mass.

• To look for a first-order differential equation for the time evolution, we go back to (VI-77) with $M = 0$. Since the terms in M disappear from (VI-78), the only remaining conditions are:

$$\alpha^i\alpha^j + \alpha^j\alpha^i = 2\delta_{ij} \qquad \text{for } i,j = 1,2,3 \tag{VI-106}$$

Since we only need three matrices, we can choose either the Pauli matrices, or their opposite:

$$\alpha^i = \pm\,\sigma_i \qquad \text{with} \quad \sigma_1 = \sigma_x \ ; \ \sigma_2 = \sigma_y \ ; \ \sigma_3 = \sigma_z \tag{VI-107}$$

This leads to the "Weyl equation", that takes two different forms depending on the sign of the right-hand side:

$$i\hbar\frac{\partial}{\partial t}\tilde{\psi}_W(\boldsymbol{p},t) = \eta\, c\, \boldsymbol{\sigma}\cdot\boldsymbol{p}\ \tilde{\psi}_W(\boldsymbol{p},t) \qquad \text{with } \eta = \pm 1 \tag{VI-108}$$

where $\tilde{\psi}_W(\boldsymbol{p},t)$ designates the two-component "Weyl spinor":

$$\tilde{\psi}_W(\boldsymbol{p},t) = \begin{pmatrix} \tilde{\psi}_1(\boldsymbol{p},t) \\ \tilde{\psi}_2(\boldsymbol{p},t) \end{pmatrix} \tag{VI-109}$$

We show below that the two versions of the equation are not equivalent. They generate "right" and "left" solutions where the spin is either parallel or antiparallel to the propagation direction.

Taking the Fourier transform of equation (VI-108), we get:

$$\frac{\partial}{\partial t}\psi_W(\boldsymbol{r},t) = -\eta\, c\, \boldsymbol{\sigma}\cdot\boldsymbol{\nabla}\psi_W(\boldsymbol{r},t) \qquad \text{with } \eta = \pm 1 \tag{VI-110}$$

(this is a rare case in quantum mechanics where a first-order dynamical equation does not contain an imaginary factor i). We now set:

$$\tilde{\sigma}^0 = (\mathbb{1}) \ ; \ \tilde{\sigma}^1 = \eta\sigma_x \ ; \ \tilde{\sigma}^2 = \eta\sigma_y \ ; \ \tilde{\sigma}^3 = \eta\sigma_z \tag{VI-111}$$

The Weyl equation (VI-110) can then be written in a concise form:

$$\tilde{\sigma}^\mu\partial_\mu\psi_W(\boldsymbol{r},t) = 0 \tag{VI-112}$$

with the standard notation $\partial_\mu = \partial/\partial x^\mu$, and the usual Einstein's convention: any repeated dummy index (here μ) is summed from 0 to 3.

B-3-b. Relativistic invariance of the equation

As the Weyl spinor is two-dimensional, we can directly use the SL$(2, C)$ representation of the Lorentz group (two-valued, non-unitary representation) to write the relativistic transformation rules for its evolution equation.

Consider a Lorentz transformation Λ defined, as in relation (VI-71), by the elements of a 4×4 matrix noted Λ:

$$x'^{\nu} = \Lambda^{\nu}_{\ \mu}\, x^{\mu} \qquad ; \qquad x^{\mu} = (\Lambda^{-1})^{\mu}_{\ \nu}\, x'^{\nu} \tag{VI-113}$$

We then have:

$$\partial'_{\nu} \equiv \frac{\partial}{\partial x'^{\nu}} = \frac{\partial}{\partial x^{\mu}}\frac{\partial x^{\mu}}{\partial x'^{\nu}} = (\Lambda^{-1})^{\mu}_{\ \nu}\, \partial_{\mu} \tag{VI-114}$$

But we saw in § 1-d of complement $\mathrm{A_V}$ that any Lorentz transformation Λ could be represented by two opposite (L) matrices of $\mathrm{SL}(2,C)$, noted $\pm L(\Lambda)$. Equation (28) of that complement defines the relation between Λ and L. These matrices act on the (Q) matrices associated with each four-vector according to $(Q') = L(\Lambda)(Q)L^{\dagger}(\Lambda)$, so that:

$$\begin{aligned}(Q') &= x'^{\nu}\sigma_{\nu} = \Lambda^{\nu}_{\ \mu}\, x^{\mu}\, \sigma_{\nu} \\ &= L(\Lambda)(Q)L^{\dagger}(\Lambda) = L(\Lambda)\,(x^{\mu}\,\sigma_{\mu})\,L^{\dagger}(\Lambda)\end{aligned} \tag{VI-115}$$

with $\sigma_0 = (\mathbb{1})$ and the definitions (VI-107) of $\sigma_{1,2,3}$. Since this equality is valid for any four-vector, we identify the x^{μ} terms and write:

$$\Lambda^{\nu}_{\ \mu}\, \sigma_{\nu} = L(\Lambda)\,\sigma_{\mu}\,L^{\dagger}(\Lambda) \tag{VI-116}$$

We shall use this result to, successively, study the transformations of the Weyl equation under rotations, then under pure Lorentz transformations, and finally under translations.

α. *Rotations*

We start with the simple case where Λ is a rotation $\mathscr{R}$ transforming an arbitrary position $\boldsymbol{r}$ into $\boldsymbol{r}' = \mathscr{R}\,\boldsymbol{r}$. The three components x^i of the position $\boldsymbol{r}$ on orthonormal axes become the x'^j:

$$x'^j = (\mathscr{R})_{ji}\, x^i \qquad x^i = (\mathscr{R}^{-1})_{ij}\, x'^j = (\mathscr{R})_{ji}\, x'^j \tag{VI-117}$$

where $(\mathscr{R})_{ji}$ is the element of the jth row and the ith column of the 3-dimensional rotation matrix $(\mathscr{R})$; as this matrix is orthogonal, we have $(\mathscr{R}^{-1})_{ij} = (\mathscr{R})_{ji}$. The partial derivatives ∂'_j with respect to x'^i are related to the partial derivatives ∂_i with respect to x^i by:

$$\partial'_j = \frac{\partial}{\partial x'^j} = \frac{\partial}{\partial x^i}\frac{\partial x^i}{\partial x'^j} = (\mathscr{R})_{ji}\,\partial_i \qquad i,j = 1,2,3 \tag{VI-118}$$

For a rotation, the matrix $L(\Lambda)$ becomes $L(\mathscr{R})$; relation (VI-116) is then written:

$$(\mathscr{R})_{ji}\,\sigma_j = L(\mathscr{R})\,\sigma_i\,L^{\dagger}(\mathscr{R}) \tag{VI-119}$$

We set[10]:

$$\psi'_W(\boldsymbol{r}',t) = L(\mathscr{R})\,\psi_W(\boldsymbol{r},t) \qquad \text{with:} \qquad \boldsymbol{r}' = \mathscr{R}\boldsymbol{r} \tag{VI-120}$$

We multiply equation (VI-110) on the left by $L(\mathscr{R})$. Since the rotation matrices $L(\mathscr{R})$ are unitary, we have $L^{-1}(\mathscr{R}) = L^{\dagger}(\mathscr{R})$, which leads to:

$$\begin{aligned}\frac{\partial}{\partial t}\psi'_W(\boldsymbol{r}',t) &= -\eta\, c\, L(\mathscr{R})\,\sigma_i\,\partial_i\,L^{\dagger}(\mathscr{R})\;\psi'_W(\boldsymbol{r}',t)\\ &= -\eta\, c\,(\mathscr{R})_{ji}\,\sigma_j\,\partial_i\,\psi'_W(\boldsymbol{r}',t) = -\eta\, c\,\sigma_j\,\partial'_j\;\psi'_W(\boldsymbol{r}',t)\end{aligned} \tag{VI-121}$$

where we have used (VI-118) for writing the last equality. This shows that the two Weyl equation, $\eta = \pm 1$, keep the same form under a rotation of the reference frame.

β. *Pure Lorentz transformations*

We now treat separately the two cases $\eta = \pm 1$.

• Taking $\eta = +1$ in (VI-110), we simply have $\sigma_\mu = \tilde{\sigma}^\mu$. We are going to show that a transformation of the Lorentz group preserves the relativistic invariance of the Weyl equation if, in the change of Lorentz reference frame associated with the Λ transformation, the Weyl spinor becomes:

$$\begin{aligned}&\psi'_W(\boldsymbol{r}',t') = L(\Lambda)\;\psi_W(\boldsymbol{r},t) \quad \text{or:} \quad \psi_W(\boldsymbol{r},t) = L^{\dagger}(\Lambda^{-1})\psi'_W(\boldsymbol{r}',t')\\ &\text{with: } (ct',\boldsymbol{r}') = \Lambda(ct,\boldsymbol{r})\end{aligned} \tag{VI-122}$$

In the first line, the second equality used the fact that the L matrix is Hermitian for a pure Lorentz transformation, so that we can write $[L(\Lambda)]^{-1} = L(\Lambda^{-1}) = L^{\dagger}(\Lambda^{-1})$.

To demonstrate the relativistic invariance, we replace Λ by its inverse in (VI-116), and then multiply by ∂_μ (this operator takes a derivative with respect to the external variables; it therefore commutes with L, which acts only on the internal components of the spinor):

$$(\Lambda^{-1})^{\nu}{}_{\mu}\;\sigma_\nu\partial_\mu = L(\Lambda^{-1})\,\sigma_\mu\partial_\mu\,L^{\dagger}(\Lambda^{-1}) \tag{VI-123a}$$

But the matrices (Λ) associated with pure Lorentz transformations are symmetrical – *cf.* relation (28) of complement $\mathrm{A_V}$ when L is Hermitian. Equation (VI-114) then provides:

$$\sigma_\nu\,\partial'_\nu = L(\Lambda^{-1})\,\sigma_\mu\partial_\mu\,L^{\dagger}(\Lambda^{-1}) \tag{VI-123b}$$

[10] We only discuss the matrix L and ignore its opposite, which is justified since two kets that differ by their sign represent the same physical state. Furthermore, since we have $\psi'_W(\boldsymbol{r},t) = L(\mathscr{R})\,\psi_W(\mathcal{R}^{-1}\boldsymbol{r},t)$, relation (VI-120) is coherent with (VI-34a) for a spinless particle.

We now apply this operator relation to the spinor $\psi'_W(\boldsymbol{r}',t')$. According to (VI-122), on the right-hand side we obtain $\sigma_\mu \partial_\mu \psi_W(\boldsymbol{r},t)$, which vanishes if $\psi_W(\boldsymbol{r},t)$ obeys the Weyl equation. We finally get:

$$\sigma^\nu \partial'_\nu \psi'(\boldsymbol{r},t) = 0 \qquad \text{(VI-124)}$$

which shows that the Weyl equation (VI-110) with $\eta = +1$ keeps the same form in different Lorentz reference frames.

- If $\eta = -1$ in (VI-110), one can no longer replace σ_μ by $\tilde{\sigma}^\mu$ in (VI-116).

One can, nevertheless, use the metric tensor (g) of special relativity:

$$g_{\mu\nu} = \delta_{\mu\nu} \quad \text{if } \mu = 0 \quad ; \quad g_{\mu\nu} = -\delta_{\mu\nu} \quad \text{if } \mu = 1,2,3 \qquad \text{(VI-125)}$$

and write $\sigma_\mu = g_{\mu\rho}\,\tilde{\sigma}^\rho$. Equality (VI-116) then becomes:

$$(\Lambda)^\nu_{\ \mu}\, g_{\nu\rho}\,\tilde{\sigma}^\rho = L(\Lambda)\, g_{\mu\tau}\tilde{\sigma}^\tau\, L(\Lambda) \qquad \text{(VI-126)}$$

We multiply this relation by $g_{\lambda\mu}$ and sum over μ. The left-hand side becomes:

$$g_{\lambda\mu}(\Lambda)^\mu_{\ \nu}\, g_{\nu\rho}\,\tilde{\sigma}^\rho = \Big((g)(\Lambda)(g)\Big)^\lambda_{\ \rho}\,\tilde{\sigma}^\rho \qquad \text{(VI-127)}$$

It can be shown in a relativistic theory[11] (see, for example, § 19.1.3 of [13], or § 11.7 of [46]) that the matrix $(g)(\Lambda)(g)$ is equal to the inverse (Λ^{-1}) of (Λ). The left-hand side is thus equal to $(\Lambda^{-1})^\lambda_{\ \rho}\,\tilde{\sigma}^\rho$. As for the right-hand side, after multiplication by $g_{\lambda\mu}$ and summation over μ, since $g_{\lambda\mu}g_{\mu\tau} = \delta_{\lambda\tau}$, it contains the matrix $\tilde{\sigma}^\lambda$ taken in between two operators L. We thus get:

$$(\Lambda^{-1})^\lambda_{\ \rho}\,\tilde{\sigma}^\rho = L(\Lambda)\,\tilde{\sigma}^\lambda\, L^\dagger(\Lambda) \qquad \text{(VI-128a)}$$

We multiply this relation by ∂_λ and once again use (VI-114) to obtain:

$$\tilde{\sigma}^\rho\,\partial'_\rho = L(\Lambda)\,\tilde{\sigma}^\lambda\,\partial_\lambda\, L^\dagger(\Lambda) \qquad \text{(VI-128b)}$$

We now set:

$$\psi'_W(\boldsymbol{r}',t') = [L^\dagger(\Lambda)]^{-1}\psi_W(\boldsymbol{r},t) \ ; \ \psi_W(\boldsymbol{r},t) = [L^\dagger(\Lambda)]\psi'_W(\boldsymbol{r}',t') \qquad \text{(VI-129)}$$

If we apply the operator relation (VI-128b) to the spinor $\psi'_W(\boldsymbol{r}',t')$, on the right-hand side we obtain the expression $\tilde{\sigma}_\lambda\partial_\lambda\psi_W(\boldsymbol{r},t)$, which vanishes if $\psi_W(\boldsymbol{r},t)$ obeys the Weyl equation. We finally get:

$$\tilde{\sigma}^\rho\partial'_\rho\psi'_W(\boldsymbol{r}',t') = 0 \qquad \text{(VI-130)}$$

which establishes the invariance of the second Weyl equation under a change of Lorentz reference frame.

---

[11] A possible definition of the (Λ) matrices is that they are the 4×4 matrices such that $(g)(\Lambda^T)(g) = (\Lambda)^{-1}$, where (Λ^T) is the transposed of (Λ).

γ. *Translations*

Space translations result in the simple replacement of the spinor variable $\boldsymbol{r}$ by $\boldsymbol{r} - \boldsymbol{b}$, where $\boldsymbol{b}$ is the translation vector. The invariance of equation (VI-110) is then obvious. The same is true for the time translation, where the variable t must be replaced by $t + \tau$.

δ. *Right and left spinors*

We note $\psi^r(\boldsymbol{r}, t)$ the solutions of the Weyl equation for $\eta = +1$, which we call the "right spinors"; the $\psi^l(\boldsymbol{r}, t)$ are the "left spinors", solutions of the equation for $\eta = -1$ (the reason for these denominations will be explained in § B-3-c). For a right spinor, equalities (VI-120) and (VI-122) show that the Λ transformations of the Lorentz group (combinations of rotations and changes of Lorentz reference frame) result in:

$$\psi_W'^{\,r}(\boldsymbol{r}', t') = L(\Lambda)\, \psi_W^r(\boldsymbol{r}, t) \quad \text{with:} \quad (ct', \boldsymbol{r}') = \Lambda(ct, \boldsymbol{r}) \tag{VI-131a}$$

For a left spinor, since the rotations correspond to unitary operators L, relations (VI-120) and (VI-129) are equivalent (the Hermitian conjugation and the inverse operation cancel each other) and both can be expressed as[12]:

$$\psi_W'^{\,l}(\boldsymbol{r}', t') = [L^\dagger(\Lambda)]^{-1}\, \psi_W^l(\boldsymbol{r}, t) = L^\dagger(\Lambda^{-1})\, \psi_W^l(\boldsymbol{r}, t) \tag{VI-131b}$$

These spinors form the representation of the Lorentz group already obtained in relation (29) of complement $\mathrm{A_V}$.

The transformation rules (VI-131a) and (VI-131b) ensure the relativistic invariance of the Weyl equation for all the transformations of the connected Poincaré group.

Comment:

It follows from the two previous relations that the scalar product of the right and left spinors obeys:

$$\begin{aligned}[\psi_W'^{\,l}(\boldsymbol{r}', t')]^\dagger\, \psi_W'^{\,r}(\boldsymbol{r}', t') &= [\psi_W^l(\boldsymbol{r}, t)]^\dagger L(\Lambda^{-1}) L(\Lambda)\, \psi_W^r(\boldsymbol{r}, t) \\ &= [\psi_W^l(\boldsymbol{r}, t)]^\dagger \psi_W^r(\boldsymbol{r}, t)\end{aligned} \tag{VI-132}$$

This scalar product is thus a Lorentz scalar.

[12]The inverse of the Hermitian conjugate of a matrix is the Hermitian conjugate of its inverse. Take, for example, the Hermitian conjugate of relation $AA^{-1} = 1$; we get $(A^{-1})^\dagger A^\dagger = 1$, which shows that $(A^{-1})^\dagger$ is the inverse $(A^\dagger)^{-1}$ of $A^\dagger$. In addition, $L^{-1}(\Lambda) = L(\Lambda^{-1})$.

B-3-c. Plane waves, helicity

Consider a Weyl spinor with momentum $\boldsymbol{p} = \hbar \boldsymbol{k}$ as the one described in (VI-109). In the position representation, $\psi_W(\boldsymbol{p}, t)$ is proportional to $e^{i\boldsymbol{k}\cdot\boldsymbol{r}}$. Choosing an axis Oz parallel to $\boldsymbol{k}$, only the Pauli matrix σ_z will play a role in equation (VI-110).

• Taking $\eta = +1$ in this equation, the right spinor solution of this equation can be written (in the eigenbasis of σ_z) as a function of two constants c_1 and c_2:

$$\psi_W^r(\boldsymbol{r}, t) = \begin{pmatrix} c_1\, e^{i(kz - ck\,t)} \\ c_2\, e^{i(kz + ck\,t)} \end{pmatrix} \tag{VI-133}$$

Now we saw in (VI-120) that a rotation by an angle θ around Oz corresponds for the spinor to the operator $L(\Lambda)$, which is written as in relation (22a) of complement A_V:

$$(L) = \begin{pmatrix} e^{-i\theta/2} & 0 \\ 0 & e^{+i\theta/2} \end{pmatrix} \tag{VI-134}$$

The first row of the spinor (VI-133) corresponds to an eigenvector of the J_z component of the (spin) angular momentum, with eigenvalue $+\hbar/2$, the second row to an eigenvector with eigenvalue $-\hbar/2$. Furthermore, the first row corresponds to a wave propagating in the direction of the Oz axis, the second in the opposite direction. In both cases, the spin's angular momentum has the same direction as the wave's propagation (following the right-hand rule). The ψ_W^r spinors are said to have a right-hand chirality.

We defined helicity not in terms of the propagation direction of a wave packet, but rather as the scalar product of momentum and spin. Now the positive and negative energy components of the spinor we just considered have the same momentum $\boldsymbol{p} = \hbar \boldsymbol{k}$, but opposite spin directions. This means that a right spinor does not have a fixed helicity: its positive energy component has a $+1/2$ helicity, whereas its negative energy component, a $-1/2$ helicity.

• Taking $\eta = -1$, we get the spinor:

$$\psi_W^l(\boldsymbol{r}, t) = \begin{pmatrix} c_1\, e^{i(kz + ck\,t)} \\ c_2\, e^{i(kz - ck\,t)} \end{pmatrix} \tag{VI-135}$$

As rotations on right and left spinors have the same effect, the eigenvalues of the spin angular momentum are the same as before. However, the propagation directions of the wave are inverted, so that the spin direction is now the opposite of the propagation direction (it follows the left-hand rule). This so-called "left spinor" has a left-hand chirality, opposite to that of the right spinor. As for the helicities, they are inverted compared to the right spinor: the positive energy component of a left spinor has a $-1/2$ helicity, and its negative energy component, a $+1/2$ helicity.

A parity operation (complement $\mathrm{D_V}$) reverses the polar vector $\boldsymbol{p}$ but not the axial vector of the angular momentum. It thus transforms a right spinor into a left spinor, and vice versa. Neither the state space spanned by the ψ_W^r nor the one spanned by the ψ_W^l is parity invariant. The parity operation transforms each of these spaces into the other. The Dirac state space is actually the direct sum of two Weyl spaces, one for the right spinors, the other for the left spinors.

The standard model of particle physics was initially constructed assuming the neutrinos had a zero mass and obeyed the Weyl equation. Following the discovery of "neutrino oscillation", it is now believed that neutrinos have a non-zero mass, so that the Weyl equation does not apply to them. This equation is still a useful tool in other domains of field theory.

Complement A_{VI}

Relativistic invariance of Dirac equation and non-relativistic limit

We studied in chapter VI the relativistic invariance of the Weyl equation, which is much simpler than the Dirac equation since it deals with 2-dimensional, rather than 4-dimensional, spinors. In § 1 of this complement, we return to these invariance calculation for the Dirac equation. The results obtained for the Weyl equation will prove to be useful, and will actually greatly simplify our task. In § 2, we shall discuss the non-relativistic limit of the Dirac equation, which is the Pauli equation (*cf.* for example § C of chapter IX in [9]). This will allow us to provide this equation with a less phenomenological foundation.

1. Relativistic invariance

1-a. Changing the base of the internal variables

Expressions (VI-85) for the α matrices, as well as relation (VI-86) for the Dirac equation, have been written using a certain basis for the internal states, whose kets will be noted $|1\rangle$, $|2\rangle$, $|3\rangle$, and $|4\rangle$. It will prove useful, for the following reasoning, to introduce a new basis of so-called "right" and "left" kets $|a_r\rangle$, $|a_l\rangle$, $|b_r\rangle$ and $|b_l\rangle$, defined as:

$$|a_{r,l}\rangle = \frac{1}{\sqrt{2}}[|1\rangle \pm |3\rangle] \qquad\qquad |b_{r,l}\rangle = \frac{1}{\sqrt{2}}[|2\rangle \pm |4\rangle] \tag{1}$$

It is easy to show that none of the three operators $\alpha^{1,2,3}$ couples the $|a_r\rangle$ or $|b_r\rangle$ kets to those with the index l, and vice versa. For example:

$$\alpha^1|a_{r,l}\rangle = \frac{1}{\sqrt{2}}\Big[|4\rangle \pm |2\rangle\Big] = \pm|b_{r,l}\rangle \tag{2}$$

In the $|a_r\rangle$, $|b_r\rangle$, $|a_l\rangle$, $|b_l\rangle$ basis, the matrices (α^i) of the $\boldsymbol{\alpha}$ components become the block diagonal matrices $(\tilde{\alpha}^i)$:

$$(\tilde{\alpha}^i) = \left(\begin{array}{c|c} +\sigma_i & 0 \\ \hline 0 & -\sigma_i \end{array}\right) \tag{3}$$

which clearly shows that the subspace $\mathscr{E}_r$ spanned by $|a_r\rangle$ and $|b_r\rangle$, and the subspace $\mathscr{E}_l$ spanned by $|a_l\rangle$ and $|b_l\rangle$, are not coupled by the $\boldsymbol{\alpha}$ components. On the other hand, this is not the case for the α^0 matrix, which now becomes:

$$(\tilde{\alpha}^0) = \left(\begin{array}{c|c} 0 & \mathbb{1} \\ \hline \mathbb{1} & 0 \end{array}\right) \tag{4}$$

This matrix sends all the kets of $\mathscr{E}_r$ into $\mathscr{E}_l$, and vice versa.

1-b. Demonstration of the invariance

The Dirac equation (VI-88) is written:

$$i\hbar\left[\frac{\partial}{\partial t} + c\,\boldsymbol{\alpha}\cdot\boldsymbol{\nabla}\right]\psi_D(\boldsymbol{r},t) = \alpha^0\, Mc^2\, \psi_D(\boldsymbol{r},t) \tag{5}$$

where, as we have just seen, the left-hand side of the equation does not couple the components $\psi^r_D(\boldsymbol{r},t)$ on $\mathscr{E}_r$ with the components $\psi^l_D(\boldsymbol{r},t)$ on $\mathscr{E}_l$. This left-hand side actually contains two independent Weyl's equations; in chapter VI, they are associated with opposite values $\eta = \pm 1$, as can be seen on the form (3) of the matrix giving the components of $\boldsymbol{\alpha}$ in the right and left bases. Their relativistic invariance has been demonstrated in § B-3-b of chapter VI. We had to apply to the right and left Weyl spinors, each of dimension 2, two different operators $L(\Lambda)$ and $L^\dagger(\Lambda^{-1})$. For the 4-dimensional Dirac kets, these two operators can be summarized into a single one, written[1]:

$$S_D(\Lambda) = L(\Lambda)\,P_r + L^\dagger(\Lambda^{-1})\,P_l \tag{6}$$

with:

$$\psi'_D(\boldsymbol{r}',t') = S_D(\Lambda)\,\psi_D(\boldsymbol{r},t) \quad \text{and:} \quad (ct',\boldsymbol{r}') = \Lambda(ct,\boldsymbol{r}) \tag{7}$$

In relation (6), the $P_{r,l}$ are the projectors onto the internal subspaces $\mathscr{E}_{r,l}$. As an example, for an infinitesimal transformation (rotation by an angle $\delta\theta$ around the unit vector $\boldsymbol{u}$, Lorentz transformation of velocity δv in the direction of the unit vector $\boldsymbol{v}$), relations (18) and (22b) of complement A_V show that:

$$L[\Lambda(\boldsymbol{u}\delta\theta,\delta\boldsymbol{v})] = \mathbb{1} - i\frac{\delta\theta}{2}\,\boldsymbol{u}\cdot\boldsymbol{\sigma} + \frac{\delta\boldsymbol{v}}{2c}\cdot\boldsymbol{\sigma} \tag{8}$$

[1]Remember that $L(\Lambda^{-1}) = L^{-1}(\Lambda)$. If Λ is a rotation, $L^\dagger(\Lambda^{-1}) = L(\Lambda)$.

and that the 4×4 matrix associated with $S_D(\Lambda)$ is written (in the $|1\rangle$, $|2\rangle$, $|3\rangle$, $|4\rangle$ basis):

$$(S_D(\Lambda)) = \left(\begin{array}{c|c} 1 - i\dfrac{\delta\theta}{2}\boldsymbol{u}\cdot\boldsymbol{\sigma} + \dfrac{\delta \boldsymbol{v}}{2c}\cdot\boldsymbol{\sigma} & 0 \\ \hline 0 & 1 - i\dfrac{\delta\theta}{2}\boldsymbol{u}\cdot\boldsymbol{\sigma} - \dfrac{\delta \boldsymbol{v}}{2c}\cdot\boldsymbol{\sigma} \end{array} \right) \tag{9}$$

The expression for the spin operator $\boldsymbol{\Sigma}$ showing the effect of rotations on the internal variables of the Dirac equation is therefore:

$$\boldsymbol{\Sigma} = \frac{\hbar}{2}\left(\begin{array}{c|c} \boldsymbol{\sigma} & 0 \\ \hline 0 & \boldsymbol{\sigma} \end{array} \right) \tag{10}$$

• We see that the effect of rotations is the same on all spinors, whether they have positive or negative energy, or whether they are right or left. We also remark on the definitions (VI-85) of these operators in chapter VI that α^0 is a rotation scalar, while $\boldsymbol{\alpha}$ is a vector operator. The scalar product $\boldsymbol{\alpha}\cdot\boldsymbol{\nabla}$ is then rotation invariant, and the rotational invariance of the Dirac equations (5) follows from the scalar properties of the operator on the two sides of the equation.

• For changes of Lorentz reference frames, we use the results[2] of § B-3-b of chapter VI. If we combine relations (VI-123b) and (VI-128b), inside which we use the relation $L(\Lambda) = L^\dagger(\Lambda)$ for the Lorentz transformations, we can write:

$$S_D(\Lambda^{-1})\left[\frac{\partial}{\partial t} + c\,\boldsymbol{\alpha}\cdot\boldsymbol{\nabla}\right] S_D^\dagger(\Lambda^{-1}) = \frac{\partial}{\partial t'} + c\,\boldsymbol{\alpha}\cdot\boldsymbol{\nabla}' \tag{11a}$$

This relation indicates how the left-hand side of equation (5) is transformed; we then have to examine the transformation of the right-hand side. We have to calculate the spin operator α'^0 defined as:

$$\alpha'^0 = S_D(\Lambda^{-1})\,\alpha^0\,S_D^\dagger(\Lambda^{-1}) \tag{11b}$$

But we will show (just below) that α'^0 is merely equal to α^0, so that relation (11a) can be generalized as:

$$\begin{aligned} S_D(\Lambda^{-1})\left[i\hbar\Big(\frac{\partial}{\partial t} + c\,\boldsymbol{\alpha}\cdot\boldsymbol{\nabla}\Big) - \alpha^0\,Mc^2\right] S_D^\dagger(\Lambda^{-1}) \\ = i\hbar\Big(\frac{\partial}{\partial t'} + c\,\boldsymbol{\alpha}\cdot\boldsymbol{\nabla}'\Big) - \alpha^0\,Mc^2 \end{aligned} \tag{11c}$$

The action of α'^0 on a ket of $\mathscr{E}_r$ is first an application of $L^\dagger(\Lambda^{-1})$, then a transfer into the subspace $\mathscr{E}_l$, and finally, an application of $L^\dagger(\Lambda)$. But the

[2]We saw in § 1-b that, here, the matrix $\boldsymbol{\alpha}$ plays the role played by $\eta\boldsymbol{\sigma}$ in chapter VI.

transfer by α^0 from $\mathscr{E}_r$ into $\mathscr{E}_l$ does not change the coordinates of the ket that are simply transferred from one subspace to the other. Consequently, the effects on these coordinates of the operator $L^\dagger(\Lambda^{-1})$ and its inverse $L^\dagger(\Lambda)$ cancel each other. The final result is that the L operators may be ignored in (11b), leaving only the effect of α^0.

If we now detail the action of operator α'^0 on a ket of $\mathscr{E}_l$, it is first an application of the Hermitian conjugate of $L^\dagger(\Lambda)$, which is $L(\Lambda)$, then a transfer of the ket into $\mathscr{E}_r$, and finally, an application of $L(\Lambda^{-1})$. As before, the effects of the two operators $L(\Lambda)$ and $L(\Lambda^{-1})$ cancel each other, leaving only the effect of α^0, showing that in both cases, $\alpha'^0 = \alpha^0$.

Once the invariance of the Dirac equation has been demonstrated by using the basis of the internal variables $|a_\pm\rangle$ and $|b_\pm\rangle$, we can return to the basis of the 4 internal kets $|1\rangle$, $|2\rangle$, $|3\rangle$, and $|4\rangle$ in both reference frames, showing that the invariance is valid in this basis or in any other one. Since the conservation relation (VI-103) of the probability fluid comes from the Dirac equation, this conservation relation remains valid in all the reference frames.

Comments:

(i) The Dirac state space is the direct sum of two Weyl state spaces, with opposite right and left chirality. The only term in the Dirac equation that couples the components of ψ_D in these two spaces is the α^0 mass term.

(ii) Relation (11b) shows that expression:

$$\psi^\dagger(\boldsymbol{r},t)\alpha^0\psi^\dagger(\boldsymbol{r},t) \tag{12}$$

is a relativistic scalar, often written $\overline{\psi}(\boldsymbol{r},t)\psi(\boldsymbol{r},t)$ with the notation $\overline{\psi}(\boldsymbol{r},t) = \psi^\dagger(\boldsymbol{r},t)\alpha^0$. However, the simple product $\psi^\dagger(\boldsymbol{r},t)\psi(\boldsymbol{r},t)$ of the row and column matrices in the spin space is not a relativistic invariant. It yields a probability density that is not invariant because of the Lorentz length contraction.

2. Non-relativistic limit of the Dirac equation

2-a. Decomposition of the Dirac spinor into two spinors of dimension 2

A 4-component Dirac spinor ψ_D can be decomposed into two spinors, each having two components. We regroup to that effect the first two positive energy components into a 2-component spinor ψ_+, and the last two negative energy components into another 2-component spinor ψ_-:

$$\psi_+(\boldsymbol{r},t) = \begin{pmatrix} \psi_1(\boldsymbol{r},t) \\ \psi_2(\boldsymbol{r},t) \end{pmatrix} \qquad \psi_-(\boldsymbol{r},t) = \begin{pmatrix} \psi_3(\boldsymbol{r},t) \\ \psi_4(\boldsymbol{r},t) \end{pmatrix} \tag{13}$$

Taking into account expression (VI-85) for the α components as a function of the Pauli matrices σ, the Dirac equation can be decomposed into two coupled equations:

$$
\begin{aligned}
i\hbar\frac{\partial}{\partial t}\psi_+(\boldsymbol{r},t) &= Mc^2\psi_+(\boldsymbol{r},t) - i\hbar c\,\boldsymbol{\sigma}\cdot\boldsymbol{\nabla}\psi_-(\boldsymbol{r},t)\\
i\hbar\frac{\partial}{\partial t}\psi_-(\boldsymbol{r},t) &= -Mc^2\psi_-(\boldsymbol{r},t) - i\hbar c\,\boldsymbol{\sigma}\cdot\boldsymbol{\nabla}\psi_+(\boldsymbol{r},t)
\end{aligned}
\tag{14}
$$

The non-relativistic limit is obtained when the momenta are small compared to Mc, hence when the energies are small compared to Mc^2. The dominant terms on the right-hand side are the $\pm Mc^2$ terms, so that the coupling between the two spinors ψ_+ and ψ_- can be treated as perturbations. We set:

$$
\psi_+(\boldsymbol{r},t) = \mathrm{e}^{-iMc^2t/\hbar}\chi_+(\boldsymbol{r},t) \qquad\qquad \psi_-(\boldsymbol{r},t) = \mathrm{e}^{iMc^2t/\hbar}\chi_-(\boldsymbol{r},t) \tag{15}
$$

and the coupled evolution equations are written:

$$
\begin{aligned}
i\hbar\frac{\partial}{\partial t}\chi_+(\boldsymbol{r},t) &= -i\hbar c\,\mathrm{e}^{2iMc^2t/\hbar}\,\boldsymbol{\sigma}\cdot\boldsymbol{\nabla}\chi_-(\boldsymbol{r},t)\\
i\hbar\frac{\partial}{\partial t}\chi_-(\boldsymbol{r},t) &= -i\hbar c\,\mathrm{e}^{-2iMc^2t/\hbar}\,\boldsymbol{\sigma}\cdot\boldsymbol{\nabla}\chi_+(\boldsymbol{r},t)
\end{aligned}
\tag{16}
$$

In the presence of an electromagnetic field, equality (VI-104) of chapter VI shows that these two equations become:

$$
\begin{aligned}
i\hbar\frac{\partial}{\partial t}\chi_+(\boldsymbol{r},t) &= qU(\boldsymbol{r},t)\,\chi_+(\boldsymbol{r},t) - c\,\mathrm{e}^{2iMc^2t/\hbar}\,\boldsymbol{\sigma}\cdot\left[i\hbar\boldsymbol{\nabla}+q\boldsymbol{A}(\boldsymbol{r},t)\right]\chi_-(\boldsymbol{r},t)\\
i\hbar\frac{\partial}{\partial t}\chi_-(\boldsymbol{r},t) &= qU(\boldsymbol{r},t)\,\chi_-(\boldsymbol{r},t) - c\,\mathrm{e}^{-2iMc^2t/\hbar}\,\boldsymbol{\sigma}\cdot\left[i\hbar\boldsymbol{\nabla}+q\boldsymbol{A}(\boldsymbol{r},t)\right]\chi_+(\boldsymbol{r},t)
\end{aligned}
\tag{17}
$$

We shall assume that the U and $\boldsymbol{A}$ potentials are weak enough for the electromagnetic energy terms to remain small compared to the rest energy Mc^2. Under these conditions, the entire right-hand sides of these equations can be treated as a perturbation.

2-b. Reconstruction of the Pauli Hamiltonian

Imagine we start from a solution where only χ_+ is non-zero to zeroth order, and let us see how this solution is modified by its coupling with χ_-. To first-order, $\chi_+(\boldsymbol{r},t)$ remains constant in time, and $\chi_-(\boldsymbol{r},t)$ is given by:

$$
\chi_-(\boldsymbol{r},t) \simeq -\frac{1}{2Mc}\,\mathrm{e}^{-2iMc^2t/\hbar}\,\boldsymbol{\sigma}\cdot\left[i\hbar\boldsymbol{\nabla}+q\boldsymbol{A}(\boldsymbol{r},t)\right]\chi_+(\boldsymbol{r},t) \tag{18}
$$

Inserting this result in the first equation (17), we get:

$$i\hbar\frac{\partial}{\partial t}\chi_+(\boldsymbol{r},t) = qU(\boldsymbol{r},t)\chi_+(\boldsymbol{r},t) + \frac{1}{2M}\left[\boldsymbol{\sigma}\cdot\Big(i\hbar\boldsymbol{\nabla}+q\boldsymbol{A}(\boldsymbol{r},t)\Big)\right]\left[\boldsymbol{\sigma}\cdot\Big(i\hbar\boldsymbol{\nabla}+q\boldsymbol{A}(\boldsymbol{r},t)\Big)\right]\chi_+(\boldsymbol{r},t) \quad (19)$$

On the right-hand side of this equation, the two scalar products introduce σ_i and σ_j components of $\boldsymbol{\sigma}$ with a summation over the indices i and j. These Pauli matrices, which only act on the internal variables, commute with the operators in the parentheses that only act on the external variables. The Pauli matrices can thus be regrouped to form their product. When $i = j$, since the squares of the Pauli matrices are equal to unity, we get a first term in:

$$\frac{1}{2M}\Big(i\hbar\boldsymbol{\nabla}+q\boldsymbol{A}(\boldsymbol{r},t)\Big)^2 \quad (20)$$

which is simply the usual kinetic energy in the Hamiltonian, proportional to the square of the velocity $\boldsymbol{V} = (\boldsymbol{P} - q\boldsymbol{A})/M$. For the $i \neq j$ terms, relation $\sigma_i\sigma_j = i\varepsilon_{ijk}\sigma_k$ shows that we can write them as:

$$\frac{1}{2M}\left[i\varepsilon_{ijk}\Big(i\hbar\partial_i + qA_i(\boldsymbol{r},t)\Big)\Big(i\hbar\partial_j + qA_j(\boldsymbol{r},t)\Big)\sigma_k\right] \quad (21)$$

The terms in $\partial_i\partial_j$ and those in A_iA_j, symmetric in i and j, cancel out once they are multiplied by ε_{ijk} and summed over i and j. We are then left with:

$$-\frac{q\hbar}{2M}\varepsilon_{ijk}\Big[\partial_i A_j(\boldsymbol{r},t) - A_j(\boldsymbol{r},t)\partial_i\Big]\sigma_k = -\frac{q\hbar}{2M}(\boldsymbol{\nabla}\times\boldsymbol{A})_k\,\sigma_k = -\frac{q\hbar}{2M}\boldsymbol{B}(\boldsymbol{r},t)\cdot\boldsymbol{\sigma} \quad (22)$$

(in the first line, there is cancellation between the two terms where the derivation only acts on the spinor on the right of the operator).

We finally get:

$$i\hbar\frac{\partial}{\partial t}\chi_+(\boldsymbol{r},t) = H\,\chi_+(\boldsymbol{r},t) \quad (23a)$$

with:

$$H = \left[\frac{1}{2M}\Big(\frac{\hbar}{i}\boldsymbol{\nabla} - q\boldsymbol{A}(\boldsymbol{r},t)\Big)^2 + qU(\boldsymbol{r},t) - \frac{q\hbar}{2M}\boldsymbol{B}(\boldsymbol{r},t)\cdot\boldsymbol{\sigma}\right]\chi_+(\boldsymbol{r},t) \quad (23b)$$

This is the well-known Pauli Hamiltonian H, which is the evolution operator for the $\chi_+(\boldsymbol{r},t)$ spinor. The first term in H describes the kinetic energy. The second corresponds to the electric energy of a charge q placed in a potential $U(\boldsymbol{r})$. The third one describes the coupling between the magnetic field and the spin magnetic moment $\boldsymbol{S} = \hbar\boldsymbol{\sigma}/2$ of the particle. For an electron of mass $M = m_e$ and charge q, it is written:

$$H_{\text{spin}} = -\frac{q}{m_e}\boldsymbol{B}(\boldsymbol{r},t)\cdot\boldsymbol{S} = -2\mu_B\,\boldsymbol{B}(\boldsymbol{r},t)\cdot(\boldsymbol{S}/\hbar) \tag{24}$$

where the Bohr magneton μ_B is defined as:

$$\mu_B = \frac{q\hbar}{2m_e} \tag{25}$$

Exercise: Assuming an homogeneous magnetic field, we set $\boldsymbol{A}(\boldsymbol{r}) = (\boldsymbol{B}\times\boldsymbol{r})/2$. Show that the Pauli Hamiltonian H can be written as:

$$H = -\frac{\hbar^2}{2M}\Delta + qU - \frac{q}{2M}\boldsymbol{B}\cdot(\boldsymbol{L}+2\boldsymbol{S}) + \frac{q^2\boldsymbol{A}^2}{2M} \tag{26}$$

where $\boldsymbol{L} = (\hbar/i)\boldsymbol{r}\times\boldsymbol{\nabla}$ is the orbital angular momentum.

The gyromagnetic ratio g of an angular momentum with a magnetic moment $\boldsymbol{M}$ is defined as $\boldsymbol{M} = g\boldsymbol{J}/\hbar$. For an orbital angular momentum $\boldsymbol{L}$, we see that $g = 1$, whereas $g = 2$ for a spin[3]; this is an important result of the Dirac equation. The last term of (26), proportional to $\boldsymbol{A}^2$, is called the "diamagnetic term".

2-c. Probability current

We now compute the non-relativistic limit of the probability current given in relation (VI-103b) of chapter VI. In terms of the ψ_+ and ψ_- spinors, this current is written:

$$\boldsymbol{j}(\boldsymbol{r},t) = c\,\psi_+^\dagger(\boldsymbol{r},t)\,\boldsymbol{\sigma}\,\psi_-(\boldsymbol{r},t) + \text{c.c.} = \mathrm{e}^{2iMc^2/\hbar}\,\chi_+^\dagger(\boldsymbol{r},t)\,\boldsymbol{\sigma}\,\chi_-(\boldsymbol{r},t) + \text{c.c.} \tag{27}$$

where c.c. means "complex conjugate". Inserting now the first-order expression (18) for $\chi_-^\dagger(\boldsymbol{r},t)$, the $i = x, y, z$ current component becomes:

$$j_i(\boldsymbol{r},t) = -\left(\frac{1}{2M}\right)\,\chi_+^\dagger(\boldsymbol{r},t)\Big[\sigma_i\,\boldsymbol{\sigma}\cdot\Big(i\hbar\boldsymbol{\nabla} + q\boldsymbol{A}(\boldsymbol{r},t)\Big)\Big]\chi_+(\boldsymbol{r},t) + \text{c.c.} \tag{28}$$

The term inside the brackets on the right-hand side can be written:

$$\sigma_i\sigma_j\Big(i\hbar\partial_j + qA_j(\boldsymbol{r},t)\Big) = \Big(i\hbar\partial_i + qA_i(\boldsymbol{r},t)\Big) + i\varepsilon_{ijk}\Big(i\hbar\partial_j + qA_j(\boldsymbol{r},t)\Big)\sigma_k \tag{29}$$

The first term yields a contribution to the probability current:

$$-\frac{1}{2M}\,\chi_+^\dagger(\boldsymbol{r},t)\,\Big(i\hbar\partial_i + qA_i(\boldsymbol{r},t)\Big)\,\chi_+(\boldsymbol{r},t) + \text{c.c.} \tag{30a}$$

[3]The value $g = 2$ predicted by Dirac's theory for the electron (or muon) spin is slightly modified (by roughly a thousandth) when taking into account the "radiative corrections" due to the quantization of the electromagnetic field.

analogous to the usual expression of the current obtained from the Schrödinger equation. The second term yields the i-component of the vector:

$$-\frac{1}{2M}\,\chi_+^\dagger(\boldsymbol{r},t)\Big[\big(-\hbar\boldsymbol{\nabla}+iq\boldsymbol{A}\big)\times\boldsymbol{\sigma}\Big]\,\chi_+(\boldsymbol{r},t)+\text{c.c.} \tag{30b}$$

where the $\boldsymbol{A}$ terms disappear in the complex conjugation.

Finally, coming back to the ψ_+ spinor, we get:

$$\boldsymbol{j}(\boldsymbol{r},t)=\frac{1}{2M}\Big[\psi_+^\dagger(\boldsymbol{r},t)\Big(\frac{\hbar}{i}\boldsymbol{\nabla}-q\boldsymbol{A}(\boldsymbol{r},t)\Big)\,\psi_+(\boldsymbol{r},t)+\text{c.c.}+\hbar\boldsymbol{\nabla}\times\langle\boldsymbol{\sigma}\rangle_{(\boldsymbol{r},t)}\Big] \tag{31a}$$

with:

$$\langle\boldsymbol{\sigma}\rangle_{(\boldsymbol{r},t)}=\psi_+^\dagger(\boldsymbol{r},t)\,\boldsymbol{\sigma}\,\psi_+(\boldsymbol{r},t) \tag{31b}$$

In this expression for the probability current $\boldsymbol{j}$, the first term is the one usually obtained when dealing with the Schrödinger or Pauli equation. The second term only appears when this equation is deduced from the non-relativistic limit of the Dirac equation. As it is the curl of a vector, its divergence is zero, and does not change the local variation of the probability density. It does, however, modify the streamlines of the probability fluid [35], which may have an influence on the measurement theory of particles arrival times [36]. The sum of the two terms on the right-hand side of (31a) is called the "Gordon decomposition".

Complement B_{VI}

Finite Poincaré transformations and Dirac state space

In chapter V, we mainly discussed infinitesimal transformations of the Poincaré group, mentioning occasionally that the finite transformations could be obtained via an exponentiation of its Lie algebra operators. The study of the finite transformations is nevertheless important as it enables a better understanding of the way different physical quantities behave upon a change of relativistic reference frame. This complement discusses in more detail these unitary finite transformations, the way they can modify the operators of the Lie algebra, and how they make it possible to build the Dirac state (spinor) space.

1. Displacement group

Displacement operators in the state space have already been studied in § 2 of complement C_V. We only give here a brief review, within the framework of all the generators of the Poincaré group.

1-a. Rotations

Consider an infinitesimal rotation by an angle $\delta\varphi$ around the unit vector $\boldsymbol{u}$. The general relation (IV-55) of chapter IV yields the expression of the corresponding operator acting in the state space. Choosing in this relation the value of the parameter $\boldsymbol{a} = \varphi\boldsymbol{u}$, we get:

$$R_{\boldsymbol{u}}(\delta\varphi) = 1 - \frac{i}{\hbar}\delta\varphi\,(\boldsymbol{u}\cdot\boldsymbol{J}) + \ldots = 1 - \frac{i}{\hbar}\delta\varphi\; J_u \tag{1}$$

where $J_u = \boldsymbol{u} \cdot J$. Integrating over φ yields the finite rotation operator:

$$R_{\boldsymbol{u}}(\varphi) = \exp(-i\varphi\, J_u/\hbar) \tag{2}$$

The commutation relations of operators $\boldsymbol{P}$ and $\boldsymbol{K}$ with $\boldsymbol{J}$ show that $\boldsymbol{P}$ and $\boldsymbol{K}$ are vector operators. In § A of chapter VIII, we define the vector operators by this type of commutation relations and show that the components of such operators are transformed under a rotation as the components of a vector in ordinary three-dimensional space, *cf.* relation (VIII-17). The auto-commutation relations of $\boldsymbol{J}$ show that it, too, is a vector operator.

As for H, since it commutes with all the components of $\boldsymbol{J}$, it is a scalar operator, hence a rotation invariant operator:

$$R_{\boldsymbol{u}}(\varphi)\, H\, R_{\boldsymbol{u}}^{\dagger}(\varphi) = H \tag{3}$$

1-b. Space translations

The operator $T(\boldsymbol{b})$ corresponding, in the state space, to a translation by a vector $\boldsymbol{b}$ is also written in the exponential form:

$$T(\boldsymbol{b}) = \exp(-i\boldsymbol{b} \cdot \boldsymbol{P}/\hbar) \tag{4}$$

Both $\boldsymbol{P}$ and H, which commute with the three components of $\boldsymbol{P}$, remain unchanged under this unitary transformation.

As for $\boldsymbol{J}$, it is transformed into operator $\tilde{\boldsymbol{J}}$ given by:

$$\tilde{\boldsymbol{J}}(\boldsymbol{b}) = \exp(-i\boldsymbol{b} \cdot \boldsymbol{P}/\hbar)\, \boldsymbol{J}\, \exp(i\boldsymbol{b} \cdot \boldsymbol{P}/\hbar) \tag{5}$$

We take an $Oxyz$ reference frame in which one unit vector, $\boldsymbol{e}_j$, is parallel to the vector $\boldsymbol{b}$. Taking into account the second relation (V-60) that gives the commutator of $\boldsymbol{J}$ and $\boldsymbol{P}$, the $\tilde{J}_i(\boldsymbol{b})$ component of $\tilde{\boldsymbol{J}}(\boldsymbol{b})$ obeys equation:

$$\frac{\mathrm{d}}{\mathrm{d}b_j}\tilde{J}_i(b_j\boldsymbol{e}_j) = \frac{-i}{\hbar}\exp(-ib_jP_j/\hbar)\,(-i\hbar\varepsilon_{ijk}P_k)\,\exp(ib_jP_j/\hbar) = -\varepsilon_{ijk}\,P_k \tag{6}$$

We thus have:

$$\tilde{J}_i(b_j\boldsymbol{e}_j) = J_i - \varepsilon_{ijk}b_jP_k \tag{7}$$

or, more generally:

$$\tilde{J}(\boldsymbol{b}) = \boldsymbol{J} - \boldsymbol{b} \times \boldsymbol{P} \tag{8}$$

This is the quantum form of the classical relation for the change of angular momentum when measured with respect to a point displaced by vector $+\boldsymbol{b}$.

Consider finally the transformation of operator $\boldsymbol{K}$, which becomes $\tilde{\boldsymbol{K}}(\boldsymbol{b})$ under a unitary transformation like the one for $\boldsymbol{J}$ in (5). Relation (V-61) shows that the commutator that plays a role is proportional to H, an invariant operator under the unitary operation. The same type of reasoning as for $\boldsymbol{J}$ leads us to:

$$\tilde{\boldsymbol{K}}(\boldsymbol{b}) = \boldsymbol{K} + \frac{\boldsymbol{b}}{c^2}H \tag{9}$$

2. Lorentz transformations

In the state space, to describe unitary operators corresponding to a change of Lorentz reference frame, it is helpful to use as a parameter the rapidity q defined in (III-114).

2-a. Rapidity

Consider the operator that changes the Lorentz reference frame:

$$T_K(\boldsymbol{q}) = \exp(-i\boldsymbol{q} \cdot \boldsymbol{K}/\hbar) \tag{10}$$

where $\boldsymbol{K}$ is the Lie algebra operator of the Poincaré group introduced in chapter V, and $\boldsymbol{q}$ a vector parameter to be defined. We can write:

$$T_K(\boldsymbol{q} + \mathrm{d}\boldsymbol{q}) = T_K(\boldsymbol{q})\, T_K(\mathrm{d}\boldsymbol{q}) \tag{11}$$

Operator $T_K(\boldsymbol{q})$ induces a change of reference frame with a velocity we call $\boldsymbol{v}(\boldsymbol{q})$. We assume that the vectors $\boldsymbol{q}$ and $\mathrm{d}\boldsymbol{q}$ to be parallel. Operator $\boldsymbol{K}$ has been introduced in chapter IV following the general definition (IV-55) of infinitesimal operators, taking the group parameter $\boldsymbol{a}$ equal to the velocity $\boldsymbol{v}$, as shown, for example, in relation (III-107). For an infinitesimal variation $\mathrm{d}\boldsymbol{q}$, operator $T_K(\mathrm{d}\boldsymbol{q})$ corresponds to a velocity change simply equal to $\mathrm{d}\boldsymbol{v} = \mathrm{d}\boldsymbol{q}$. But when $\boldsymbol{q}$ has a finite value, we must use the usual relation for adding relativistic velocities, written as:

$$v(q + \mathrm{d}q) = \frac{v + \mathrm{d}q}{1 + \frac{v\mathrm{d}q}{c^2}} \tag{12}$$

which leads to:

$$\frac{\mathrm{d}v}{\mathrm{d}q} = 1 - \frac{v^2}{c^2} \tag{13}$$

Integrating this equation yields:

$$\frac{v}{c} = \tanh\frac{q}{c} \tag{14}$$

This relation is simply the definition (III-114) of the rapidity q as a function of the velocity v. It is not surprising that the rapidity appears in the argument of the exponential operator (10): we know that, in relativity, it is the rapidities (and not the velocities) which can be added, just as we can add the arguments of a product of exponentials.

2-b. Momentum-energy couple

We consider a Lorentz transformation along the Oz direction, and examine the transformations of the $\boldsymbol{P}$ and H operators:

$$\begin{aligned}\tilde{\boldsymbol{P}}(q) &= \exp(-iqK_z/\hbar)\,\boldsymbol{P}\,\exp(iqK_z/\hbar)\\ \tilde{H}(q) &= \exp(-iqK_z/\hbar)\,H\,\exp(iqK_z/\hbar)\end{aligned} \tag{15}$$

The commutation relations of K_z with the P_z and H operators are given in (V-61) and (V-62), and lead to:

$$\begin{aligned}\frac{\mathrm{d}}{\mathrm{d}q}\tilde{P}_z(q) &= \frac{-i}{\hbar}\exp(-iqK_z/\hbar)\,(\frac{-i\hbar}{c^2}H)\,\exp(iqK_z/\hbar) = -\frac{1}{c^2}\tilde{H}(q)\\ \frac{\mathrm{d}}{\mathrm{d}q}\tilde{H}(q) &= \frac{-i}{\hbar}\exp(-iqK_z/\hbar)\,(-i\hbar P_z)\,\exp(iqK_z/\hbar) = -\tilde{P}_z(q)\end{aligned} \tag{16}$$

As for the Ox and Oy components of $\boldsymbol{P}$, they are invariant under the unitary transformation under study:

$$\tilde{P}_x(q) = P_x \qquad\qquad \tilde{P}_y(q) = P_y \tag{17}$$

Adding and subtracting relations (16), we get:

$$\frac{\mathrm{d}}{\mathrm{d}q}[\tilde{H}(q) \pm c\tilde{P}_z(q)] = \mp\frac{1}{c}[\tilde{H}(q) \pm c\tilde{P}_z] \tag{18}$$

and therefore:

$$[\tilde{H}(q) \pm c\tilde{P}_z(q)] = \exp(\mp\frac{q}{c})\,[H \pm cP_z] \tag{19}$$

Finally:

$$c\tilde{P}_z(q) = \cosh(\frac{q}{c})\,cP_z - \sinh(\frac{q}{c})\,H \tag{20a}$$

$$\tilde{H}(q) = -\sinh(\frac{q}{c})\,cP_z + \cosh(\frac{q}{c})\,H \tag{20b}$$

Using relation (14) to express q as a function of the velocity, we see that relations (17) and (20) are those of a Lorentz transformation. This means that, in a Lorentz transformation, the four operators H and $c\boldsymbol{P}$ are transformed as the components of a four-vector. We find again, for finite transformations, the definition of a four-vector operator introduced in chapter V in (V-64) for infinitesimal transformations.

2-c. $J - K$ couple

We now transform both operators $\boldsymbol{J}$ and $\boldsymbol{K}$ with the same unitary operation:

$$\tilde{\boldsymbol{J}}(q) = \exp(-iqK_z/\hbar)\, \boldsymbol{J}\, \exp(iqK_z/\hbar)$$
$$\tilde{\boldsymbol{K}}(q) = \exp(-iqK_z/\hbar)\, \boldsymbol{K}\, \exp(iqK_z/\hbar) \tag{21}$$

The commutation relations (V-61) lead to:

$$\frac{\mathrm{d}}{\mathrm{d}q}\tilde{J}_x(q) = \frac{-i}{\hbar}\exp(-iqK_z/\hbar)\,(i\hbar K_y)\,\exp(iqK_z/\hbar) = \tilde{K}_y(q)$$
$$\frac{\mathrm{d}}{\mathrm{d}q}\tilde{K}_y(q) = \frac{-i}{\hbar}\exp(-iqK_z/\hbar)\,(\frac{i\hbar}{c^2}J_x)\,\exp(iqK_z/\hbar) = \frac{1}{c^2}\tilde{J}_x(q) \tag{22}$$

as well as:

$$\frac{\mathrm{d}}{\mathrm{d}q}\tilde{J}_y(q) = \frac{-i}{\hbar}\exp(-iqK_z/\hbar)\,(-i\hbar K_x)\,\exp(iqK_z/\hbar) = -\tilde{K}_x(q)$$
$$\frac{\mathrm{d}}{\mathrm{d}q}\tilde{K}_x(q) = \frac{-i}{\hbar}\exp(-iqK_z/\hbar)\,(\frac{-i\hbar}{c^2}J_y)\,\exp(iqK_z/\hbar) = -\frac{1}{c^2}\tilde{J}_y(q) \tag{23}$$

Concerning the $\tilde{J}_z$ and $\tilde{K}_z$ operators, it is easy to see that the commutation relations show that they are independent of q:

$$\tilde{J}_z(q) = J_z \qquad\qquad \tilde{K}_z(q) = K_z \tag{24}$$

By a similar calculation as for H and $\boldsymbol{P}$, we get:

$$\tilde{J}_x(q) = \cosh(\frac{q}{c})\, J_x + \sinh(\frac{q}{c})\, cK_y \tag{25a}$$
$$c\tilde{K}_y(q) = \sinh(\frac{q}{c})\, J_x + \cosh(\frac{q}{c})\, cK_y \tag{25b}$$

and:

$$\tilde{J}_y(q) = \cosh(\frac{q}{c})\, J_y - \sinh(\frac{q}{c})cK_x \tag{26a}$$
$$c\tilde{K}_x(q) = -\sinh(\frac{q}{c})\, J_y + \cosh(\frac{q}{c})\, cK_x \tag{26b}$$

or:

$$\tilde{\boldsymbol{J}}_\perp(q) = \cosh(\frac{q}{c})\boldsymbol{J}_\perp - \sinh(\frac{q}{c})\,\hat{\boldsymbol{q}}\times c\boldsymbol{K}_\perp \qquad\qquad \tilde{\boldsymbol{J}}_\parallel(q) = \boldsymbol{J}_\parallel$$
$$c\tilde{\boldsymbol{K}}_\perp(q) = \sinh(\frac{q}{c})\,\hat{\boldsymbol{q}}\times \boldsymbol{J}_\perp + \cosh(\frac{q}{c})\, c\boldsymbol{K}_\perp \qquad\qquad \tilde{\boldsymbol{K}}_\parallel(q) = \boldsymbol{K}_\parallel \tag{27}$$

where $\hat{\boldsymbol{q}}$ is the unit vector in the direction of the Lorentz transformation; $\boldsymbol{J}_\parallel$ and $\boldsymbol{K}_\parallel$ are the projections of $\boldsymbol{J}$ and $\boldsymbol{K}$ onto this vector, and $\boldsymbol{J}_\perp$ and $\boldsymbol{K}_\perp$ are the perpendicular components. These transformation relations are the finite form of the infinitesimal equalities obtained in § 2-a of complement $\mathrm{B_V}$, and are similar to those obtained for the electric and magnetic fields in classical electromagnetism.

2-d. Spin–helicity couple

We finally discuss the transformations of the two operators $\boldsymbol{S}$ and $\boldsymbol{J} \cdot \boldsymbol{P} = \boldsymbol{P} \cdot \boldsymbol{J}$. We showed in § C-3-c of chapter V that operator $\boldsymbol{J} \cdot \boldsymbol{P}$ commutes with $\boldsymbol{J}$ (it is a scalar), with $\boldsymbol{P}$ (it is translation invariant) and H (it is a constant of motion for a free particle). Remember that helicity is defined as the operator $\boldsymbol{J} \cdot \boldsymbol{P}$ whose eigenvalues have been divided by the momentum modulus.

We set:

$$\tilde{\boldsymbol{S}}(q) = \exp(-iqK_z/\hbar)\, \boldsymbol{S}\, \exp(iqK_z/\hbar)$$
$$(\tilde{\boldsymbol{J}} \cdot \tilde{\boldsymbol{P}})(q) = \exp(-iqK_z/\hbar)\, (\boldsymbol{J} \cdot \boldsymbol{P})\, \exp(iqK_z/\hbar) \tag{28}$$

The commutation relations of K_z with S_z and $\boldsymbol{J} \cdot \boldsymbol{P}$ are written in (V-81) and (V-83). We thus get:

$$\frac{\mathrm{d}}{\mathrm{d}q}\tilde{S}_z(q) = \frac{-i}{\hbar} \exp(-iqK_z/\hbar)\, (\frac{-i\hbar}{Mc^2}\boldsymbol{J} \cdot \boldsymbol{P})\, \exp(iqK_z/\hbar) = -\frac{1}{Mc^2}(\tilde{\boldsymbol{J}} \cdot \tilde{\boldsymbol{P}})(q)$$
$$\frac{\mathrm{d}}{\mathrm{d}q}(\tilde{\boldsymbol{J}} \cdot \tilde{\boldsymbol{P}})(q) = \frac{-i}{\hbar} \exp(-iqK_z/\hbar)\, (-i\hbar M S_z)\, \exp(iqK_z/\hbar) = -M\tilde{S}_z(q) \tag{29}$$

As for the Ox and Oy components of $\boldsymbol{S}$, they are invariant under the unitary transformation:

$$\tilde{S}_x(q) = S_x \qquad \tilde{S}_y(q) = S_y \tag{30}$$

This finally yields:

$$Mc\,\tilde{S}_z(q) = \cosh(\frac{q}{c})\, Mc\, S_z - \sinh(\frac{q}{c})\, \boldsymbol{J} \cdot \boldsymbol{P} \tag{31a}$$

$$(\tilde{\boldsymbol{J}} \cdot \tilde{\boldsymbol{P}})(q) = -\sinh(\frac{q}{c})\, Mc\, S_z + \cosh(\frac{q}{c})\, \boldsymbol{J} \cdot \boldsymbol{P} \tag{31b}$$

These relations show that the couple $(\boldsymbol{J} \cdot \boldsymbol{P}, Mc\boldsymbol{S})$ is transformed as a four-vector (the so-called Pauli–Lubanski four-vector, *cf.* § 2 of complement B_V).

Comment:

If the mass is equal to zero, the operator $\boldsymbol{S}$ is no longer defined, but we can introduce the operator $\boldsymbol{\Sigma} = M\boldsymbol{S}$, as we did in § C-4 of chapter V. The commutation relations established in § C-3-c of that chapter remain valid, provided we multiply the right-hand side by M. For example, equality (V-81) becomes:

$$[\Sigma_i\,,\, K_j] = \frac{i\hbar}{c^2}\, \delta_{ij}\, \boldsymbol{P} \cdot \boldsymbol{J} \tag{32}$$

We see that, once MS_z is replaced by Σ_z, relations (31) remain valid when the mass is equal to zero.

3. State space and Dirac operators

3-a. Construction of the state space

We start from an eigenvector $|\boldsymbol{p}, E, \nu\rangle$ common to the momentum $\boldsymbol{P}$ and the Hamiltonian H, with eigenvalues $\boldsymbol{p}$ and E. We choose an axis Oz parallel to $\boldsymbol{p}$ and apply to this eigenvector the Lorentz unitary transformation $T_K(\boldsymbol{q})$ defined in (10), where $\boldsymbol{q}$ is also parallel to the Oz axis. This yields a new eigenvector of $\boldsymbol{P}$ and H, with different eigenvalues, since using relations (20) we can write:

$$\boldsymbol{P}\, T_K(q)|\boldsymbol{p}, E, \nu\rangle = T_K(q)\tilde{\boldsymbol{P}}(-q)|\boldsymbol{p}, E, \nu\rangle = \left[\cosh(\frac{q}{c})\,\boldsymbol{p} + \sinh(\frac{q}{c})\,\hat{\boldsymbol{q}}\frac{E}{c}\right] T_K(q)|\boldsymbol{p}, E, \nu\rangle$$
$$H\, T_K(q)|\boldsymbol{p}, E, \nu\rangle = T_K(q)\tilde{H}(-q)|\boldsymbol{p}, E, \nu\rangle = \left[\sinh(\frac{q}{c})\, cp + \cosh(\frac{q}{c})\, E\right] T_K(q)|\boldsymbol{p}, E, \nu\rangle \tag{33}$$

The terms inside the brackets on the right-hand side yield the new eigenvalues, which are transformed as the components of a four-vector.

α. *Non-zero mass*

If M is different from zero, we have $|E| > c|p|$ and the eigenvalues are the components of a time-like four-vector. Changing the Lorentz reference frame, one can annul its spatial component: we simply have to take $\tanh(q/c) = -cp/E$ in relation (33). This means that there necessarily exists, in the state space, an eigensubspace $\mathscr{E}(\boldsymbol{p} = 0)$, common to $\boldsymbol{P}$ and H, in which the three components of $\boldsymbol{P}$ have a zero eigenvalue; the eigenvalue of H is $E = \pm Mc^2$.

We can use the subspace $\mathscr{E}(\boldsymbol{p} = 0)$ as a starting point to build the Dirac state space $\mathscr{E}_D$. The form (VI-77) of the Dirac equation and that of the matrices $\boldsymbol{\alpha}$ show that $\mathscr{E}(\boldsymbol{p} = 0)$ is a 4-dimensional subspace, in which we can introduce an orthonormal basis of the kets $|\boldsymbol{p} = 0, \nu\rangle$, where ν takes four values. The form (VI-85) of α^0 shows that the energy E can take two values:

$$E_\nu = Mc^2 \text{ for } \nu = 1, 2 \qquad ; \qquad E_\nu = -Mc^2 \text{ for } \nu = 3, 4 \tag{34}$$

Applying the operator $T_K(\boldsymbol{q})$, we can then obtain basis kets that have an arbitrary momentum $\boldsymbol{p}$, as we now show. We introduce the vector $\boldsymbol{q}(\boldsymbol{p}, \nu)$, parallel to $\boldsymbol{p}$, and whose modulus obeys:

$$\tanh\left[\frac{q(p, \nu)}{c}\right] = \frac{cp}{E_\nu} \tag{35}$$

Relation (33) shows that the kets:

$$|\boldsymbol{p}, E, \nu\rangle = T_K[\boldsymbol{q}(\boldsymbol{p}, \nu)]\; |\boldsymbol{p} = 0, \nu\rangle \tag{36}$$

are eigenvectors of $\boldsymbol{P}$ and H with eigenvalues $\boldsymbol{p}$ and E (from now on we do not write in the ket the eigenvalue E, no longer necessary). For $\nu = 1, 2, 3, 4$, we get the four kets that span the subspace $\mathscr{E}(\boldsymbol{p})$ with arbitrary momentum $\boldsymbol{p}$. If now $\boldsymbol{p}$ varies in the entire momentum space, these kets span the entire state space $\mathscr{E}_D$ of the Dirac equation.

β. Zero mass

If M is zero, we have $E = \pm cp$, and the brackets in equations (33) reduce to $e^{\pm q/c}$, exponentials whose sign cannot change. A pure Lorentz transformation can no longer change the sign of the momentum.

To build the state space, we cannot start from a zero-momentum subspace, since such a subspace does not exist. Starting from a non-zero momentum $\boldsymbol{p}$, we can use rotations to obtain arbitrary directions for $\boldsymbol{p}$, then Lorentz transformations parallel to the momentum to simultaneously change E and $\boldsymbol{p}$, while preserving the relation $E^2 = p^2c^2$.

3-b. Operators

α. Internal and external variables

With the Dirac equation (VI-86), and as in the Pauli spin theory (*cf.* for example § C of chapter IX in [9]), we associate the eigenvalue $\boldsymbol{p}$ with an "external variable" of the system, whereas the quantum numbers ν that distinguish the four spinor components (VI-87) correspond to spin "internal variables" . Consequently, operator $\boldsymbol{P}$ only acts on the external variables. This is not the case for the Hamiltonian written inside the brackets on the right-hand side of the Dirac equation (VI-86): as it includes the product $\boldsymbol{\alpha} \cdot \boldsymbol{p}$, it acts on both types of variables. We computed in § B-2-b its eigenvectors with well-defined momentum $\boldsymbol{p}$. Unless $\boldsymbol{p}$ is zero, they have both a component on the first two basis eigenkets ($\nu = 1, 2$), called the positive energy component, and a component on the last two basis kets ($\nu = 3, 4$), called the negative energy component.

To obtain the form (VI-88) of the Dirac equation, we have performed a Fourier transform on the momentum variables, which did not affect the internal variables. Operator $\boldsymbol{R}$ is thus, by definition, the operator whose eigenvectors are the Fourier transforms:

$$|\boldsymbol{r}, \nu\rangle = \frac{1}{(2\pi\hbar)^{3/2}} \int \mathrm{d}^3p \; \mathrm{e}^{i\boldsymbol{p}\cdot\boldsymbol{r}/\hbar} \; |\boldsymbol{p}, \nu\rangle \tag{37}$$

The $\boldsymbol{R}$ operator only acts on the external variables, but not on the quantum numbers ν.

On the other hand, since the commutator of $\boldsymbol{K}$ and $\boldsymbol{P}$ is proportional to H, which acts on the internal variables, this commutator also acts on the internal variables. This means that, contrary to what we saw for the Galileo group, $-\boldsymbol{K}/M$ is not an acceptable position operator for the Poincaré group. Finally, since $\boldsymbol{J}$ is proportional to the commutator of two components of $\boldsymbol{K}$, we should expect $\boldsymbol{J}$ to also act on the two types of variables.

The orbital angular momentum operator $\boldsymbol{L} = -\boldsymbol{K} \times \boldsymbol{P}/M$ annuls any ket inside the eigensubspace $\mathscr{E}(\boldsymbol{p} = 0)$. Relation (V-76) then shows that the action of the total angular momentum $\boldsymbol{J}$ is simply the action of the spin $\boldsymbol{S}$. By analogy with the Pauli theory (a more precise argument is given in complement A_{VI}, *cf.* relation (10)), we assume that the first two eigenkets $\nu = 1$ and $\nu = 2$ correspond to the eigenkets of the S_z spin component , with respective eigenvalues $+\hbar/2$ and $-\hbar/2$; we make the same hypothesis for the last two eigenkets $\nu = 3$ and $\nu = 4$. It means that the 4 spin vectors span the direct sum of two spin 1/2 spaces having, respectively, a positive and negative energy. In the subspace $\mathscr{E}(\boldsymbol{p} = 0)$, the spin operator is represented by the matrices:

$$\boldsymbol{S} = \frac{\hbar}{2}\left(\begin{array}{c|c} \boldsymbol{\sigma} & 0 \\ \hline 0 & \boldsymbol{\sigma} \end{array}\right) \tag{38}$$

Using relations (30) and (31), we can obtain the action of any spin component in an eigenspace with an arbitrary momentum $\boldsymbol{p}$, knowing that $\boldsymbol{J} \cdot \boldsymbol{P}$ yields zero in the initial space $\mathscr{E}(\boldsymbol{p} = 0)$.

β. *Velocity operator*

The definition of the eigenvectors of $\boldsymbol{R}$ by the Fourier transform (37) implies the standard commutation relations:

$$[R_i\,,\,P_j] = i\hbar\,\delta_{ij} \tag{39}$$

which, eventually, lead to:

$$[\boldsymbol{R}, H] = i\hbar c\,\boldsymbol{\alpha} \tag{40}$$

In the Heisenberg point of view, the operator yielding the time derivative of operator $\boldsymbol{R}$ is thus simply $c\boldsymbol{\alpha}$. This is coherent with relation (VI-99) of chapter VI, where it is the three components of $\boldsymbol{\alpha}$ that yield the spatial components of the probability current, and hence the local velocity.

Note that, even for a free particle, the time derivative of the position is not an operator acting only on external variables, which is contrary to expectations (and to what happens with the Pauli equations). With the Dirac equation, the velocity operator is actually an operator acting only on the internal variables, coupling the positive and negative energy components.

Besides, expression (VI-85) for the α^i components shows that they do not commute with each other: the different velocity components are incompatible quantities. Moreover, in the internal variable state, the eigenvalues of these operators are equal to $\pm c$, each being doubly degenerate. If one could measure a single component of the particle's velocity, one could only find the light velocity; it is only by superposing two internal components with opposite energies that one can obtain an average lower value.

A more precise computation (see, for example, § 37 of chapter XX in [6]) shows that the evolution of the mean value $\langle \boldsymbol{R} \rangle$ for a free particle is the sum of two terms:

– a $\langle \boldsymbol{P} \rangle t/m$ term , the usual one for a free particle,

– a term oscillating at high frequency, corresponding to the difference in energy between the positive and negative energy states. It is an interference effect between the corresponding quantum states. This rapid oscillation movement is generally called "Zitterbewegung" (from the German: "shivering motion"). These properties of the Dirac particle's position are not very intuitive.

γ. *Helicity*

(i) If the mass M is different from zero, relations (36) lead to (for $\boldsymbol{q}$ parallel to the Oz axis):

$$\begin{aligned} S_z|\boldsymbol{p},\nu\rangle &= S_z T_K[\boldsymbol{q}(\boldsymbol{p},\nu)]\; |\boldsymbol{p}=0,\nu\rangle &&= T_K[\boldsymbol{q}(\boldsymbol{p},\nu)]\; \tilde{S}_z(-\boldsymbol{q})|\boldsymbol{p}=0,\nu\rangle \\ (\boldsymbol{J}\cdot\boldsymbol{P})|\boldsymbol{p},\nu\rangle &= (\boldsymbol{J}\cdot\boldsymbol{P})T_K[\boldsymbol{q}(\boldsymbol{p},\nu)]\; |\boldsymbol{p}=0,\nu\rangle &&= T_K[\boldsymbol{q}(\boldsymbol{p},\nu)](\tilde{\boldsymbol{J}}\cdot\tilde{\boldsymbol{P}})\; |\boldsymbol{p}=0,\nu\rangle \end{aligned} \tag{41}$$

Equalities (30) and (31) show how the spin operators and the $\boldsymbol{J}\cdot\boldsymbol{P}$ operator are transformed under a Lorentz transformation. Applying these operators to a ket of $\mathscr{E}(\boldsymbol{p}=0)$, where operator $\boldsymbol{J}\cdot\boldsymbol{P}$ has a zero eigenvalue, these equalities simplify to:

$$\tilde{S}_z(q) \Rightarrow \cosh(\frac{q}{c})\, S_z \qquad (\tilde{\boldsymbol{J}}\cdot\tilde{\boldsymbol{P}})(q) \Rightarrow -\sinh(\frac{q}{c})\, Mc\, S_z \tag{42}$$

Similarly, equations (20) simplify to:

$$c\tilde{P}_z(q) \Rightarrow -\sinh(\frac{q}{c})\, H \qquad \tilde{H}(q) \Rightarrow \cosh(\frac{q}{c})\, H \tag{43}$$

The equalities first show that, in the subspace $\mathscr{E}(\boldsymbol{p})$, the eigenvalues of S_z and H both vary (increase) with the same proportionality coefficient $\cosh(q/c)$. They also show that the eigenvalues of $\boldsymbol{J}\cdot\boldsymbol{P}$ and P_z vary in a similar way, proportionally to $\sinh(q/c)$; their ratio is thus constant (excluding the singular value $q=0$). The helicity is obtained by dividing the eigenvalue of $\boldsymbol{J}\cdot\boldsymbol{P}$ by the modulus of the eigenvalue of P_z (helicity and $\boldsymbol{J}\cdot\boldsymbol{P}$ always have the same sign). Consequently, the helicity sign is changed by inverting the momentum in a Lorentz transformation, but its absolute value remains constant. Since $\boldsymbol{P}$, but not $\boldsymbol{J}$, changes sign in a parity transformation (complement D_V), a parity operation yields the same result.

(ii) If the mass M is zero, we saw that a reference frame where the particle is at rest cannot be obtained by a Lorentz transformation. Consequently, the particle's momentum cannot be inverted. We also noted that relations (31) remain valid, provided we replace MS_z by Σ_z. For a massless particle ($E = cp$) with a momentum parallel to Oz, relation (V-104) shows that $\Sigma_z = (\boldsymbol{J} \cdot \boldsymbol{P})/c$, so that the hyperbolic functions of (31) regroup into a single exponential with a fixed sign. As a result, a pure Lorentz transformation cannot change the signs of Σ_z or of the helicity. Since operator $\boldsymbol{J} \cdot \boldsymbol{P}$ commutes with $\boldsymbol{J}$, $\boldsymbol{P}$ and H, no other transformation of the Poincaré group can change that operator: helicity cannot change sign under any transformation of this group.

δ. *Non-relativistic limit*

Several difficulties of the Dirac equation come from the existence of the two negative energy components. In the non-relativistic limit of velocities much lower than c, it is possible to eliminate these two components and go back to a two-component theory. To lowest order in v/c, this approach leads to the Pauli theory (see, for example, § 29 of chapter XX in reference [6]).

One can then go to a higher order in v/c, thanks to a unitary transformation that bears the name of "Foldy–Wouthuysen transformation". This transformation approaches Dirac's theory via a two-component theory, to any order in v/c. For example, it enables transforming operator $\boldsymbol{R}$ into a "mean position" operator that no longer couples the positive and negative energy components, and is thus free of Zitterbewegung.

The Foldy–Wouthuysen transformation also enables calculating the correction terms to the Pauli equation Hamiltonian. For an electron in a central potential (hydrogen atom in the approximation of an infinitely heavy proton), it introduces two new terms:

– the so-called "spin-orbit coupling" term, containing the scalar product $\boldsymbol{L} \cdot \boldsymbol{S}$. Its physical interpretation involves the coupling between the electron spin and the magnetic field created, in the electron reference frame, by the electric field of the nucleus.

– the "Darwin term", which contains the Laplacian of the central potential under study. The interpretation of this term involves the Zitterbewegung, which causes the electron to average the values of the central potential over the amplitude of its own very rapid spatial motion.

These two terms are used in atomic physics to compute the relativistic effects in atoms and ions (the so-called "fine structure" effects).

3-c. Interpretation difficulties

We already underlined in § B-2-f of chapter VI several intrinsic difficulties of the Dirac equation. They come from the presence of states with negative energies that have no lower boundary. All the positive energy states, i.e. all those attributed to a spin 1/2 particle, should then be considered as unstable. We mentioned how Dirac proposed to get around this difficulty by postulating that all the negative energy states are already populated, and hence not accessible to a positive energy particle because of the Pauli principle. He furthermore suggested that a hole in this "Dirac sea" could be interpreted as the presence of an antiparticle, a positron, with a charge opposed to that of the electron. In spite of Anderson's spectacular later discovery (1933) of the positron, this idea remains delicate. In particular, it assumes that space is filled with a Dirac sea negatively charged and with an infinite mass, while having no other effect, be it electromagnetic or gravitational.

In this complement, we discussed some additional difficulties. Constructing the Dirac state space led us to study two operators, momentum $\boldsymbol{P}$ and velocity $c\boldsymbol{\alpha}$, with no apparent relation (although momentum and velocity are proportional vectors for a classical relativistic particle). The first operator acts only on the external variables, and the second on the internal variables. We also noted that, for any Lorentz reference frame, the eigenvalues of the velocity could take only two values $\pm c$, with no other possible intermediate values.

Furthermore, relations (20) predict singular properties for the momentum: a reference frame where the particle's momentum is zero can be changed by a Lorentz transformation into a frame where the momentum is parallel to the classical velocity (hence opposed to the velocity boost of the new reference frame), but only for the positive energy components. For the negative energy components, the momentum has an opposite sign: the larger the velocity boost in a given direction, the larger the momentum in that direction! It is as if the mass of the particle became negative. Finally, in Dirac's theory, electrons and positrons do not play a symmetrical role. All these difficulties are no longer present in quantum field theory.

Complement C_{VI}

Lagrangians and conservation laws for wave equations

In this complement, we show how the wave equations established in chapter VI can be derived using a variation principle with a Lagrangian. We then discuss how Noether's theorem (complement B_I) shows that certain invariance properties of this Lagrangian lead to local conservation laws of the type:

$$\frac{\partial}{\partial t}\rho(\boldsymbol{r},t) + \boldsymbol{\nabla}\cdot\boldsymbol{J}(\boldsymbol{r},t) = 0 \tag{1}$$

linking the time evolution of the local density ρ of a certain physical quantity (probability, energy, or momentum density) to the associated current $\boldsymbol{J}$.

For a wave equation, the Lagrangian depends on all the values of the wave function, and is the spatial integral of a Lagrangian density $\mathcal{L}$. Such a Lagrangian and the resulting Lagrange equations have already been introduced in § 1 of complement B_I, but we limited our study to fields with real components. We shall first generalize this discussion to the case where the field components can be complex.

1. Complex fields

1-a. Stationarity of the action

For a complex field $\boldsymbol{\Phi}(\boldsymbol{r},t)$, each component is written:

$$\phi_j(\boldsymbol{r},t) = \phi_j^R(\boldsymbol{r},t) + i\phi_j^I(\boldsymbol{r},t) \tag{2}$$

where $\phi_j^R(\boldsymbol{r},t)$ and $\phi_j^I(\boldsymbol{r},t)$ are two real functions. It is equivalent to consider the system as having n complex components, or $2n$ real components, which can be considered as independent dynamical variables.

The Lagrangian density $\mathcal{L}$ is expressed as a function of $\boldsymbol{\Phi}(\boldsymbol{r},t)$ and $\boldsymbol{\Phi}^*(\boldsymbol{r},t)$. We asssume it is invariant under a change of the global phase of the function $\boldsymbol{\Phi}(\boldsymbol{r},t)$, which amounts to saying that each term of $\mathcal{L}$ contains $\boldsymbol{\Phi}$ as many times as $\boldsymbol{\Phi}^*$. As we vary $\phi_j^R(\boldsymbol{r},t)$ and $\phi_j^I(\boldsymbol{r},t)$ to obtain the stationarity conditions for the action $\mathscr{A}[\Gamma]$, the variations of $\boldsymbol{\Phi}$ and $\boldsymbol{\Phi}^*$ are not independent as they are complex conjugates of each other. As we now show, we can still obtain the stationarity conditions of $\mathscr{A}[\Gamma]$ as if these variations were independent.

If $\mathscr{A}$ is stationary, its variation is zero when $\boldsymbol{\Phi}$ varies by $\delta\boldsymbol{\Phi}$, or by $e^{i\chi}\delta\boldsymbol{\Phi}$ as well. Adding the variations introduced by the simultaneous variations of $\boldsymbol{\Phi}$ and $\boldsymbol{\Phi}^*$, we get:

$$\delta\mathscr{A} = e^{i\chi}\,\delta c_1 + e^{-i\chi}\,\delta c_2 = 0 \tag{3}$$

where δc_1 is the variation induced by $\delta\boldsymbol{\Phi}$ for $\chi = 0$, and δc_2 the variation induced by $\delta\boldsymbol{\Phi}^*$. This function of χ will remain equal to zero for any value of χ only if both coefficients δc_1 and δc_2 are zero. This shows that the variation of the action must go to zero when one varies independently the field and its complex conjugate.

The previous arguments can be generalized to physical systems composed of several complex fields, independent or coupled.

1-b. Computation of densities and Noether's currents

We now examine how relations (13) and (16) of complement B_I should be modified when the field become complex. Its real $\phi_j^R(\boldsymbol{r},t)$ and imaginary $\phi_j^I(\boldsymbol{r},t)$ parts then play the role of independent dynamical variables, and we have:

$$\begin{aligned}
\frac{\partial\mathcal{L}}{\partial\dot{\phi}_j^R} &= \frac{\partial\mathcal{L}}{\partial\dot{\phi}_j}\frac{\partial\dot{\phi}_j}{\partial\dot{\phi}_j^R} + \frac{\partial\mathcal{L}}{\partial\dot{\phi}_j^*}\frac{\partial\dot{\phi}_j^*}{\partial\dot{\phi}_j^R} = \frac{\partial\mathcal{L}}{\partial\dot{\phi}_j} + \frac{\partial\mathcal{L}}{\partial\dot{\phi}_j^*} \\
\frac{\partial\mathcal{L}}{\partial\dot{\phi}_j^I} &= \frac{\partial\mathcal{L}}{\partial\dot{\phi}_j}\frac{\partial\dot{\phi}_j}{\partial\dot{\phi}_j^I} + \frac{\partial\mathcal{L}}{\partial\dot{\phi}_j^*}\frac{\partial\dot{\phi}_j^*}{\partial\dot{\phi}_j^I} = i\left[\frac{\partial\mathcal{L}}{\partial\dot{\phi}_j} - \frac{\partial\mathcal{L}}{\partial\dot{\phi}_j^*}\right]
\end{aligned} \tag{4}$$

Since $f_j = \delta\phi_j/\delta\varepsilon$, the local density ρ contains the sum over j of the expression:

$$\begin{aligned}
\delta\phi_j^R\frac{\partial\mathcal{L}}{\partial\dot{\phi}_j^R} + \delta\phi_j^I\frac{\partial\mathcal{L}}{\partial\dot{\phi}_j^I} &= \left(\frac{\delta\phi_j + \delta\phi_j^*}{2}\right)\left[\frac{\partial\mathcal{L}}{\partial\dot{\phi}_j} + \frac{\partial\mathcal{L}}{\partial\dot{\phi}_j^*}\right] \\
&\quad + \left(\frac{\delta\phi_j - \delta\phi_j^*}{2i}\right)i\left[\frac{\partial\mathcal{L}}{\partial\dot{\phi}_j} - \frac{\partial\mathcal{L}}{\partial\dot{\phi}_j^*}\right] \\
&= \delta\phi_j\frac{\partial\mathcal{L}}{\partial\dot{\phi}_j} + \delta\phi_j^*\frac{\partial\mathcal{L}}{\partial\dot{\phi}_j^*}
\end{aligned} \tag{5}$$

For the calculation of the current, it is sufficient to replace the time derivative $\dot{\phi}_j$ with a space derivative, and the calculation proceeds in the same way. We therefore check that, to calculate a local density or a current, it is equivalent to chose dynamical variables that are the real and imaginary parts of a field, or this field and its complex conjugate.

2. Schrödinger equation

We note ψ the wave function, and $\dot{\psi}$ its partial time derivative. Consider the Lagrangian density[1]:

$$\mathcal{L} = i\hbar\psi^*\dot{\psi} - \frac{\hbar^2}{2m}\boldsymbol{\nabla}\psi^* \cdot \boldsymbol{\nabla}\psi - V\psi^*\psi \tag{6}$$

where V is an external potential, and m the particle's mass.

2-a. Deriving the equation

We first vary the function ψ^*. Since:

$$\frac{\partial\mathcal{L}}{\partial\psi^*} = i\hbar\dot{\psi} - V\psi \quad ; \quad \frac{\partial\mathcal{L}}{\partial\dot{\psi}^*} = 0 \quad ; \quad \frac{\partial\mathcal{L}}{\partial\boldsymbol{\nabla}\psi^*} = \boldsymbol{\nabla}_{\boldsymbol{\nabla}\psi^*}(\mathcal{L}) = -\frac{\hbar^2}{2m}\boldsymbol{\nabla}\psi \tag{7}$$

the corresponding Lagrange equation is written:

$$i\hbar\dot{\psi} - V\psi + \frac{\hbar^2}{2m}\Delta\psi = 0 \tag{8a}$$

or:

$$i\hbar\dot{\psi} = V\psi - \frac{\hbar^2}{2m}\Delta\psi \tag{8b}$$

which is the standard Schrödinger equation.

We now vary the function ψ. We get:

$$\frac{\partial\mathcal{L}}{\partial\psi} = -V\psi^* \quad ; \quad \frac{\partial\mathcal{L}}{\partial\dot{\psi}} = i\hbar\psi^* \quad ; \quad \frac{\partial\mathcal{L}}{\partial\boldsymbol{\nabla}\psi} = \boldsymbol{\nabla}_{\boldsymbol{\nabla}\psi}(\mathcal{L}) = -\frac{\hbar^2}{2m}\boldsymbol{\nabla}\psi^* \tag{9}$$

The Lagrange equation is now written:

$$-V\psi^* - i\hbar\frac{\mathrm{d}}{\mathrm{d}t}\psi^* + \frac{\hbar^2}{2m}\Delta\psi^* = 0 \tag{10}$$

which is the complex conjugate of (8b). This leads to the same equation of motion.

[1]An equivalent Lagrangian density is obtained by replacing the first term $i\hbar\psi^*\dot{\psi}$ by the more symmetric form $(i\hbar/2)(\psi^*\dot{\psi} - \dot{\psi}^*\psi)$, so that $\mathcal{L}$ becomes real.

The previous discussion can be easily generalized to the case where the wave function describes several particles, and where the potential includes an interaction component depending on the mutual positions of the particles. We must, however, replace the second term on the right-hand side of (6) (the kinetic energy term) by a summation over the particles, with mass m_k and position $\boldsymbol{r}_k$:

$$-\frac{\hbar^2}{2m}\boldsymbol{\nabla}\psi^* \cdot \boldsymbol{\nabla}\psi \Rightarrow -\sum_k \frac{\hbar^2}{2m_k}\boldsymbol{\nabla}_{\boldsymbol{r}_k}\psi^* \cdot \boldsymbol{\nabla}_{\boldsymbol{r}_k}\psi \tag{11}$$

2-b. Local conservation equations

Applying Noether's theorem leads to a number of local conservation laws contained in the Schrödinger equation.

• Phase invariance, conservation of probability density

The Lagrangian density (6) is invariant under a change of ψ into $\mathrm{e}^{-i\alpha}\psi$. When α takes on an infinitesimal value $\alpha = \delta\varepsilon$, the infinitesimal variations of ψ and ψ^* are, respectively:

$$\delta\psi = -i\delta\varepsilon\psi \quad \text{and:} \quad \delta\psi^* = i\delta\varepsilon\psi^* \tag{12}$$

In the equations (13) of complement $\mathrm{B_I}$, we now have $f = -i\psi(\boldsymbol{r}, t)$, and we obtain (dividing ρ and $\boldsymbol{J}$ by $\hbar$, which has no consequence on the conservation equation):

$$\begin{aligned} \rho(\boldsymbol{r}, t) &= |\psi(\boldsymbol{r}, t)|^2 \\ \boldsymbol{J}(\boldsymbol{r}, t) &= \frac{\hbar}{2mi}\left[\psi^*(\boldsymbol{r}, t)\boldsymbol{\nabla}\psi(\boldsymbol{r}, t) - \psi(\boldsymbol{r}, t)\boldsymbol{\nabla}\psi^*(\boldsymbol{r}, t)\right] \end{aligned} \tag{13}$$

(when computing the current, remember that ψ and ψ^* are treated as independent variables, hence associated with two different values of j whose contributions are added). This yields the usual conservation equation of the probability density for the Schrödinger equation.

• Energy conservation

Since the Lagrangian density is not explicitly time-dependent, we can use the results of § 4 in complement $\mathrm{B_I}$. This yields a conservation equation for another energy density and its associated current:

$$\begin{aligned} \rho_{\mathrm{E}}(\boldsymbol{r}, t) &= i\hbar\psi^*(\boldsymbol{r}, t)\dot{\psi}(\boldsymbol{r}, t) - \mathcal{L} \\ &= \frac{\hbar^2}{2m}\boldsymbol{\nabla}\psi^*(\boldsymbol{r}, t) \cdot \boldsymbol{\nabla}\psi(\boldsymbol{r}, t) + V(\boldsymbol{r}, t)|\psi(\boldsymbol{r}, t)|^2 \\ \boldsymbol{J}_{\mathrm{E}}(\boldsymbol{r}, t) &= -\frac{\hbar^2}{2m}\left[\dot{\psi}(\boldsymbol{r}, t)\boldsymbol{\nabla}\psi^*(\boldsymbol{r}, t) + \dot{\psi}^*(\boldsymbol{r}, t)\boldsymbol{\nabla}\psi(\boldsymbol{r}, t)\right] \end{aligned} \tag{14}$$

The energy density ρ_E is the Hamiltonian density, which is the sum of, first, a kinetic energy term and second, a potential energy term.

• Momentum conservation

Here again the Lagrangian density does not depend explicitly on the space variable $\boldsymbol{r}$, and we can use the results of § 4 in complement $\mathrm{B_I}$. Noting $\boldsymbol{e}_x$ the unit vector along the Ox axis, this yields for the Ox momentum component:

$$\rho_{p_x}(\boldsymbol{r},t) = i\hbar\,\psi^*(\boldsymbol{r},t)\,\partial_x\psi(\boldsymbol{r},t) \tag{15}$$

$$\boldsymbol{J}_{p_x}(\boldsymbol{r},t) = -\frac{\hbar^2}{2m}\Big[(\partial_x\psi(\boldsymbol{r},t))\boldsymbol{\nabla}\psi^*(\boldsymbol{r},t) + (\partial_x\psi^*(\boldsymbol{r},t))\boldsymbol{\nabla}\psi(\boldsymbol{r},t)\Big] - \mathcal{L}\boldsymbol{e}_x$$

We get two similar expressions for the Oy and Oz momentum components.

These equalities show that neither ρ_{p_x}, nor $\boldsymbol{J}_{p_x}$, are a priori real. Their respective imaginary parts are written:

$$\mathrm{Im}(\rho_{p_x}) = \frac{\hbar}{2}[\psi^*\partial_x\psi + \psi\partial_x\psi^*]$$

$$\mathrm{Im}(\boldsymbol{J}_{p_x}) = \frac{1}{2i}[\mathcal{L} - \mathcal{L}^*]\boldsymbol{e}_x = \frac{\hbar}{2}[\psi^*\dot{\psi} + \dot{\psi}^*\psi]\boldsymbol{e}_x \tag{16}$$

We see directly from these expressions that, regardless of the evolution of ψ, these imaginary parts satisfy a separate conservation law:

$$\frac{\mathrm{d}}{\mathrm{d}t}\mathrm{Im}(\rho_{p_x}) + \boldsymbol{\nabla}\cdot\mathrm{Im}(\boldsymbol{J}_{p_x}) = 0 \tag{17}$$

These imaginary parts do not yield any additional information on the dynamics, and can be simply ignored. They can be subtracted from the local conservation equation, keeping only the real parts of relations (15).

The same conclusion can be reached by noticing that $\mathcal{L}$ and $\mathcal{L}^*$ are equivalent Lagrangian densities, since their difference is equal to the total time derivative of $i\hbar\psi^*\psi$. Their half-sum, which is real, is another Lagrangian that directly yields the real parts of the quantities written in (15).

Exercise: Using Schrödinger equation, calculate the time derivatives of the three ρ densities discussed above, and show that they are the opposite of the divergence of the corresponding currents.

3. Klein-Gordon equation

3-a. Expression for the Lagrangian

We start from the Lagrangian density:

$$\mathcal{L} = \frac{\hbar^2}{mc^2}\dot{\psi}^*\dot{\psi} - \frac{\hbar^2}{m}\boldsymbol{\nabla}\psi^*\cdot\boldsymbol{\nabla}\psi - mc^2\psi^*\psi \tag{18}$$

It leads to:

$$\frac{\partial\mathcal{L}}{\partial\psi^*} = -mc^2\psi \quad ; \quad \frac{\partial\mathcal{L}}{\partial\dot{\psi}^*} = \frac{\hbar^2}{mc^2}\dot{\psi} \quad ; \quad \frac{\partial\mathcal{L}}{\partial\boldsymbol{\nabla}\psi^*} = \boldsymbol{\nabla}_{\boldsymbol{\nabla}\psi^*}(\mathcal{L}) = -\frac{\hbar^2}{m}\boldsymbol{\nabla}\psi \tag{19}$$

The Lagrange equation is then written:

$$\frac{\hbar^2}{mc^2}\ddot{\psi} = -mc^2\psi + \frac{\hbar^2}{m}\Delta\psi = 0 \tag{20}$$

which is identical to the Klein–Gordon equation written in (VI-65b), except for the notation of the mass (here in minuscule).

It is not necessary to discuss a variation of ψ, since ψ and ψ^* play a symmetric role in the Lagrangian. The result would necessarily be an equation that is the complex conjugate of (20).

3-b. Relativistic invariance of the Lagrangian density

Consider a change of Lorentz reference frame that transforms the four-vector $(ct, \boldsymbol{r})$ with components x^μ into a four-vector $(ct', \boldsymbol{r}')$ with components x'^ν. One goes from these (contravariant) components with upper indices to (covariant) components with lower indices using the metric tensor g of special relativity, and writing:

$$x_\mu = g_{\mu\rho}x^\rho \qquad x'_\nu = g_{\nu\rho}x^\rho \tag{21}$$

We use the standard notation:

$$\partial^\mu = \frac{\partial}{\partial x_\mu} \qquad \partial_\mu = \frac{\partial}{\partial x^\mu} = g_{\mu\nu}\partial^\nu \qquad \partial'^\mu = \frac{\partial}{\partial x'_\mu} \qquad \partial'_\mu = \frac{\partial}{\partial x'^\mu} \tag{22}$$

For a partial derivative, moving a lower index to an upper index (or conversely) does not change the derivative if $\mu = 0$, and changes its sign if $\mu = 1, 2, 3$. We thus have:

$$\frac{1}{c^2}\frac{\partial^2}{\partial t^2} - \Delta = \partial^\mu\partial_\mu \tag{23}$$

This expression is a relativistic invariant, and hence equal to $\partial'^\mu\partial'_\mu$. If we assume a relativistic invariance for ψ:

$$\psi'(\boldsymbol{r}', t') = \psi(\boldsymbol{r}, t) \tag{24}$$

we can finally write:

$$\begin{aligned}\mathcal{L} &= \hbar^2c^2[\partial^\mu\psi^*(\boldsymbol{r},t)][\partial_\mu\psi(\boldsymbol{r},t)] - m^2c^4\psi^*(\boldsymbol{r},t)\psi(\boldsymbol{r},t)\\ &= \hbar^2c^2[\partial'^\mu\psi'^*(\boldsymbol{r}',t')][\partial'_\mu\psi'(\boldsymbol{r}',t')] - m^2c^4\psi'^*(\boldsymbol{r}',t')\psi'(\boldsymbol{r}',t')\end{aligned} \tag{25}$$

which shows that the Lagrangian density is a relativistic invariant. Since the $d^3r dt$ space–time volume element is also an invariant, the action $\mathscr{A}$ is invariant. Its stationarity can thus be written in any inertial reference frame, as they all play an equivalent role.

Comment:

$|\psi(\boldsymbol{r},t)|^2$ is an invariant, whereas we know that in relativity the local probability density cannot be invariant since d^3r is not a Lorentz transformation invariant. Actually, this probability density is given by relation (VI-68), different from $|\psi|^2$.

3-c. Local conservation equations

As above, we examine successively the conservation relations coming from the invariance of phase, time, and position.

• Local conservation of probability

The density and current associated with the invariance of the Lagrangian (18) under a phase change of the wave function are written, once divided by $2\hbar$ (from now on and for the sake of concision, we no longer write explicitly the obvious $\boldsymbol{r}$ and t dependence):

$$\begin{aligned}\rho &= \frac{i\hbar}{2mc^2}\left[\psi^*\dot{\psi} - \dot{\psi}^*\psi\right]\\ \boldsymbol{J} &= \frac{\hbar}{2mi}\left[\psi^*\boldsymbol{\nabla}\psi - \psi\boldsymbol{\nabla}\psi^*\right]\end{aligned} \tag{26}$$

We find again relations (VI-67) and (VI-68) of chapter VI; as already mentioned, this density ρ is not always positive.

• Energy conservation

Since the Lagrangian density is not explicitly time-dependent, we get:

$$\begin{aligned}\rho_{\mathrm{E}} &= \frac{\hbar^2}{mc^2}\left[2\,\dot{\psi}^*\dot{\psi}\right] - \mathcal{L} = \frac{\hbar^2}{mc^2}\dot{\psi}^*\dot{\psi} + \frac{\hbar^2}{m}\boldsymbol{\nabla}\psi^*\cdot\boldsymbol{\nabla}\psi + mc^2|\psi|^2\\ \boldsymbol{J}_{\mathrm{E}} &= -\frac{\hbar^2}{m}\left[\dot{\psi}\boldsymbol{\nabla}\psi^* + \dot{\psi}^*\boldsymbol{\nabla}\psi\right]\end{aligned} \tag{27}$$

The energy density is a Hamiltonian density, which, as for the Schrödinger equation, is the sum of a first kinetic energy term and of two other potential energy terms. As opposed to ρ, the energy density ρ_E is always positive or zero.

• Momentum conservation

As above, since the Lagrangian density does not depend explicitly on the space variable $\boldsymbol{r}$, we get for the Ox momentum component:

$$\begin{aligned}\rho_{p_x} &= \frac{\hbar^2}{mc^2}\left[\dot{\psi}^*\left(\partial_x\psi\right) + \left(\partial_x\psi^*\right)\dot{\psi}\right]\\ \boldsymbol{J}_{p_x} &= -\frac{\hbar^2}{m}\Big[\left(\partial_x\psi\right)\boldsymbol{\nabla}\psi^* + \left(\partial_x\psi^*\right)\boldsymbol{\nabla}\psi\Big] - \mathcal{L}\boldsymbol{e}_x\end{aligned} \tag{28}$$

with two similar expressions for the Oy and Oz momentum components.

Exercise: Compute the time derivative of the above three densities ρ, and verify that they yield the three local conservation relations.

4. Dirac equation

4-a. Expression for the Lagrangian

We now show that the Dirac equation can be obtained from the Lagrangian density:

$$\mathcal{L} = i\hbar\,\psi_D^\dagger\dot{\psi}_D + i\hbar c\,\psi_D^\dagger\,\boldsymbol{\alpha}\cdot\boldsymbol{\nabla}\psi_D - mc^2\,\psi_D^\dagger\,\alpha^0\,\psi_D \tag{29a}$$

or, using the Einstein summation convention:

$$\mathcal{L} = i\hbar c\,\overline{\psi}_D\,\gamma^\mu\,\partial_\mu\psi_D - mc^2\,\overline{\psi}_D\,\psi_D \qquad \text{with:} \qquad \overline{\psi}_D = \psi_D^\dagger\alpha^0 \tag{29b}$$

Taking the derivative[2] of expression (29a) with respect to $\psi_D^\dagger$, we get:

$$\frac{\partial\mathcal{L}}{\partial\psi_D^\dagger} = i\hbar\dot{\psi}_D + i\hbar c\,\boldsymbol{\alpha}\cdot\boldsymbol{\nabla}\psi_D - mc^2\alpha^0\,\psi_D \;\;;\;\; \frac{\partial\mathcal{L}}{\partial\dot{\psi}_D^\dagger} = 0 \;\;;\;\; \frac{\partial\mathcal{L}}{\partial\boldsymbol{\nabla}\psi_D^\dagger} = 0 \tag{30}$$

which leads to the Dirac equation (VI-88):

$$i\hbar\dot{\psi}_D + i\hbar c\,\boldsymbol{\alpha}\cdot\boldsymbol{\nabla}\psi_D - mc^2\alpha^0\psi_D = 0 \tag{31}$$

We can also consider variations of ψ_D and write the derivatives as:

$$\frac{\partial\mathcal{L}}{\partial\psi_D} = -mc^2\psi_D^\dagger\,\alpha^0 \;\;;\;\; \frac{\partial\mathcal{L}}{\partial\dot{\psi}_D} = i\hbar\,\psi_D^\dagger \;\;;\;\; \frac{\partial\mathcal{L}}{\partial\boldsymbol{\nabla}\psi_D} = i\hbar c\,\psi_D^\dagger\,\boldsymbol{\alpha} \tag{32}$$

To write the Lagrange equation, we need the expression for the spatial divergence of the third of these derivatives. Since:

$$\sum_{i=1}^{3}\frac{\partial}{\partial x_i}\psi_D^\dagger\alpha^i = \boldsymbol{\nabla}\psi_D^\dagger\cdot\boldsymbol{\alpha} \tag{33}$$

we get:

$$-mc^2\psi_D^\dagger\,\alpha^0 - i\hbar\,\dot{\psi}_D^\dagger - i\hbar c\,\boldsymbol{\nabla}\psi_D^\dagger\boldsymbol{\alpha} = 0 \tag{34}$$

which is the Hermitian conjugate of (31). This shows that the variations of ψ_D, compared to those of $\psi_D^\dagger$, do not impose any additional constraint.

---

[2]The derivative of $\mathscr{L}$ with respect to $\psi_D^\dagger$ actually contains 4 (functional) derivatives with respect to the 4 elements of the row matrix representing $\psi_D^\dagger$. These 4 derivatives can be written in the column of a spinor, since their product by the row matrix expressing the variation $\mathrm{d}\psi_D^\dagger$ yields the variation of the Lagrangian density.

4-b. Relativistic invariance of the Lagrangian density

The Lagrangian density (29a) can be written:

$$\mathcal{L} = \psi_D^\dagger \left[i\hbar \frac{\partial}{\partial t} \psi_D + i\hbar c \boldsymbol{\alpha} \cdot \boldsymbol{\nabla} \psi_D - mc^2 \alpha^0 \psi_D \right] \tag{35}$$

As in complement A_{VI} we assume that, in a change of reference frame, the transformation of the spinor ψ_D is:

$$\psi'_D(\boldsymbol{r}', t') = S_D(\Lambda)\, \psi_D(\boldsymbol{r}, t) \quad \text{with:} \qquad (ct', \boldsymbol{r}') = \Lambda(ct, \boldsymbol{r}) \tag{36}$$

If Λ is a rotation, relation (35) directly shows that $\mathcal{L}$ is invariant, since the scalar product $\boldsymbol{\alpha} \cdot \boldsymbol{\nabla}$ of two vectors is a rotation scalar, and since α^0 is also a scalar. The space or time translation invariance can be checked in a similar way.

If the transformation is a change of Lorentz reference frame, $S_D(\Lambda)$ is Hermitean, and relation (11c) of complement A_{VI} allows us to write:

$$\begin{aligned} \mathcal{L} &= \psi_D^\dagger(\boldsymbol{r}, t)\, S_D^\dagger(\Lambda) \left[i\hbar \frac{\partial}{\partial t'} + i\hbar c \boldsymbol{\alpha} \cdot \boldsymbol{\nabla}' - mc^2 \alpha^0 \right] S_D(\Lambda)\, \psi_D(\boldsymbol{r}, t) \\ &= \psi'^\dagger_D(\boldsymbol{r}', t') \left[i\hbar \frac{\partial}{\partial t'} + i\hbar c \boldsymbol{\alpha} \cdot \boldsymbol{\nabla}' - mc^2 \alpha^0 \right] \psi'_D(\boldsymbol{r}', t') = \mathcal{L}' \end{aligned} \tag{37}$$

This proves the relativistic invariance of the Lagrangian density.

Exercise: Generalize the previous calculations to the Dirac equation (VI-104) containing an electromagnetic field.

Comment:

The complex conjugate of the Lagrangian density reads:

$$\mathcal{L}^* = \left[-i\hbar \frac{\partial}{\partial t} \psi_D^\dagger - i\hbar c \boldsymbol{\nabla} \psi_D^\dagger \cdot \boldsymbol{\alpha} - mc^2\, \psi_D^\dagger \alpha^0 \right] \psi_D \tag{38}$$

When integrated over space–time, $\mathcal{L}$ and $\mathcal{L}^*$ lead to complex conjugate actions. If one of them is stationnary, so is the other, and the two densities are equivalent. We can obtain the same conclusion by noting that:

$$\mathcal{L} = \mathcal{L}^* + i\hbar \frac{\partial}{\partial t} (\psi_D^\dagger \psi_D) + i\hbar c \boldsymbol{\nabla} \cdot (\psi_D^\dagger \boldsymbol{\alpha} \psi_D) \tag{39}$$

The difference $\mathcal{L}^* - \mathcal{L}$ is the sum of a total time derivative and of a gradient, both disappearing in the integration that yields the action; this difference does not contribute to the stationnarity of the action. The Dirac equation can be obtained from a real $\mathcal{L} + \mathcal{L}^*$ as well from a complex Lagrangian density.

4-c. Local conservation equations

Like for the Schrödinger equation, we examine successively the conservation relations coming from the invariance of phase, time, and position.

• Local conservation of probability

The density and current associated with the invariance of the Lagrangian (18) under a phase change of the wave function are written:

$$\begin{aligned} \rho &= \psi_D^\dagger \psi_D \\ \boldsymbol{J} &= c\,\psi_D^\dagger \boldsymbol{\alpha} \psi_D \end{aligned} \tag{40}$$

(the 4 spinor components correspond to 4 values of the index j; as above, we have divided ρ and $\boldsymbol{J}$ by $\hbar$). This yields again relations (VI-99) of chapter VI.

• Energy conservation

Since the Lagrangian density is not explicitly time-dependent, we get:

$$\begin{aligned} \rho_{\text{E}} &= i\hbar\,\psi_D^\dagger \dot{\psi} - \mathcal{L} = -i\hbar c\,\psi_D^\dagger\, \boldsymbol{\alpha}\cdot\boldsymbol{\nabla}\psi_D + mc^2\,\psi_D^\dagger \alpha^0 \psi_D \\ \boldsymbol{J}_{\text{E}} &= i\hbar c\,\psi_D^\dagger\, \boldsymbol{\alpha}\,\dot{\psi}_D \end{aligned} \tag{41}$$

The energy density is merely equal to the local Hamiltonian density. The energy current is similar to the probability current, but with an additional time derivative multiplied by $\hbar$.

• Momentum conservation

Here again the Lagrangian density does not depend explicitly on the space variable $\boldsymbol{r}$. We thus get, for the Ox momentum component:

$$\begin{aligned} \rho_{p_x} &= i\hbar\psi_D^\dagger\,(\partial_x \psi) \\ \boldsymbol{J}_{p_x} &= i\hbar c\,\psi_D^\dagger\,\boldsymbol{\alpha}\,(\partial_x\psi_D) - \mathcal{L}\boldsymbol{e}_x \end{aligned} \tag{42}$$

with two similar expressions for the Oy and Oz momentum components.

Note that the right-hand sides of expressions (41) and (42) are not necessarily real. However, as we did for the Schrödinger equation, we can use the fact that $\mathcal{L}$ and $\mathcal{L}^*$ are equivalent Lagrangian densities to obtain a real Lagrangian density; we can then replace the right-hand sides of these relations by their real parts.

Exercise: Compute the time derivative of the above three densities ρ, and verify that they yield the three local conservation relations.

Chapter VII

Rotation group, angular momenta, spinors

In this chapter, we study and build the rotation operators using the properties of the $R_{(3)}$ rotation group, and the commutation relations of its infinitesimal generators. These rotation operators, infinitesimal or finite, act in the state space of an arbitrary system. We first give in § A a simple review of angular momentum operators and of the construction of a standard basis. This will suggest the general form of the unitary irreducible representation of $R_{(3)}$, enabling us to obtain the explicit form of the corresponding representation matrices. These results will enable us to complete in § B the studies of chapter VI concerning the non-zero spin representations, and discuss some properties of two-component spinors. In § C, we will discuss the problem of angular momenta addition, which amounts to

Introduction to Continuous Symmetries: From Space–Time to Quantum Mechanics, First Edition. Franck Laloë.

the decomposition into irreducible representations of a representation of the $R_{(3)}$ group obtained by a tensor product of irreducible representations.

A. General properties of rotation operators

A-1. Elementary theory of angular momentum

Most basic quantum mechanics books present an elementary theory of angular momentum, often unconnected with the concept of group representation. We give here a short review of that theory, and the reader unfamiliar with angular momentum may consult the corresponding chapter in one the books cited in the bibliography, for example, chapter VI of [9], whose notations we systematically use.

We obtained in chapter III the commutation relations between the J_{x_i} ($x_i = x, y, z$) components of an operator $\boldsymbol{J}$ corresponding to the rotation in the state space of an arbitrary physical system:

$$\left[J_{x_i}, J_{x_j}\right] = i\hbar\, \varepsilon_{ijk}\, J_{x_k} \tag{VII-1}$$

Operator:

$$\boldsymbol{J}^2 = J_x^2 + J_y^2 + J_z^2 \tag{VII-2}$$

commutes[1] with all the components of $\boldsymbol{J}$:

$$\left[\boldsymbol{J}^2, J_{x_i}\right] = 0 \tag{VII-3}$$

The problem we address in this § A is the construction, using these commutation relations (i.e. the structure constants of the rotation group), of matrices describing explicitly the action of operators J_x, J_y, and J_z in the state space $\mathscr{E}$ of an arbitrary system. This will show us how to construct the form of rotation operators.

Comment:

The following discussion is obviously not limited to quantum mechanics: the space $\mathscr{E}$ can be any representation space (a space for the action of the operators associated with the representation matrices). As an example, $\mathscr{E}$ could be the ordinary 3-dimensional space.

[1]Operator $\boldsymbol{J}^2$ is, within a factor $-1/2$, the Casimir operator of the rotation group (§ 4 of complement A_{III}).

A-1-a. A few useful reminders

We shall state, without demonstration, a certain number of standard results. The two J_+ and J_- operators are defined as:

$$J_{\pm} = J_x \pm i\, J_y \tag{VII-4}$$

We have :

$$[J_z, J_{\pm}] = \pm\hbar\, J_{\pm} \tag{VII-5a}$$

$$J_+\, J_- = \boldsymbol{J}^2 - J_z^2 + \hbar\, J_z \tag{VII-5b}$$

$$J_-\, J_+ = \boldsymbol{J}^2 - J_z^2 - \hbar\, J_z \tag{VII-5c}$$

$$\boldsymbol{J}^2 = \frac{1}{2}\,(J_+\, J_- + J_-\, J_+) + J_z^2 \tag{VII-5d}$$

The following properties can be established:

(i) The eigenvalues of $\boldsymbol{J}^2$ are real, positive, or zero. They are noted $j(j+1)\,\hbar^2$, where $j \geq 0$. As for the eigenvalues of J_z, they are written $m\hbar\,(m \gtrless 0)$.

(ii) In the state space, we can construct an orthonormal basis with eigenvectors common to $\boldsymbol{J}^2$ and J_z. These vectors are noted $|k, j, m\rangle$, where the index k is used to distinguish between eigenkets corresponding to the same values of j and m (k is useful if $\boldsymbol{J}^2$ and J_z do not form alone a CSCO).

(iii) We have:

$$-j \leq m \leq +j \tag{VII-6a}$$

These inequalities come from the fact that the squared norm of the kets $J_{\pm}\,|k, j, m\rangle$ must be positive or zero.

(iv) $J_-\,|k, j, m\rangle$ is a (non-zero) eigenket common to $\boldsymbol{J}^2$, with the (unchanged) eigenvalue $j(j+1)\hbar^2$, and J_z, with the eigenvalue $(m-1)\hbar$ (lowered by $\hbar$), except when $m = -j$ (the ket is then equal to zero).

(v) Symmetric property for the operator J_+ where the sign $-$ is replaced by $+$: $J_+\,|k, j, m\rangle$ is a (non-zero) eigenket common to $\boldsymbol{J}^2$, with the (unchanged) eigenvalue $j(j+1)\hbar^2$, and J_z, with the eigenvalue $(m+1)\hbar$ (increased by $\hbar$), except when $m = +j$ (the ket is then equal to zero).

(vi) The repeated action of operator J_+ on a ket $|k, j, m\rangle$ must always, after a finite number of operations, yield a zero ket [otherwise inequality (VII-6a) would not be respected]. According to (v), this leads to

$j - m$ is an integer.

The same reasoning with J_- shows that

$j + m$ is an integer.

(vii) As a result, both numbers $2j$ and $2m$ are integers. Or else:

- j is an integer, as well as m;
- j is a "half-integer" (ratio of the type $p/2$, where p is an odd integer), as well as m.

(viii) With any ket $|k, j, m\rangle$ we can associate, by the repeated action of operators J_+ and J_-, a family of $(2j+1)$ orthogonal kets that correspond to the same value of j and to the $(2j+1)$ values of:

$$m = j,\ j-1,\ j-2,\ \ldots,\ -j+1,\ -j \tag{VII-6b}$$

The action of J_z, J_+, or J_- on one of these kets yields another ket of the same family, to within a proportionality constant (that can be zero for J_z if $m = 0$, for J_+ if $m = j$ and for J_- if $m = -j$). The subspace spanned by this family of $(2j+1)$ kets is stable under the action of any J_u component of operator $\boldsymbol{J}$.

A-1-b. Standard basis

α. Constructing the basis

In the state space $\mathscr{E}$, operator $\boldsymbol{J}^2$ can have, in general, several eigenvalues, each corresponding to a given value of the quantum number j. We choose one, and consider the eigensubspace common to $\boldsymbol{J}^2$ and J_z, with respective eigenvalues $j(j+1)\hbar^2$ and $j\hbar$; this means that the quantum number m has taken its maximum value. In this eigensubspace, we consider an orthonormal basis $|\tau, j, j\rangle$ whose vectors are distinguished by the value of τ. We then have:

$$\langle \tau, j, j | \tau', j, j\rangle = \delta_{\tau,\tau'} \tag{VII-7}$$

We can apply operator J_- to one of the $|\tau, j, j\rangle$, and obtain a ket with quantum numbers j and $m = j-1$, according to property (iv) recalled above. This operation can be repeated to obtain a ket with quantum numbers j and $m = j-2$, etc. Successive applications of J_- will finally yield the last ket of the series, with quantum numbers j and $m = -j$ (this last ket would be annulled by an additional application of the operator). None of these $2j+1$ kets is zero; we assume they have somehow been normalized. They span a subspace $\mathscr{E}(\tau, j)$ of dimension $2j+1$, in which they form an orthonormal basis:

$$\langle \tau, j, m | \tau, j, m'\rangle = \delta_{mm'} \tag{VII-8}$$

Consider now two sets of $2j+1$ kets, $|\tau, j, m\rangle$ and $|\tau', j', m'\rangle$, obtained by the previous construction, but corresponding to two different values of τ. They are orthogonal if $j \neq j'$ or $m \neq m'$, since the indices j and m label the eigenvalues of Hermitian operators. Let us show that they are also orthogonal if $j = j'$ and $m = m'$, but $\tau \neq \tau'$. This is the case by definition for the kets $|\tau, j, j\rangle$, i.e. when m takes its maximum value $m = j$. Using (VII-5b), we can write:

$$\langle \tau, j, j | J_+ J_- | \tau', j, j \rangle = 2j\hbar^2 \langle \tau, j, j | \tau', j, j \rangle = 0 \tag{VII-9}$$

Since $|\tau, j, j-1\rangle$ is by definition proportional to $J_- |\tau, j, j\rangle$, this equality proves the orthogonality of $|\tau, j, j-1\rangle$ and $|\tau', j, j-1\rangle$. By recurrence, one can show the orthogonality of the kets $|\tau, j, m\rangle$ and $|\tau', j, m\rangle$ for the $2j+1$ possible values of m. It follows that:

$$\langle \tau, j, m' | \tau', j, m \rangle = \delta_{\tau,\tau'} \delta_{mm'} \tag{VII-10}$$

The previous procedure can be repeated for all the possible values of j in the entire space state $\mathscr{E}$. This yields a basis of this space where, since j labels the eigenvalue of a Hermitian operator, two kets with different values of j are orthogonal. Relation (VII-10) then includes an additional $\delta_{jj'}$ if j and j' are different.

To sum up, the properties of the kets $|\tau, j, m\rangle$ we have defined are[2]:

$$\boxed{\begin{aligned} \boldsymbol{J}^2 |\tau, j, m\rangle &= j(j+1)\hbar^2 |\tau, j, m\rangle \\ J_z |\tau, j, m\rangle &= m\hbar |\tau, j, m\rangle \\ J_\pm |\tau, j, m\rangle &= \hbar\sqrt{j(j+1) - m(m \pm 1)}\, |\tau, j, m \pm 1\rangle \end{aligned}} \tag{VII-11}$$

In additon, we have the orthogonality relation:

$$\boxed{\langle \tau, j, m | \tau', j', m' \rangle = \delta_{\tau\tau'}\, \delta_{jj'}\, \delta_{mm'}} \tag{VII-12}$$

and the closure relation:

$$\boxed{\sum_{\tau,j,m} |\tau, j, m\rangle \langle \tau, j, m| = \mathbb{1}} \tag{VII-13}$$

expressing that, by construction, the sum of the projectors onto the spaces $\mathscr{E}(\tau, j)$ is the identity operator.

The set of $|\tau, j, m\rangle$ forms the so-called "*standard basis*" of angular momentum in the state space of the system. This type of basis is particularly well adapted to the study of the angular momentum of any physical system. The

[2]Strictly speaking, one should also verify that while the third relation (VII-11) allows defining the ket $|\tau, j, m-1\rangle$ by the action of J_- on $|\tau, j, m\rangle$, it also allows coming back from $|\tau, j, m-1\rangle$ toward $|\tau, j, m\rangle$ by the action of J_+. This property is easily established using again (VII-5b) or (VII-5c).

matrix elements of J_x, J_y, and J_z are easily found from relation (VII-11), taking into account definition (VII-4) for J_+ and J_-. Regardless of the nature of the physical system under study, the spaces $\mathscr{E}(\tau, j)$ always have the same dimension $2j+1$ (where $j = 0, 1/2, 1, 3/2, \ldots$ etc.) and the same properties in relation to the total angular momentum $\boldsymbol{J}$. Finally, the space $\mathscr{E}(\tau, j)$ is invariant, or stable, under the action of any function $F(\boldsymbol{J})$ of the angular momentum[3]; this function is represented in each $\mathscr{E}(\tau, j)$ by matrices that only depend on F and not on the physical system under study.

The following discussion applies whenever a physical system is studied in quantum mechanics, without having to specify the nature of the system.

β. *Another way of constructing a standard basis*

Imagine we have a set of observables $K_1, K_2, \ldots, K_i$ forming with $\boldsymbol{J}^2$ and J_z a CSCO. We assume that these operators K_i commute, not only with each other and J_z, but also with the two other components J_x and J_y (as the K_i commute with any component J_u of $\boldsymbol{J}$ they are "scalar operators"). The eigenkets:

$$|k_1, k_2, \ldots, k_i, \ldots ; j, m\rangle$$

common to $K_1, K_2, \ldots, K_i$, $\boldsymbol{J}^2$ and J_z form a standard basis under the condition that their relative phases be fixed by:

$$J_\pm \, |k_1, k_2, \ldots, k_i, \ldots ; j, m\rangle = \sqrt{j(j+1) - m(m\pm 1)} \; |k_1, k_2, \ldots, k_i, \ldots ; j, m\pm 1\rangle \quad \text{(VII-14)}$$

Note that to be able to write this equation, it is not enough that $[J_z, K_i] = 0$, but we must also have $[J_\pm, K_i] = 0$. Operator K could be, for example, the Hamiltonian of system invariant under rotation.

Comment:

If $\boldsymbol{J}^2$ and J_z form a CSCO by themselves, which amounts to assuming that the degeneracy of each eigenvalue of $\boldsymbol{J}^2$ takes its minimum value $2j+1$, the indices k or τ are not necessary, and there is only one way of constructing the $\mathscr{E}(\tau, j)$. However, in the general case of a larger degeneracy of the eigenvalues of $\boldsymbol{J}^2$, these indices are necessary. To construct the $\mathscr{E}(\tau, j)$ we started from an orthogonal basis in an eigensubspace of j and $m = j$, and the choice of that basis was arbitrary. Each choice leads to different spaces $\mathscr{E}(\tau, j)$, and to a change of unitary basis in each subspaces spanned by the $|\tau, j, m\rangle$ with a given value of j and m.

[3]A subspace $\mathscr{F}$ is invariant, or stable, under the action of an operator R if $|\psi\rangle \in \mathscr{F}$ implies that $R|\psi\rangle \in \mathscr{F}$, regardless of the ket $|\psi\rangle$. This does not necessarily imply an invariance ket by ket: in general, we do not have $R|\psi\rangle \propto |\psi\rangle$.

γ. *Irreducible spaces*

We now show that the spaces $\mathscr{E}(\tau, j)$ are irreducible: they do not contain any smaller subspace that remains invariant under the action of all the angular momentum operators, and hence under the action of all the rotation operators.

An infinitesimal rotation operator by an angle $\delta\varphi$ around a vector $\boldsymbol{u}$, acting in the space state, can be written using the general relation (IV-55) of chapter IV (where we set $\boldsymbol{a} = \varphi\boldsymbol{u}$):

$$R_{\boldsymbol{u}}(\varphi) = 1 - \frac{i}{\hbar}\delta\varphi\,(\boldsymbol{u}\cdot\boldsymbol{J}) + \ldots \tag{VII-15}$$

Integrating over φ, we get a finite rotation operator:

$$R_{\boldsymbol{u}}(\varphi) = \exp(-i\varphi\, J_u/\hbar) \qquad \text{with:} \qquad J_u = \boldsymbol{u}\cdot J \tag{VII-16}$$

A subspace invariant under the action of all the components of the angular momentum is invariant under the action of all the rotation operators, and vice versa. When a subspace is invariant under the action of J_x and J_y, it is also invariant under the action of iJ_y, and hence of the operators J_+ and J_-.

It follows that $\mathscr{E}(\tau, j)$ is invariant under the action of the rotation operators (VII-15) and (VII-16). Let us start from an arbitrary non-zero vector of $\mathscr{E}(\tau, j)$. Repeated application of operator J_+ yields kets that must belong to the same irreducible subspace as the initial ket. At some point they will reach the ket $|\tau, j, j\rangle$ with the maximal value of m. Applying then repeatedly operator J_-, we can transform this ket into a ket $|\tau, j, m\rangle$ with any value of m. We can thus span the entire basis that defines $\mathscr{E}(\tau, j)$. This shows that there exists no subspace of $\mathscr{E}(\tau, j)$ that remains invariant under the action of all the rotations. The subspace $\mathscr{E}(\tau, j)$ is thus irreducible with respect to the entire set of rotations.

A-2. Explicit construction of the rotation matrices

We explicitly construct the irreducible sets of rotation matrices associated with the subspaces $\mathscr{E}(\tau, j)$. They are matrices with $(2j+1)$ rows and $(2j+1)$ columns, associated in these subspaces with the operators $R_{\boldsymbol{u}}(\varphi)$. We choose an integer or half-integer value of j, and then use relations (VII-11) to construct the $(2j+1)\times(2j+1)$ matrices associated with J_x, J_y, and J_z (or with J_u, which is simply a linear combination of those three). By exponentiation, one obtains the matrices associated with the operators $R_{\boldsymbol{u}}(\varphi)$.

These matrices will be noted $(R_{\boldsymbol{u}}^{[j]}(\varphi))$, and their elements in a standard basis $R_{\boldsymbol{u}}^{[j]}(\varphi)_{mm'}$. These notations will often be simplified into $(R^{[j]})$ and $(R^{[j]})_{mm'}$, when the $\boldsymbol{u}$ and φ dependance will not be essential:

$$\left(R_{\boldsymbol{u}}^{[j]}(\varphi)\right) = \left(R^{[j]}\right) = \text{matrix of } R_{\boldsymbol{u}}(\varphi) \text{ in } \mathscr{E}(\tau, j) \tag{VII-17a}$$

$$\left(R^{[j]}\right)_{mm'} = \langle\tau, j, m|\exp\{-i\varphi\, J_u/\hbar\}\,|\tau, j, m'\rangle \tag{VII-17b}$$

In what follows, the quantum number τ does not play any role, and can thus be omitted.

A-2-a. Zero $j = 0$ angular momentum

The value $j = 0$ leads to the trivial one-dimensional representation. The quantum number m can only be zero [*cf.* (VII-6b)] and relations (VII-11) show that:

$$(J_u) = (0) \tag{VII-18}$$

As for the rotation operators, it is easy to show that:

$$\left(R_{\boldsymbol{u}}^{[0]}(\varphi)\right) = (\mathbb{1}) \tag{VII-19}$$

regardless of $\boldsymbol{u}$ and φ.

A-2-b. Angular momentum $j = 1/2$, spin $1/2$

The rotation matrices are 2×2 matrices. We easily get[4]:

$$(J_z) = \frac{\hbar}{2}\begin{pmatrix} 1 & 0 \\ 0 & -1 \end{pmatrix} \tag{VII-20}$$

$$(J_+) = \hbar\begin{pmatrix} 0 & 1 \\ 0 & 0 \end{pmatrix} \qquad (J_-) = \hbar\begin{pmatrix} 0 & 0 \\ 1 & 0 \end{pmatrix} \tag{VII-21}$$

According to the definition of $J_\pm$, this leads to:

$$(J_x) = \frac{\hbar}{2}\begin{pmatrix} 0 & 1 \\ 1 & 0 \end{pmatrix} \qquad (J_y) = \frac{\hbar}{2}\begin{pmatrix} 0 & -i \\ i & 0 \end{pmatrix} \tag{VII-22}$$

These are simply, to within a factor $\hbar/2$, the three Pauli matrices. As for operator $\boldsymbol{J}^2$, it is associated with the 2×2 identity matrix, multiplied by $3\hbar^2/4$.

Consider now the finite rotation operators. For example, by exponentiation of (VII-20), we easily get:

$$\left(R_z(\varphi)\right) = \begin{pmatrix} \mathrm{e}^{-i\varphi/2} & 0 \\ 0 & \mathrm{e}^{i\varphi/2} \end{pmatrix} \tag{VII-23}$$

---

[4]By convention, and whatever the value of j, we always rank the basis vectors in descending order of their m value:

$$|j,j\rangle,\ |j,j-1\rangle,\ |j,j-2\rangle,\ \ldots,\ |j,-j\rangle$$

Similarly, for a rotation around Ou, the infinitesimal operator is:

$$J_u = u_x \; J_x + u_y \; J_y + u_z \; J_z \tag{VII-24}$$

We can use the properties of the Pauli matrices ($\sigma_u^2 = 1$; *cf.* chapter III, § B-3) to regroup as in (III-56) the even and odd powers of $J_{\boldsymbol{u}}$ in an exponential expansion. This yields:

$$\begin{aligned}\left(R_{\boldsymbol{u}}^{[1/2]}(\varphi)\right) &= \left(\mathrm{e}^{-i\varphi \, J_u/\hbar}\right) = \cos\frac{\varphi}{2} - \frac{2i}{\hbar}\sin\frac{\varphi}{2}\; \boldsymbol{u}\cdot\boldsymbol{J} \\ &= \begin{pmatrix} \cos\frac{\varphi}{2} - i\,u_z\,\sin\frac{\varphi}{2} & (-i\,u_x - u_y)\,\sin\frac{\varphi}{2} \\ (-i\,u_x + u_y)\,\sin\frac{\varphi}{2} & \cos\frac{\varphi}{2} + i\,u_z\,\sin\frac{\varphi}{2}\end{pmatrix}\end{aligned} \tag{VII-25}$$

These are the SU(2) matrices already introduced in chapter III; replacing in (VII-25) φ by a we obtain exactly the same matrix as in (III-77). We shall see in § A-3-d to what extent these matrices constitute a representation of the rotation group.

Comment:

Relation (16) of complement C_V expresses a rotation $\mathscr{R}(\alpha, \beta, \gamma)$, defined by its Euler angles α, β, and γ, as a function of rotations around the Oy and Oz axes. This relation allows expressing $(R^{[1/2]})$ as a function of the Euler angles:

$$\begin{aligned}\left(R^{[1/2]}(\alpha,\beta,\gamma)\right) &= \pm\left(R_{\boldsymbol{u}_z}^{[1/2]}(\alpha)\right)\;\left(R_{\boldsymbol{u}_y}^{[1/2]}(\beta)\right)\;\left(R_{\boldsymbol{u}_z}^{[1/2]}(\gamma)\right) \\ &= \pm\begin{pmatrix} \mathrm{e}^{-i(\alpha+\gamma)/2}\,\cos\,\beta/2 & -\,\mathrm{e}^{i(\gamma-\alpha)/2}\,\sin\,\beta/2 \\ \mathrm{e}^{i(\alpha-\gamma)/2}\,\sin\,\beta/2 & \mathrm{e}^{i(\gamma+\alpha)/2}\,\cos\,\beta/2\end{pmatrix}\end{aligned} \tag{VII-26}$$

Somewhat anticipating what follows (§ A-3-d), we have inserted a $\pm$ in these equations to account for the fact the representation of the rotation group may be projective (or, in this case, double-valued).

A-2-c. Angular momentum $j = 1$, rotation matrices in ordinary space

The dimension of the representation is 3 in this case (*cf.* note 4 for the ranking of the basis vectors).

α. *Expressions of the matrices*

Using relations (VII-11) with the equalities $J_x = (J_+ + J_-)/2$ and $J_y = (J_+ - J_-)/2i$, we get:

$$(J_z) = \hbar \begin{pmatrix} 1 & 0 & 0 \\ 0 & 0 & 0 \\ 0 & 0 & -1 \end{pmatrix} \quad (J_x) = \frac{\hbar}{\sqrt{2}} \begin{pmatrix} 0 & 1 & 0 \\ 1 & 0 & 1 \\ 0 & 1 & 0 \end{pmatrix} \quad (J_y) = \frac{i\hbar}{\sqrt{2}} \begin{pmatrix} 0 & -1 & 0 \\ 1 & 0 & -1 \\ 0 & 1 & 0 \end{pmatrix} \tag{VII-27}$$

Following a similar reasoning as for $j = 1/2$, we obtain:

$$\left(R^{[1]}_{\boldsymbol{e}_z}(\varphi)\right) = \begin{pmatrix} \mathrm{e}^{-i\varphi} & 0 & 0 \\ 0 & 1 & 0 \\ 0 & 0 & \mathrm{e}^{i\varphi} \end{pmatrix} \tag{VII-28}$$

and, for example:

$$\left(R_{\boldsymbol{u}_y}(\varphi)\right) = \begin{pmatrix} (1+\cos\varphi)/2 & -\sin\varphi/\sqrt{2} & (1-\cos\varphi)/2 \\ \sin\varphi/\sqrt{2} & \cos\varphi & -\sin\varphi/\sqrt{2} \\ (1-\cos\varphi)/2 & \sin\varphi/\sqrt{2} & (1+\cos\varphi)/2 \end{pmatrix} \tag{VII-29}$$

As in (VII-26), we get :

$$\begin{aligned} \left(R^{[1]}(\alpha,\beta,\gamma)\right) &= \left(R^{[1]}_{\boldsymbol{u}_z}(\alpha)\right)\ \left(R^{[1]}_{\boldsymbol{u}_y}(\beta)\right)\ \left(R^{[1]}_{\boldsymbol{u}_z}(\gamma)\right) \\ &= \begin{pmatrix} \mathrm{e}^{-i(\alpha+\gamma)}\,\frac{1+\cos\beta}{2} & -\mathrm{e}^{-i\alpha}\,\frac{\sin\beta}{\sqrt{2}} & \mathrm{e}^{i(\gamma-\alpha)}\,\frac{1-\cos\beta}{2} \\ \mathrm{e}^{-i\gamma}\,\frac{\sin\beta}{\sqrt{2}} & \cos\beta & -\mathrm{e}^{i\gamma}\,\frac{\sin\beta}{\sqrt{2}} \\ \mathrm{e}^{i(\alpha-\gamma)}\,\frac{1-\cos\beta}{2} & \mathrm{e}^{i\alpha}\,\frac{\sin\beta}{\sqrt{2}} & \mathrm{e}^{i(\gamma+\alpha)}\,\frac{1+\cos\beta}{2} \end{pmatrix} \end{aligned} \tag{VII-30}$$

Note that the factor $\pm$ is not needed here since, as we shall see in the next paragraph A-2-c-β, the 3×3 matrices we have obtained are the usual rotation matrices, though in a different basis (equivalent representations). We won't go into any more detail concerning the properties of the rotation matrices $\left(R^{[j]}_{\boldsymbol{u}}(\varphi)\right)$, some of which are discussed in the appendix C-IV , volume II of the book on quantum mechanics by A. Messiah [6].

β. *Relation to the rotation matrices in ordinary space*

We wrote in § C-1 of chapter III the expression for infinitesimal rotations in ordinary space. A rotation by an angle $\delta\varphi$ around the unit vector $\boldsymbol{u}$ is written as:

$$\mathscr{R}_{\boldsymbol{u}}(\delta\varphi) = 1 + \delta\varphi\left[u_x \mathscr{M}_x + u_y \mathscr{M}_y + u_z \mathscr{M}_z\right] \tag{VII-31}$$

where the expressions of the matrices $\mathscr{M}$ are given by relations (III-88) and (III-90). These expressions are valid in the basis formed by the three basis vectors $\boldsymbol{e}_x$, $\boldsymbol{e}_y$, and $\boldsymbol{e}_z$ corresponding to the three axes of an orthogonal reference frame $Oxyz$. The matrices $(\mathscr{R})$ in ordinary space are real and orthogonal, a special case of the complex and unitary matrices such as the $(R^{[j]})$.

We now carry out a change of basis:

$$\begin{cases} \boldsymbol{e}_{+1} = -\dfrac{1}{\sqrt{2}}\left[\boldsymbol{e}_x + i\,\boldsymbol{e}_y\right] \\ \boldsymbol{e}_0 \;\;= \boldsymbol{e}_z \\ \boldsymbol{e}_{-1} = \dfrac{1}{\sqrt{2}}\left[\boldsymbol{e}_x - i\,\boldsymbol{e}_y\right] \end{cases} \tag{VII-32}$$

Knowing the expressions of the $\mathscr{M}$ matrices in the initial basis $\boldsymbol{e}_x$, $\boldsymbol{e}_y$, $\boldsymbol{e}_z$ (chapter III, § C-1), it is easy to compute their action in the new basis. As an example:

$$\begin{aligned} \mathscr{M}_x\, \boldsymbol{e}_{+1} &= -\frac{1}{\sqrt{2}}\,\mathscr{M}_x\left[\boldsymbol{e}_x + i\,\boldsymbol{e}_y\right] = -\frac{i}{\sqrt{2}}\,\boldsymbol{e}_z = -\frac{i}{\sqrt{2}}\,\boldsymbol{e}_0 \\ \mathscr{M}_x\, \boldsymbol{e}_0 &= \mathscr{M}_x\, \boldsymbol{e}_z = -\boldsymbol{e}_y = -\frac{i}{\sqrt{2}}\left[\boldsymbol{e}_{+1} + \boldsymbol{e}_{-1}\right] \\ \mathscr{M}_x\, \boldsymbol{e}_{-1} &= \frac{i}{\sqrt{2}}\,\mathscr{M}_x\left[\boldsymbol{e}_x - i\,\boldsymbol{e}_y\right] = -\frac{i}{\sqrt{2}}\,\boldsymbol{e}_0 \end{aligned} \tag{VII-33}$$

The $\mathscr{M}_{x,y,z}$ matrices then become, in the new basis:

$$\tilde{\mathscr{M}}_x = -\frac{i}{\sqrt{2}}\begin{pmatrix} 0 & 1 & 0 \\ 1 & 0 & 1 \\ 0 & 1 & 0 \end{pmatrix} \quad \tilde{\mathscr{M}}_y = \frac{1}{\sqrt{2}}\begin{pmatrix} 0 & -1 & 0 \\ 1 & 0 & -1 \\ 0 & 1 & 0 \end{pmatrix} \quad \tilde{\mathscr{M}}_z = \begin{pmatrix} -i & 0 & 0 \\ 0 & 0 & 0 \\ 0 & 0 & i \end{pmatrix} \tag{VII-34}$$

These expressions, multiplied by $i\hbar$, are the same as the matrices of the three components of $\boldsymbol{J}$ written in (VII-27). By exponentiation [real exponentials for the rotation matrices $\mathscr{R}$, purely imaginary for the matrices $(R^{[j=1]})$], we go from infinitesimal to finite rotations.

This establishes that, on the $\boldsymbol{e}_{+1}$, $\boldsymbol{e}_0$, $\boldsymbol{e}_{-1}$ basis, the rotation matrices $(R_{\boldsymbol{u}}(\varphi))$ become identical to the $R_{\boldsymbol{u}}^{[j=1]}(\varphi)$ matrices. The unitary matrix (S) for the change of basis (VII-32) is written:

$$(S) = \begin{pmatrix} -\frac{1}{\sqrt{2}} & 0 & \frac{1}{\sqrt{2}} \\ -\frac{i}{\sqrt{2}} & 0 & -\frac{i}{\sqrt{2}} \\ 0 & 1 & 0 \end{pmatrix} \tag{VII-35}$$

The standard relations for a change of basis become the equivalence relations between the two representations:

$$(R_{\boldsymbol{u}}^{[j=1]}(\varphi)) = (S)^{-1}\ (\mathscr{R}_{\boldsymbol{u}}(\varphi))\ (S) \tag{VII-36a}$$

$$(\mathscr{R}_{\boldsymbol{u}}(\varphi)) = (S)\ (R_{\boldsymbol{u}}^{[j=1]}(\varphi))\ (S)^{-1} \tag{VII-36b}$$

In the $j = 1$ case, our general construction of irreducible representations yielded the rotation matrices themselves, i.e. the SO(3) group.

γ. *Rotation of vector fields*

When rotations are performed on vector fields, the rotation matrices $j = 1$ operate twice. A vector field $\boldsymbol{v}(\boldsymbol{r})$ is defined at every point of space $\boldsymbol{r}$ by its three components $v_i(\boldsymbol{r})$ $(i = x, y, z)$ as functions of $\boldsymbol{r}$. Let us perform a rotation $\mathscr{R}$ that transforms the position variables $\boldsymbol{r}$ into $\boldsymbol{r}' = \mathscr{R}\boldsymbol{r}$, either directly (active rotation), or by rotating the coordinate axes with $\mathscr{R}^{-1}$ (passive rotation). We then have to rotate, not only the variable $\boldsymbol{r}$, but also the components $v_i(\boldsymbol{r})$ of the vector field $\boldsymbol{v}(\boldsymbol{r})$. We then obtain the field $\boldsymbol{v}'(\boldsymbol{r}')$ given by:

$$\boldsymbol{v}'(\boldsymbol{r}') = \mathscr{R}\boldsymbol{v}(\boldsymbol{r}) \qquad \text{with:} \qquad \boldsymbol{r}' = \mathscr{R}\boldsymbol{r} \tag{VII-37a}$$

or:

$$\boldsymbol{v}'(\boldsymbol{r}) = \mathscr{R}\boldsymbol{v}(\mathscr{R}^{-1}\boldsymbol{r}) \quad ; \quad v_i'(\boldsymbol{r}) = \sum_j (\mathscr{R})_{ij}\ v_j(\mathscr{R}^{-1}\boldsymbol{r}) \tag{VII-37b}$$

We can then use the form (VII-36b) of the rotation matrix, or the components v_m of $\boldsymbol{v}$ in the $\boldsymbol{e}_m$ basis (VII-32), to write:

$$v_m'(\boldsymbol{r}) = \sum_{m'} R_{mm'}^{[1]}\ v_{m'}(\mathscr{R}^{-1}\boldsymbol{r}) \tag{VII-38}$$

A-3. Irreducible unitary representations in state space

A-3-a. Rotation matrix

In the state space $\mathscr{E}$ of any physical state taken as a whole, to each operator $R_{\boldsymbol{u}}(\varphi)$ is associated a matrix describing its action in an orthonormal basis $\{|u_p\rangle\}$ of $\mathscr{E}$:

$$\Big(R_{\boldsymbol{u}}(\varphi)\Big) = \begin{pmatrix} \langle u_1|\mathrm{e}^{-i\varphi\, J_u/\hbar}|u_1\rangle & \dots & \langle u_1|\mathrm{e}^{-i\varphi\, J_u/\hbar}|u_N\rangle \\ \langle u_2|\mathrm{e}^{-i\varphi\, J_u/\hbar}|u_1\rangle & & \\ \vdots & & \\ \langle u_N|\mathrm{e}^{-i\varphi\, J_u/\hbar}|u_1\rangle & \dots & \langle u_N|\mathrm{e}^{-i\varphi\, J_u/\hbar}|u_N\rangle \end{pmatrix} \tag{VII-39}$$

If the dimension N of $\mathcal{E}$ is finite, this is an $N \times N$ matrix; actually, N is often infinite, and so is the associated matrix $R_{\boldsymbol{u}}(\varphi)$. By construction, these matrices obey the pertinent commutation relations for infinitesimal rotations. However, for finite rotations, we should verify if these matrices form representations in the strict sense of the term, or if they provide only projective representations.

α.   *Block decomposition of matrices*

The matrices (VII-39) constitute a representation of the rotation group. Since the $R_{\boldsymbol{u}}(\varphi)$ operators are unitary and the $\{|u_k\rangle\}$ basis is orthonormal, these matrices as well as the representation are unitary. One can try to reduce this representation to a sum of representations with dimensions lower than N. This amounts to looking for a new basis $\{|u'_k\rangle\}$, in which the $R_{\boldsymbol{u}}(\varphi)$ operators are represented by block diagonal matrices:

$$\left(R_{\boldsymbol{u}}(\varphi)\right)' = \begin{pmatrix} \langle u'_1|e^{-i\varphi\, J_u/\hbar}|u'_1\rangle & \cdots & \langle u'_1|e^{-i\varphi\, J_u/\hbar}|u'_N\rangle \\ \vdots & & \\ \langle u'_N|e^{-i\varphi\, J_u/\hbar}|u'_1\rangle & \cdots & \langle u'_N|e^{-i\varphi\, J_u/\hbar}|u'_N\rangle \end{pmatrix}$$

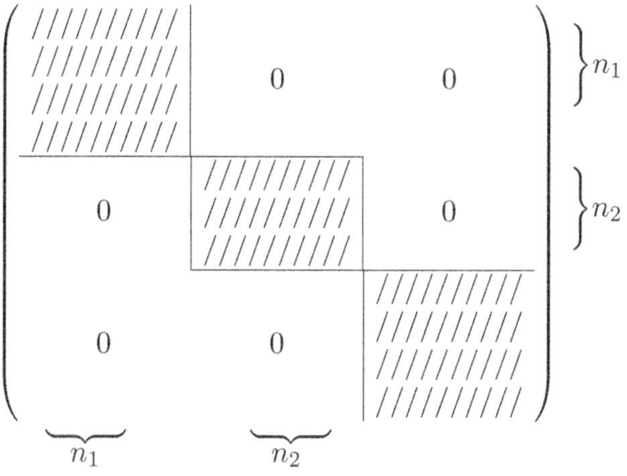

$$= \begin{pmatrix} \boxed{/\!/\!/\!/\!/} & 0 & 0 \\ 0 & \boxed{/\!/\!/\!/\!/} & 0 \\ 0 & 0 & \boxed{/\!/\!/\!/\!/} \end{pmatrix} \begin{matrix} \left.\right\}n_1 \\ \left.\right\}n_2 \\ \ \end{matrix} \qquad \text{(VII-40)}$$

$$\underbrace{}_{n_1} \quad \underbrace{}_{n_2}$$

Comment:

As any unitary matrix is diagonalizable one can always, for a single matrix, reduce each n_i to unity. As we already pointed out in chapter II, the problem is not to diagonalize *a single* given matrix, but *all* the matrices of the group, using *the same basis* $\{|u'_k\rangle\}$; a complete diagonalization is not possible in general, and we must limit ourselves to a block diagonalization.

β. The irreducible blocks correspond to the $\mathscr{E}(\tau, j)$

We saw that the rotation invariance of a subspace is equivalent to the invariance under the action of the three components J_x, J_y, and J_z of the angular momentum operator. The form of the matrix on the right-hand side of (VII-40) indicates that a first subspace $\mathscr{F}_1$, associated with the first block of n_1 rows and columns, is invariant under the action of the whole set of rotation operators. In the same way, the second $n_2 \times n_2$ block corresponds to an invariant subspace $\mathscr{F}_2$ with dimension n_2, etc.

Having decomposed the rotation matrices into diagonal blocks as in (VII-40), we could try to apply the same procedure to each block in order to obtain smaller blocks. This operation could continue until we obtain *irreducible* representations of the rotation group (chapter II, § B-5) for which, by definition, no further decomposition into a direct sum is possible (still considering, of course, the whole set of rotation matrices).

At this stage, each subspace $\mathscr{F}_n$ in which the blocks act is an irreducible subspace. Let us show that this subspace necessarily coincides with one of the irreducible subspaces $\mathscr{E}(\tau, j)$ constructed in § A-1-b. Since $\mathscr{F}_n$ is stable under the action of the angular momentum operators, one can diagonalize $\boldsymbol{J}^2$ and J_z in this subspace and obtain a common eigenvector $|j, m\rangle$. Applying $J - m$ times operator J_+ to this ket we obtain, still remaining inside $\mathscr{F}_n$, the vector $|j, j\rangle$. We can then obtain the $2j + 1$ basis vectors of a $\mathscr{E}(\tau, j)$ subspace by successive applications of operator J_-. This shows that $\mathscr{F}_n$ contains the space $\mathscr{E}(\tau, j)$. However, since $\mathscr{F}_n$ is irreducible, it cannot contain an invariant subspace: both subspaces must coincide.

We now know the general form of the block diagonal matrix (VII-40): it is constructed by placing along the diagonal a set of $\left(R^{[j]}\right)$ matrices. The state space of the physical system is the direct sum of $\mathscr{E}(\tau, j)$ spaces.

A-3-b. True representations and double-valued representations

We now examine whether the rotation operators acting in the state space $\mathscr{E}$ constitute a representation in the strict sens of the term (without a phase factor) of the rotation group $R_{(3)}$ or a projective representation.

Answering this question is not a priori obvious. The structure of the rotation group enabled us to *locally* eliminate the phase factors, but there is no evidence that one can do the same, globally, for the entire rotation group. Until now, our reasoning was based on necessary conditions: starting from the commutation relations between infinitesimal generators, we established relations, such as (VII-11), that must necessarily be satisfied, with the proper choice of basis, in any space associated with an irreducible representation of $R_{(3)}$. However, we did not verify whether these conditions were sufficient; strictly speaking it is not obvious that, for any integer or half-integer value of j, relations (VII-11) actually lead to representations of $R_{(3)}$, projective or not. In addition, we will see that the two cases (integer or half-integer values of j) must be distinguished. We thus go back to the matrices we obtained and discuss whether they actually constitute representations of the $R_{(3)}$ group.

Two simple cases come up immediately. First of all, the $j = 0$ case yields the non-faithful trivial representation, with dimension 1. The second, corresponding to $j = 1$, we now discuss.

A-3-c. Case where $j = 1$

This is a simple case since, as we saw in § A-2-c-β, the $R^{[1]}$ matrices we obtained are none other than the rotation matrices $(\mathscr{R})$ in ordinary space, to within a unitary basis change. This was to be expected: we have been looking systematically for all the irreducible representations of the rotation group, and for each dimension $(2j + 1)$ we only found one, to within a basis change, in each representation space $\mathscr{E}(\tau, j)$. Now the rotation matrices themselves constitute an irreducible 3-dimensional representation[5]. Consequently, the two representations must be equivalent.

It is clear that this representation is faithful (one-to-one correspondence) and that it is a true (non-projective) representation in the strict sense of the term: the product of the matrices $R^{[j=1]}$ automatically corresponds to the product of the matrices $(\mathscr{R})$. In this case we obtained a representation where all the phase factors have been eliminated.

A-3-d. Case where $j = 1/2$

It is easy to see, in (VII-25) or (VII-26) for example, that this case is less straightforward as the previous one. The expressions for the $(R^{[1/2]})$ matrices include trigonometric functions of the half-angles $\varphi/2$, $\alpha/2$, etc. This means that, a-priori, at least two distinct matrices $(R^{[1/2]})$ can correspond to a rotation $\mathscr{R}$: nothing changes in the rotation when φ is replaced by $\varphi + 2\pi$.

The matrices described in (VII-25) belong to the SU(2) group, as can easily been seen by comparison with expression (III-77) of chapter III, where a plays the role of the parameter φ. As the parameter $\boldsymbol{a}$ of a rotation equals $\boldsymbol{u}\varphi$, we must associate with each rotation the matrix of SU(2) corresponding to the same value of the parameter $\boldsymbol{a}$. The difficulty comes from the fact that to span the entire SU(2) group, $\boldsymbol{a}$ must vary for each direction $\boldsymbol{u}$ within a 4π interval (*cf.* figure 11 of chapter III); on the other hand, in the rotation group $R_{(3)}$, a 2π interval is

[5]If the representation was reducible, it could be decomposed into a sum of either two representations (with respective dimensions 1 and 2) or of three (with dimension 1). In both cases, there would be a direction of space (one-dimensional subspace of R^3), invariant under the action of all the rotations. We know that this is not the case: we only have to consider a rotation by a non-zero angle around an axis different from that direction, to see that this direction is not invariant.

sufficient for the parameter φ (any 2π rotation is the identity rotation). Under these conditions, what will be the precise correspondence between the elements of $R_{(3)}$ and SU(2)?

• A first point of view is to consider that, mathematically, a single matrix must be associated with each rotation. One can restrict[6] φ in (VII-25) to a 2π interval:

$$-\pi < \varphi < +\pi \tag{VII-41}$$

This means that there is a correspondence between the whole set of rotations and half of SU(2) , the part that corresponds, in figure 13 of chapter III, to the left sphere, centered around the identity matrix. This subset of SU(2) has been noted[7] $SU(2)_+$ in § B-3 of chapter III.

This point of view will not lead stricto sensu to a representation of the group $R_{(3)}$, since the representation matrices do not form a group as the set is no longer closed under multiplication. Take as an example the rotations around Oz:

$$\mathscr{R}_{\boldsymbol{u}_z}(\varphi_1) \qquad -\pi < \varphi_1 < \pi \quad ; \quad \mathscr{R}_{\boldsymbol{u}_z}(\varphi_2) \qquad -\pi < \varphi_2 < \pi \tag{VII-42}$$

Their product yields:

$$\left(R^{[1/2]}(\varphi_1)\right)\ \left(R^{[1/2]}(\varphi_2)\right) = \begin{pmatrix} \mathrm{e}^{-i(\varphi_1+\varphi_2)/2} & 0 \\ 0 & \mathrm{e}^{i(\varphi_1+\varphi_2)/2} \end{pmatrix} \tag{VII-43}$$

If $|\varphi_1 + \varphi_2| < \pi$, this simply yields the matrix associated with the product rotation, by the angle $\varphi_1 + \varphi_2$. However, if $|\varphi_1 + \varphi_2| \geq \pi$, which is perfectly compatible with inequalities (VII-42), this product does not belong to the set of chosen matrices. Taking into account the angle convention, this product rotation is by an angle $\varphi_1 + \varphi_2 - 2\pi$; it corresponds to the matrix opposite to (VII-43).

In this first point of view, we have obtained a projective representation where the phase factor $\mathrm{e}^{-i\xi}$ can only take the two discrete values ± 1. There exists a *local* isomorphism between the rotations and the $R^{[1/2]}$ matrices: as long as the rotation angles are small enough for the product to remain in the set of representations (positive trace matrices), the matrix product corresponds to the rotation product. But if one considers the whole set of rotations, it may be the opposite of the matrix product that corresponds to the rotation product: the isomorphism is not *global*.

[6]Even restricting the accessible domain of φ, each rotation can be described two ways, by changing φ and $\boldsymbol{u}$ into their opposite. This is of no consequence since the matrix (VII-25) is unchanged in this operation.

[7]Remember that the $SU(2)_+$ matrices are those having a positive trace, and the $SU(2)_-$ (right sphere in figure 13 of chapter III) are the ones with negative traces.

Comment:

The situation we just mentioned is not surprising, since we have only used the properties of the Lie algebra of the $R_{(3)}$ group in the vicinity of a given rotation. We thus lost a global property of $R_{(3)}$, namely that two rotations around the same vector $\boldsymbol{u}$ and by an angle φ that tends respectively toward $+\pi$ and $-\pi$, tend toward the same rotation (or in other words we lost the fact that a 2π rotation is the identity rotation).

• A second point of view is to associate with each rotation of $R_{(3)}$ not one but a pair of two SU(2) matrices: the one discussed in the above point of view [SU$(2)_+$ matrix] and the opposite matrix [SU$(2)_-$ matrix]. These two matrices are obtained by taking, in (VII-20), the angle φ and the angle $\varphi + 2\pi$. This second point of view is schematized in figure 1, where a point A (whose position yields the parameter $\boldsymbol{a}$ corresponding to a rotation) is associated with two points A' and A'' (with respective positions $\boldsymbol{a}$ and $\boldsymbol{a} \pm 2\pi\, \boldsymbol{a}/a$). This results in a *double-valued* correspondence between the elements of $R_{(3)}$ and SU(2); conversely, this correspondence induces a "surjection" of SU(2) into $R_{(3)}$. In this way, the representation set is the entire SU(2) group. Furthermore, with the product $\mathscr{R}_2\,\mathscr{R}_1$ of two rotations is always associated the couple of the two matrices $\pm U_2\, U_1$ obtained by multiplication of the unitary matrices $\pm U_1$ and $\pm U_2$ corresponding respectively to $\mathscr{R}_1$ and $\mathscr{R}_2$. This more convenient point of view is generally used in quantum mechanics. There is no longer a representation (even projective)

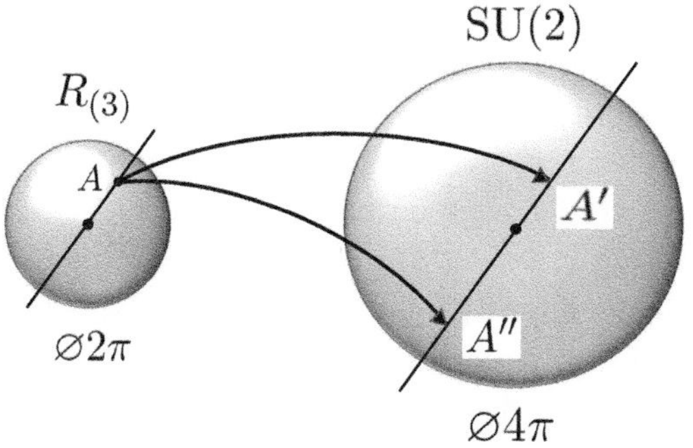

Figure 1: The left sphere represents the rotation group $R_{(3)}$, the one on the right the SU(2) group of the unitary 2×2 matrices with a determinant equal to one. Instead of associating with each rotation (point A) a single matrix of SU(2), two opposite matrices (points A′ and A″) are associted with it, leading to a double-valued representation.

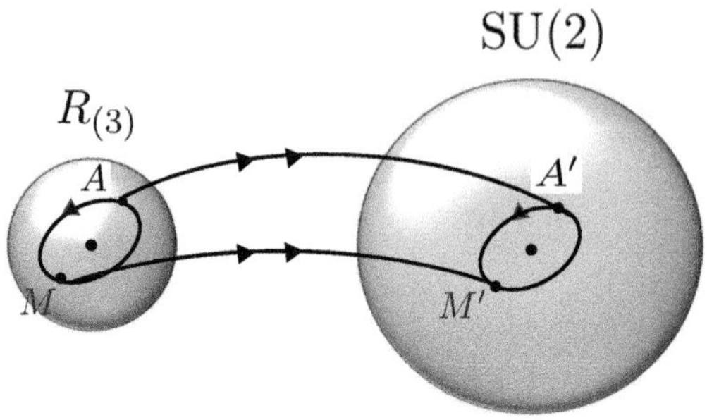

Figure 2: The continuous image in SU(2) of a closed loop homotopic to zero in the rotation group is necessarily a closed loop in SU(2) (as the representation is double-valued, there are two such images, only one of them being represented on the figure).

of $R_{(3)}$ in the usual sense of the term, but the concept of representation can certainly be enlarged to matrices that are multi-valued functions of the elements of the group.

The possibility of a double-valued correspondence between the rotations and the SU(2) matrices, albeit preserving a local homomorphism between the two groups, is related to their topology: $R_{(3)}$ is 2-connected whereas SU(2) is simply connected. Consider a closed loop homotopic to zero in $R_{(3)}$, starting from point A (*cf.* figure 2). We choose, for example, A' as the image of A and, by continuity, M' as the image of M. When the loop closes and M tends toward A, M' necessarily tends toward A' (and not A''): one comes back to the same matrix of SU(2) as long as the closed loop is homotopic to zero. This is because such a loop can be continuously deformed into a zero path and, by continuity, the matrix associated with the rotation cannot jump from its initial value to an opposite value. More generally, consider two closed loops starting and ending in A. If these two paths are homotopic with each other, they necessarily start and end at the same image of A (either A', or A'').

How can the image of a closed loop start in A' and end in A'' instead of A'? As on figure 3, we must take in $R_{(3)}$ a closed path non-homotopic to zero, that crosses once the sphere with the π radius. The image of that closed path in $R_{(3)}$ is an open path in SU(2). This explains why the 2-connected structure of $R_{(3)}$ allows double-valued representations, but not higher multi-valued representations.

Figure 4 shows how the continuous image of a path in $R_{(3)}$, homotopic to zero but more complex (crossing twice the sphere of the π rotations) is a closed

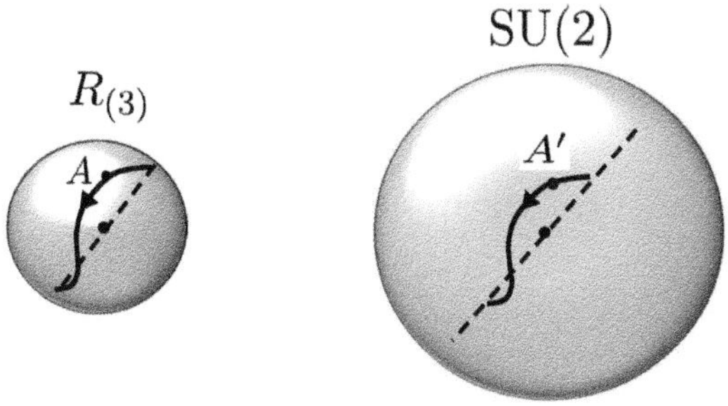

Figure 3: The image of a closed path non-homotopic to zero in the rotation group is an open path in SU(2), since it ends at the starting point of the second image of that path. Following that second image by continuity brings us back to the departure point in SU(2); following the two images in a row yields a closed path (not represented on the figure).

path in SU(2), and hence homotopic to zero since SU(2) is simply connected. There are, however, two possible closed paths, depending on the departure point in SU(2).

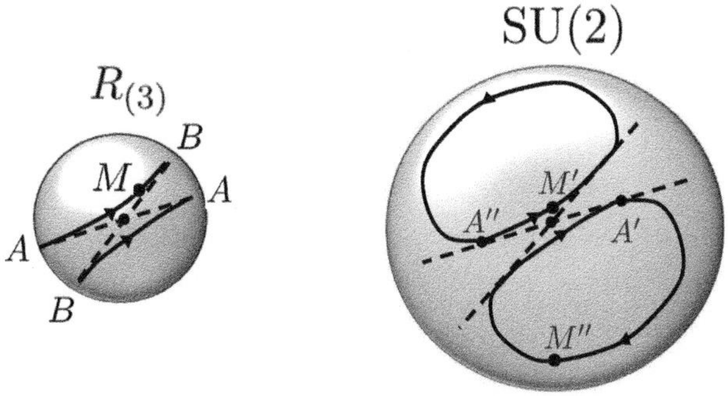

Figure 4: The figure on the left-hand side shows a closed path in the rotation group which, although more complex than the one in figure 2, is homotopic to zero (cf. figure 16 of complement C_V). Its image in the double-valued representation is a set of two closed paths in SU(2).

• From a physical point of view, can we accept these double-valued representations in quantum mechanics? A priori yes: since a global phase factor of a state vector is of no consequence on its physical properties, nothing prevents associating two opposite operators with the same geometric transformation.

To make it clear that the state vector depends not only on the initial and final positions of the system, but also on the path it took to get there, we take as an example the simple case of a rotation around a given axis. If the angle φ increases from 0 to 2π, the phase of the state vector is multiplied by -1; if the angle increases from 0 to 4π, by $+1$; from 0 to 6π, by -1, etc. Since a global phase factor does not change the physical properties of the system, the effect we just described plays no role in most experiments. It may, however, appear in certain interference experiments (*cf.* § B-4).

No fundamental physics principle is violated by accepting the double-valued representations obtained for $j = 1/2$. This mathematical possibility offered by the formalism of quantum mechanics is actually used by Nature, and numerous particles[8] such as the electron, the proton, the neutron, etc. have a spin $1/2$.

A-3-e. Any value of j

The results we have obtained can be generalized to any value of j: the integer values lead to "true" representations of the rotation group, the half-integer values to "double-valued" representations.

Consider, for example, rotations around the Oz axis, by an angle φ defined in the interval $[-\pi, +\pi]$. In the standard basis $\{|\tau, j, m\rangle\}$, each rotation leads to a simple multiplication by the phase factor $e^{im\varphi}$. When dealing with the product of two rotations, the phases add up, hence leading to a final phase outside the interval $[-\pi, +\pi]$. If m is an integer, this has no consequence as we can subtract 2π from the phase without changing the final ket: the representation matrices form a group under the internal composition law. On the other hand, if m is a half-integer, we could end up with the opposite matrix. Since φ is by definition between $-\pi$ and $+\pi$, the representation matrices do not form a group, and the discussion around the $j = 1/2$ case can be transposed to the case where j is an arbitrary half-integer.

We will not develop this argument further and examine what happens for rotations around different axes. In the general theory of the addition of angular momenta, in § C, we will see that a half-integer angular momentum j can be obtained by adding

[8] Historically, the possibility of half-integer angular momenta led to an amusing controversy at the beginning of quantum mechanics. Strong reservations were expressed when Uhlenbeck and Goudsmit put forward the hypothesis of the electron spin to explain certain particularities of the hydrogen spectrum. The then very recent quantization of the orbital variables had led to the belief that an angular momentum had to be a multiple of $\hbar$. Accepting the additional value $\hbar/2$ opened the way to other values like $\hbar/3, \hbar/4, \ldots, \hbar/n$, which amounted to letting $\hbar$ tend toward zero, hence giving up any quantization! We now understand how the 2-connected structure of $R_{(3)}$ does not allow fractions of $\hbar$ lower than 1/2.

a 1/2 moment and another integer moment $j - 1/2$. This yields the tensor product of a true representation and a double-valued one and the double-valued nature of the final representation comes from that of the spin 1/2.

B. Spin 1/2 particule; spinors

We now extend the results of § A-1 in chapter VI for a free particle. Starting from general considerations on the geometric transformations that can be performed on a physical system, we build a state space $\mathscr{E}$ in which this transformation group can be linearly represented. However, instead of choosing as before the simplest possibility, we shall consider the next case in the order of increasing complexity. This will yield the state space of a spin 1/2 particle.

B-1. Construction of the state space

As in § A-1 of chapter VI and in the framework of the Galilean group, we assume the existence of a position operator $\boldsymbol{R}$ of the system, where $\boldsymbol{R}$ symbolizes three Hermitian operators X, Y, and Z. However, we do not limit ourselves to the case where X, Y, Z form a CSCO. An orthonormal basis of eigenvectors common to these three operators will be written:

$$|x, y, z, \mu\rangle = |\boldsymbol{r}, \mu\rangle \tag{VII-44}$$

where μ is an index that distinguishes between the various eigenvectors associated with the same eigenvalues x, y, z. At the moment, the possible values of the quantum number μ have not been specified (they can a priori be discrete or continuous).

We use the translation operators $T(\boldsymbol{0}, \boldsymbol{b}) = \exp(-i\boldsymbol{b} \cdot \boldsymbol{P})/\hbar$ to define the relative phase of the kets $|\boldsymbol{r}, \mu\rangle$ by:

$$|\boldsymbol{r}, \mu\rangle = T\left(\boldsymbol{0}, \boldsymbol{b} = \boldsymbol{r}\right)\, |\boldsymbol{r} = 0, \mu\rangle \tag{VII-45a}$$

This equality defines the $|\boldsymbol{r}, \mu\rangle$ kets at any point with respect to the ket associated with $\boldsymbol{r} = \boldsymbol{0}$. In particular, it implies[9] that the number of accessible values of μ and their values themselves are independent of the specific point $\boldsymbol{r}$ (space homogeneity). This leads to:

$$|\boldsymbol{r}, \mu\rangle = T\left(\boldsymbol{0}, \boldsymbol{b} = \boldsymbol{r} - \boldsymbol{r}'\right)\, |\boldsymbol{r}', \mu\rangle \tag{VII-45b}$$

meaning that the origin $\boldsymbol{r} = 0$ does not play any particular role in the definition of the basis $\{|\boldsymbol{r}, \mu\rangle\}$.

[9]Since the T operators are unitary, the orthogonality of the kets $|\boldsymbol{r} = \boldsymbol{0}, \mu\rangle$ associated with various values of μ implies that the same is true for the kets $|\boldsymbol{r}, \mu\rangle$.

We can then consider that the state space $\mathscr{E}$ spanned by this basis is the tensor product of a space $\mathscr{E}_{\boldsymbol{r}}$, spanned by the kets $\{|\boldsymbol{r}\rangle\}$, and a space spanned by the $\{|\mu\rangle\}$ that we shall note $\mathscr{E}_S$:

$$|\boldsymbol{r},\mu\rangle = |\boldsymbol{r}\rangle \otimes |\mu\rangle \qquad |\boldsymbol{r}\rangle \in \mathscr{E}_{\boldsymbol{r}} \qquad |\mu\rangle \in \mathscr{E}_S$$

$$\mathscr{E} = \mathscr{E}_{\boldsymbol{r}} \otimes \mathscr{E}_S \tag{VII-46}$$

Relation (VII-45b) implies that the index μ is invariant under the action of any translation operator, i.e. of any component of $\boldsymbol{P}$. Consequently, by definition, the translation operator $\boldsymbol{P}$ does not act in $\mathscr{E}_S$, but only in $\mathscr{E}_{\boldsymbol{r}}$. This means that, in $\mathscr{E}_S$, we have a trivial representation of the translation group (the $n_j \times n_j$ identity matrix, where n_j is the dimension of $\mathscr{E}_S$). The observables acting only in $\mathscr{E}_S$ are all invariant under a translation, and μ appears as an index related to the *internal variables* of the system – as opposed to $\boldsymbol{r}$, which is related to its *external variables.*

If we want the internal variables of the system to not remain invariant under any displacement, we must choose for the $R_{(3)}$ group a non-trivial representation. As a given representation can be decomposed into a direct sum of irreducible representations and following a maximum simplicity criteria, we shall demand the representation of the $R_{(3)}$ group $\mathscr{E}_S$ to be irreducible. Excluding the trivial representation of dimension one (that would bring us back to the spinless particle case studied in chapter VI), the representation with the lowest dimension corresponds to the quantum number $j = 1/2$.

The operator generating the rotations in $\mathscr{E}_S$ is noted $\boldsymbol{S}$, whereas $\boldsymbol{J}$ denotes the operator generating the rotations in the entire space $\mathscr{E}$. We thus have in the space $\mathscr{E}_S$ of the internal variables:

$$\begin{aligned} &S_z\,|\mu\rangle = \mu\hbar\,|\mu\rangle \\ &S_\pm\,|\mu\rangle = \hbar\sqrt{S(S+1)-\mu(\mu\pm 1)}\;|\mu\pm 1\rangle \end{aligned} \tag{VII-47a}$$

where S_x, S_y, and S_z are the components of $\boldsymbol{S}$ and $S_\pm = S_x \pm i\,S_y$; the quantum number j has been replaced by S, and we keep the notation $j(j+1)\hbar^2$ for the eigenvalues of $\boldsymbol{J}^2$. In the entire space $\mathscr{E}$, these relations become:

$$\begin{aligned} &S_z\,|\boldsymbol{r},\mu\rangle = \mu\hbar\,|\boldsymbol{r},\mu\rangle \\ &S_\pm\,|\boldsymbol{r},\mu\rangle = \hbar\sqrt{S(S+1)-\mu(\mu\pm 1)}\;|\boldsymbol{r},\mu\pm 1\rangle \end{aligned} \tag{VII-47b}$$

since, by hypothesis, $\boldsymbol{S}$ does not act in the space $\mathscr{E}_{\boldsymbol{r}}$.

From now on, we shall call orbital (or external) variables the variables acting in $\mathscr{E}_{\boldsymbol{r}}$, and spin (or internal) variables those acting in $\mathscr{E}_S$. The name "spin" (turning about itself) comes from the fact that the operator $\boldsymbol{S}$, called spin of the system, is a rotation operator acting on the internal variables of the system. It is thus reminiscent of the spinning of a top about itself, its spin being the internal angular momentum. This classical analogy must be handled with care, as the spin is of a quantum nature: it does not correspond to the rotation of a finite size particle, but of a point particle (*cf.* § B-4).

B-2. State vector of the particle

It is equivalent to define the state vector $|\psi\rangle \in \mathscr{E}$ of the system or its components in the basis $\{|\boldsymbol{r}, \mu\rangle\}$:

$$\psi_\mu(\boldsymbol{r}) = \langle \boldsymbol{r}, \mu | \psi \rangle \tag{VII-48}$$

The state of the system is not just defined by a single wave function $\psi(\boldsymbol{r})$, but by $(2S+1)$ fonctions $\psi_\mu(\boldsymbol{r})$, where $\mu = S, S-1, \ldots, -S$.

In what follows, we shall discuss the $S = 1/2$ case (the $S = 0$ case brings us back to the already studied spinless particle, and the generalization to higher values of S is done without difficulty). The system is now called a "spin 1/2 particle" and we note $+$ or $-$ the respective values $\mu = +1/2$ and $-1/2$. The two wave functions $\psi_\pm(\boldsymbol{r})$ are often regrouped into a column matrix:

$$[\psi(\boldsymbol{r})] = \begin{pmatrix} \psi_+(\boldsymbol{r}) \\ \psi_-(\boldsymbol{r}) \end{pmatrix} \tag{VII-49}$$

This mathematical object is called a "spinor". At any point $\boldsymbol{r}$ in space, a spinor is defined by two components ψ_+ and ψ_-. This is quite similar to the definition of a vector field $\boldsymbol{V}(\boldsymbol{r})$ in space, except that a vector $\boldsymbol{V}$ has three components and not two (see exercise below). The adjoint spinor is the row matrix:

$$[\psi(\boldsymbol{r})]^\dagger = \begin{pmatrix} \psi_+^\star(\boldsymbol{r}) & \psi_-^\star(\boldsymbol{r}) \end{pmatrix} \tag{VII-50}$$

The product of the row and column matrices:

$$[\psi(\boldsymbol{r})]^\dagger \; [\psi(\boldsymbol{r})] = |\psi_+(\boldsymbol{r})|^2 + |\psi_-(\boldsymbol{r})|^2 \tag{VII-51}$$

yields the probability density of finding the particle at point $\boldsymbol{r}$, regardless of its spin state.

Since the basis $\{|\boldsymbol{r}, \mu\rangle\}$ is orthonormal, we have:

$$\langle \boldsymbol{r}, \mu | \boldsymbol{r}', \mu' \rangle = \delta\left(\boldsymbol{r} - \boldsymbol{r}'\right) \delta_{\mu\mu'} \tag{VII-52}$$

The closure relation in this basis is written:

$$\int d^3r \sum_{\mu=+1/2,-1/2} |\boldsymbol{r}, \mu\rangle \; \langle \boldsymbol{r}, \mu| = \mathbb{1} \tag{VII-53}$$

This last equality easily shows that:

$$|\psi\rangle = \int d^3r \sum_{\mu=+,-} |\boldsymbol{r},\mu\rangle\,\langle \boldsymbol{r},\mu|\psi\rangle$$
$$= \int d^3r\,\Big\{\psi_+(\boldsymbol{r})|\boldsymbol{r},+\rangle + \psi_-(\boldsymbol{r})|\boldsymbol{r},-\rangle\Big\} \tag{VII-54}$$

In the same way, considering two kets $|\varphi\rangle$ and $|\psi\rangle$, we easily get:

$$\langle\varphi|\psi\rangle = \int d^3r\,\Big\{\varphi_+^\star(\boldsymbol{r})\,\psi_+(\boldsymbol{r}) + \varphi_-^\star(\boldsymbol{r})\,\psi_-(\boldsymbol{r})\Big\}$$
$$= \int d^3r\,\left[\varphi(\boldsymbol{r})\right]^\dagger\,\left[\psi(\boldsymbol{r})\right] \tag{VII-55}$$

B-3. Operators

All the results we obtained in § A-1 of chapter VI remain valid in $\mathscr{E}_{\boldsymbol{r}}$: the operator $\boldsymbol{P}$ acts as $(\hbar/i)\boldsymbol{\nabla}$, the angular momentum operator in $\mathscr{E}_{\boldsymbol{r}}$ is:

$$\boldsymbol{L} = \boldsymbol{R}\times\boldsymbol{P} \tag{VII-56}$$

etc. We can follow the same procedure as in § 2-c of complement $\mathrm{C_V}$, where the state space $\mathscr{E}$ had a tensor product structure $\mathscr{E} = \mathscr{E}_1\otimes\mathscr{E}_2$ [*cf.* (VI-39)]. We showed that the total linear and angular momenta of the system were the sum of the corresponding operators in the spaces $\mathscr{E}_1$ and $\mathscr{E}_2$. Here, $\mathscr{E}_{\boldsymbol{r}}$ plays the role of $\mathscr{E}_1$ and $\mathscr{E}_S$ that of $\mathscr{E}_2$. In addition we have already noted that $\mathscr{E}_S$ is invariant under any translation, which amounts to considering that the momentum operator is zero in $\mathscr{E}_S$. The total momentum is thus equal to the operator $\boldsymbol{P}$ defined in $\mathscr{E}_{\boldsymbol{r}}$, whereas the total angular momentum $\boldsymbol{J}$ equals:

$$\boldsymbol{J} = \boldsymbol{L}+\boldsymbol{S}$$
$$= \boldsymbol{R}\times\boldsymbol{P}+\boldsymbol{S} \tag{VII-57}$$

The rotation operator $R_{\boldsymbol{u}}(\varphi)$ is expressed as:

$$R_{\boldsymbol{u}}(\varphi) = \mathrm{e}^{-i\varphi\,J_u/\hbar} = \mathrm{e}^{-i\varphi\,(L_u+S_u)/\hbar}$$
$$= \mathrm{e}^{-i\varphi\,L_u/\hbar}\,\mathrm{e}^{-i\varphi\,S_u/\hbar} \tag{VII-58}$$

We can write:

$$\mathrm{e}^{-i\varphi\,L_u/\hbar}\,|\boldsymbol{r}\rangle = |\boldsymbol{r}'\rangle \quad \text{with} \quad \boldsymbol{r}' = \mathscr{R}_{\boldsymbol{u}}(\varphi)\,\boldsymbol{r} \tag{VII-59a}$$

and:

$$\mathrm{e}^{-i\varphi\,S_u/\hbar}\,|\mu\rangle = \sum_{\mu'} \langle\mu'|\mathrm{e}^{-i\varphi\,S_u/\hbar}\,|\mu\rangle \times |\mu'\rangle$$
$$= \sum_{\mu'} \left(R_{\boldsymbol{u}}^{[1/2]}(\varphi)\right)_{\mu'\mu}\,|\mu'\rangle \tag{VII-59b}$$

where the $R^{[1/2]}_{\mu'\mu}$ have been defined in (VII-17b) and (VII-25). Combining these equalities, we get:

$$R_{\boldsymbol{u}}(\varphi)\,|\boldsymbol{r},\mu\rangle = \sum_{\mu'} \left(R^{[1/2]}_{\boldsymbol{u}}(\varphi)\right)_{\mu'\mu} |\boldsymbol{r}',\mu'\rangle \tag{VII-60}$$

The notation $[\psi(\boldsymbol{r})]$ in (VII-49) enabled us to symbolize a two-components column matrix (spinor) representing a ket. In the same way, the notation $[\![A]\!]$ will symbolize a 2×2 matrix acting on the column matrices associated with the spinors. As an example, consider an operator A acting only in $\mathscr{E}_S$. We then have:

$$\begin{aligned} A|\psi\rangle &= \int d^3r\ \{\psi_+(\boldsymbol{r})A|\boldsymbol{r}\rangle|+\rangle + \psi_-(\boldsymbol{r})A|\boldsymbol{r}\rangle|-\rangle\} \\ &= \int d^3r\ |\boldsymbol{r}\rangle\ \{\psi_+(\boldsymbol{r})\ [A_{++}|+\rangle + A_{-+}|-\rangle] + \psi_-(\boldsymbol{r})\ [A_{+-}|+\rangle + A_{--}|-\rangle]\} \end{aligned} \tag{VII-61}$$

If $\psi'_+(\boldsymbol{r})$ and $\psi'_-(\boldsymbol{r})$ are the components of the spinor associated with $|\psi'\rangle = A|\psi\rangle$, we have:

$$\begin{pmatrix} \psi'_+(\boldsymbol{r}) \\ \psi'_-(\boldsymbol{r}) \end{pmatrix} = \begin{pmatrix} A_{++} & A_{+-} \\ A_{-+} & A_{--} \end{pmatrix} \begin{pmatrix} \psi_+(\boldsymbol{r}) \\ \psi_-(\boldsymbol{r}) \end{pmatrix} \tag{VII-62}$$

This amounts to writing:

$$[\![A]\!] = \begin{pmatrix} A_{++} & A_{+-} \\ A_{-+} & A_{--} \end{pmatrix} \tag{VII-63}$$

As A only acts in $\mathscr{E}_S$, the elements of $[\![A]\!]$ do not act on the variable $\boldsymbol{r}$. On the other hand, if we had chosen an operator acting only in $\mathscr{E}_{\boldsymbol{r}}$, as P_x for example, we would get:

$$[\![P_x]\!] = \frac{\hbar}{i} \begin{pmatrix} \partial/\partial x & 0 \\ 0 & \partial/\partial x \end{pmatrix} \tag{VII-64}$$

which is the product of an operator acting in $\boldsymbol{r}$ by the 2×2 identity matrix. One can also consider mixed operators acting both in $\mathscr{E}_{\boldsymbol{r}}$ and $\mathscr{E}_S$. Rotation operators $R_{\boldsymbol{u}}(\varphi)$ are such an example, as shown in (VII-58). We have:

$$|\psi'\rangle = R_{\boldsymbol{u}}(\varphi)|\psi\rangle = \int d^3r \sum_{\mu=+,-} \psi_\mu(\boldsymbol{r})\ R_{\boldsymbol{u}}(\varphi)\ |\boldsymbol{r},\mu\rangle \tag{VII-65}$$

or, using (VII-60):

$$|\psi'\rangle = \int d^3r \sum_{\mu\mu'} \psi_\mu(\boldsymbol{r})\ \left(R^{[1/2]}\right)_{\mu'\mu} |\boldsymbol{r}',\mu'\rangle \tag{VII-66}$$

where $\boldsymbol{r}'$ is defined in (VII-59a).

Finally, replacing $\boldsymbol{r}$ by $\mathscr{R}^{-1}\boldsymbol{r}'$, and changing in (VII-66) the integration variable $\boldsymbol{r}$ into $\boldsymbol{r}'$:

$$\begin{pmatrix} \psi'_+(\boldsymbol{r}) \\ \psi'_-(\boldsymbol{r}) \end{pmatrix} = \begin{pmatrix} \left(R^{[1/2]}\right)_{++} & \left(R^{[1/2]}\right)_{+-} \\ \left(R^{[1/2]}\right)_{-+} & \left(R^{[1/2]}\right)_{--} \end{pmatrix} \begin{pmatrix} \psi_+\left(\mathscr{R}^{-1}\boldsymbol{r}\right) \\ \psi_-\left(\mathscr{R}^{-1}\boldsymbol{r}\right) \end{pmatrix} \tag{VII-67}$$

This relation is similar to the one used in the rotation of a vector field, *cf.* (VII-38): to rotate a spinor, we first consider its components at point $\mathscr{R}^{-1}\boldsymbol{r}$, then take a linear combination of its components.

Exercise : Write the formulas for the rotation transformation of the components of a vector field in 3-dimensional space. Considering infinitesimal rotations, define orbital and spin angular momenta, $\boldsymbol{L}$ and $\boldsymbol{S}$. Write explicitly the S_x, S_y, and S_z matrices and calculate $\boldsymbol{S}^2 = S_x^2 + S_y^2 + S_z^2$. Does this yield an irreducible representation of the rotation group? Which one? Show that there is a perfect analogy with the spin $S = 1$ particle (*cf.* [6], § 21 in chapter XIII)].

B-4. Particular properties of a half-integer spin

In § A-3-b, we noted and discussed an important particularity of representations associated with half-integer momenta j. These representations are double-valued and, depending on the way the system is rotated to reach its final position, two different state vectors $\pm|\psi\rangle$ may be obtained. In particular, a 2π rotation results in a sign change of $|\psi\rangle$. From a physical point of view, two kets $\pm|\psi\rangle$ yield identical measurement results (since they only differ by a global phase factor). A superficial examination of the system could lead us to believe that this mathematical curiosity has no physical consequence. This is not the case, as we now explain with an optics analogy. We know that a light detector is not sensitive to the phase of an optical wave. If such a detector is used to measure the intensity of a wave that has gone through a phase object of unknown index, it is not possible to determine the phase change of the optical wave. But we also know how to go around this difficulty thanks to interference experiments. We superpose to the wave that has gone through the phase object a reference wave, whose phase is known in all points in space. The interference pattern obtained depends on the phase variations of the unknown wave.

In quantum mechanics, interference experiments can also be performed. They actually underlie this whole field (linear superposition principle, concept of probability amplitude, etc). The basic idea of the experiment is simple (figure 5): one builds an interferometer for spin 1/2 particles (for example electrons, protons, or neutrons), and on one of the arms of the interferometer one imposes a $2n\pi$ rotation of the internal (spin) variables; after recombination of the two waves, one examines[10] the interference pattern as a function of the value of n. If the particle has a half-integer spin and if $n = 1$ (or, in a more general way, if n is odd) the phase of one of the interfering wave is changed by π (since $e^{i\pi} = -1$)

[10]Considering the operation imposed on the particle's spin from a global point of view (modification of the spinor at each point $\boldsymbol{r}$), it appears more complex than the simple action of operator $e^{-i\varphi S_u/\hbar}$. In the experiment, the rotation operation depends on $\boldsymbol{r}$ (the particle's position), and its action on the orbital and spin variables is therefore correlated.

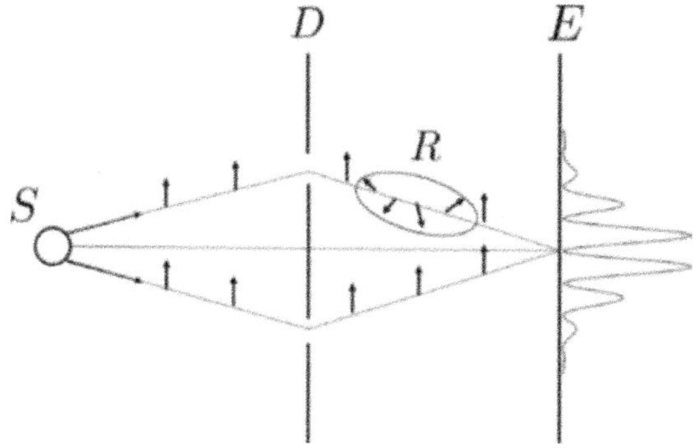

Figure 5: A source S emits neutrons (spin 1/2 particles) with polarized spins. They go through a diaphragm D that has two holes. Because of diffraction, the beams recombine on a screen E where an interference pattern is observed. One then applies in the region R, traversed by one of the beams, a magnetic field that rotates the spins by a 2π angle, so that the spins recover their initial direction at the end. Nevertheless, one observes a complete modification of the interference pattern on the screen, the central bright fringe becoming for example a dark fringe.

and the interference pattern is modified. On the other hand, if $n = 2, 4, 6, \ldots$, the pattern is the same as if $n = 0$ (no rotation).

The experiment has been performed on neutrons [42, 43], whose charge is zero (no Lorentz force, hence no magnetic deviation of the trajectories), but whose non-zero spin magnetic moment undergoes the Larmor precession in a magnetic field $\boldsymbol{B}$. It was verified that a 2π rotation does indeed change the interference pattern as expected. This remarkable experiment shows that in quantum physics, a 2π rotation of an object may change certain of its properties.

C. Addition of angular momenta

In § B we have already seen cases where the system's state space is the tensor product of two (orbital and spin) spaces, each associated with a representation of the rotation group. This occurs very often as, for example, when the system is composed of two spinless particles, with respective state spaces $\mathscr{E}_1$ and $\mathscr{E}_2$. The state space of the total system is:

$$\mathscr{E} = \mathscr{E}_1 \otimes \mathscr{E}_2 \tag{VII-68}$$

In both spaces $\mathscr{E}_1$ and $\mathscr{E}_2$, one can define subspaces that generate irreducible representations of the rotation group $R_{(3)}$. Taking their product we obtain, in $\mathscr{E}$, representations of $R_{(3)}$ that are the tensor products of the individual representations. We are going to discuss the characteristics of these new representations.

C-1. Statement of the problem

Consider two physical systems $\mathscr{S}_1$ and $\mathscr{S}_2$ (or two kinds of different variables $\mathscr{S}_1$ and $\mathscr{S}_2$ of the same system, for example, the orbital and spin variables). The corresponding state spaces $\mathscr{E}_1$ and $\mathscr{E}_2$ can be decomposed into direct sums of irreducible spaces:

$$\mathscr{E}_1 = \sum_{\oplus\, \tau_1 j_1} \mathscr{E}(\tau_1, j_1) \qquad \mathscr{E}_2 = \sum_{\oplus\, \tau_2 j_2} \mathscr{E}(\tau_2, j_2) \tag{VII-69}$$

and the corresponding standard bases are noted $\{|\tau_1, j_1, m_1\rangle\}$ and $\{|\tau_2, j_2, m_2\rangle\}$. The total space $\mathscr{E}$, expressed in (VII-68) as a tensor product, is written:

$$\mathscr{E} = \sum_{\oplus\, \tau_1\tau_2\, j_1 j_2} \mathscr{E}(\tau_1, \tau_2\,; j_1, j_2) \tag{VII-70a}$$

where:

$$\mathscr{E}(\tau_1, \tau_2\,; j_1, j_2) = \mathscr{E}(\tau_1, j_1) \otimes \mathscr{E}(\tau_2, j_2) \tag{VII-70b}$$

This subspace $\mathscr{E}(\tau_1\tau_2\,; j_1 j_2)$ of $\mathscr{E}$ is generated by the kets:

$$|\tau_1\tau_2\,; j_1 j_2\; m_1 m_2\rangle = |\tau_1 j_1 m_1\rangle \otimes |\tau_2 j_2 m_2\rangle \tag{VII-70c}$$

where τ_1, τ_2, j_1, j_2 are fixed, whereas the indices m_1 and m_2 have, respectively, $(2j_1+1)$ and $(2j_2+1)$ possible values. The space $\mathscr{E}(\tau_1\tau_2\,; j_1 j_2)$ is invariant under the action of any function of $\boldsymbol{J}_1$ and $\boldsymbol{J}_2$, and in particular under the action of the total angular momentum $\boldsymbol{J} = \boldsymbol{J}_1 + \boldsymbol{J}_2$.

C-1-a. Independent rotations of the two sub-systems

The operator associated with the rotation of system 1 by an angle φ_1 around $\boldsymbol{u}_1$ and with the rotation of system 2 by an angle φ_2 around $\boldsymbol{u}_2$ is written:

$$R^1_{\boldsymbol{u}_1}(\varphi_1)\; R^2_{\boldsymbol{u}_2}(\varphi_2) = \exp\left\{-\frac{i}{\hbar}\varphi_1\, \boldsymbol{J}_1 \cdot \boldsymbol{u}_1\right\}\; \exp\left\{-\frac{i}{\hbar}\varphi_2\, \boldsymbol{J}_2 \cdot \boldsymbol{u}_2\right\} \tag{VII-71a}$$

The set of this type of operators form a group, tensor product of the group of rotations of system 1 around O and the group of rotations of system 2 around the same point [the elements of $R_{(3)} \otimes R_{(3)}$ depend on 6 continuous parameters]. In the subspace $\mathscr{E}(\tau_1, \tau_2\,; j_1, j_2)$, these operators are represented by matrices whose elements are:

$$\begin{aligned}&\langle \tau_1\tau_2; j_1 j_2\; m_1 m_2| \mathrm{e}^{-i\varphi_1\, \boldsymbol{J}_1\cdot \boldsymbol{u}_1/\hbar}\; \mathrm{e}^{-i\varphi_2\, \boldsymbol{J}_2\cdot \boldsymbol{u}_2/\hbar}\; |\tau_1\tau_2; j_1 j_2\; m_1' m_2'\rangle \\ &\quad = \left(R^{[j_1]}_{\boldsymbol{u}_1}(\varphi_1)\right)_{m_1 m_1'} \left(R^{[j_2]}_{\boldsymbol{u}_2}(\varphi_2)\right)_{m_2 m_2'}\end{aligned} \tag{VII-71b}$$

The group product is represented by matrices that are the tensor product of $(R^{[j_1]})$ and $(R^{[j_2]})$.

This representation is irreducible as we now show. A subspace $\mathscr{E}(\tau_1, \tau_2\,;\, j_1, j_2)$ that is globally invariant under the action of all the operators of the product group should be, in particular, invariant under the action of the infinitesimal operators (associated with any $\delta\varphi_1$ and $\delta\varphi_2$). It shoud therefore be invariant under the action of $J_{1_\pm}$ and $J_{2_\pm}$. Now, any ket of the basis $|\tau_1\tau_2\,;\, j_1 j_2\; m_1 m_2\rangle$ that generates $\mathscr{E}(\tau_1\tau_2\,;\, j_1 j_2)$ can be obtained by the repeated action, on any ket of the same basis, of $J_{1_\pm}$ and $J_{2_\pm}$. In a general way, any ket $|\psi_{\tau_1\tau_2 j_1 j_2}\rangle$ of $\mathscr{E}(\tau_1\tau_2\,;\, j_1 j_2)$ can yield $|\tau_1\tau_2\,;\, j_1 j_2 j_1 j_2\rangle$ by the repeated action of operators J_{1_+} and J_{2_+}, and then any basis ket can be obtained by the repeated action of J_{1_-} and J_{2_-}. Consequently, $\mathscr{E}(\tau_1\tau_2\,;\, j_1 j_2)$ does not contain any invariant subspace under the action of the tensor product group whose elements are given in (VII-71).

C-1-b. Same rotations of the two sub-systems

Instead of discussing independent rotations of the two sub-systems $\mathscr{S}_1$ and $\mathscr{S}_2$, we are now interested in the rotation of the global system formed by $\mathscr{S}_1$ and $\mathscr{S}_2$ considered as a whole. The relevant operators (VII-71) are those where:

$$\boldsymbol{u}_1 = \boldsymbol{u}_2 = \boldsymbol{u} \qquad \text{and} \quad \varphi_1 = \varphi_2 = \varphi \tag{VII-72}$$

and they are written :

$$\begin{aligned} R_{\boldsymbol{u}}(\varphi) &= \exp\left\{-\frac{i}{\hbar}\,\varphi\,\boldsymbol{J}_1\cdot\boldsymbol{u}\right\}\,\exp\left\{-\frac{i}{\hbar}\,\varphi\,\boldsymbol{J}_2\cdot\boldsymbol{u}\right\} \\ &= \exp\left\{-\frac{i}{\hbar}\,\varphi\,\boldsymbol{J}\cdot\boldsymbol{u}\right\} \end{aligned} \tag{VII-73}$$

where:

$$\boldsymbol{J} = \boldsymbol{J}_1 + \boldsymbol{J}_2 \tag{VII-74}$$

They form a subgroup of the tensor product group considered above. Taking into account equalities (VII-72), it is clear that the matrices (VII-71a) yield a representation of the group of operators (VII-73). We thus have directly, in space $\mathscr{E}_1 \otimes \mathscr{E}_2$, new representations of the $R_{(3)}$ group, with dimension $(2j_1+1)(2j_2+1)$:

$$\begin{aligned} &R_{\boldsymbol{u}}(\varphi)|\tau_1\tau_2\,;\, j_1 j_2\; m_1 m_2\rangle = \\ &\sum_{m_1' m_2'} \left(R_{\boldsymbol{u}}^{[j_1]}(\varphi)\right)_{m_1' m_1} \left(R_{\boldsymbol{u}}^{[j_2]}(\varphi)\right)_{m_2' m_2} |\tau_1\tau_2\,;\, j_1 j_2\; m_1' m_2'\rangle \end{aligned} \tag{VII-75}$$

The matrix elements of the representations are given by:

$$\left(R_{\boldsymbol{u}}(\varphi)\right)_{m_1' m_2'\,;\, m_1 m_2} = \left(R_{\boldsymbol{u}}^{[j_1]}(\varphi)\right)_{m_1' m_1} \left(R_{\boldsymbol{u}}^{[j_2]}(\varphi)\right)_{m_2' m_2} \tag{VII-76}$$

The question is whether this representation is irreducible or not. The answer is not obvious since the problem is no longer to diagonalize simultaneously by block all the operators (VII-71a) (we already know it is not possible) but only those obeying relation (VII-72). If we can reduce this representation, it means there exists another basis of $\mathscr{E}(\tau_1\tau_2\,;\,j_1j_2)$ where the matrix can be written:

$$(R_{\boldsymbol{u}}(\varphi))' = \underbrace{\left(\begin{array}{c|c|c|c} \text{/////} & 0 & 0 & 0 \\ \hline 0 & \text{/////} & 0 & 0 \\ \hline 0 & 0 & \text{/////} & 0 \\ \hline 0 & 0 & 0 & \text{/////} \end{array}\right)}_{\mathscr{E}(\tau,\, J)} \qquad \text{(VII-77)}$$

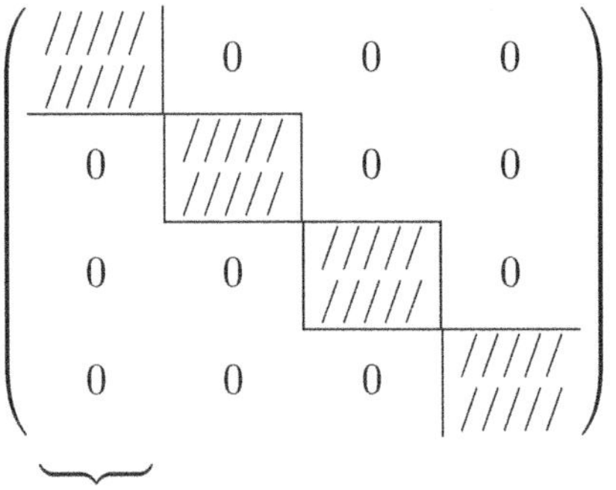

with :

$$\mathscr{E}(\tau_1\tau_2\,;\,j_1j_2) = \sum_{\substack{\oplus \\ \tau,\, J}} \mathscr{E}(\tau, J) \qquad \text{(VII-78)}$$

[we simplify the notation $\mathscr{E}(\tau_1\tau_2\,;\,\tau J)$ into $\mathscr{E}(\tau, J)$].

We are going to show that such a decomposition is actually possible. The problem will be to obtain the new representations (characterized by the values of J), to determine how many times each value of J appears, how to perform the basis change, etc.

C-2. Reduction of the product representation

We won't give any details on the demonstrations of this paragraph as they can be found in elementary quantum mechanics books [see [9], for example, chapter X, § C].

C-2-a. Eigenvalues of J^2 and $|J, M\rangle$ kets

It readily appears that the product representation is in general reducible. If this was not the case, J could take only one value such that $2J+1 = (2j_1+1)(2j_2+1)$ and each eigenvalue $M\hbar$ of the operator:

$$J_z = J_{1_z} + J_{2_z} \qquad \text{(VII-79a)}$$

would occur once and only once (between $+J$ and $-J$). Now we know (unless j_1 or j_2 is zero) that one can find different combinations of m_1 and m_2 leading to the same sum:

$$M = m_1 + m_2 \tag{VII-79b}$$

The same reasoning easily gives the possible values of J. With each irreducible representation in $\mathscr{E}(\tau, J)$ is associated once and only once any values of M equal to $J, J-1, J-2, \ldots, -J$. As a result the value $J = j_1 + j_2$ occurs once and only once, since there is only one ket such that $M = m_1 + m_2 = j_1 + j_2$. As for the value $M = j_1 + j_2 - 1$, it is 2-fold degenerate $[m_1 = j_1 - 1,\ m_2 = j_2$ or $m_1 = j_1,\ m_2 = j_2 - 1]$. Since the maximum value of $J = j_1 + j_2$ implies the existence of a ket $M = j_1 + j_2 - 1$, the value $J = j_1 + j_2 - 1$ necessarily exists. The argument is continued in a similar way, leading to the well-known result that the possible values of J are:

$$J = \begin{cases} j_1 + j_2 \\ j_1 + j_2 - 1 \\ j_1 + j_2 - 2 \\ \vdots \\ \\ |j_1 - j_2| + 1 \\ |j_1 - j_2| \end{cases} \tag{VII-80}$$

Each value of J occurring only once, the index τ in (VII-78) is no longer necessary[11]. We can check that the sum of dimensions of the spaces $\mathscr{E}(J)$ is indeed equal to that of the initial space:

$$\sum_{J=|j_1-j_2|}^{j_1+j_2} (2J+1) = (2j_1+1)\,(2j_2+1) \tag{VII-81}$$

As a result $\boldsymbol{J}^2$ and J_z form a CSCO in $\mathscr{E}(\tau_1\tau_2\,;\,j_1j_2)$. We already have the set of the two operators J_{1_z} and J_{2_z}. How should we go from the basis $|\tau_1\tau_2\,;\,j_1j_2m_1m_2\rangle$ associated with the second CSCO (from now on, the kets will be noted $|j_1j_2m_1m_2\rangle$, ignoring the indices no longer useful) to the $|J, M\rangle$ basis associated with the first one?

The above reasoning shows that we can set (we no longer write the indices τ_1 and τ_2 in the eigenkets, since they don't play any role):

$$|J = j_1 + j_2,\ M = m_1 + m_2\rangle = |j_1, j_2, m_1 = j_1, m_2 = j_2\rangle \tag{VII-82}$$

[11]It so happens that, in the case of the rotation group, the decomposition of a tensor product of irreducible representations yields each irreducible representation only once. This is not always the case for other groups.

Through the action of:

$$J_- = J_{1-} + J_{2-} \tag{VII-83}$$

we can build all the other kets:

$$|J = j_1 + j_2, M\rangle$$

of the same "family" labeled by the value $j_1 + j_2$ of J. This yields the first $2j_1 + 2j_2 + 1$ kets of the new basis spanning $\mathscr{E}(J = j_1 + j_2)$.

We then reason in the complement of this subspace where the eigenvalue $M = j_1 + j_2 - 1$ occurs once (it is non-degenerate) instead of twice in the initial space. The reasoning is just the same, except that we now have to choose the phase of one the basis kets (the action of $J_\pm$ then fixing the phase of the others). By convention we set:

$$\langle j_1 j_2 j_1\, J - j_1 | JJ\rangle \text{ real } > 0 \tag{VII-84}$$

It is clear that we can continue the reasoning and obtain all the $|J, M\rangle$ kets in this way, forming what is called the "coupled" basis.

To summarize, we have written the decomposition (VII-78) of the subspace $\mathscr{E}(\tau_1\tau_2\,;\, j_1 j_2)$ in the form:

$$\mathscr{E}\,(\tau_1\tau_2\,;\, j_1 j_2) = \sum_{J=|j_1-j_2|}^{j_1+j_2} \mathscr{E}(J) \tag{VII-85a}$$

Each $\mathscr{E}(J)$ is generated by the $2J+1$ kets $|J, M\rangle$ of the standard basis associated with the corresponding value of J and to $M = J, J-1, J-2, \ldots, -J$. The index τ of (VII-78) is no longer needed since each allowed value of J appears only once in the decomposition of the tensor product representation (VII-76). In a symbolic way, one can write:

$$\left(R^{[j_1]}\right) \otimes \left(R^{[j_2]}\right) = \sum_{J=|j_1-j_2|}^{j_1+j_2} \left(R^{[J]}\right) \tag{VII-85b}$$

C-2-b. Clebsch–Gordan coefficients

The Clebsch–Gordan coefficients are, by definition, the scalar products:

$$\langle j_1 j_2 m_1 m_2 | JM\rangle$$

They are the $[(2j_1+1)(2j_2+1)]^2$ coefficients of a unitary matrix of a basis change (the matrix is actually real, hence orthogonal).

(i) By construction, these coefficients are totally independent of the nature of the system under study, as well as of τ_1 and τ_2: they reflect the properties of the rotation group $R_{(3)}$.

(ii) They are real:

$$\langle j_1 j_2 m_1 m_2 | JM \rangle = \langle JM | j_1 j_2 m_1 m_2 \rangle \tag{VII-86}$$

(since the action of $J_\pm$, J_z yields real coefficients).

(iii) Selection rules:

$\langle j_1 j_2 m_1 m_2 | JM \rangle$ is non-zero only if:

- $M = m_1 + m_2$;
- $|j_1 + j_2| \geq J \geq |j_1 - j_2|$ (triangle law *cf.* figure 6);
- J is an integer if j_1 and j_2 are both integers or both half-integers;
- J is a half-integer if one and only one of the two quantum numbers (j_1, j_2) is a half-integer.

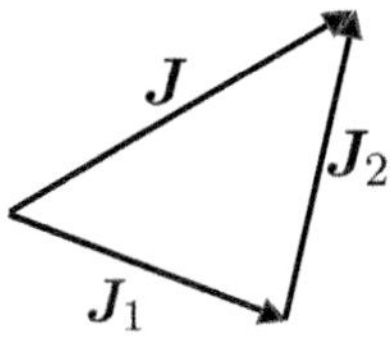

Figure 6: Triangle law: for a Clebsch–Gordan coefficient $\langle j_1 j_2 m_1 m_2 | JM \rangle$ to be different from zero, one must be able to build a triangle with the lengths of the angular momenta J_1, J_2, and J.

(iv) Orthogonality relations:

Since the two bases $\{|J, M\rangle\}$ and $\{|j_1 j_2 m_1 m_2\rangle\}$ of the space $\mathscr{E}(\tau_1 \tau_2 \,;\, j_1 j_2)$ are orthonormal, we can write:

$$\sum_{m_1, m_2} \langle JM | j_1 j_2 m_1 m_2 \rangle \, \langle j_1 j_2 m_1 m_2 | J' M' \rangle = \delta_{JJ'} \, \delta_{MM'} \tag{VII-87a}$$

and:

$$\sum_{J,M} \langle j_1 j_2 m_1 m_2 | JM \rangle \, \langle JM | j_1 j_2 m_1' m_2' \rangle = \delta_{m_1 m_1'} \, \delta_{m_2 m_2'} \tag{VII-87b}$$

(v) Relation to the rotation matrices. We have:

$$\left(R^{[j_1]}\right)_{m_1 m_1'} \left(R^{[j_2]}\right)_{m_2 m_2'} = \tag{VII-88a}$$

$$\sum_{J=|j_1-j_2|}^{J=j_1+j_2} \sum_{M,M'} \langle j_1 j_2 m_1 m_2 | JM \rangle \langle j_1 j_2 m_1' m_2' | JM' \rangle \left(R^{[J]}\right)_{MM'} \tag{VII-88b}$$

and, conversely:

$$\left(R^{[J]}\right)_{MM'} = \sum_{m_1 m_1'} \sum_{m_2 m_2'} \langle j_1 j_2 m_1 m_2 | JM \rangle \langle j_1 j_2 m_1' m_2' | JM' \rangle$$

$$\left(R^{[j_1]}\right)_{m_1 m_1'} \left(R^{[j_2]}\right)_{m_2 m_2'} \tag{VII-88c}$$

Exercise: To prove the above equalities, compute the matrix element:

$$\langle j_1 j_2 m_1 m_2 | \mathrm{e}^{-i\varphi \boldsymbol{J}\cdot\boldsymbol{u}/\hbar} | j_1 j_2 m_1' m_2' \rangle$$

by introducing rotation matrices and using the relation $\boldsymbol{J} = \boldsymbol{J}_1 + \boldsymbol{J}_2$. Developing the kets $|j_1 j_2 m_1 m_2\rangle$ and $|j_1 j_2 m_1' m_2'\rangle$ on the basis $|J, M\rangle$, establish equation (VII-87a). To prove equation (VII-87b), start from:

$$\langle J, M | \mathrm{e}^{-i\varphi \boldsymbol{J}\cdot\boldsymbol{u}/\hbar} | J, M' \rangle$$

and develop the kets $|J, M\rangle$ on the basis $|j_1 j_2 m_1 m_2\rangle$.

Comment:

The following relations can be established (see for example reference [9], complement B_X) – note that the first one has been set by convention in (VII-84):

$$\langle j_1 j_2 j_1 \, (J - j_1) \, | J, J \rangle \quad \text{real} > 0$$
$$\langle j_1 j_2 j_1 \, (M - j_1) \, | J, M \rangle \quad \text{real} > 0$$
$$\langle j_2 j_1 m_2 m_1 | J, M \rangle = (-1)^{j_1+j_2-J} \langle j_1 j_2 m_1 m_2 | J, M \rangle$$
$$\langle j_1 j_2 - m_1 - m_2 | J, -M \rangle = (-1)^{j_1+j_2-J} \langle j_1 j_2 m_1 m_2 | J, M \rangle \tag{VII-89}$$

to which we add the recurrence relations:

$$\sqrt{J(J+1) - M(M \pm 1)} \, \langle j_1 j_2 m_1 m_2 | J, M \pm 1 \rangle =$$
$$\sqrt{j_1(j_1+1) - m_1(m_1 \mp 1)} \, \langle j_1 j_2 (m_1 \mp 1) \, m_2 | JM \rangle$$
$$+ \sqrt{j_2(j_2+1) - m_2(m_2 \mp 1)} \, \langle j_1 j_2 \, m_1 (m_2 \mp 1) \, | JM \rangle \tag{VII-90}$$

Using these relations, we get:

$$\langle j\, j\, m - m|0,0\rangle = \frac{(-1)^{j-m}}{\sqrt{2j+1}} \tag{VII-91}$$

This last equality gives the coefficients corresponding to an angular momentum $J = 0$ (with $j_1 = j_2 = j$). It will be useful in what follows.

Exercises:

(i) Establish relations (VII-90) and (VII-91).

(ii) Compute the coupling coefficients of two 1/2 spins and show that:

$$\begin{aligned}&\langle \tfrac{1}{2}\tfrac{1}{2}{+}\tfrac{1}{2}{+}\tfrac{1}{2}|11\rangle = 1 \qquad ; \qquad \langle \tfrac{1}{2}\tfrac{1}{2}{-}\tfrac{1}{2}{-}\tfrac{1}{2}|1-1\rangle = 1\\ &\langle \tfrac{1}{2}\tfrac{1}{2}{+}\tfrac{1}{2}{-}\tfrac{1}{2}|10\rangle = \langle \tfrac{1}{2}\tfrac{1}{2}{-}\tfrac{1}{2}{+}\tfrac{1}{2}|10\rangle = \frac{1}{\sqrt{2}}\end{aligned} \tag{VII-92}$$

(iii) Compute explicitly the Clebsch–Gordan coefficients:

$\langle j_1 = 1,\, j_2 = 1\, m_1 m_2 | J, M\rangle$.

Complement A_{VII}

Rotation of a spin 1/2 and SU(2) matrices

The density matrix of a spin 1/2 is a 2×2 matrix, which is Hermitian and has a trace equal to unity. The unitary transformations acting on such matrices are described by 2×2 unitary matrices, with a determinant equal to 1. These matrices are those of the 3-dimensional SU(2) group (§ B-3 in chapter III). This complement will show that these unitary transformations amount to simple rotations of the spin enabling us to associate a simple physical image with each SU(2) matrix.

More precisely, we are going to show that there is an homomorphic mapping of the SU(2) group onto the SO(3) group – remember that this last group is formed by the rotation matrices in ordinary space [$R_{(3)}$ group], i.e. the 3×3 matrices that are orthogonal (unitary and real) and unimodular (determinant +1). In other words, it is possible to establish a correspondence between each (U) matrix of SU(2) and a SO(3) matrix so that to the matrix product of the (U) in SU(2) corresponds the matrix product in SO(3). This is not a one-to-one correspondence since two opposite (U) matrices correspond to the same rotation: there is simply an homomorphism between the two groups, and not an isomorphism.

Comment:

The SU(2) group is a subgroup of the matrices of the SL(2, C) group, studied in complement A_V. These matrices are not necessarily unitary, and form a 6-dimensional group. As we saw in that complement, they are associated with the transformations of the Lorentz group, which is also 6-dimensional. We also noted that the SL(2, C) matrices corresponding to rotations were unitary, hence belonging to SU(2). The aim of this complement is to discuss this correspondence in more detail in the context of the properties of a spin 1/2.

1. Modification of a spin 1/2 polarization induced by an SU(2) matrix

Consider any vector $\boldsymbol{P}$ in the real 3-dimensional space, and note P_x, P_y, and P_z its components. We can associate with it a 2×2 Hermitian ρ matrix, defined as:

$$\begin{aligned}\rho &= \frac{1}{2}\left(\mathbb{1} + \boldsymbol{\sigma}\cdot\boldsymbol{P}\right)\\ &= \frac{1}{2}\left(\mathbb{1} + \sigma_x P_x + \sigma_y P_y + \sigma_z P_z\right)\end{aligned} \tag{1}$$

where $\mathbb{1}$ is the 2×2 identity matrix, and σ_x, σ_y, σ_z are the three Pauli matrices [definitions (III-72)]. The ρ matrix can be interpreted[1] as the density matrix of a spin 1/2 (complement E_{IV} in reference [9]), subject to the condition that $P = \sqrt{P^2} < 1$. The average value of this spin is $\langle \boldsymbol{S}\rangle = (\hbar/2)\boldsymbol{P}$, and $\boldsymbol{P}$ is called the "Bloch vector", or the "spin polarization". More generally, ρ can describe the state of any two-level quantum system.

The Pauli matrices have well-known properties:

$$\sigma_j^2 = 1 \qquad j = x, y, z \tag{2a}$$

$$\text{Tr}\,\{\sigma_j\} = 0 \tag{2b}$$

$$(\boldsymbol{\sigma}\cdot\boldsymbol{A})\,(\boldsymbol{\sigma}\cdot\boldsymbol{B}) = \boldsymbol{A}\cdot\boldsymbol{B}\,\mathbb{1} + i\boldsymbol{\sigma}\cdot(\boldsymbol{A}\times\boldsymbol{B}) \tag{2c}$$

where $\boldsymbol{A}$ and $\boldsymbol{B}$ are two given constant vectors. It follows that:

$$\boldsymbol{P} = \langle\boldsymbol{\sigma}\rangle = \text{Tr}\,\{\rho\boldsymbol{\sigma}\} \tag{3}$$

which shows that the vector $\boldsymbol{P}$ indicates the direction of the spin. In addition:

$$\rho^2 = \frac{1}{4}\left(1 + 2\boldsymbol{\sigma}\cdot\boldsymbol{P} + \boldsymbol{P}^2\right) \tag{4}$$

so that:

$$\text{Tr}\left\{\rho^2\right\} = \frac{1}{2}\left(1 + \boldsymbol{P}^2\right) \tag{5}$$

For a spin, the density matrix ρ represents a pure state if the modulus of $\boldsymbol{P}$ equals one; at the other extreme, if this modulus is zero, the spin is not polarized. It is easy to show that when $\boldsymbol{P}$ is real and arbitrary, equality (1) represents the most general 2×2 Hermitian matrix having a trace equal to one.

[1]This interpretation of ρ as a density matrix is the reason for the presence of the term $\mathbb{1}$ in (1). In what follows, one could easily suppress that diagonal term without changing the reasoning; the set of ρ matrices would then be the set of 2×2 Hermitian matrices with a trace equal to zero (which is still a 3-dimensional vector space).

The general expression of the matrices $M(\boldsymbol{a})$ of the SU(2) group is given in (III-77). Any $M(\boldsymbol{a})$ matrix of SU(2) allows associating with any ρ matrix another ρ' matrix through the relation:

$$\rho' = M(\boldsymbol{a})\,\rho M^{\dagger}(\boldsymbol{a}) \tag{6}$$

Like the ρ matrix, ρ' is Hermitian and has a trace equal to one. It can thus be decomposed as:

$$\rho' = \frac{1}{2}\left(1 + \boldsymbol{\sigma}\cdot\boldsymbol{P}'\right) \tag{7}$$

where $\boldsymbol{P}'$ is real. The correspondence between ρ and ρ' given by the matrix $M(\boldsymbol{a})$ can also be considered as a correspondence $\mathscr{R}$ between two vectors $\boldsymbol{P}$ and $\boldsymbol{P}'$ in 3-dimensional space:

$$\begin{array}{ccccccc} \boldsymbol{P} & \Longrightarrow & \rho & \overset{M(\boldsymbol{a})}{\Longrightarrow} & \rho' = M(\boldsymbol{a})\,\rho M^{\dagger}(\boldsymbol{a}) & \Longrightarrow & \boldsymbol{P}' \\ \downarrow & & & \mathscr{R} & & & \uparrow \end{array} \tag{8}$$

In other words, with any matrix M of SU(2) is associated a transformation of the vector $\boldsymbol{P}$.

2. The transformation is a rotation

$\boldsymbol{P}'$ can be expressed as a function of $\boldsymbol{P}$:

$$\boldsymbol{P}' = \text{Tr}\left\{\rho'\boldsymbol{\sigma}\right\} = \text{Tr}\left\{M\left(\frac{1+\boldsymbol{\sigma}\cdot\boldsymbol{P}}{2}\right)M^{\dagger}\boldsymbol{\sigma}\right\} \tag{9}$$

which shows that $\boldsymbol{P}'$ is a linear function of $\boldsymbol{P}$. Since the traces of the Pauli matrices are zero, $\boldsymbol{P} = 0$ corresponds to $\boldsymbol{P}' = 0$. We can thus write:

$$\boldsymbol{P}' = (\mathscr{R})\boldsymbol{P}$$

where $(\mathscr{R})$ is a 3×3 matrix. In addition:

$$\begin{aligned} \text{Tr}\left\{\rho'^2\right\} &= \text{Tr}\left\{M\rho M^{\dagger}\, M\rho M^{\dagger}\right\} \\ &= \text{Tr}\left\{M\rho^2 M^{\dagger}\right\} = \text{Tr}\left\{\rho^2\right\} \end{aligned} \tag{10}$$

(since the trace is invariant under a circular permutation of operators), and, using relation (5):

$$\text{Tr}\left\{\rho'^2\right\} = \frac{1}{2}\left(1 + \boldsymbol{P}^2\right) \tag{11}$$

Consequently:

$$\boldsymbol{P}'^2 = \boldsymbol{P}^2 \tag{12}$$

meaning that the linear transformation of $\boldsymbol{P}$ into $\boldsymbol{P}'$ preserves the vector length[2].

The transformation is thus, either a rotation around the origin, or the product of a rotation and a symmetry with respect to the origin. Actually, this second possibility must be rejected as we now explain. The two possibilities can be distinguished by the fact that the determinant of $(\mathscr{R})$ equals $+1$ in the first case (pure rotation), and -1 in the second (product of a rotation and a point symmetry). Choosing for M the identity matrix $\mathbb{1}$, we have $\boldsymbol{P}' = \boldsymbol{P}$ and the corresponding determinant equals 1. Now, any matrix M of SU(2) can be obtained from matrix $\mathbb{1}$ by a continuous variation of the parameters, and the determinant of $(\mathscr{R})$ cannot jump discontinuously between $+1$ and -1. Consequently, $(\mathscr{R})$ is a rotation matrix, i.e. a SO(3) matrix.

3. Homomorphism

Consider two matrices M and M' of SU(2). Applying to ρ first the unitary transformation M, followed by the transformation M', we get:

$$\rho'' = M'\rho' M'^{\dagger} = M'M\rho M^{\dagger} M'^{\dagger} \tag{13}$$

Since:

$$(M'M)^{\dagger} = M^{\dagger} M'^{\dagger} \tag{14}$$

ρ'' is the density matrix obtained directly by the unitary transformation associated with the product $M'M$ of the matrices. In addition, we have:

$$\boldsymbol{P}'' = (\mathscr{R})'\boldsymbol{P}' = (\mathscr{R}')(\mathscr{R})\boldsymbol{P} \tag{15}$$

which shows that the product $(\mathscr{R}')(\mathscr{R})$ is associated with the product of the corresponding M matrices. This establishes that the correspondence:

$$M \Longrightarrow (\mathscr{R}) \tag{16}$$

is an homomorphism.

It is not a one-to-one correspondence (isomorphism) as we now explain. It is easy to see on (6) that two opposite matrices, M and $-M$, yield the same matrices ρ', and hence the same transformation $\boldsymbol{P} \Longrightarrow \boldsymbol{P}'$. We are going to show

[2]If ρ is interpreted as the density matrix of a spin 1/2, this corresponds to the fact that the unitary transformation M does not change the degree of polarization of the spin.

that, conversely, if two matrices M_1 and M_2 correspond to the same rotation $(\mathscr{R})$, they are either equal or opposite.

Demonstration: Imagine that two matrices M_1 and M_2 are associated with the same rotation $(\mathscr{R})$ and transform any matrix ρ (Hermitian and with a trace equal to unity) into the same matrix ρ':

$$M_1\rho M_1^\dagger = M_2\rho M_2^\dagger \tag{17}$$

This leads to:

$$M_2^\dagger M_1 \, \rho M_1^\dagger M_2 = \rho \tag{18}$$

regardless of the chosen ρ matrix. We call M_0 the most general matrix of SU(2) that corresponds to the identity operation for the vector $\boldsymbol{P}$. We see that:

$$M_2^\dagger M_1 = M_0 \tag{19a}$$

or:

$$M_1 = M_2 M_0 \tag{19b}$$

The set of matrices M_1, associated with the same rotation $(\mathscr{R})$ as a given matrix M_2, is obtained by multiplying that matrix by the set of matrices M_0 associated with the identity rotation.

The unitary transformation M changes the system of the two orthonormal eigenvectors $|\chi_1\rangle$ and $|\chi_2\rangle$ of ρ into the system of the two orthonormal eigenvectors $|\chi_1'\rangle$ and $|\chi_2'\rangle$ of ρ'. We can write:

$$\rho\,|\chi_\ell\rangle = p_\ell\,|\chi_l\rangle \qquad \ell = 1,2 \tag{20}$$

where p_ℓ is real. This leads to:

$$\begin{aligned} \rho' M\,|\chi_\ell\rangle &= M\,\rho M^\dagger\, M\,|\chi_\ell\rangle \\ &= M\rho\,|\chi_\ell\rangle = p_\ell\, M\,|\chi_\ell\rangle \end{aligned} \tag{21}$$

If the two eigenvalues p_1 and p_2 are non-degenerate, we necessarily have:

$$|\chi_l'\rangle = \mathrm{e}^{i\alpha_l} M\,|\chi_l\rangle \qquad l = 1,2 \tag{22}$$

(where α_l is real) and the eigenvalues of ρ' are the same as those of ρ.

We now show that the set of the M_0 matrices only includes the identity matrix $(\mathbb{1})$ and its opposite. The eigenvectors $|\chi_1\rangle$ and $|\chi_2\rangle$ of ρ form an arbitrary orthonormal basis of the 2-dimensional space where the M matrices act; only a scalar matrix [proportional to $(\mathbb{1})$] can accept any of these arbitrary bases as an eigenbasis. Consequently:

$$M_0 = c\,(\mathbb{1}) \tag{23}$$

where c is a complex constant. As M_0 is unitary, $|c| = 1$ and $c = \mathrm{e}^{i\beta}$, where β is real. This leads to:

$$\det\, M_0 = \mathrm{e}^{2i\beta} \det\,(\mathbb{1}) = \mathrm{e}^{2i\beta} \tag{24}$$

which is equal to 1 only if $\beta = n\pi$ (where n is an integer). We thus finally get:

$$M_0 = \pm(\mathbb{1}) \tag{25}$$

In conclusion, only equal or opposite matrices of SU(2) correspond to the same rotation ($\mathscr{R}$). Conversely, a given rotation corresponds to two matrices, M and $-M$, of SU(2). These two matrices belong, respectively, to the subsets noted $\mathrm{SU}(2)_+$ and $\mathrm{SU}(2)_-$ in § B-3 of chapter III.

4. Link with the chapter VII discussion

We now show that the correspondence (16) between unitary matrices M and rotations $\mathscr{R}$ is the same as the one we obtained in chapter VII, § A-2-b. This will enable us to show that this correspondence allows obtaining all the rotation matrices, which is likely since all the SU(2) as well as the SO(3) matrices depend on three parameters.

We first take an infinitesimal parameter $\delta\boldsymbol{a} = \boldsymbol{u}\delta a$ in relation (III-74), which yields:

$$M(\delta\boldsymbol{a}) = (\mathbb{1}) - i\frac{\delta a}{2}\left[u_x\sigma_x + u_y\sigma_y + u_z\sigma_z\right] \tag{26}$$

This equality allowed introducing σ_x, σ_y, σ_z as operators of the Lie algebra of SU(2). Relation (6) becomes:

$$\begin{aligned}\rho' = M(\delta\boldsymbol{a})\,\rho\,M^\dagger(\delta\boldsymbol{a}) &= \rho - i\frac{\delta a}{2}\left[u_x\sigma_x + u_y\sigma_y + u_z\sigma_z\ ,\ \rho\right]\\ &= \rho - i\frac{\delta a}{4}\left[u_x\sigma_x + u_y\sigma_y + u_z\sigma_z\ ,\ \boldsymbol{\sigma}\cdot\boldsymbol{P}\right]\end{aligned} \tag{27}$$

Expanding the product $\boldsymbol{\sigma}\cdot\boldsymbol{P}$ and using relation:

$$[\sigma_x, \sigma_y] = 2i\sigma_z \tag{28}$$

(as well as those obtained by circular permutation of the x, y, and z indices), we get:

$$\begin{aligned}\rho' - \rho = \frac{\delta a}{2}\Big[(u_yP_z - u_zP_y)\,\sigma_x + (u_zP_x - u_xP_z)\,\sigma_y&\\ + (u_xP_y - u_yP_x)\,\sigma_z\Big]&\end{aligned} \tag{29}$$

which leads to:

$$\boldsymbol{P}' - \boldsymbol{P} = \delta a\,\boldsymbol{u}\times\boldsymbol{P} \tag{30}$$

This is indeed the relation yielding an infinitesimal rotation of vector $\delta\boldsymbol{a}$. Consequently, to each infinitesimal matrix (VII-25), represented by the parameter $\delta\boldsymbol{a}$, corresponds a rotation. of the same parameter.

Integrating these infinitesimal operations in SU(2) and SO(3), we can check that there is a correspondence between any vector $\boldsymbol{a}$ (with $|a| < \pi$) defining a matrix M and a rotation with the same parameter $\boldsymbol{a}$. Note in particular that all the rotation matrices can be obtained that way.

Comment:

It is easy to reason on matrices associated with finite values of $\boldsymbol{a}$. Choosing in (III-68) a vector $\boldsymbol{u}$ parallel to Oz, we get the matrix:

$$M = \begin{pmatrix} e^{-ia/2} & 0 \\ 0 & e^{ia/2} \end{pmatrix} = e^{-ia\sigma_z/2} \tag{31}$$

so that (6) becomes:

$$\rho' = M\rho M^\dagger = \frac{1}{2}\left[1 + e^{-ia\sigma_z/2}\, \boldsymbol{P}\cdot\boldsymbol{\sigma}\; e^{ia\sigma_z/2}\right] \tag{32}$$

An elementary calculation on 2×2 matrices leads to:

$$\begin{aligned} e^{-ia\sigma_z/2}\,\sigma_x\, e^{ia\sigma_z/2} &= \cos a\,\sigma_x + \sin a\,\sigma_y \\ e^{-ia\sigma_z/2}\,\sigma_y\, e^{ia\sigma_z/2} &= -\sin a\,\sigma_x + \cos a\,\sigma_y \\ e^{-ia\sigma_z/2}\,\sigma_z\, e^{ia\sigma_z/2} &= \sigma_z \end{aligned} \tag{33}$$

Inserting these results in ρ', we get:

$$\begin{cases} P'_x = \cos a\, P_x - \sin a\, P_y \\ P'_y = \sin a\, P_x + \cos a\, P_y \\ P'_z = P_z \end{cases} \tag{34}$$

which corresponds to a rotation around Oz by an angle $\varphi = a$.

Choosing now $\boldsymbol{u}$ parallel to Oy, the M matrix is real:

$$M = \begin{pmatrix} \cos(a/2) & -\sin(a/2) \\ \sin(a/2) & \cos(a/2) \end{pmatrix} = e^{-ia\sigma_y/2} \tag{35}$$

A similar calculation yields:

$$\begin{aligned} P'_x &= \sin a\, P_z + \cos a\, P_y \\ P'_y &= P_y \\ P'_z &= \cos a\, P_z - \sin a\, P_y \end{aligned} \tag{36}$$

This corresponds to a rotation ($\mathscr{R}$) around Oy by an angle $\theta = a$.

Finally, we just have to use relation (16) to establish that any rotation $(\mathscr{R})$ can be obtained by the product of rotations around Oz and Oy.

5. Link with double-valued representations

We have established a correspondence (not bijective) between each matrix M of SU(2) and each rotation matrix $(\mathscr{R})$. If $\boldsymbol{a}$ belongs to the sphere of radius π accessible in the case of rotations [the matrix M belongs to $\text{SU}(2)_+$], the correspondence is made through the same value of a. On the other hand, if $\pi < a < 2\pi$ [the matrix M is in $\text{SU}(2)_-$], we set:

$$\boldsymbol{a}' = \boldsymbol{a} - 2\pi\frac{\boldsymbol{a}}{a} \tag{37}$$

which brings back the parameter in the π radius sphere, changes M into $-M$, and yields the parameter of the associated rotation. From the point of view of figure 13 of chapter III, it amounts to bringing back all the points of the right sphere [set $\text{SU}(2)_-$] on points having the same position with respect to the center of the left sphere [set $\text{SU}(2)_+$]. In this SU(2) $\Longrightarrow$ SO(3) correspondence, the product of M matrices is always associated with the product of the rotations. This correspondence is schematized by a reversal of the direction of the arrows on figure 1 of chapter VII.

Let us go back to the point of view of the double-valued correspondence SO(3) $\Longrightarrow$ SU(2). Each rotation is the image of a couple of M matrices. Two rotations $\mathscr{R}_1$ and $\mathscr{R}_2$ correspond to two couples $\pm M_1$ and $\pm M_2$ of unitary matrices. It is clear that the product of two rotations corresponds to the couple obtained by the product of the $\pm M_1$ and $\pm M_2$ matrices. This is indeed what we found in § A-3-d of chapter VII. However, if one takes only one of the two M matrices of the couple, the one belonging to $\text{SU}(2)_+$ for example, it may very well happen that the product rotation is associated with the opposite of the product of the corresponding M matrices.

Complement B_{VII}

Addition of more than two angular momenta

This complement introduces two kinds of coefficients that will prove useful in many computations involving rotations: the 3-j and 6-j coefficients. The 3-j coefficients come about when adding three angular momenta so that the resultant is zero. They are equivalent to the Clebsch–Gordan coefficients, but have interesting symmetry properties. The 6-j coefficients characterize different ways of adding three angular momenta, so that the resultant is different from zero; they can be expressed as a function of the 3-j coefficients.

1. Zero total angular momentum; 3-j coefficients

Consider a system $\mathscr{S}_1$ with an angular momentum $\boldsymbol{J}_1$ and call $\mathscr{E}_1$ the (irreducible) space spanned by the $(2j_1+1)$ eigenkets of J_z, noted $|1:m_1\rangle$. For a second system $\mathscr{S}_2$, with an angular momentum $\boldsymbol{J}_2$, $\mathscr{E}_2$ is spanned by the $(2j_2+1)$ kets $|1:m_2\rangle$. In the space $\mathscr{E}_1 \otimes \mathscr{E}_2$ of dimension $(2j_1+1)(2j_2+1)$, the minimum value of the quantum number J of the total angular momentum is $|j_1-j_2|$; it can only be zero if $j_1=j_2=j$ (figure 1).

Using (VII-91), the state with a zero total angular momentum is written:

$$|\psi_0\rangle = \sum_{m=j}^{+j} \frac{(-1)^{j-m}}{\sqrt{2j+1}}\,|1:m\rangle \otimes |2:-m\rangle \tag{1}$$

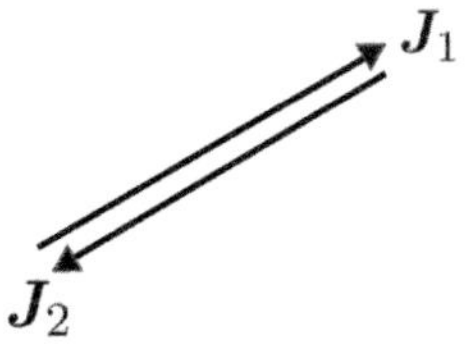

Figure 1: Two angular momenta with equal lengths but opposite directions add up to a zero total angular momentum.

What will happen if we now consider a set of three systems, $\mathscr{S}_1$, $\mathscr{S}_2$, and $\mathscr{S}_3$ with angular momenta $\boldsymbol{J}_1$, $\boldsymbol{J}_2$, and $\boldsymbol{J}_3$? Starting by adding $\boldsymbol{J}_1$ and $\boldsymbol{J}_2$ to obtain $\boldsymbol{J}'$:

$$\boldsymbol{J}_1 + \boldsymbol{J}_2 = \boldsymbol{J}' \tag{2}$$

then adding $\boldsymbol{J}'$ to $\boldsymbol{J}_3$ to obtain the total angular momentum $\boldsymbol{J}$:

$$\boldsymbol{J}' + \boldsymbol{J}_3 = \boldsymbol{J}$$

leads to the following table:

$$\begin{array}{ccl}
J' & & \qquad J \\[1em]
(j_1+j_2) & \longrightarrow & \begin{cases} j_1+j_2+j_3 \\ j_1+j_2+j_3-1 \\ \vdots \\ |j_1+j_2-j_3| \end{cases} \\[1em]
(j_1+j_2-1) & \longrightarrow & \begin{cases} j_1+j_2+j_3-1 \\ \vdots \\ |j_1+j_2-j_3-1| \end{cases} \\[1em]
\vdots & & \\[1em]
|j_1-j_2| & \longrightarrow & \begin{cases} |j_1-j_2|+j_3 \\ \vdots \\ \big||j_1-j_2|-j_3\big| \end{cases}
\end{array}$$

It is clear that most values of J occur several times: $\boldsymbol{J}^2$ and J_z do not form a CSCO. We must also specify the value of J' (hence $\boldsymbol{J}^2$, J_z, and $\boldsymbol{J}'^2$ form a CSCO). There are, however, some values of J that occur only once, for example the maximum value $J = j_1+j_2+j_3$. One cannot state a priori that the minimum value occurs only once[1]. Let us discuss in which cases this minimum value is zero.

For J to be equal to zero, we must have $J' = j_3$, and hence (triangle law):

$$\begin{aligned} &(j_1+j_2) \geq j_3 \geq |j_1-j_2| \\ &j_1+j_2+j_3 \qquad \text{integer} \end{aligned} \tag{3}$$

When the intermediate quantum number J' is fixed, there exists only one possible ket associated with the value $J = 0$ (defined to within a phase factor, assuming the ket is normalized). The situation is schematized in figure 2.

[1] In the case of 3 spins 1/2's for example, the minimum value is $J = 1/2$, and it occurs twice (either for $J' = 0$ or for $J' = 1$).

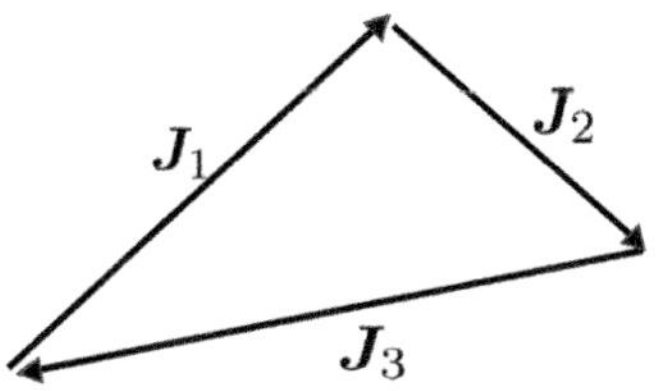

Figure 2: Three angular momenta that add up to zero.

We now build this ket, which we call $|\psi_0\rangle$. Assuming that conditions (3) are met, we add $\boldsymbol{J}_1$ and $\boldsymbol{J}_2$ to obtain the kets:

$$|J' = j_3, M'\rangle = \sum_{m_1,m_2} \langle j_1 j_2 m_1 m_2 | j_3 M'\rangle \; |1 : m_1\rangle \otimes |2 : m_2\rangle \tag{4}$$

where M takes one of the values j_3, $j_3 - 1$, ..., $-j_3$. The ket $|\psi_0\rangle$ is then written:

$$|\psi_0\rangle = (-1)^{j_1 - j_2 - j_3} \sum_{M' + m_3 = 0} \langle j_3 j_3 m_3 M' | 0\, 0\rangle \; |J' = j_3, M'\rangle \otimes |3 : m_3\rangle \tag{5}$$

where the factor $(-1)^{j_1 - j_2 - j_3}$ has been introduced for convenience. Using relations (VII-91) and (4), we get:

$$|\psi_0\rangle = \sum_{m_1 m_2 m_3} \frac{(-1)^{j_1 - j_2 - m_3}}{\sqrt{2 j_3 + 1}} \langle j_1 j_2 m_1 m_2 | j_3 - m_3\rangle$$
$$|1 : m_1\rangle \otimes |2 : m_2\rangle \otimes |3 : m_3\rangle \tag{6}$$

with the constraint $m_1 + m_2 + m_3 = 0$ in the summation. Defining the "$3 - j$ coefficients" as:

$$\begin{pmatrix} j_1 j_2 j_3 \\ m_1 m_2 m_3 \end{pmatrix} = \frac{(-1)^{j_1 - j_2 - m_3}}{\sqrt{2 j_3 + 1}} \langle j_1 j_2 m_1 m_2 | j_3 - m_3\rangle \tag{7}$$

we can write:

$$|\psi_0\rangle = \sum_{\substack{m_1 m_2 m_3 \\ (m_1 + m_2 + m_3 = 0)}} \begin{pmatrix} j_1 j_2 j_3 \\ m_1 m_2 m_3 \end{pmatrix} |1 : m_1\rangle \otimes |2 : m_2\rangle \otimes |3 : m_3\rangle \tag{8}$$

The choice of phase we made for $|\psi_0\rangle$ gives the $3 - j$ coefficients useful symmetry properties. A coefficient $3 - j$ is:

- invariant under a circular permutation of its columns:

$$\begin{pmatrix} j_1 j_2 j_3 \\ m_1 m_2 m_3 \end{pmatrix} = \begin{pmatrix} j_2 j_3 j_1 \\ m_2 m_3 m_1 \end{pmatrix} = \begin{pmatrix} j_3 j_1 j_2 \\ m_3 m_1 m_2 \end{pmatrix} \tag{9a}$$

- multiplied by $(-1)^{j_1+j_2+j_3}$ when two columns are exchanged:

$$\begin{pmatrix} j_2 j_1 j_3 \\ m_2 m_1 m_3 \end{pmatrix} = (-1)^{j_1+j_2+j_3} \begin{pmatrix} j_1 j_2 j_3 \\ m_1 m_2 m_3 \end{pmatrix} \tag{9b}$$

Demonstration: Consider the ket:

$$\begin{aligned} |\psi'_0\rangle &= \sum_{m_1 m_2 m_3} \begin{pmatrix} j_2 j_3 j_1 \\ m_2 m_3 m_1 \end{pmatrix} |1:m_1\rangle \otimes |2:m_2\rangle \otimes |3:m_3\rangle \\ &= \sum_{m_1 m_2 m_3} \frac{(-1)^{j_2-j_3-m_1}}{\sqrt{2j_1+1}} \langle j_2 j_3 m_2 m_3 | j_1 - m_1\rangle \; |1:m_1\rangle \otimes |2:m_2\rangle \otimes |3:m_3\rangle \end{aligned} \tag{10}$$

$|\psi'_0\rangle$ corresponds to the addition of $\boldsymbol{J}_2$ and $\boldsymbol{J}_3$, yielding $\boldsymbol{J}''$:

$$\boldsymbol{J}_2 + \boldsymbol{J}_3 = \boldsymbol{J}'' \tag{11}$$

then to the addition of $\boldsymbol{J}''$ and $\boldsymbol{J}_1$, to form $\boldsymbol{J}$. As the eigenvalue $J = 0$ is non-degenerate, we necessarily have:

$$|\psi'_0\rangle = \lambda |\psi_0\rangle \tag{12}$$

Furthermore, since the two kets are normalized by construction we have $|\lambda| = 1$ and since their coefficients on the kets of the initial basis are real (products and sums of Clebsch–Gordan coefficients) we can only have $\lambda = \pm 1$.

The only remaining problem when comparing $|\psi_0\rangle$ and $|\psi'_0\rangle$ is a question of sign. Consider for example the coefficients of these two kets on the basis vector:

$$|1:j_1\rangle \otimes |2:j_3-j_1\rangle \otimes |3:-j_3\rangle$$

For $|\psi_0\rangle$, equality (6) and convention (VII-84) show that this coefficient has the sign of $(-1)^{j_1-j_2+j_3}$. For $|\psi'_0\rangle$, according to (10) the sign is:

$$(-1)^{j_2-j_3-j_1} \times \text{ sign of coefficient } \langle j_2 j_3 (j_3-j_1)(-j_3) | j_1 - j_1\rangle$$

Now the last two relations (VII-89), and the fact that $j_1+j_2+j_3$ is an integer according to (3), show that the sign of the above mentioned Clebsch–Gordan coefficient is positive. As a result:

$$|\psi_0\rangle = |\psi'_0\rangle \tag{13}$$

and relation (9a) is established.

We now examine what happens if we change the order of j_1 and j_2 (permutation of 1 and 2). This leads to the ket:

$$\begin{aligned}|\psi_0''\rangle &= \sum_{m_1,m_2,m_3} \begin{pmatrix} j_2 j_1 j_3 \\ m_2 m_1 m_3 \end{pmatrix} |1:m_1\rangle \otimes |2:m_2\rangle \otimes |3:m_3\rangle \\ &= \sum_{m_1,m_2,m_3} \frac{(-1)^{j_2-j_1-m_3}}{\sqrt{2j_3+1}} \langle j_2 j_1 m_2 m_1 | j_3 - m_3\rangle\ |1:m_1\rangle \otimes |2:m_2\rangle \otimes |3:m_3\rangle\end{aligned} \tag{14}$$

As before, we know a priori that $|\psi_0''\rangle = \pm|\psi_0\rangle$. But the sign of the coefficient of $|\psi_0''\rangle$ on the ket $|1:j_1\rangle \otimes |2:j_3-j_1\rangle \otimes |3:-j_3\rangle$ is:

$$\begin{aligned}&(-1)^{j_2-j_1+j_3} \times \text{ sign of coefficient } \langle j_2 j_1\ (j_3-j_1)\ j_1 | j_3 j_3\rangle \\ &\qquad = (-1)^{j_2-j_1+j_3}\ (-1)^{j_1+j_2-j_3} = (-1)^{2j_2}\end{aligned} \tag{15}$$

[*cf.* [9], complement B_X, formula (23)]. We thus have:

$$|\psi_0''\rangle = (-1)^{j_1+j_2+j_3}\ |\psi_0\rangle \tag{16}$$

which establishes equality (9b).

These symmetry relations allow the permutation in a Clebsch–Gordan coefficient of either j_1 and m_1, or j_2 and m_2, with J and M. This yields, using (9a), (9b), and definition (7):

$$\begin{aligned}\langle j_1 j_2 m_1 m_2 | J, M\rangle &= (-1)^{j_1-J+m_2} \sqrt{\frac{2J+1}{2j_1+1}}\ \langle J j_2 M - m_2\, | j_1 m_1\rangle \\ &= (-1)^{j_2-J-m_1} \sqrt{\frac{2J+1}{2j_2+1}}\ \langle j_1 J - m_1 M\, | j_2 m_2\rangle\end{aligned} \tag{17}$$

For the $3-j$ coefficients, the orthogonality relations (VII-87) become:

$$\begin{aligned}&\sum_{j_3 m_3} (2j_3+1) \begin{pmatrix} j_1 j_2 j_3 \\ m_1 m_2 m_3 \end{pmatrix} \begin{pmatrix} j_1 j_2 j_3 \\ m_1' m_2' m_3 \end{pmatrix} = \delta_{m_1 m_1'}\, \delta_{m_2 m_2'} \\ &\qquad \sum_{m_1 m_2} \begin{pmatrix} j_1 j_2 j_3 \\ m_1 m_2 m_3 \end{pmatrix} \begin{pmatrix} j_1 j_2 j_3' \\ m_1 m_2 m_3' \end{pmatrix} = \frac{1}{2j_3+1}\, \delta_{j_3 j_3'}\, \delta_{m_3 m_3'}\end{aligned} \tag{18}$$

2. 6-j Wigner coefficients

Given three angular momenta $\boldsymbol{J}_1$, $\boldsymbol{J}_2$, and $\boldsymbol{J}_3$, consider the two ways of adding them together:

$$\begin{cases} \boldsymbol{J}_1 + \boldsymbol{J}_2 = \boldsymbol{G} \\ \boldsymbol{G} + \boldsymbol{J}_3 = \boldsymbol{J} \end{cases} \qquad \begin{cases} \boldsymbol{J}_2 + \boldsymbol{J}_3 = \boldsymbol{G}' \\ \boldsymbol{J}_1 + \boldsymbol{G}' = \boldsymbol{J} \end{cases} \tag{19}$$

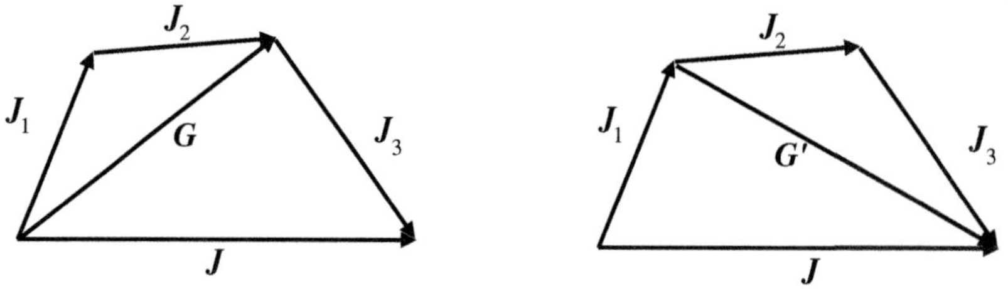

Figure 3: Different ways of adding together three angular momenta.

These two ways are schematized in figure 3, and to each corresponds a different basis of kets, noted respectively:

$$\begin{aligned} &|\,(j_1, j_2)\; g, j_3\,;\; J\,M\rangle = |\varphi^g_{JM}\rangle \\ &|j_1, (j_2, j_3)\; g'\,;\; J'\,M'\rangle = |\psi^{g'}_{J'M'}\rangle \end{aligned} \tag{20}$$

The $|\varphi^g_{JM}\rangle$ and $|\psi^{g'}_{J'M'}\rangle$ kets are necessarily orthogonal, unless $J = J'$ and $M = M'$ (eigenvectors of $\boldsymbol{J}^2$ and J_z with different eigenvalues are orthogonal). As we now show, their scalar product depends on J, g, and g', but is independent of M. Between the bra $\langle\varphi^g_{JM}|$ and the ket $|\psi^{g'}_{JM}\rangle$ we introduce relation (VII-5c):

$$J_- J_+ = \boldsymbol{J}^2 - J_z^2 - \hbar\, J_z$$

Using the fact that, by construction:

$$\begin{aligned} J_\pm|\varphi^g_{JM}\rangle &= \hbar\sqrt{J\,(J+1) - M\,(M\pm 1)}\,|\varphi^g_{JM\pm 1}\rangle \\ \left[\boldsymbol{J}^2 - J_z^2 - J_z\right]|\varphi^g_{JM}\rangle &= \hbar^2\left[J\,(J+1) - M^2 - M\right]|\varphi^g_{JM}\rangle \end{aligned} \tag{21}$$

(with a similar relation for the kets $|\psi^{g'}_{JM}\rangle$), we get:

$$\langle\varphi^g_{JM}|\psi^{g'}_{JM}\rangle = \langle\varphi^g_{JM\pm 1}|\psi^{g'}_{JM\pm 1}\rangle \tag{22}$$

The coefficient $6-j$, noted $\begin{Bmatrix} j_1 & j_2 & g \\ j_3 & J & g' \end{Bmatrix}$, is defined as:

$$\langle\varphi^g_{JM}|\psi^{g'}_{J'M'}\rangle = \delta_{JJ'}\,\delta_{MM'}\sqrt{(2g+1)(2g'+1)}\;(-1)^{j_1+j_2+j_3+J}\begin{Bmatrix} j_1 & j_2 & g \\ j_3 & J & g' \end{Bmatrix} \tag{23}$$

where the factor $(-1)^{j_1+j_2+j_3+J}$ and the square root have been, here again, introduced by convenience to give the coefficients a greater symmetry.
The $6-j$ coefficients have the following properties:

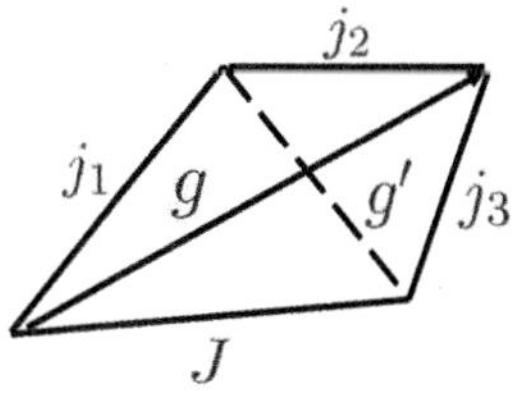

Figure 4: Tetrahedron associated with a $6-j$ coefficient.

(i) they are real;

(ii) they are invariant under any permutation of the columns;

(iii) they are invariant when replacing two elements of the first row by the corresponding elements of the second;

(iv) each of them can be associated with a tetrahydron in space (figure 4). The $6-j$ coefficients depend only on the angular momenta and their position on the tetrahydron;

(v) they are equal to zero unless, for each face of the tetrahedron:

- the sum of the angular momenta is an integer,
- the triangle laws are verified.

To demonstrate these relations, we expand the two vectors (20) on the basis: $|1:m_1\rangle \otimes |2:m_2\rangle \otimes |3:m_3\rangle$. For example:

$$|\varphi^g_{JM}\rangle = \sum_{m_1m_2m_3} (-1)^{j_1-j_2-j_3+m_{12}+g+M'}$$
$$\sqrt{(2g+1)(2J+1)} \begin{pmatrix} j_1 j_2 g \\ m_1 m_2 - m_{12} \end{pmatrix} \begin{pmatrix} g\, j_3\, J \\ m_{12} m_3 - M' \end{pmatrix}$$
$$|1:m_1\rangle \otimes |2:m_2\rangle \otimes |3:m_3\rangle \tag{24}$$

where $m_{12} = m_1 + m_2$. Using this equality in the scalar product on the left-hand side of (23), we get the relation:

$$\delta_{j_3 j_3'}\,\delta_{m_3 m_3'}\,\frac{1}{2j_3+1}\begin{Bmatrix} j_1 & j_2 & j_3 \\ J_1 & J_2 & J_3 \end{Bmatrix} = \sum_{\substack{M_1 M_2 M_3 \\ m_1 m_2}} (-1)^{J_1+J_2+J_3+M_1+M_2+M_3}$$

$$\begin{pmatrix} J_1 & J_2 & j_3 \\ M_1 & -M_2 & m_3 \end{pmatrix}\begin{pmatrix} J_2 & J_3 & j_1 \\ M_2 & -M_3 & m_1 \end{pmatrix}\begin{pmatrix} J_3 & J_1 & j_2 \\ M_3 & -M_1 & m_2 \end{pmatrix}\begin{pmatrix} j_1 & j_2 & j_3' \\ m_1 & m_2 & m_3' \end{pmatrix} \tag{25}$$

This equality (where the summation over 5 indices is reduced to a double summation as a result of the selection rules) leads to the symmetry relations mentioned above.

A great number of useful relations can be established using the various properties of the $3-j$ and $6-j$ coefficients . These relations are not given in the present complement but can be found in the appendix C of volume II, reference [6], or in the book of A.R. Edmonds [44]. There are also numerical tables for the $3-j$ and $6-j$ coefficients [45]. The procedure discussed above can be generalized, allowing the definition of $9-j$, $12-j$, etc. coefficients.

Chapter VIII

Transformation of observables under rotation

Introduction

In the previous chapter, we discussed various properties of the operators $R_{\boldsymbol{u}}(\varphi)$ describing the rotation of the state vector of any physical system. The same operators $R_{\boldsymbol{u}}(\varphi)$ can describe the rotation of the measurement instruments associated with the various observables A of the system (Hermitian operators with a

Introduction to Continuous Symmetries: From Space–Time to Quantum Mechanics, First Edition. Franck Laloë.

complete spectrum). We saw in § C of chapter IV that the operator A' associated with measurement instruments having undergone a rotation by an angle φ around the unit vector $\boldsymbol{u}$ is written:

$$A' = R_{\boldsymbol{u}}(\varphi)\, A\, R^{\dagger}_{\boldsymbol{u}}(\varphi) \tag{VIII-1}$$

where, as we recall:

$$R_{\boldsymbol{u}}(\varphi) = \exp\left\{-\frac{i}{\hbar}\varphi J_u\right\} \tag{VIII-2}$$

In particular, for an infinitesimal rotation (by an angle $\delta\varphi$):

$$A' = A - \frac{i}{\hbar}\delta\varphi\,[J_u\,, A] \tag{VIII-3}$$

It is the commutator of A with the various components J_u of the total angular momentum $\boldsymbol{J}$ that describes the effect of an infinitesimal rotation on each observable.

This chapter is devoted to the study of rotation of observables. Compared with the previous chapter, the main difference is that transformation of observables instead of state vectors involves replacing formulas of the type:

$$|\psi'\rangle = R_{\boldsymbol{u}}(\varphi)|\psi\rangle \tag{VIII-4}$$

by relation (VIII-1), where $R_{\boldsymbol{u}}(\varphi)$ appears twice rather than once. In a similar way, for infinitesimal rotations, instead of having J_u simply act on $|\psi\rangle$, it is the commutator (VIII-3) that is involved. For the rest, we shall simply use the previous reasoning (as well as the previous notations) of chapter VII.

Instead of presenting important and/or new ideas, this chapter discusses applications and useful calculation methods. The concept of vector (§ A) or tensor (see an elementary review of classical tensors in complement A_{VIII}) operators will be clarified. It will lead to introducing (§ B) the definition of "*irreducible tensor operators*". In quantum mechanics, these operators help simplify numerous calculations where angular momentum algebra and rotation invariance play a role. The simplification mainly comes from the *Wigner–Eckart theorem*, which will be demonstrated and discussed (§ C). An important application of these types of computations concerns the electric or magnetic multipolar operators of a system of charges (complement C_{VIII}).

Comment:

Confusing rotation operations on a physical system or on its observables can lead to calculation errors, particularly on the sign of the result. In

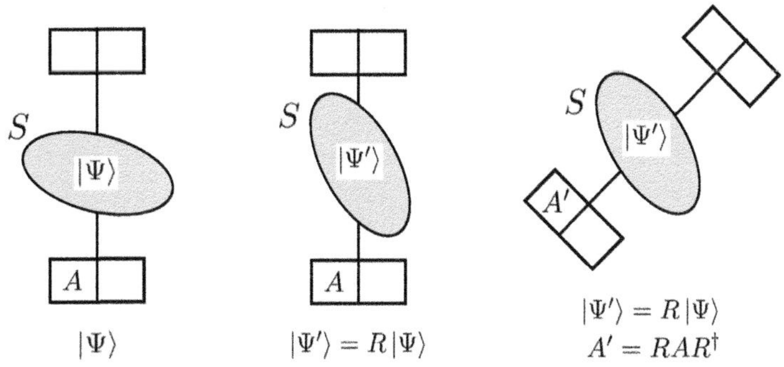

Figure 1: The left-hand side of the figure schematizes a physical system S described by a ket $|\Psi\rangle$ undergoing a measurement associated with operator A. The middle part represents a situation where S has been rotated and is described by the ket $|\Psi'\rangle$. The right-hand side of the figure finally assumes that the physical system and the measurement instrument have undergone the same rotation; the measurements are performed along different axes. Neither the measurement results nor their probabilities can be different from the initial situation.

quantum mechanics, the operation $|\psi'\rangle = R\,|\psi\rangle$ is the analog of a rotation $\mathscr{R}$ of the positions $\boldsymbol{r}$ of a classical system, whereas transforming the observables according to (VIII-1) amounts to changing the axes along which the measurements are performed; it is thus the analog of a change of the coordinate axes. It is not equivalent to perform a rotation $\mathscr{R}$ of each point $\boldsymbol{r}$ in space or to perform the same rotation of the coordinate axes (for the coordinates of $\boldsymbol{r}$ to be transformed according to the rotation $\mathscr{R}$, one must rotate the axes by the inverse operation $\mathscr{R}^{-1}$).

Figure 1 symbolizes the way we obtained (VIII-1), following the reasoning of §§ A and C of chapter IV. We built the observable A', the rotation transform of A, so that all the measurement results are unchanged if one rotates both the physical system and the measurement instruments; this implies that $\langle\psi'|A|\psi'\rangle = \langle\psi|A|\psi\rangle$. The idea is that the possible measurement results and their probabilities depend only on the *relative position* of the measurement instruments and the system under study.

To take a specific example, consider a Stern and Gerlach-type experiment. The measured quantity is the component of a particle's spin along a given direction $\boldsymbol{v}$, and the observable A is written $\boldsymbol{S}\cdot\boldsymbol{v}$ ($\boldsymbol{v}$ is supposed to be a unit vector). Imagine that initially the spin is in the eigenstate $|+\rangle_{\boldsymbol{v}}$ of this observable (eigenvalue $+\hbar/2$). The measurement result is then certain, $+\hbar/2$. If the system undergoes a rotation $R_{\boldsymbol{u}}(\varphi)$, its state becomes $|+\rangle_{\boldsymbol{v}'}$, a state where the spin points in the direction $\boldsymbol{v}'$

given by:

$$\boldsymbol{v}' = \mathscr{R}_{\boldsymbol{u}}(\varphi)\boldsymbol{v} \tag{VIII-5}$$

The ket $|+\rangle_{\boldsymbol{v}'}$ is an eigenket of the component of $\boldsymbol{S}$ on $\boldsymbol{v}'$ (eigenvalue $+\hbar/2$), a component that is precisely the observable A', the transform of A under the rotation of the measurement instruments:

$$(\boldsymbol{S}\cdot\boldsymbol{v})' = \boldsymbol{S}\cdot\boldsymbol{v}' = R_{\boldsymbol{u}}(\varphi)\ (\boldsymbol{S}\cdot\boldsymbol{v})\ R^{\dagger}_{\boldsymbol{u}}(\varphi) \tag{VIII-6}$$

This example illustrates how the measurement instruments define the space axes along which the physical properties of the system are studied – this can also be seen in figure 1. Of course, if one rotates both the physical system (spin) and the measurement instruments (Stern Gerlach magnets), nothing changes: the measurement result remains certain, equal to $+\hbar/2$.

A. Scalar and vector operators

In § B-5 of chapter V, we defined the *scalar operators*, which commute with all the components of the total angular momentum $\boldsymbol{J}$ of the physical system under study. These operators are rotation invariant, and the most common example is the Hamiltonian of an isolated system.

In terms of simplicity concerning their transformations under rotation, the vector operators come directly after the scalar operators. In the Lie algebra of the Galilean and Poincaré groups, all the non-scalar operators are vector operators.

A-1. Definition of a vector operator

In both Galilean and Poincaré groups, the three infinitesimal generators for translation $\boldsymbol{P}$, rotation $\boldsymbol{J}$ and change of Galilean reference frame $\boldsymbol{K}$ obey the same commutation relations:

$$[J_x\ ,\ V_y] = i\hbar\, V_z \qquad \text{and a circular permutation of } x, y, \text{ and } z \tag{VIII-7}$$

or, in a more concise notation:

$$[J_{x_i}\ ,\ V_{x_j}] = i\hbar\, \varepsilon_{ijk} V_{x_k} \quad \text{with: } i,j,k = 1,2,3 \text{ and: } x_{i,j,k} = x,y,z \tag{VIII-8}$$

By definition, any set of three operators obeying these commutation relations with the total angular momentum $\boldsymbol{J}$ is called a vector operator. We noted in (V-28) in chapter V that the scalar product $\boldsymbol{V}\cdot\boldsymbol{W}$ of two vector operators $\boldsymbol{V}$ and $\boldsymbol{W}$ is a scalar operator. We now discuss in more detail the link between the definition of a vector operator and geometric rotations.

A-1-a. Classical vectors, reminders and notations

When the coordinate axes rotate by an angle φ around a vector $\boldsymbol{u}$, the components of the new basis vectors as a function of the old ones are written in the columns of the rotation matrix [*cf.* for example relation (18) of complement C_V]:

$$\left(\boldsymbol{e}'_x \;\; \boldsymbol{e}'_y \;\; \boldsymbol{e}'_z\right) = \left(\boldsymbol{e}_x \;\; \boldsymbol{e}_y \;\; \boldsymbol{e}_z\right) \left(\mathscr{R}_{\boldsymbol{u}}(\varphi)\right) \tag{VIII-9}$$

A given vector $\boldsymbol{r}$ can be expanded on the two bases:

$$\boldsymbol{r} = x\boldsymbol{e}_x + y\boldsymbol{e}_y + z\boldsymbol{e}_z = x'\boldsymbol{e}'_x + y'\boldsymbol{e}'_y + z'\boldsymbol{e}'_z \tag{VIII-10}$$

or:

$$\boldsymbol{r} = \left(\boldsymbol{e}_x \;\; \boldsymbol{e}_y \;\; \boldsymbol{e}_z\right) \begin{pmatrix} x \\ y \\ z \end{pmatrix} = \left(\boldsymbol{e}'_x \;\; \boldsymbol{e}'_y \;\; \boldsymbol{e}'_z\right) \begin{pmatrix} x' \\ y' \\ z' \end{pmatrix} \tag{VIII-11}$$

Inserting relation (VIII-9) on the right-hand side, we can identify the components on the three vectors $\boldsymbol{e}_x$, $\boldsymbol{e}_y$, and $\boldsymbol{e}_z$. This yields:

$$\begin{pmatrix} x \\ y \\ z \end{pmatrix} = \left(\mathscr{R}_{\boldsymbol{u}}(\varphi)\right) \begin{pmatrix} x' \\ y' \\ z' \end{pmatrix} \quad \text{or:} \quad \begin{pmatrix} x' \\ y' \\ z' \end{pmatrix} = \left(\mathscr{R}_{\boldsymbol{u}}(\varphi)\right)^{-1} \begin{pmatrix} x \\ y \\ z \end{pmatrix} \tag{VIII-12}$$

This shows that matrix $(\mathscr{R})$ allows going from the new coordinates to the old ones, and the inverse matrix $(\mathscr{R})^{-1}$ from the old coordinates to the new ones.

Comments:

(i) We have computed the transformation of the coordinates of a given vector $\boldsymbol{v}$ from one basis to a second one, which had undergone a rotation $(\mathscr{R})$ with respect to the first.

Another point of view is to keep the same basis. Relation (VIII-12) then defines the coordinates x, y, and z of a vector $\boldsymbol{v}$, which, under the rotation $(\mathscr{R})^{-1}$, becomes a new vector $\boldsymbol{v}'$ with coordinates x', y', and z':

$$\boldsymbol{v}' = \mathscr{R}_{\boldsymbol{u}}^{-1}(\varphi)\, \boldsymbol{v} \tag{VIII-13}$$

This illustrates again the fact that the coordinates of a vector are transformed the same way, whether the axes undergo a rotation $\mathscr{R}$, or the vector undergoes the inverse rotation $\mathscr{R}^{-1}$.

In the particular case of an infinitesimal rotation by an angle $\delta\varphi$, the vector $\boldsymbol{r}'$ becomes [*cf.* relation (28) of complement C_V]:

$$\boldsymbol{v}' = \boldsymbol{v} + \delta\boldsymbol{v} = \boldsymbol{v} - \delta\varphi\, \boldsymbol{u} \times \boldsymbol{v} \tag{VIII-14}$$

(ii) It is sometimes more convenient to work with the matrix $(\mathscr{R})$ rather than with its inverse. Since the rotation matrices $(\mathscr{R})$ are orthogonal (the inverse matrix is equal to the transpose of the matrix), the second equation (VIII-12) can be written in an equivalent form:

$$(x'\ y'\ z') = (x\ y\ z)\,(\mathscr{R}_{\boldsymbol{u}}(\varphi)) \tag{VIII-15}$$

or:

$$x'_i = \sum_j (\mathscr{R})^{-1}_{ij}\, x_j = \sum_j (\mathscr{R})_{ji}\, x_j \qquad x_i, x_{i'} = x, y, z \tag{VIII-16}$$

A-1-b. Vector operator and geometric rotation

Relation (VIII-8) is not the only way to define a vector operator, we can also use finite rotations.

• A vector operator is a set of three components V_x, V_y, V_z that are transformed, under a rotation $\mathscr{R}$ of the axes of the measurement instrument, as the components of a 3-dimensional classical vector upon a change of coordinate axes. As in relation (VIII-6), rotation $\mathscr{R}$ acts on the directions along which the measurements are performed, and the transformations of the components of a vector operator are given, as in (VIII-12), by:

$$\begin{pmatrix} V'_x \\ V'_y \\ V'_z \end{pmatrix} = \Big(\mathscr{R}_{\boldsymbol{u}}(\varphi)\Big)^{-1} \begin{pmatrix} V_x \\ V_y \\ V_z \end{pmatrix} \tag{VIII-17}$$

with:

$$V'_{x_i} = R_{\boldsymbol{u}}(\varphi)\, V_{x_i}\, R^{\dagger}_{\boldsymbol{u}}(\varphi) \tag{VIII-18}$$

As we saw in (VIII-15), this relation is equivalent to:

$$\Big(V'_x\ V'_y\ V'_z\Big) = \Big(V_x\ V_y\ V_z\Big)\Big(\mathscr{R}_{\boldsymbol{u}}(\varphi)\Big) \tag{VIII-19}$$

The three operators V_x, V_y, and V_z can be regrouped in the vector notation $\boldsymbol{V} = V_x \boldsymbol{e}_x + V_y \boldsymbol{e}_y + V_z \boldsymbol{e}_z$. The components $V'_{x,y,z}$ are not the components of a new operator but, as in (VIII-10), the components of the same vector[1] on axes having undergone a rotation $\mathscr{R}$:

$$\boldsymbol{V} = \sum_i V_{x_i} \boldsymbol{e}_{x_i} = \sum_i V'_{x_i} \boldsymbol{e}'_{x_i} \tag{VIII-20}$$

[1] It is easy to show, using (VIII-9) and (VIII-17) that:

$$\sum_i V'_{x_i} \boldsymbol{e}'_{x_i} = \sum_{ijk} \Big(\mathscr{R}^{-1}_{\boldsymbol{u}}(\varphi)\Big)_{ij} V_{x_j} \Big(\mathscr{R}_{\boldsymbol{u}}(\varphi)\Big)_{ki} \boldsymbol{e}_{x_k} = \sum_{jk} \delta_{jk}\, V_{x_j} \boldsymbol{e}_{x_k} = \boldsymbol{V}$$

where each x_i, x_j, and x_k represents one of the three indices x, y, or z. In the example of the spin measurement in a Stern and Gerlach magnet, it is not the spin that rotates, but the magnet that defines the measurement axes.

• One can also, as in (VIII-13), leave the axes unchanged and consider the V_i' as the components of a new vector $\boldsymbol{V}'$. Taking (VIII-17) and (VIII-18) into account, $\boldsymbol{V}'$ is written:

$$\boldsymbol{V}' = \mathscr{R}_{\boldsymbol{u}}^{-1}(\varphi)\boldsymbol{V} = R_{\boldsymbol{u}}(\varphi)\,\boldsymbol{V}\,R_{\boldsymbol{u}}^{\dagger}(\varphi) \tag{VIII-21}$$

As in (VIII-13), the inverse rotation appears in this relation.

In the particular case of an infinitesimal rotation by an angle $\delta\varphi$, $\boldsymbol{V}'$ becomes, *cf.* (VIII-14):

$$\boldsymbol{V}' = \boldsymbol{V} + \delta\boldsymbol{V} = \boldsymbol{V} - \delta\varphi\,\boldsymbol{u}\times\boldsymbol{V} \tag{VIII-22}$$

The condition for $\boldsymbol{V}$ to be a vector operator is therefore:

$$[J_u, \boldsymbol{V}] = -i\hbar\,\boldsymbol{u}\times\boldsymbol{V} \tag{VIII-23}$$

or else, after taking the scalar product of this relation with a vector $\boldsymbol{w}$:

$$\Big[(\boldsymbol{J}\cdot\boldsymbol{u})\,,\,(\boldsymbol{V}\cdot\boldsymbol{w})\Big] = i\hbar\,(\boldsymbol{u}\times\boldsymbol{w})\cdot\boldsymbol{V} \tag{VIII-24}$$

Projecting onto the three axes, we find again relations (VIII-7), i.e. our initial definition of a vector operator. Conversely, integrating infinitesimal rotations into finite rotations, we obtain (VIII-17). These various definitions of a vector operator are thus equivalent.

A-2. Standard components

In § A-3-c of chapter VII, we saw that the rotation matrices $(R^{[j=1]})$ are simply the usual rotation matrices $(\mathscr{R})$ written in another basis. We are going to look for a basis change that modifies the components of a vector operator $\boldsymbol{V}$ so that they transform as the kets $|j=1,m\rangle$, with $m = 0, \pm 1$.

Starting from the basis $\boldsymbol{e}_x$, $\boldsymbol{e}_y$, and $\boldsymbol{e}_z$ in which the matrices $(\mathscr{R})$ act, we introduce the three vectors:

$$\begin{cases} \boldsymbol{e}_{+1} = -\dfrac{1}{\sqrt{2}}\,[\boldsymbol{e}_x + i\boldsymbol{e}_y] \\ \boldsymbol{e}_0 = \boldsymbol{e}_z \\ \boldsymbol{e}_{-1} = \dfrac{1}{\sqrt{2}}\,[\boldsymbol{e}_x - i\boldsymbol{e}_y] \end{cases} \tag{VIII-25}$$

which form the basis in which the $(\mathscr{R})$ are identical to the $(R^{[j=1]})$. The matrix (S) of this basis change (whose columns contain the components of the new base vectors onto the old ones) is unitary and its inverse is the Hermitian conjugate matrix:

$$(\mathrm{S}) = \begin{pmatrix} -\frac{1}{\sqrt{2}} & 0 & \frac{1}{\sqrt{2}} \\ -\frac{i}{\sqrt{2}} & 0 & -\frac{i}{\sqrt{2}} \\ 0 & 1 & 0 \end{pmatrix} \qquad (S)^{-1} = \begin{pmatrix} -\frac{1}{\sqrt{2}} & \frac{i}{\sqrt{2}} & 0 \\ 0 & 0 & 1 \\ \frac{1}{\sqrt{2}} & \frac{i}{\sqrt{2}} & 0 \end{pmatrix} \tag{VIII-26}$$

The relations for a basis change are given in (VII-36):

$$(R_{\boldsymbol{u}}^{[j=1]}(\varphi)) = (S)^{-1}\,(\mathscr{R}_{\boldsymbol{u}}(\varphi))\,(S) \tag{VIII-27a}$$

$$(\mathscr{R}_{\boldsymbol{u}}(\varphi)) = (S)\,(R_{\boldsymbol{u}}^{[j=1]}(\varphi))\,(S)^{-1} \tag{VIII-27b}$$

Inserting the second of these equalities in (VIII-19), we get:

$$\left(V_x'\;V_y'\;V_z'\right) = (V_x\;V_y\;V_z)\,(S)\left(R_{\boldsymbol{u}}^{[j=1]}(\varphi)\right)(S)^{-1} \tag{VIII-28}$$

The relations for a basis change of a vector components are:

$$(V_{+1}\;V_0\;V_{-1}) = (V_x\;V_y\;V_z)\,(S) \tag{VIII-29a}$$

$$(V_{+1}'\;V_0'\;V_{-1}') = \left(V_x'\;V_y'\;V_z'\right)(S) \tag{VIII-29b}$$

where V_{+1}', V_0', and V_{-1}' are the transformed operators of V_{+1}, V_0, and V_{-1}:

$$V_m' = R_{\boldsymbol{u}}(\varphi)\;V_m\;R_{\boldsymbol{u}}^{\dagger}(\varphi) \quad \text{with} \quad m = +1, 0, -1 \tag{VIII-30}$$

Equality (VIII-28) can be simplified into:

$$(V_{+1}'\;V_0'\;V_{-1}') = (V_{+1}\;V_0\;V_{-1})\left(R_{\boldsymbol{u}}^{[j=1]}(\varphi)\right) \tag{VIII-31a}$$

or:

$$V_{m'}' = \sum R_{m'm}^{[1]} V_m \tag{VIII-31b}$$

Expliciting relation (VIII-29a), we get:

$$\begin{cases} V_{+1} = -\dfrac{1}{\sqrt{2}}\,(V_x + iV_y) \\ V_0 = V_z \\ V_{-1} = \dfrac{1}{\sqrt{2}}\,(V_x - iV_y) \end{cases} \tag{VIII-32}$$

The operators V_m thus defined (with $m = +1, 0, -1$) are called the "standard components" of the vector operator $\boldsymbol{V}$. Relation (VIII-31a) shows that these standard components are transformed in a rotation as the kets of a standard basis with angular momentum $j = 1$. For rotations around the Oz axis, relation (VIII-31b) shows that the transformations are purely diagonal. Introducing these standard components V_m allows us to apply straightforwardly the angular momentum theory and the concept of standard basis of the previous chapter.

Comment:

Expanding the vector operator $\boldsymbol{V}$ on the basis (VIII-25), one should note that V_{+1} is associated with $\boldsymbol{e}_{-1}$ and V_{-1} with $\boldsymbol{e}_{+1}$, and pay attention to the signs:

$$\boldsymbol{V} = V_x\,\boldsymbol{e}_x + V_y\,\boldsymbol{e}_y + V_z\,\boldsymbol{e}_z = -V_{-1}\,\boldsymbol{e}_{+1} + V_0\,\boldsymbol{e}_0 - V_{+1}\,\boldsymbol{e}_{-1} \tag{VIII-33}$$

This comes from the fact that the three basis kets (VIII-25) are orthonormal with a scalar product $\boldsymbol{e}_m^* \cdot \boldsymbol{e}_{m'} = \delta_{mm'}$ and that $\boldsymbol{e}_{\pm 1}^* = -\boldsymbol{e}_{\mp 1}$.

B. Tensor operators

B-1. Definition

There are operators whose transformation laws under rotation are neither those of scalar nor those of vector operators. As an example, operator

$$T_{XX} = XP_X + P_X X \tag{VIII-34}$$

follows more complicated transformation laws even though it is defined in terms of the components of vector operators.

Any observable A can be obtained as a function F of the fundamental observables

$$A = F\,(\boldsymbol{R}, \boldsymbol{P}, \boldsymbol{S}) \tag{VIII-35}$$

A standard prodedure to obtain such a function is to apply the canonical quantization rules to one (or a set of) spinless particle(s). Starting with a classical function $F(\boldsymbol{r}_q, \boldsymbol{p}_q)$ of the positions $\boldsymbol{r}_q$ and momenta $\boldsymbol{p}_q$ of the various particles q of the system, one obtains the quantum observable by replacing $\boldsymbol{r}_q$ by the operator $\boldsymbol{R}_q$, and $\boldsymbol{p}_q$ by the operator $\boldsymbol{P}_q$, adding if needed the proper symmetrization (to account for the non-commutativity between operators).

Generally, the function F can be expanded as a Taylor series, which leads to terms such as:

$$C \times V_{x_i}\, W_{x_j} \ldots Z_{x_\ell} \tag{VIII-36}$$

In this expression, C is a constant (proportional to the nth derivative of the function F evaluated at the origin) and V_{x_i}, $W_{x_j}, \ldots,$ Z_{x_ℓ} are the Cartesian components of n vector operators ($x_i, x_j, \ldots, x_\ell = x, y,$ or z). For a single particle, operators $\boldsymbol{V}, \boldsymbol{W}, \ldots, \boldsymbol{Z}$ each represent one of the three operators $\boldsymbol{R}$, $\boldsymbol{P}$, or $\boldsymbol{S}$. However, what follows remains valid if they represent different vector operators ($\boldsymbol{L} = \boldsymbol{R} \times \boldsymbol{P}$, for example). For several particles, these operators depend on the particle and should be labeled by an index q; since this index plays no role in what follows, we shall omit it.

Consider all the terms of the type (VIII-36) corresponding to a given choice of the n vector operators (the number of times each operator $\boldsymbol{R}$, $\boldsymbol{P}$, or $\boldsymbol{S}$ appears and in which order, is specified). These terms differ only by the choice of the components $i, j, \ldots, \ell$, and there are 3^n of them. These 3^n terms are regrouped in an operator table (setting $C = 1$) called a tensor operator $T^{(n)}$ of order n (or of rank n, depending on the authors). The components of this operator are:

$$T^{(n)}_{x_i x_j \ldots x_\ell} = V_{x_i} W_{x_j} \ldots Z_{x_\ell} \qquad x_i, x_j, \ldots, x_\ell = x, y, \text{ or } z \tag{VIII-37}$$

and we note:

$$\boldsymbol{T}^{(n)} = \boldsymbol{V} \otimes \boldsymbol{W} \otimes \ldots \otimes \boldsymbol{Z} \tag{VIII-38}$$

For example, if $n = 1$, we find the already studied case of a vector operator, represented by a row matrix with 3 elements. If $n = 2$, the table becomes a 3×3 matrix:

$$\overline{\overline{T}}^{(2)} = \begin{pmatrix} V_x W_x & V_x W_y & V_x W_z \\ V_y W_x & V_y W_y & V_y W_z \\ V_z W_x & V_z W_y & V_z W_z \end{pmatrix} \tag{VIII-39}$$

(a tensor of order two is often written under a double line). Similarly, $n = 3$ leads to a $3 \times 3 \times 3$ cubic table (set of three 3×3 matrices), etc. It is clear that any function F can be expressed, through a Taylor series, as a linear combinations of such tensors.

B-2. Transformation under rotation

The question remains of how the components of these tensor operators $T^{(n)}$ transform under a rotation. Expliciting the components of relation (VIII-19), we get:

$$V'_{x_i} = R\, V_{x_i}\, R^\dagger = \sum_j (\mathscr{R})_{ji}\, V_{x_j} \tag{VIII-40}$$

This relation applies to $\boldsymbol{V}$, $\boldsymbol{W}, \ldots,$ $\boldsymbol{Z}$, which are vector operators. Under a rotation, the component (VIII-37) of the tensor operator becomes:

$$\begin{aligned} R\, T^{(n)}_{x_i x_j \ldots x_\ell} R^\dagger &= R\, V_{x_i} R^\dagger\; R\, W_j R^\dagger \ldots R\, Z_{x_\ell} R^\dagger \\ &= \sum_{i'\, j' \ldots \ell'} (\mathscr{R})_{i'i}\, (\mathscr{R})_{j'j} \ldots (\mathscr{R})_{\ell'\ell}\, T^{(n)}_{x_{i'} x_{j'} \ldots x_{\ell'}} \end{aligned} \tag{VIII-41}$$

We shall call tensor operator $T^{(n)}$ any set of 3^n components (operators) that obey this law in a rotation transformation. It is clear that not only the tensor product operators (VIII-41), but all their linear combinations (for a given value of n) generate[2] tensor operators $\boldsymbol{T}^{(n)}$.

The linear transformation laws (VIII-41) for the components of the tensor operators form a linear representation of the rotation group. Applying successively two rotations R_1 then R_2 to a vector operator $\boldsymbol{V}$, its V_{x_i} component is transformed as:

$$V_{x_i} \Rightarrow R_1 V_{x_i} R_1^\dagger \Rightarrow R_2 R_1 V_{x_i} R_1^\dagger R_2^\dagger \tag{VIII-42}$$

Using relation (VIII-40) twice, we get:

$$\begin{aligned} R_2 R_1 V_{x_i} R_1^\dagger R_2^\dagger &= \sum_j (\mathscr{R}^{(1)})_{ji}\ R_2 V_{x_j} R_2^\dagger = \sum_{jk} (\mathscr{R}^{(1)})_{ji} (\mathscr{R}^{(2)})_{kj} V_{x_k} \\ &= \sum_k \left(\mathscr{R}^{(2)} \mathscr{R}^{(1)}\right)_{ki} V_{x_k} \end{aligned} \tag{VIII-43}$$

where $\mathscr{R}^{(1)}$ and $\mathscr{R}^{(2)}$ are two rotations in the usual space associated respectively with R_1 and R_2. This means that for a tensor operator of order $n = 1$, the matrix product does correspond to the transformation product. For a tensor operator of higher order n, we apply the same relation to each term of product (VIII-41), and we get a tensor product representation.

We should then wonder whether this representation is reducible or not. In the $n = 1$ case (vector operators), the representation is obviously irreducible since the standard components, introduced thanks to a basis change, are directly transformed with the $(R^{[j=1]})$ matrices. In the $n \geq 2$ case, the answer is not obvious a priori. Once again, we shall define new components so as to directly use the results of chapter VII on the irreducible representations of the rotation group.

B-3. Spherical components

We now introduce the "spherical components" (or, more explicitly, the "uncoupled spherical components") of a tensor $T^{(n)}$. Their advantage is that they are transformed in a rotation just as a tensor product of kets with angular momenta $j = 1$.

[2]If the state space $\mathscr{E}$ has an infinite dimension, the set of $\boldsymbol{T}^{(n)}$ for a fixed value of n forms in general another vector space of infinite dimension. For example, the number of scalar (or vector) operators that are linearly independent is infinite if one can define independently reduced matrix elements within all the $\mathscr{E}(\tau, j)$, *cf.* § C-2. One should not confuse the operator space and the state space $\mathscr{E}$ (in which the elements of the operator space act).

B-3-a. Definition and transformation properties

For a vector operator $\boldsymbol{V}$, the spherical components are identical[3] to the standard components V_m defined by relation (VIII-32). Once expanded, relation (VIII-29a) is written:

$$V_{m_i} = \sum_{i'} (S)_{i'\,m_i}\, V_{x_{i'}} \tag{VIII-44}$$

where $m_i = 1, 0, -1$ is the column index of matrix (S), and $x_{i'} = x, y, z$ for $i = 1, 2, 3$ is the row index. For a tensor $\boldsymbol{T}^{(n)} = \boldsymbol{V} \otimes \boldsymbol{W} \otimes \ldots \otimes \boldsymbol{Z}$, the uncoupled spherical components are the $T^{(n)}_{m_i\, m_j\, \ldots\, m_\ell}$ defined by applying this relation to each component of the tensor. This yields the following expression:

$$T^{(n)}_{m_i\, m_j\, \ldots\, m_\ell} = \sum_{i'\, j'\, \ldots\, \ell'} (S)_{i'\, m_i}\, (S)_{j'\, m_j}\, \ldots\, (S)_{\ell'\, m_\ell}\, T^{(n)}_{x_{i'}\, x_{j'}\, \ldots\, x_{\ell'}} \tag{VIII-45a}$$

Conversely, since the matrix (S) is unitary, we have:

$$T^{(n)}_{x_i, x_j\, \ldots\, x_\ell} = \sum_{m_{i'}\, m_{j'}\, \ldots\, m_{\ell'}} (S)^\star_{i\, m_{i'}}\, (S)^\star_{j\, m_{j'}}\, \ldots\, (S)^\star_{\ell\, m_{\ell'}}\, T^{(n)}_{m_{i'}\, m_{j'}\, \ldots\, m_{\ell'}} \tag{VIII-45b}$$

These two equalities, established for a tensor of the form $\boldsymbol{V} \otimes \boldsymbol{W} \otimes \ldots \otimes \boldsymbol{Z}$, remain valid for any tensor (they are linear with respect to T).

How are the spherical components transformed under a rotation? Relation (VIII-31a) enables calculating the transformation under rotation of each factor V_{m_i}, W_{m_j}, ..., Z_{m_ℓ} of the tensor. Inserting as many times as needed $R^\dagger R \equiv 1$ in between the operators V_{m_i}, W_{m_j}, ..., we get:

$$R\, T^{(n)}_{m_i\, m_j\, \ldots\, m_\ell} R^\dagger = \sum_{i'\, j'\, \ldots\, \ell'} \left(R^{[1]}\right)_{m_i'\, m_i} \left(R^{[1]}\right)_{m_j'\, m_j} \ldots \left(R^{[1]}\right)_{m_\ell' m_\ell} T^{(n)}_{m_i'\, m_j'\, \ldots\, m_\ell'} \tag{VIII-46}$$

Here again, starting from the case where $\boldsymbol{T}^{(n)}$ is a simple tensor product, one can extend by linearity the validity of this formula to the general case of any tensor operator of order n.

B-3-b. Link with a tensor product of states

Consider n physical systems, with angular momenta $j = 1$: system $\mathcal{S}_1$, with (3-dimensional) state space $\mathscr{E}_1$ generated by the kets $|1\colon m_1\rangle$ ($m_1 = +1, 0$, or -1); system $\mathcal{S}_2$, with state space $\mathscr{E}_2$, generated by the three kets $|2\colon m_2\rangle$ ($m_2 = +1, 0,$

[3]If $n = 1$, the standard components and the uncoupled spherical components are the same, but this is no longer the case when $n \geq 2$, as will be shown below.

or −1), etc. A basis ("uncoupled" basis) of the state space of the total system $\mathcal{S}$ formed by the gathering of $\mathcal{S}_1, \mathcal{S}_2, \ldots, \mathcal{S}_n$, is given by the set of kets:

$$|1\colon m_1\rangle \otimes |2\colon m_2\rangle \otimes \ldots \otimes |n\colon m_n\rangle$$

where all the m equal +1, 0, or −1. Under a rotation of the entire system $\mathcal{S}$, these kets are transformed as:

$$\begin{aligned} R_{\boldsymbol{u}}(\varphi)\left[|1\colon m_1\rangle \otimes |2\colon m_2\rangle \otimes \ldots \otimes |n\colon m_n\rangle\right] \\ = \left[\mathrm{e}^{-\mathrm{i}\varphi \boldsymbol{J}_1\cdot\boldsymbol{u}/\hbar}|1\colon m_1\rangle\right] \otimes \left[\mathrm{e}^{-\mathrm{i}\varphi \boldsymbol{J}_2\cdot\boldsymbol{u}/\hbar}|2\colon m_2\rangle\right] \ldots \otimes \left[\mathrm{e}^{-\mathrm{i}\varphi \boldsymbol{J}_n\cdot\boldsymbol{u}/\hbar}|n\colon m_n\rangle\right] \\ = \sum_{m_1'\, m_2' \ldots m_n'} \left(R^{[1]}\right)_{m_1'\, m_1} \left(R^{[1]}\right)_{m_2'\, m_2} \cdots \left(R^{[1]}\right)_{m_n'\, m_n} \\ |1\colon m_1'\rangle \otimes |2\colon m_2'\rangle \otimes \ldots \otimes |n\colon m_n'\rangle \end{aligned} \quad \text{(VIII-47)}$$

where $\boldsymbol{J}_1$, $\boldsymbol{J}_2$, ... represent the angular momentum operators associated respectively with $\mathcal{S}_1$, $\mathcal{S}_2, \ldots$ We thus obtain, for the kets of the uncoupled basis, the same transformation law as with relation (VIII-46) for the uncoupled spherical components. In other words, the elements $T^{(n)}_{m_i\, m_j \ldots m_\ell}$ are transformed according to matrices that are the tensor product of $(R^{[1]})$ matrices.

This property will enable us to apply the general theory for the coupling of angular momenta, going from uncoupled spherical components to coupled (or standard) components. Applying n times rotation matrices $R^{[j]}$ with $j = 1$, we know that we obtain the following values of J:

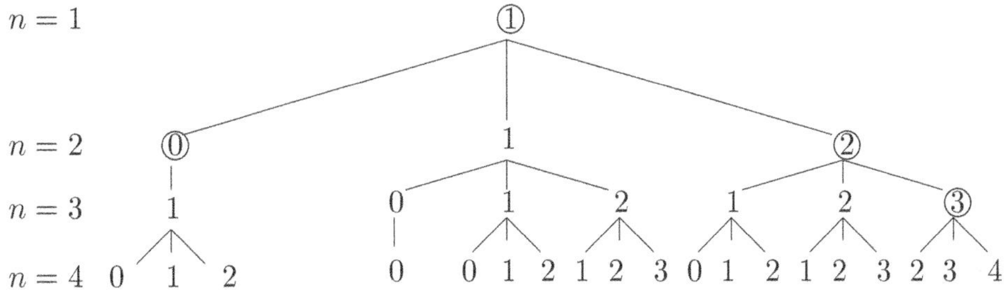

This brings up irreducible representations of the rotation group with J taking its maximum value $J = n$ (values circled on the table[4]). Note that this maximum value is only reached once (only one coupling scheme), as opposed to the other values. The operators obtained after this angular momenta coupling scheme for building irreducible representations are called "irreducible tensor operators". Their more precise definition as well as certain of their properties is discussed in the next § B-4.

[4]The value $J = 0$ is also circled for $n = 2$ as it is the first time it appears. It corresponds to a scalar operator (rotation invariant).

B-4. Irreducible tensor operators

We start discussing the $n = 2$ case, the smallest value of n that does not bring us back to the initial vector operators ($n = 1$). It will help introducing the main ideas, as well as the notations used in the general case for any value of n.

B-4-a. Standard components of an $n = 2$ tensor

When $n = 2$, the tensor $\boldsymbol{T}^{(2)}$ is often noted $\overline{\overline{T}}$, where the double bar explicits its tensor character. It has 9 Cartesian components $T_{x_i x_j}$, as well as 9 (uncoupled) spherical components $T_{m_i m_j}$. Under a rotation, they are transformed as the 9 kets:

$$|1\ 1\ m_i\, m_j\rangle = |1\colon m_i\rangle \ \otimes\ |2\colon m_j\rangle \tag{VIII-48}$$

associated with two systems with angular momentum 1.

Coupling the states (VIII-48) yields the kets $|J, M\rangle$ whose explicit expression is (second exercise of § C-2-b in chapter VII):

$$\begin{cases} |2,2\rangle = |1\ 1\ 1\ 1\rangle \\ |2,1\rangle = \dfrac{1}{\sqrt{2}}|1\ 1\ 1\ 0\rangle + \dfrac{1}{\sqrt{2}}|1\ 1\ 0\ 1\rangle \\ |2,0\rangle = \dfrac{1}{\sqrt{6}}|1\ 1\ 1\ -1\rangle + \dfrac{2}{\sqrt{6}}|1\ 1\ 0\ 0\rangle + \dfrac{1}{\sqrt{6}}|1\ 1\ -1\ 1\rangle \\ |2,-1\rangle = \dfrac{1}{\sqrt{2}}|1\ 1\ 0\ -1\rangle + \dfrac{1}{\sqrt{2}}|1\ 1\ -1\ 0\rangle \\ |2,-2\rangle = |1\ 1\ -1\ -1\rangle \end{cases} \tag{VIII-49a}$$

$$\begin{cases} |1,1\rangle = \dfrac{1}{\sqrt{2}}|1\ 1\ 1\ 0\rangle - \dfrac{1}{\sqrt{2}}|1\ 1\ 0\ 1\rangle \\ |1,0\rangle = \dfrac{1}{\sqrt{2}}|1\ 1\ 1\ -1\rangle - \dfrac{1}{\sqrt{2}}|1\ 1\ -1\ 1\rangle \\ |1,-1\rangle = \dfrac{1}{\sqrt{2}}|1\ 1\ 0\ -1\rangle - \dfrac{1}{\sqrt{2}}|1\ 1\ -1\ 0\rangle \end{cases} \tag{VIII-49b}$$

$$|0,0\rangle = \frac{1}{\sqrt{3}}|1\ 1\ 1\ -1\rangle - \frac{1}{\sqrt{3}}|1\ 1\ 0\ 0\rangle + \frac{1}{\sqrt{3}}|1\ 1\ -1\ 1\rangle \tag{VIII-49c}$$

The fractions appearing in these equalities are the various Clebsch–Gordan coefficients $\langle 1\,1\,m_1\,m_2|J\,M\rangle$. Just as these coefficients allow building the states $|J, M\rangle$ of the coupled basis as a function of the states $|1\,1\,m_1\,m_2\rangle$, one can build the "*standard components*" (or coupled spherical components) $T_Q^{[K]}$ of the tensor operator $\overline{\overline{T}}$ as:

$$\boxed{T_Q^{[K]} = \sum_{m_1\,m_2} \langle 1\,1\,m_1\,m_2|K\,Q\rangle \; T_{m_1\,m_2}} \tag{VIII-50a}$$

In this definition, K is the order[5] of the tensor that replaces the index J of the total angular momentum, equal here to 2, 1, or 0. The index Q that replaces M varies by integer values between $+K$ and $-K$. Conversely, we have:

$$T_{m_1\,m_2} = \sum_{K,Q} \langle 1\,1\,m_1\,m_2|K\,Q\rangle \; T_Q^{[K]} \tag{VIII-50b}$$

Consequently, just as:

$$R\,|J,M\rangle = \sum_{M'} \left(R^{[j]}\right)_{M'M} |J,M'\rangle \tag{VIII-51a}$$

the rotation transforms of operators $T_Q^{[K]}$ are given by:

$$\boxed{R\,T_Q^{[K]}\,R^\dagger = \sum_{Q'} \left(R^{[K]}\right)_{Q'Q} \; T_{Q'}^{[K]}} \tag{VIII-51b}$$

This result comes from the fact that the transformation laws for the $T_{m_1\,m_2}$ and for the kets $|1\,1\,m_1\,m_2\rangle$ are identical, and that the $T_Q^{[K]}$ are defined as a function of the $T_{m_1\,m_2}$ in exactly the same way as the $|J,M\rangle$ are defined as a function of the $|1\,1\,m_1\,m_2\rangle$.

Exercise: Check this equality explicitly using the properties of the Clebsch–Gordan coefficients and of the rotation matrices.

The $\boldsymbol{T}^{[K]}$ are called "*irreducible tensor operators*"; each of them has $2K+1$ standard components $T_Q^{[K]}$. Their explicit values are given in § 1 of complement $\mathrm{B_{VIII}}$. Using relation (VIII-50b), one can come back to the 9 uncoupled sperical components associated with each $\boldsymbol{T}^{[K]}$. One must perform a summation over Q, but not over K, so as to obtain separately the $T_{m_1 m_2}$ associated with each value of K. One then uses equality (VIII-45b) to get the corresponding Cartesian components. This calculation is carried out in § 2 of complement $\mathrm{B_{VIII}}$.

For an infinitesimal rotation, we know that:

$$R_{\boldsymbol{u}}(\delta\varphi) = 1 - i\frac{\delta\varphi}{\hbar}J_u \tag{VIII-52}$$

[5]The same word is used for the order n of a Cartesian tensor and the order K of an irreducible tensor operator; the context indicates which of these two indices is meant.

so that (VIII-51b) becomes, to first-order in $\delta\varphi$:

$$\left[J_u, T_Q^{[K]}\right] = \sum_{Q'} \langle J = K, M = Q' | J_u | J = K, M = Q \rangle \; T_{Q'}^{[K]} \tag{VIII-53}$$

Taking $\boldsymbol{u}$ successively parallel to Ox, Oy, and Oz, and combining linearly the first two equalities to obtain $J_\pm = J_x \pm iJ_y$, we get:

$$\boxed{\begin{cases} \left[J_z, T_Q^{[K]}\right] = Q\hbar \; T_Q^{[K]} \\ \left[J_\pm, T_Q^{[K]}\right] = \hbar\sqrt{K(K+1) - Q(Q\pm 1)} \; T_{Q\pm 1}^{[K]} \end{cases}} \tag{VIII-54}$$

These relations may constitute another definition of an irreducible tensor operator $T_Q^{[K]}$. By integration, equalities (VIII-54) that concern the transformations of operators under an infinitesimal rotation can become the transformation laws under a finite rotation; we then get relation (VIII-51b).

Comment:

The sum over K in (VIII-50b) shows that the initial tensor $\overline{\overline{T}}$ is the sum of three contributions coming from the values $K = 0$, 1, 2:

$$\overline{\overline{T}}^{(n)} = \overline{\overline{T}}^{[K=2]} + \overline{\overline{T}}^{[K=1]} + \overline{\overline{T}}^{[K=0]} \tag{VIII-55}$$

This decomposition is performed in § 2 of complement $\mathrm{B_{VIII}}$.

B-4-b. General case

Starting with the uncoupled basis described above:

$$|1\!: m_1\rangle \otimes |2\!: m_2\rangle \otimes \ldots \otimes |n\!: m_n\rangle \tag{VIII-56}$$

where each ket is associated with the same value $j = 1$, one can change for a new coupled basis with eigenvectors common to $\boldsymbol{J}^2$ and J_z (operator $\boldsymbol{J}$ is the total angular momentum $\boldsymbol{J}_1 + \boldsymbol{J}_2 + \ldots + \boldsymbol{J}_n$). For example, one can add $\boldsymbol{J}_1$ and $\boldsymbol{J}_2$ and get a quantum number $J_{1\,2}$ (we set $\boldsymbol{J}_{1\,2} = \boldsymbol{J}_1 + \boldsymbol{J}_2$), then add their sum to $\boldsymbol{J}_3$ which yields the quantum number $J_{1\,2\,3}$ (we set $\boldsymbol{J}_{1\,2\,3} = \boldsymbol{J}_{1\,2} + \boldsymbol{J}_3$), etc. In this way, we get the states:

$$|J_{1\,2}, J_{1\,2\,3}, \ldots ; J, M\rangle = \sum_{m_1+\ldots+m_n=M} C_{m_1\ldots m_n}\,[J_{1\,2}, J_{1\,2\,3}, \ldots ; J\, M] \; |1\!: m_1\rangle \otimes |2\!: m_2\rangle \ldots \otimes |n\!: m_n\rangle \tag{VIII-57a}$$

which form another basis of the space generated by the uncoupled basis (VIII-56), as we now check by recurrence. The coupling of the first two $j = 1$ angular momenta introduces, in their 9-dimensional space, the orthonormal basis described in (VIII-49) where $J_{12} = 0, 1, 2$. The coupling of J_{12} with the third spin leads (*cf.* table on page 367) to a 27-dimensional space, including a subspace $J_{123} = 0$, three subspaces $J_{123} = 1$ that differ by the value of J_{12}, two subspaces $J_{123} = 1$ that also differ by the value of J_{12}. We thus obtain 27 new kets forming a new orthonormal basis. The reasoning continues until the total angular momentum J is obtained.

The coefficients C, products of Clebsch–Gordan coefficients that are real, are thus real constants. These coefficients define a $3^n \times 3^n$ orthogonal matrix, which makes it easy to invert relation (VIII-57a) and write:

$$|1: m_1\rangle \otimes |2: m_2\rangle \ldots \otimes |n: m_n\rangle = \sum_{J_{12}\ J_{123} \ldots ; J} C_{m_1 \ldots m_n} \left[J_{12}, J_{123}, \ldots ; J\, M\right] |J_{12}, J_{123}, \ldots ; J, M\rangle \tag{VIII-57b}$$

where $M = m_1 + m_2 + \ldots + m_n$.

When M varies from $+J$ to $-J$ in (VIII-57a) (the other quantum numbers J_{12}, J_{123}, ... remaining constant), we get $2J+1$ kets that generate an irreducible subspace with $2J+1$ dimensions[6]. Unlike what happened in the $n = 2$ case, these subspaces are specified not only by the value of J, but also through the values of the intermediate angular momenta J_{12}, J_{123}, etc.

Copying relation (VIII-57a), one defines the "*standard components*" and we set:

$$T_Q^{[K]}\left[J_{12}, J_{123}, \ldots\right] = \sum_{m_1+m_2+\ldots+m_n=Q} C_{m_1\, m_2 \ldots m_n} \left[J_{12}, J_{123} \ldots ;\ K\, Q\right] T^{(n)}_{m_1\, m_2 \ldots m_n} \tag{VIII-58}$$

These operators are transformed under a rotation according to equality:

$$R\, T_Q^{[K]}\left[J_{12}, J_{123}, \ldots\right] R^{\dagger} = \sum_{Q'} \left(R^{[j=K]}\right)_{Q'Q} T_{Q'}^{[K]}\left[J_{12}, J_{123}, \ldots\right] \tag{VIII-59}$$

which is completely identical to relation (VIII-51b) for second-order tensors.

Conversely, since the coefficients C define a real orthogonal matrix, we can copy (VIII-57b) and write an equation giving the uncoupled spherical components $T_{m_1\, m_2 \ldots m_n}$ as a function of the irreducible tensor components.

[6]The irreducible subspaces thus obtained, as well as the standard basis that can be built, obviously depend on the coupling scheme chosen for the individual angular momenta.

Comments:

(i) To express the characteristic property (VIII-59) of the irreducible tensor operators, it is sometimes said that operators $T_Q^{[K]}$ are transformed under a rotation as spherical harmonics of order $\ell = K$.

(ii) The number K is an integer, never a half-integer, as can be seen in the table on page 367. Contrary to what happens for state vectors, only "true" representations of $R_{(3)}$ may come into play, and not the "double-valued" ones (chapter VII, § A-3-b). Note that an operator with a half-integer order would change into its opposite under a 2π rotation: this would imply that a global rotation of all the corresponding measurement instruments would change all the possible results into their opposite (a change of sign of an observable has immediate physical consequences, which is not the case for a state vector).

(iii) The maximum value $K = n$ occurs only once ($J_{12} = 2$, $J_{123} = 3$, etc.). One sometimes keeps the name of irreducible tensor operators for the operators $T^{[K]}$ where $K = n$ (they correspond to the circled values on the right-hand side of the table at the end of § B-3). They are simply noted $T^{[K]}$ since the values of the intermediate angular momenta J_{12}, J_{123}, ... need no longer be specified.

(iv) One should not confuse the three types of tensor components

$$\boxed{\text{Cartesian}} \underset{\leftarrow}{\rightarrow} \boxed{\begin{array}{c}\text{spherical}\\ \text{uncoupled}\end{array}} \underset{\leftarrow}{\rightarrow} \boxed{\text{standard}}$$

which are all different, unless $n = 1$ (vector operator). Many authors call "spherical" (without specifying "coupled") the standard components.

B-5. Properties of irreducible tensor operators

B-5-a. Commutation relations with J

The commutation relations of an irreducible tensor operator $T_Q^{[K]}$ with the total angular momentum $\boldsymbol{J}$ of the system under study can be obtained in the same way as in the previous § B-4-a, equalities (VIII-52), (VIII-53), and (VIII-54). Relations (VIII-54) can thus be generalized to tensors of any order.

These equalities are a necessary and sufficient condition for the $2K + 1$ operators $T_Q^{[K]}$ to be considered as the components of an irreducible tensor operator of order K. They show that the $T_Q^{[K]}$ are transformed under a rotation like the kets $|J = K,\ M = Q\rangle$, and they can be used as a definition for irreducible tensor operators.

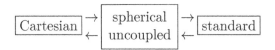

B-5-b. Symmetries of the $T^{[K]}$ with maximum order

In this § B-5-b, we assume that K takes on its maximum value, $K = n$.

α. *Symmetries*

Returning to the Cartesian components of $T^{[K]}$, we get a *symmetric* tensor with a *trace equal to zero*.

In the case where $K = 2$, one can see the symmetry of the tensor written in relation (13) of complement B_{VIII}. In the case where $K = 3$ for example, we expect:

$$T^{[K=3]}_{x\,x\,y} = T^{[K=3]}_{x\,y\,x} = \ldots$$

a relation that is obviously not obeyed for any (reducible) tensor operator.

Demonstration: We first establish the symmetry of the tensor, and show that the uncoupled spherical components $T_{m_1\,m_2 \ldots m_n}$ are invariant under a permutation of two m indices. This property is obviously satisfied if all the indices m equal $+1$. Since:

$$|J = K, M = K\rangle = |1\colon m = 1\rangle \otimes |2\colon m = 1\rangle \otimes \ldots \otimes |n\colon m = 1\rangle \tag{VIII-60a}$$

we have:

$$T^{[K=n]}_{Q=n} = T^{(n)}_{1,1,1\ldots 1} \tag{VIII-60b}$$

We apply to equality (VIII-60a) the operator J_- :

$$J_- = J_{1-} + J_{2-} + \ldots + J_{n-} \tag{VIII-61}$$

We get an expression for the ket $|J = K,\; M = K - 1\rangle$ that is symmetric with respect to the permutation of any two angular momenta 1, 2, ..., n. The expression of operator $T^{[K=n]}_{Q=K-1}$ as a function of the spherical components is symmetric with respect to the permutation of two of the coordinates m_i:

$$T^{[K=n]}_{Q=K-1} = \frac{1}{\sqrt{K}} \left[T^{(n)}_{0\,1\,1\ldots 1} + T^{(n)}_{1\,0\,1\ldots 1} + \ldots + T^{(n)}_{1\,1\,1\ldots 0} \right] \tag{VIII-62a}$$

One continues by recurrence, applying repeatedly the operator written in (VIII-61). This operator is symmetric with respect to the indices 1, 2, ..., n, so that the initial symmetry is conserved after each application. Each $T^{[K=n]}_{Q}$ can thus be expressed in a symmetrical way as a function of the Cartesian components of the tensor.

To obtain the spherical components of the $T^{(n)}_{m_i\,m_j\ldots m_\ell}$, we must invert these relations, assuming that among all the tensors $T^{[K']}$ that will appear, the only non-zero one corresponds to $K = n$. As we have seen that the matrix of the coefficients $C_{m_1\,m_2\ldots m_n}[K = n, Q = n - 1]$ is orthogonal, no computation is necessary. For example, relation (VIII-62a) yields simply:

$$T^{(n)}_{0\,1\,1\ldots 1} = T^{(n)}_{1\,0\,1\ldots 1} = \ldots = T^{(n)}_{1\,1\,1\ldots 0} = \frac{1}{\sqrt{K}}\, T^{[K]}_{Q=K-1} \tag{VIII-62b}$$

which shows that the components $T^{(n)}_{m_1 m_2 \dots m_2}$ are invariant under the permutation of two of their indices[7].

To finally obtain the Cartesian coordinates, we use formula (VIII-45b). The same symmetry property is obeyed by the Carsesian components: the same matrix (S) is used to change each spherical index into a Cartesian index.

The symmetry property has thus been established for $Q = n - 1$. The reasoning is repeated by successive applications of the symmetric operator J_-, so as to reach all the values of Q.

β. *Zero trace*

We now show that $T^{[K=n]}$ is a tensor with a trace equal to zero:

$$\sum_{x_i = x,y,z} T^{[K=n]}_{x_i\, x_i\, x_j \dots x_\ell} = 0 \tag{VIII-63}$$

We have chosen to contract the first two indices, taking them equal and summing over their 3 possible values. One could obviously do the same for any two indices, adjacent or not, and the reasoning would be the same.

Demonstration: The announced property comes from the fact that if we start by coupling the first two angular momenta $\boldsymbol{J}_1$ and $\boldsymbol{J}_2$ to obtain $J_{1\,2} = 0$, and if we take the product of the resulting ket by completely arbitrary kets of the systems 3, 4, ..., K, one can only obtain states orthogonal to $|J = n, M\rangle$. This is because, if $J_{1\,2} = 0$, the maximum value of J is $n - 2$ (eigenkets of $\boldsymbol{J}^2$ with different eigenvalues are orthogonal).

The ket associated with this coupling of $\boldsymbol{J}_1$ and $\boldsymbol{J}_2$ is written, according to relation (VIII-49c):

$$|J_{1\,2} = 0\rangle = \frac{1}{\sqrt{3}} \left[|1\colon m_1 = 1\rangle\ |2\colon m_2 = -1\rangle \right.$$
$$\left. - |1\colon m_1 = 0\rangle\ |2\colon m_2 = 0\rangle + |1\colon m_1 = -1\rangle\ |2\colon m_2 = 1\rangle \right] \tag{VIII-64}$$

The fact that:

$$|J_{1\,2} = 0\rangle \otimes |3\colon m_3\rangle \otimes |4\colon m_4\rangle \otimes \dots |n\colon m_n\rangle$$

is orthogonal to $|J = n, M\rangle$ leads to:

$$C_{1\,-1\,m_3\,m_4 \dots m_n} - C_{0\,0\,m_3\,m_4 \dots m_n} + C_{-1\,1\,m_3\,m_4 \dots m_n} = 0 \tag{VIII-65a}$$

where the coefficients C have been defined in (VIII-57a). This equality implies that:

$$T^{[K=n]}_{1\,-1\,m_3\,m_4 \dots m_n} - T^{[K=n]}_{0\,0\,m_3\,m_4 \dots m_n} + T^{[K=n]}_{-1\,1\,m_3\,m_4 \dots m_n} = 0 \tag{VIII-65b}$$

[7]The equality of all the components, as seen in (VIII-62a) and (VIII-62b), is not a general property, as can be observed in the second relation (VIII-49a). It only occurs in the particular case where the sum Q of the m indices equals K or $K - 1$, or their opposite.

We simply have to change basis with the matrix (S) to get the Cartesian components [formulas (VIII-45)]. Starting with the transformation of the first two indices into Cartesian indices, we get:

$$T^{[K=n]}_{x\,x\,m_3\,m_4\,...\,m_n} + T^{[K=n]}_{y\,y\,m_3\,m_4\,...\,m_n} + T^{[K=n]}_{z\,z\,m_3\,m_4\,...\,m_n} = 0 \qquad \text{(VIII-66)}$$

(the trace of the tensor over two indices only appears in the tensor where the corresponding angular momenta are coupled to yield a zero momentum). One can also change the other indices from spherical to Cartesian, and we get equality (VIII-63).

γ. *Hermiticity*

A tensor operator $\overline{\overline{T}}$ is considered to be Hermitian if its Cartesian components are Hermitian operators. The necessary and sufficient condition to be obeyed is:

$$\boxed{\left[T^{[K]}_{Q}\right]^{\dagger} = (-1)^{Q}\, T^{[K]}_{-Q}} \qquad \text{(VIII-67)}$$

For example, if $\boldsymbol{V}$ is Hermitian ($V_x = V_x^{\dagger}, ...$), we have:

$$\begin{cases} [V_{+1}]^{\dagger} = -\dfrac{1}{\sqrt{2}}\,(V_x - iV_y) = -V_{-1} \\ [V_0]^{\dagger} = V_0 \\ [V_{-1}]^{\dagger} = \dfrac{1}{\sqrt{2}}\,(V_x + iV_y) = -V_{+1} \end{cases} \qquad \text{(VIII-68)}$$

and condition (VIII-67) is fulfilled. This result can be generalized to any value of K.

Demonstration: Assuming the tensor to be Hermitian, we write the Hermitian conjugate relation of (VIII-45a). On the right-hand side, and since the Cartesian components of the tensor are Hermitian, we simply take the complex conjugates of all the elements of the matrix (S) for the basis change. We start by taking the complex conjugate of the first one, $(S)_{i'\,i}$. Expression (VIII-26) for this matrix, where $i = 1$, 2, and 3 correspond respectively to $m_i = +1$, 0, and -1, shows that $[(S)_{i'i}]^* = (-1)^{m_i}(S)_{i'-i}$. This leads to the expression of the following component of the tensor:

$$(-1)^{m_i}\, T^{(n)}_{-m_i\,m_j\,...\,m_\ell}$$

Repeating the operation for all the tensor indices, we get on the right-hand side the expression of the spherical component of the tensor:

$$(-1)^{m_i+m_j+...+m_\ell}\, T^{(n)}_{-m_i\,-m_j\,...\,-m_\ell}$$

The hermiticity condition for these spherical components is thus:

$$\left[T^{(n)}_{m_i\,m_j\,...\,m_\ell}\right]^{\dagger} = (-1)^{m_i+m_j+...+m_\ell}\, T_{-m_i\,-m_j\,...\,-m_\ell} \qquad \text{(VIII-69)}$$

For the irreducible components, since the coefficients C are real and the sum of the m coefficients is equal to Q, relation (VIII-58) yields:

$$\left[T_Q^{[K]}\right]^\dagger = (-1)^Q \sum_{m_i+m_j+\ldots+m_\ell} C_{m_i\,m_j\,\ldots\,m_\ell}\; T^{(n)}_{-m_i\,-m_j\,\ldots\,-m_\ell}$$

Using the last of the formulas (VII-89) with $J = j_1 + j_2$ (maximum value of the angular momentum), it is easy to show by recurrence that:

$$C_{m_i\,m_j\,\ldots\,m_\ell} = C_{-m_i\,-m_j\,\ldots\,-m_\ell}$$

and we obtain the result (VIII-67).

Comment:

Using relation (VII-89) limits the validity of the demonstration to tensors of maximum order. For arbitrary tensors, this relation shows that an additional sign is needed:

$$(-1)^{1+1-J_{12}+1+J_{12}-J_{123}+\cdots-K} = (-1)^{n-K} \tag{VIII-70}$$

The factor $(-1)^Q$ of (VIII-67) must be replaced by $(-1)^{Q+n-K}$. It is not surprising that this coefficient may vary from one tensor operator to another, since the transformation conditions of an operator under rotation or Hermitian conjugation are independent. An irreducible tensor operator will remain irreducible when multiplied by i, but it will change the hermicity conditions.

B-5-c. Product of two tensor operators

Consider two irreducible tensor operators , operator $T^{[K_1]}$ of order K_1 and operator $W^{[K_2]}$ of order K_2. We shall discuss the products of their components:

$$T_{Q_1}^{[K_1]}\, W_{Q_2}^{[K_2]} \quad \text{with:} -K_1 \le Q_1 \le K_1 \quad , \quad -K_2 \le Q_2 \le K_2 \tag{VIII-71}$$

This yields a set of $(2K_1+1)\,(2K_2+1)$ operators that transform linearly into each other under rotation. In other words, they form an operator vector space, with dimension $(2K_1+1)\,(2K_2+1)$, globally invariant under rotation. In general, this space is not irreducible, but can be decomposed into invariant subspaces with lower dimensions:

$$\begin{aligned} R\,T_{Q_1}^{[K_1]}\,W_{Q_2}^{[K_2]}\,R^\dagger &= R\,T_{Q_1}^{[K_1]}\,R^\dagger\,R\,W_{Q_2}^{[K_2]}\,R^\dagger \\ &= \sum_{Q_1'Q_2'} \left(R^{[K_1]}\right)_{Q_1'Q_1} \left(R^{[K_2]}\right)_{Q_2'Q_2} T_{Q_1'}^{[K_1]}\,W_{Q_2'}^{[K_2]} \end{aligned} \tag{VIII-72}$$

The operators written in (VIII-71) are transformed under rotation as tensor products of kets $|J = K_1,\ M = Q_1\rangle \otimes |J = K_2,\ M = Q_2\rangle$. We are facing once again the problem of the addition (coupling) of angular momenta. Consequently, we expect the operators $Z_Q^{[K]}$ defined by[8]:

$$Z_Q^{[K]} = \sum_{Q_1+Q_2=Q} \langle K_1\, K_2\, Q_1\, Q_2 | K\, Q\rangle\; T_{Q_1}^{[K_1]}\; W_{Q_2}^{[K_2]}$$
$$K = K_1 + K_2, K_1 + K_2 - 1, \ldots, |K_1 - K_2|$$
$$Q = K,\, K-1, \ldots, \ldots \quad \ldots, \quad -K \tag{VIII-73}$$

to be transformed under rotation in a simpler way, as the kets of the coupled basis $|J = K,\; M = Q\rangle$:

$$R\, Z_Q^{[K]}\, R^\dagger = \sum_{Q'} \left(R^{[K]}\right)_{Q'Q} Z_{Q'}^{[K]} \tag{VIII-74}$$

The $2K+1$ operators $Z_Q^{[K]}$ associated with a same value of K yield an irreducible tensor operator.

Conversely, the products (VIII-71) can be expressed as a function of the new operators as:

$$T_{Q_1}^{[K_1]}\, W_{Q_2}^{[K_2]} = \sum_{K,Q} \langle K_1\, K_2\, Q_1\, Q_2 | K\, Q\rangle\; Z_Q^{[K]} \tag{VIII-75}$$

This yields a sum of irreducible tensor operators whose order goes from $|K_1 - K_2|$ to $K_1 + K_2$.

We have already seen above (§ B-4-a) how the composition of vector operators ($K_1 = K_2 = 1$) yields operators $K = 0, 1, 2$. Equalities (VIII-73) and (VIII-75) allow generalizing the calculations to any values K_1 and K_2. In particular, we see that for a product of operators to yield a *scalar* operator, one must always start from two irreducible tensor operators of the same order. One then uses (1) to write the scalar product of two operators $\boldsymbol{T}^{[K]}$ and $\boldsymbol{W}^{[K]}$:

$$\boldsymbol{T}^{[K]} \cdot \boldsymbol{W}^{[K]} = \sum_{Q} (-1)^Q\; T_Q^{[K]}\; W_{(-Q)}^{[K]} \tag{VIII-76}$$

In (VII-91), there appears a coefficient $(-1)^K\;(2K+1)^{-1/2}$ that we have left out. For $K = 1$, this will allow us to recover the usual scalar product of two vectors (*cf.* relation (2) of complement $\mathrm{B_{VIII}}$).

[8]The tensor $Z_Q^{[K]}$ is the tensor's irreducible part extracted from the tensor product $\boldsymbol{T}^{[K_1]} \otimes \boldsymbol{W}^{[K_2]}$. It can be noted:

$$Z_Q^{[K]} = \left[T^{[K_1]} \otimes W^{[K_2]}\right]_Q^{[K]}$$

Exercise: Consider two tensors $T^{[K=2]}$ and $Z^{[K=2]}$. Write their Cartesian components $\overline{\overline{T}}_{x_i x_j}$ and $\overline{\overline{Z}}_{x_i x_j}$. Calculate the "contracted product" $\sum_{ij} \overline{\overline{T}}_{x_i x_j} \overline{\overline{Z}}_{x_i x_j}$ and verify that it yields the scalar product $\boldsymbol{T}^{[K=2]} \cdot \boldsymbol{Z}^{[K=2]}$.

The rule $K_1 = K_2$ is useful whenever one tries to build scalar operators starting from tensor operators. This is the case, for example, when building the Hamiltonian of an isolated system (rotation invariant) starting from products of operators acting in the state space of the system.

Examples

(*i*) For a free particle, operators such as $\boldsymbol{P}^2$, $\boldsymbol{R} \cdot \boldsymbol{L}$, are, a priori[9], good candidates for the Hamiltonian (which must commute with $\boldsymbol{J} = \boldsymbol{L} + \boldsymbol{S}$). For a particle in a central potential, one can choose $\xi(r)\ \boldsymbol{L} \cdot \boldsymbol{S}$.

(*ii*) Consider a set of two interacting systems $\mathcal{S}_1$ and $\mathcal{S}_2$, isolated as a whole (no external potential). The interaction Hamiltonian W must be invariant under the simultaneous rotation of $\mathcal{S}_1$ and $\mathcal{S}_2$, but not necessarily under the rotation of $\mathcal{S}_1$ or $\mathcal{S}_2$ separately. Consequently, W may include terms that would be impossible in H_1 and H_2 separately: terms in $\boldsymbol{J}_1 \cdot \boldsymbol{J}_2$, for example, in $\boldsymbol{J}_1 \cdot \boldsymbol{S}_2$ (spin–other–orbit coupling between the two electrons of the helium atom), in $\delta(r)\ \boldsymbol{J}_1 \cdot \boldsymbol{S}_2$ [if $\boldsymbol{J}_1 = \boldsymbol{I}$, the nuclear spin, and if $\boldsymbol{S}_2$ is an electron spin, this is the contact term $\delta(r)\ \boldsymbol{I} \cdot \boldsymbol{S}$ of the hyperfine structure Hamiltonian]. Another term of this type is the Hamiltonian W_{dd} of the dipole-dipole magnetic coupling between two systems (two electrons, for example) with positions $\boldsymbol{R}_1$ and $\boldsymbol{R}_2$ and spins $\boldsymbol{S}_1$ and $\boldsymbol{S}_2$; this Hamiltonian is written:

$$W_{\text{dd}} = \frac{\lambda}{R^3} \left[\boldsymbol{S}_1 \cdot \boldsymbol{S}_2 - 3\, \frac{(\boldsymbol{S}_1 \cdot \boldsymbol{R})\,(\boldsymbol{S}_2 \cdot \boldsymbol{R})}{R^2} \right] \tag{VIII-77}$$

where λ is a constant and $\boldsymbol{R} = \boldsymbol{R}_1 - \boldsymbol{R}_2$.

Exercise:

Verify that W_{dd} is rotation invariant. Introduce, in the state space ${}^S\mathscr{E}(1) \otimes {}^S\mathscr{E}(2)$ associated with the spins of the two particles, a spin irreducible tensor operator ${}^S T^{[K=2]}$ and write its components. Using the relative angular position operator $\boldsymbol{R}/R$ (which is a vector), build an irreducible tensor operator ${}^O T^{[K=2]}$ in the orbital space ${}^O\mathscr{E}(1) \otimes {}^O\mathscr{E}(2)$; show that ${}^O T_0^{(2)}$ is proportional to $3\cos^2\theta - 1$

[9]Note however that $\boldsymbol{P} \cdot \boldsymbol{S}$ is not parity invariant (it is changed into its opposite under a symmetry with respect to a point, *cf.* complement D_V) and that $\boldsymbol{R} \cdot \boldsymbol{L}$ is not invariant under time reversal (*cf.* appendice).

(where θ is the angle between $\boldsymbol{r}_1 - \boldsymbol{r}_2$ and a fixed Oz axis). Taking the product of ${}^{S}T^{[K=2]}$ and ${}^{O}T^{[K=2]}$, build a scalar operator and show that it is equal to W_{dd} (to within a coefficient).

C. Wigner–Eckart theorem

The Wigner–Eckart theorem gives a very useful expression for the matrix element:

$$\langle \tau, J, M | T_Q^{[K]} | \tau', J', M' \rangle$$

of an irreducible tensor operator $T_Q^{[K]}$ in a standard basis. It shows that this matrix element is proportional to the Clebsch–Gordan coefficient $\langle JM | J'KM'Q \rangle$ with a proportionality coefficient independent of M, M', and Q. In quantum mechanics, where we most frequently use a standard basis, this leads to a significant simplification of the computations: thanks to the Wigner–Eckart theorem, knowing one matrix element necessarily means knowing many others. The presence of the Clebsch–Gordan coefficient also leads to *selection rules.* For example, one must have $M = Q + M'$ for this coefficient to be different from zero; all the matrix elements that do not obey this equality are necessarily zero.

Before giving a precise demonstration of the theorem, we first present the general idea. We start by noticing that, under a rotation, the ket:

$$T_Q^{[K]} | \tau', J', M' \rangle$$

becomes:

$$R\, T_Q^{[K]}\, R^\dagger\ R | \tau', J', M' \rangle$$

In this expression, $R\, T_Q^{[K]}\, R^\dagger$ is known (transformation of an irreducible tensor operator), as is $R\ | \tau', J', M' \rangle$ (transformation of the kets of a standard basis). The same type of reasoning we already used several times shows that the rotation transformation of the kets $T_Q^{[K]} | \tau', J', M' \rangle$ is analogous to the rotation transformation of the tensor products $| J = K, M = Q \rangle \otimes | J', M' \rangle$. This will lead us to couple $\boldsymbol{J}'$ and $\boldsymbol{K}$ to form $\boldsymbol{J}'' = \boldsymbol{J}' + \boldsymbol{K}$, and to write states $| k, J = J'', M'' \rangle$ thanks to the Clebsch–Gordan coefficients $\langle J'KM'Q | J''M'' \rangle$. We then just need to take a scalar product with $\langle \tau, J, M |$. We expect it to be zero (as for a standard basis) unless $J = J''$, $M = M''$; these equalities, inserted in the coefficient $\langle J'KM'Q | J''M'' \rangle$, yield the predicted coefficient.

C-1. Demonstration

Let us make the reasoning more explicit. We shall proceed in two steps, beginning with a lemma concerning the scalar product $\langle \tau, J, M | k, J'', M'' \rangle$ of kets from two different standard bases, labeled respectively by the indices τ and k. The theorem itself will then be demonstrated.

C-1-a. Lemma

Consider two sets of $2J+1$ and $2J''+1$ vectors:

$$|\tau, J, M\rangle \qquad M = J,\ J-1,\ \ldots, -J$$
$$|k, J'', M''\rangle \qquad M'' = J'',\ J''-1,\ \ldots, -J''$$

By hypothesis, these kets obey the rules of a standard basis[10]: the action of J_+, J_-, J_z, and hence of any component J_u of $\boldsymbol{J}$, is given by relations (VII-11). As a result, we are going to show that the scalar product $\langle \tau, J, M | k, J'', M''\rangle$

- is zero unless $J = J''$ and $M = M''$;
- is independent of M (and M'').

The demonstration of this lemma is very simple:

• two eigenkets of $\boldsymbol{J}^2$ or of J_z (both Hermitian operators) are orthogonal if they are associated with different eigenvalues.

• Using relation (VII-5b), we insert operator:

$$J_+ J_- = \boldsymbol{J}^2 - J_z^2 + \hbar\, J_z \tag{VIII-78}$$

between $\langle \tau J M|$ and $|k J M\rangle$. This yields:

$$\begin{aligned}\langle \tau, J, M | J_+ J_- | k, J, M\rangle &= \left(\langle \tau, J, M | J_+\right)\left(J_- | k, J, M\rangle\right)\\ &= \hbar^2 \left[J(J+1) - M(M-1)\right]\ \langle \tau, J, M-1 | k, J, M-1\rangle\end{aligned} \tag{VIII-79}$$

since the $|\tau, J, M\rangle$ and the $|k, J, M\rangle$ obey the relations of a standard basis. In addition, relation (VIII-78) leads to, since $|k, J, M\rangle$ is an eigenvector of $\boldsymbol{J}^2$ and J_z:

$$\langle \tau, J, M | J_+\, J_- | k, J, M\rangle = \langle \tau, J, M | k, J, M\rangle\ \hbar^2 \left[J(J+1) - M^2 + M\right] \tag{VIII-80}$$

Comparing (VIII-79) and (VIII-80), we see that:

$$\langle \tau, J, M | k, J, M\rangle = \langle \tau, J, M-1 | k, J, M-1\rangle \tag{VIII-81}$$

which proves the lemma.

[10] Different notations $|\tau, J, M\rangle$ and $|k, J'', M''\rangle$ are used to stress the fact that these kets do not necessarily belong to the same standard basis. The expansion of $|k, J'', M''\rangle$ on the $|\tau, J, M\rangle$ may involve several values of τ.

Comment:

Another way to establish this property is to use the "Schur lemma", which we only quote here without demonstration – see [15] chapter 9, theorems 2 and 3, or [16], chapter 3, § 3-14. Consider two sets of matrices $M_1(g)$ and $M_2(g)$, which form two *irreducible* representations, with respective dimensions p_1 and p_2, of the same group G (whose elements are noted g). If there exists a matrix F, with p_1 columns and p_2 rows, such that:

$$F\ M_1(g) = M_2(g)\ F \tag{VIII-82}$$

one can conclude that:

- either the representations $M_1(g)$ and $M_2(g)$ are equivalent (which is obviously possible only if $p_1 = p_2$), or F is zero;
- if the representations are not only equivalent but also identical $[M_1(g) \equiv M_2(g)$ for any $g \in G]$, F is necessarily a scalar matrix (i.e. proportional to the identity matrix).

To apply this lemma, we consider the two spaces $\mathscr{E}_{\tau,J}$ (with dimension $2J+1$), generated by the $|\tau, J, M\rangle$, and $\mathscr{E}_{k,J''}$ (with dimension $2J''+1$) generated by the $|k, J'', M''\rangle$. They are both invariant under the action of all the rotation operators, and we can write:

$$\begin{aligned}
\langle \tau, J, M|R|k, J'', M''\rangle &= \sum_{M'} \langle \tau, J, M|R|\tau, J, M'\rangle\ \langle \tau, J, M'|k, J'', M''\rangle \\
&= \sum_{M'} \left(R^{[J]}\right)_{MM'}\ (F)_{M'M''} \\
&= \sum_{M'''} \langle \tau, J, M|k, J'', M'''\rangle\ \langle k, J'', M'''|R|k, J'', M''\rangle \\
&= \sum_{M'''} (F)_{MM'''}\ \left(R^{[J'']}\right)_{M'''M''}
\end{aligned} \tag{VIII-83}$$

where $(F)_{MM''}$ is defined as:

$$(F)_{MM''} = \langle \tau, J, M|k, J'', M''\rangle \tag{VIII-84}$$

The matrix (F) is associated with a linear application of $\mathscr{E}_{k,J''}$ into $\mathscr{E}_{\tau,J}$, such that:

$$F|k, J'', M''\rangle = \sum_{M} (F)_{MM''}|\tau, J, M\rangle \tag{VIII-85}$$

In addition, equalities (VIII-83) yield:

$$\left(R^{[J]}\right)\ (F) = (F)\ \left(R^{[J'']}\right) \tag{VIII-86}$$

and, unless (F) is zero, the Schur lemma states that (F) is proportional to the identity matrix (in this case, $J = J''$).

C-1-b. The theorem

We now demonstrate the well-known Wigner–Eckart theorem, expressed as:

$$\boxed{\begin{aligned}&\langle\tau, J, M|T_Q^{[K]}|\tau', J', M'\rangle\\ &\qquad= \langle J\,M|J'\,K\,M'\,Q\rangle\,\frac{\langle\tau, J||T^{[K]}||\tau', J'\rangle}{\sqrt{2J+1}}\\ &\qquad= (-1)^{J-M}\begin{pmatrix} J & K & J'\\ -M & Q & M'\end{pmatrix}\langle\tau, J||T^{[K]}||\tau', J'\rangle\end{aligned}} \tag{VIII-87}$$

In the second equality appears a $3-j$ coefficient, whose definition is given in complement B_{VII}. The number $\langle\tau\, J||T^{[K]}||\tau'\, J'\rangle$ is called the "*reduced matrix element*" of $T^{[K]}$ between the subspaces $\mathscr{E}(\tau, J)$ and $\mathscr{E}(\tau', J')$. It depends only on these subspaces and on the operator $T^{[K]}$ under study. The value of this reduced matrix element determines, thanks to relation (VIII-87), the values of the $(2J+1)(2J'+1)(2K+1)$ matrix elements of the operators $T_Q^{[K]}$, associated with all possible values of J, M', and Q.

Demonstration: Using relation (VIII-59), we can write the transformation laws of the kets $T_Q^{[K]}|\tau', J', M'\rangle$:

$$\begin{aligned}R\,T_Q^{[K]}|\tau', J', M'\rangle &= R\,T_Q^{[K]}\,R^\dagger\,R|\tau', J', M'\rangle\\ &= \sum_{Q'}\sum_{M''}\left(R^{[K]}\right)_{Q'Q}\left(R^{[J']}\right)_{M''M'}\,T_{Q'}^{[K]}|\tau', J', M''\rangle\end{aligned} \tag{VIII-88}$$

This equality shows that these kets are transformed in the same way as the tensor products:

$$|J=K, M=Q\rangle\otimes|J', M'\rangle \tag{VIII-89}$$

Consequently, we can use the general theory of angular momentum addition and introduce the kets $|k, J'', M''\rangle$ given by:

$$|k, J'', M''\rangle = \sum_{Q\,M'}\langle J'\,K\,M'\,Q|J''\,M''\rangle\;\;T_Q^{[K]}|\tau', J', M'\rangle \tag{VIII-90}$$

where $\langle J'\,K\,M'Q|J''\,M''\rangle$ is a Clebsch–Gordan coefficient. These kets are simply transformed like the kets of a standard basis, i.e. according to:

$$R|k, J'', M''\rangle = \sum_{M'''}\left(R^{[J'']}\right)_{M'''\,M''}\,|k, J'', M'''\rangle \tag{VIII-91}$$

The demonstration is the same as the one of formula (VIII-51b), using definition (VIII-50a).

Inverting equality (VIII-90), we get (using the fact that the coefficients $\langle J'\,K\,M'''\,Q|J''M''\rangle$ define an orthogonal change of basis):

$$T_Q^{[K]}|\tau', J', M'\rangle = \sum_{J''M''} \langle J'\,K\,M'\,Q|J''\,M''\rangle|k, J'', M''\rangle \tag{VIII-92}$$

Since the $|k, J'', M''\rangle$ are transformed like the kets of a standard basis, we can apply the lemma established above:

$$\langle \tau, J, M|k, J'', M''\rangle = \delta_{JJ''}\,\delta_{MM''} \times [\text{factor independent of } M]$$

We then set:

$$\langle \tau, J, M|k, J'', M''\rangle = \delta_{JJ''}\,\delta_{MM''}\frac{1}{\sqrt{2J+1}}\,\langle \tau, J||T^{[K]}||\tau', J'\rangle \tag{VIII-93}$$

where the factor $(2J+1)^{-1/2}$ has been introduced by pure convention; the quantity $\langle \tau, J||T^{[K]}||\tau', J'\rangle$ depends only on the quantum numbers that appear in it (it does not depend on M, M', or Q). Multiplying expression (VIII-92) by the bra $\langle \tau, J, M|$, we get relation (VIII-87).

Comments:

(i) For a tensor of maximum order, the (VIII-67) hermiticity condition of $T^{[K]}$ is equivalent to[11]:

$$\langle \tau, J||T^{[K]}||\tau', J'\rangle = (-1)^{J-J'}\langle \tau,'\,J'||T^{[K]}||\tau, J\rangle^{\star} \tag{VIII-94}$$

Demonstration: If $T_Q^{[K]\dagger} = (-1)^Q\,T_{-Q}^{[K]}$, we have:

$$\begin{aligned}&\langle \tau, J, M|(T_Q^{[K]})^{\dagger}|\tau', J', M'\rangle \\ &\quad = (-1)^Q\,\frac{1}{\sqrt{2J+1}}\,\langle J, M|J'\,K\,M'\,-Q\rangle\langle \tau\,J||T^{[K]}||\tau'\,J'\rangle\end{aligned} \tag{VIII-95}$$

In addition, according to the definition of an adjoint operator, this matrix element is equal to:

$$\begin{aligned}&\langle \tau', J', M'|T_Q^{[K]}|\tau, J, M\rangle^{\star} \\ &\quad = \frac{1}{\sqrt{2J'+1}}\,\langle J', M'|J\,K\,M\,Q\rangle\langle \tau'\,J'||T^{[K]}||\tau\,J\rangle^{\star}\end{aligned} \tag{VIII-96}$$

where we have used again the Wigner–Eckart theorem to write the right-hand side. Since relations (18) of chapter VII show that:

$$\langle J'\,M'|J\,K\,M\,Q\rangle = (-1)^{J-J'+Q}\,\sqrt{\frac{2J'+1}{2J+1}}\,\langle J\,M|J'\,K\,M'\,-Q\rangle \tag{VIII-97}$$

[11] The simplicity of relation (VIII-94) explains the arbitrary introduction of the coefficient $\sqrt{2J+1}$ in the denominator of the right-hand side of (VIII-93).

equality (VIII-94) is established.

Inside a subspace $\mathscr{E}(\tau, j)$, the hermiticity condition of operator $T^{[K]}$ is reduced to:

$$\langle \tau, J || T^{[K]} || \tau, J \rangle = \langle \tau, J || T^{[K]} || \tau, J \rangle^{\star} \tag{VIII-98}$$

which simply indicates that the reduced matrix element is real.

(ii) The Wigner–Eckart theorem is not limited to the rotation group $R_{(3)}$. This group, however, exhibits a property that is not general: in the decomposition of a tensor product of representations $(R^{[j_1]}) \otimes (R^{[j_2]})$ into irreducible representations, each representation $(R^{[J]})$ will appear only once (at most). As a result, there exists only one reduced matrix element. Writing the Wigner–Eckart theorem for groups other than $R_{(3)}$ (crystal symmetry groups, for example), requires changing the right-hand side of (VIII-87) by adding a sum including several reduced matrix elements.

D. Applications and examples

D-1. Scalar operators

When $K = 0$, operator $T^{[K]}$ is said to be a *scalar* operator. Relations (VIII-54) show that it is a rotation invariant operator, commuting with all the components of the total angular momentum $\boldsymbol{J}$. The Clebsch–Gordan coefficient appearing in expression (VIII-87) of the Wigner–Eckart theorem is written:

$$\langle J\, M | J'\, 0\, M'\, 0 \rangle = \delta_{JJ'}\, \delta_{MM'} \tag{VIII-99}$$

Accordingly, the only non-zero matrix elements of a scalar operator are those associated with the same values, in the bra and the ket, of the numbers J and M. In particular, inside a subspace $\mathscr{E}(\tau, J)$, these matrix elements are diagonal elements. This property can be demonstrated without the Wigner–Eckart theorem, using simply the fact that $T^{[K=0]}$ commutes with J_z and $J_{\pm}$.

How should we write the matrix representing an operator $T^{[K=0]}$ in a standard basis of the total state space? We just saw that, in each $\mathscr{E}(\tau, J)$, an operator $T^{[0]}$ is represented by a diagonal matrix whose elements are all equal (scalar matrix). It is possible for $T^{[0]}$ to also have non-diagonal elements different from zero, but always between a bra and a ket associated with the same value of J and of M (only the value of τ is different). In other words, we have square scalar matrices between two different subspaces $\mathscr{E}(\tau, J)$ and $\mathscr{E}(\tau', J)$, just as we do inside each of these subspaces:

$$
\begin{array}{ccccc}
 & \mathcal{E}(\tau_1, J) & \mathcal{E}(\tau_1, J') & \cdots & \mathcal{E}(\tau_2, J) \\
(T^{[0]}) = & \multicolumn{4}{c}{\begin{pmatrix}
\boxed{\begin{matrix}\lambda_1 & & \\ & \lambda_1 & \\ \vdots & \ddots & \vdots \\ & \cdots & \lambda_1\end{matrix}} & \cdots & \cdots & \boxed{\begin{matrix}\lambda_4 & & \\ & \lambda_4 & \\ \vdots & \ddots & \vdots \\ & \cdots & \lambda_4\end{matrix}} \\
\vdots & \boxed{\begin{matrix}\lambda_2 & & \\ & \lambda_2 & \\ \vdots & \ddots & \vdots \\ & \cdots & \lambda_2\end{matrix}} & & \vdots \\
\vdots & & \ddots & \vdots \\
\boxed{\begin{matrix}\lambda_5 & & \\ & \lambda_5 & \\ \vdots & \ddots & \vdots \\ & \cdots & \lambda_5\end{matrix}} & \cdots & \cdots & \boxed{\begin{matrix}\lambda_3 & & \\ & \lambda_3 & \\ \vdots & \ddots & \vdots \\ & \cdots & \lambda_3\end{matrix}}
\end{pmatrix}}
\end{array}
$$

(VIII-100)

Consider now two scalar operators $T_a^{[0]}$ and $T_b^{[0]}$. Inside each (τ, J), they are represented by scalar matrices of the form $\lambda_a(\tau) \times (\mathbb{1})$ and $\lambda_b(\tau) \times (\mathbb{1})$. Setting:

$$c_{ab}(\tau) = \lambda_a(\tau)/\lambda_b(\tau) \tag{VIII-101}$$

we get:

$$T_a^{[0]} = c_{ab}(\tau)\, T_b^{[0]} \tag{VIII-102a}$$

meaning $T_a^{[0]}$ and $T_b^{[0]}$ are proportional [all their matrix elements in $\mathscr{E}(\tau, J)$ are in the same ratio $c_{ab}(\tau)$]. For any state $|\psi_\tau\rangle$ of the physical system belonging to $\mathscr{E}(\tau, J)$, we have:

$$\langle T_a^{[0]}\rangle = \langle \psi_\tau | T_a^{[0]} | \psi_\tau \rangle = c_{ab}(\tau)\ \langle T_b^{[0]}\rangle \tag{VIII-102b}$$

One should be careful using these equalities, which are *valid only inside a given subspace* $\mathscr{E}(\tau, J)$. It is generally *incorrect* to say that $T_a^{[0]}$ and $T_b^{[0]}$ are proportional in the entire state space $\mathscr{E}$, the direct sum of all the $\mathscr{E}(\tau, J)$. In addition, the proportionality coefficient $\lambda_a(\tau)/\lambda_b(\tau)$ depends on the considered subspace $\mathscr{E}(\tau, J)$.

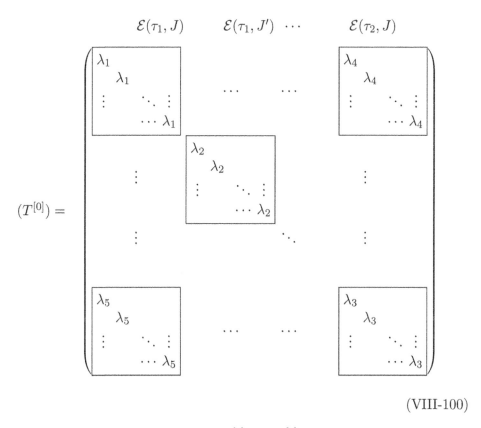

Comments:

(i) The best way to avoid possible mistakes is to never write a relation of the type (VIII-102a) and to replace it by:

$$P(\tau, J)\ T_a^{[0]}\ P(\tau, J) = c_{ab}(\tau)\ P(\tau, J)\ T_b^{[0]}\ P(\tau, J) \tag{VIII-102c}$$

where $P(\tau, J)$ is the orthogonal projector onto the subspace $\mathscr{E}(\tau, J)$. This formulation is more correct, but a bit heavier.

(ii) Here is an example of a classical mistake if one fails to remember that, in the entire space, two scalar operators are not necessarily proportional. Imagine we know the proportionality coefficient $c_{ab}(\tau)$, and consider the product operator:

$$T_a^{[0]}\ T_b^{[0]}$$

It is easy to show that this operator is also a scalar operator. Using (VIII-102a), one could be tempted to write that, in $\mathscr{E}(\tau, J)$, this operator is proportional to:

$$c_{ab}(\tau)\ \left[T_b^{[0]}\right]^2$$

This formulation is *incorrect*, and one should write:

$$\begin{aligned}&\langle \tau, J, M|T_a^{[0]}\ T_b^{(0)}|\tau, J, M\rangle \\ &\quad = \sum_{\tau'} \langle \tau, J, M|T_a^{[0]}|\tau', J, M\rangle\langle \tau', J, M|T_b^{[0]}|\tau, J, M\rangle\end{aligned} \tag{VIII-103}$$

where, in the inserted closure relation, the scalar character of the operators allows an elimination of the summations over intermediate quantum numbers J' and M'. Matrix elements of $T_a^{[0]}$ and $T_b^{[0]}$ between different $\mathscr{E}(\tau, J)$ spaces come into play explicitly (unless the summation over τ' happens to be reduced to $\tau' = \tau$; this would be case for example if we chose $T_a^{[0]} = \boldsymbol{J}^2$).

(iii) The non-diagonal matrix elements of two scalar operators $T_a^{[0]}$ and $T_b^{[0]}$ between two different subspaces $\mathscr{E}(\tau, J)$ and $\mathscr{E}(\tau', J)$ are also proportional. The proportionality coefficients depend on τ, τ', and J. We can insert these proportionality coefficients on the right-hand side of (VIII-103). However, their τ' dependence prevents the appearance of a closure relation on the basis $\{|\tau', J, M\rangle\}$ and hence of the operator $T_b^{(0)2}$.

D-2. Vector operators

A vector operator is associated with the value $K = 1$ and to the three possible values of Q, which are $+1$, 0, and -1. One can also use the Cartesian components $T_x^{[K=1]}$, $T_y^{[K=1]}$, and $T_z^{[K=1]}$ of the operator. The three values of Q, or of the Cartesian index x_i, correspond to three matrices whose elements obey the selection rules of the Clebsch–Gordan coefficient $\langle J\,M|J'\,1\,M'\,Q\rangle$. This yields, for the matrix elements $\langle\tau, J, M|T_Q^{[K=1]}|\tau, J', M'\rangle$, the following selection rules:

$$M' = M + Q \tag{VIII-104a}$$

$$J' = J + 1, \quad J, \quad J - 1 \tag{VIII-104b}$$

As for the matrix element $\langle\tau, J, M|T_i^{[K=1]}|\tau', J', M'\rangle$, it still obeys the selection rule (VIII-104b); the rule (VIII-104a) must be replaced by:

$$\begin{array}{lll} M' = M \pm 1 & \text{if } x_i = x \text{ or } y & (i = 1, 2) \\ M' = M & \text{if } x_i = z & (i = 3) \end{array} \tag{VIII-104c}$$

Inside a given subspace $\mathscr{E}(\tau, J)$, all the matrix elements of the operator $T^{[K=1]}$, denoted more conveniently $\boldsymbol{V}$, depend on a single parameter, the reduced matrix element:

$$\langle\tau, J||\boldsymbol{V}||\tau, J\rangle$$

It is a real number if $\boldsymbol{V}$ is Hermitian (*cf.* § B-5-b).

The selection rules (VIII-104c) show that the matrix representing V_z in $\mathscr{E}(\tau, J)$ is always diagonal. Its elements are proportional to the value of M (eigenvalue of J_z) appearing in the bra and the ket. This property can be verified, either directly on the coefficient $\langle J\,M|J\,1\,M\,0\rangle$, or by using the projection theorem that will be established below. As for the matrix elements of V_{+1}, they are located right above the main diagonal[12], and the matrix elements of V_{-1} right below. For the matrix elements representing V_x and V_y, the two previous types are possible; if V is Hermitian, the elements of V_x are real, those of V_y purely imaginary.

D-2-a. Link between two vector operators

Consider two vector operators $\boldsymbol{V}_a$ and $\boldsymbol{V}_b$; the ratio between their various matrix elements inside each subspace $\mathscr{E}(\tau, J)$ is, for each of their components (Cartesian or standard), always equal to the ratio of their reduced matrix elements. Setting:

$$\alpha_{ab}(\tau) = \frac{\langle\tau, J||\boldsymbol{V}_a||\tau, J\rangle}{\langle\tau, J||\boldsymbol{V}_b||\tau, J\rangle} \tag{VIII-105}$$

we can write, inside $\mathscr{E}(\tau, J)$:

$$\boldsymbol{V}_a = \alpha_{ab}(\tau)\,\boldsymbol{V}_b \tag{VIII-106a}$$

[12]The vectors of the standard basis are assumed to be arranged by decreasing order of M.

As for relation (VIII-102a), this equality should be used with caution: it concerns only operators $\boldsymbol{V}_a$ and $\boldsymbol{V}_b$ restricted to a subspace $\mathscr{E}(\tau, J)$. It is more correct to write, as in (VIII-102c):

$$P(\tau, J)\, \boldsymbol{V}_a\ P(\tau, J) = \alpha_{ab}(\tau)\, P(\tau, J)\, \boldsymbol{V}_b\, P(\tau, J) \qquad \text{(VIII-106b)}$$

D-2-b. Projection theorem

One can choose for $\boldsymbol{V}_b$ the total angular momentum operator $\boldsymbol{J}$, which commutes with $P(\tau, J)$. Relation (VIII-106b) then becomes, for any vector operator $\boldsymbol{V}$:

$$P(\tau, J)\, \boldsymbol{V}\ P(\tau, J) = \alpha_V(\tau)\, P(\tau, J)\, \boldsymbol{J} \qquad \text{(VIII-106c)}$$

To compute the constant $\alpha_V(\tau)$, we can take the scalar product of this equality with $\boldsymbol{J}$, using again the commutation of $P(\tau, J)$ and $\boldsymbol{J}$:

$$P(\tau, J)\, \boldsymbol{J} \cdot \boldsymbol{V}\ P(\tau, J) = \alpha_V(\tau)\, P(\tau, J)\, \boldsymbol{J}^2 \qquad \text{(VIII-107)}$$

In this equation, both sides are scalar operators (obtained by the product of two vector operators). The right-hand side is simply written:

$$\alpha_V(\tau)\, J(J+1)\, \hbar^2\ P(\tau, J)$$

since the subspace $\mathscr{E}(\tau, J)$ is an eigenspace of $\boldsymbol{J}$ with eigenvalue $J(J+1)\,\hbar$; the left-hand side will be noted:

$$\langle \boldsymbol{J} \cdot \boldsymbol{V} \rangle_{\tau,J}\ P(\tau, J)$$

where the constant $\langle \boldsymbol{J} \cdot \boldsymbol{V} \rangle_{\tau,J}$ is the common value of all the diagonal elements of the operator $\boldsymbol{J} \cdot \boldsymbol{V}$ in $\mathscr{E}(\tau, J)$. In other words, for any normalized ket $|\psi\rangle$ belonging to $\mathscr{E}(\tau, J)$, the average value $\langle \psi | \boldsymbol{J} \cdot \boldsymbol{V} | \psi \rangle$ is equal to the constant $\langle \boldsymbol{J} \cdot \boldsymbol{V} \rangle_{\tau,J}$. Finally, equality (VIII-107) yields:

$$\boxed{\alpha_V(\tau) = \frac{\langle \boldsymbol{J} \cdot \boldsymbol{V} \rangle_{\tau,J}}{J(J+1)\,\hbar^2}} \qquad \text{(VIII-108)}$$

The proportionality coefficient between $\boldsymbol{J}$ and $\boldsymbol{V}$ is the ratio between the scalar product of $\boldsymbol{J}$ and $\boldsymbol{V}$ and the modulus squared of $\boldsymbol{J}$. This result is often referred to as the "*projection theorem*". It can be interpreted classically by noting that, in the course of the precession of any vector $\boldsymbol{v}$ around the total (constant) angular momentum $\boldsymbol{j}$ of a physical system, the only non-zero average value over time is the projection:

$$\boldsymbol{v}_{/\!/} = \boldsymbol{j} \times \frac{\boldsymbol{v} \cdot \boldsymbol{j}}{\boldsymbol{j}^2} \qquad \text{(VIII-109)}$$

of $\boldsymbol{v}$ onto $\boldsymbol{j}$ (*cf.* figure 2); the projection $\boldsymbol{v}_\perp$ of $\boldsymbol{v}$ onto the plane perpendicular to $\boldsymbol{j}$ averages out to zero[13] after a long enough precession time of the system around $\boldsymbol{j}$.

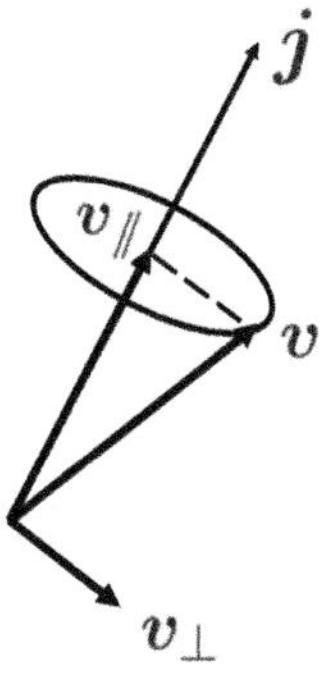

Figure 2: When the vector $\boldsymbol{v}$ rotates fast enough around the vector $\boldsymbol{j}$, its average value over time is equal to its projection $\boldsymbol{v}_{/\!/}$ onto $\boldsymbol{j}$.

Here again, care should be taken not to use this result outside its validity range. As an example, equality (VIII-108) does not allow obtaining the proportionality coefficient $\alpha_{ab}(\tau)$ between any two vector operators $\boldsymbol{V}_a$ and $\boldsymbol{V}_b$ in the form:

$$\frac{\langle \boldsymbol{V}_a \cdot \boldsymbol{V}_b \rangle_{\tau,J}}{\langle \boldsymbol{V}_b^2 \rangle_{\tau,J}}$$

This is because, in the demonstration of (VIII-108) we had to commute operators $P(\tau, J)$ and $\boldsymbol{J}$, which, in general, is no longer possible when $\boldsymbol{J}$ is replaced by another vector operator. Unless $\boldsymbol{V}_b$ commutes with $P(\tau, J)$, $\alpha_{ab}(\tau)$ is generally not equal to the above expression.

For matrix elements between two different subspaces $\mathscr{E}(\tau, J)$, thanks to relations (VIII-104), the Wigner–Eckart theorem considerably reduces the number of possible non-zero matrix elements of a vector operator $\boldsymbol{V}$. As an example, consider the V_z component of this operator. It is represented by a matrix that looks like the one written in (VIII-100), with the addition of non-zero matrix elements between subspaces $\mathscr{E}(\tau, J)$ and $\mathscr{E}(\tau, J \pm 1)$; the selection rules are slightly less strict for a vector operator than for a scalar operator.

[13]This classical analogy of the quantum mechanics projection theorem should be interpreted with caution. The theorem gives an instantaneous property of operators where time plays no role, whereas in classical mechanics $\boldsymbol{v}$ averages out to $\boldsymbol{v}_{/\!/}$ only after a sufficiently long time.

D-2-c. Application to optical dipole transitions

There are numerous applications of the Wigner–Eckart theorem in spectroscopy and atomic physics. Consider an atom being at the initial time $t = 0$ in one of the Zeeman sublevels (states $|J_e, M_e\rangle$) of an excited atomic level with angular momentum J_e. Through spontaneous emission of a photon with polarization $\boldsymbol{e}_\lambda$, it can fall back into one of the Zeeman sublevels (states $|J_f, M_f\rangle$) of its ground state. How should we evaluate the probability of the various processes, and the intensity of the emitted radiation as a function of the initial and final sublevels M_e and M_f, and what is the role of the polarization $\boldsymbol{e}_\lambda$ of the emitted radiation?

We assume that the atom is not subjected to any external magnetic field, or that this field is weak enough to be neglected[14]. All the $|J_e, M_e\rangle$ sublevels, as well as the $|J_f, M_f\rangle$ sublevels, have the same energy. They are often represented as in figure 3, by a diagram where the values of M are arranged by increasing value from left to right, and where vertical or oblique lines represent atomic transitions between excited and ground sublevels. The question is to evaluate the relative probabilities of all the processes that link a given sublevel M_e to a given sublevel M_f, and to understand the role of the polarization $\boldsymbol{e}_\lambda$.

We can use the Fermi golden rule (or the Wigner-Weisskopf theory for spontaneous emission) to compute the probability per unit time for the emission of a photon with momentum $\hbar\boldsymbol{k}$ and polarization $\boldsymbol{e}_\lambda$, by an atom going from sublevel M_e to sublevel M_f:

$$\Gamma\left(\boldsymbol{k}, \boldsymbol{e}_\lambda\right)_{M_e \to M_f} = \frac{2\pi}{\hbar}\, \left|\langle J_e, M_e; 0|H_{\text{int}}|J_f, M_f; \boldsymbol{k}, \boldsymbol{e}_\lambda\rangle\right|^2 \rho(E_f) \tag{VIII-110}$$

In this expression, $|J_e, M_e; 0\rangle$ is the state of the global system (atom + radiation) where the atom is in the state $|J_e, M_e\rangle$ and the radiation in its ground state (zero photon); $|J_f, M_f; \boldsymbol{k}, \boldsymbol{e}_\lambda\rangle$ represents the state of the global system where the atom is described by $|J_f, M_f\rangle$, whereas the radiation in the mode $\boldsymbol{k}, \boldsymbol{e}_\lambda$ is in its $n = 1$ excited state (one photon). The final state density is noted $\rho(E_f)$; conservation of energy requires $\hbar c k = E_e - E_f$. Finally, H_{int} represents the interaction Hamiltonian:

$$H_{\text{int}} = b \sum_q \boldsymbol{P}_q \cdot \boldsymbol{A}\left(\boldsymbol{R}_q\right) \tag{VIII-111}$$

where b is a constant, $\boldsymbol{P}_q$ and $\boldsymbol{R}_q$ are the momentum and position of the qth electron of the atom, and $\boldsymbol{A}$ is the vector potential of the electromagnetic field. This vector potential can be expressed as a function of the operators $a_{\boldsymbol{k},\lambda}$ that

[14] In practice, magnetic fields of the order of 0.1 tesla (1000 gauss) are necessary to significantly change the characteristics of the Zeeman lines emitted by an atomic vapor at room temperature.

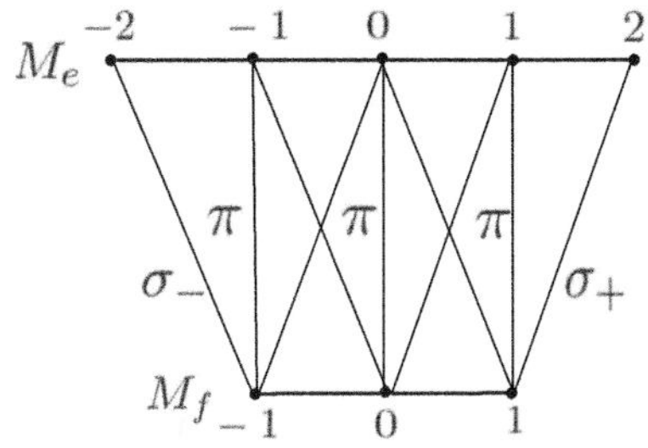

Figure 3: Polarization diagram for optical transitions between a lower level with angular momentum $J_f = 1, M_f$ and a higher level with angular momentum $J_e = 2, M_e$. The vertical lines represent a linear polarization parallel to the quantization axis Oz, called π polarization; because of the transverse character of a light wave, a beam with a pure π polarization propagates perpendicularly to Oz. The oblique lines leaning toward the right represent a right-hand circular polarization for a beam propagating in the Oz direction, and whose photons have an angular momentum $+\hbar$ (σ_+ polarization). The oblique lines leaning toward the left represent a left-hand circular polarization, with photons having an opposite angular momentum $-\hbar$ (σ_- polarization); the corresponding beams also propagate along the Oz axis.

annihilate a quantum in the mode $\boldsymbol{k}$, $\boldsymbol{e}_\lambda$ (for more details, see for example chapters XIX and XX of [9]):

$$\boldsymbol{A}(\boldsymbol{r}) = \sum_{\boldsymbol{k},\lambda} \mathcal{A}_k \, a_{\boldsymbol{k},\lambda} \, \boldsymbol{e}_\lambda \, \mathrm{e}^{\mathrm{i}\boldsymbol{k}\cdot\boldsymbol{r}} + \text{Hermitian conjugate} \tag{VIII-112}$$

where $\mathcal{A}_k$ is a function of k that we will not need here. Using the electric dipole approximation, we neglect the dimensions of the atom, on the order of the Bohr radius a_0, compared with the wavelength $\lambda_{\text{ph}} = 2\pi/k$ of the emitted photons; this approximation is well founded in the optical domain ($a_0/\lambda_{\text{ph}} \lesssim 10^{-3}$). Assuming the atom's nucleus to be fixed and at the origin of the reference frame, one can replace in (VIII-112) the exponentials $\mathrm{e}^{i\boldsymbol{k}\cdot\boldsymbol{r}}$ by 1. This leads to:

$$H_{\text{int}} = b \sum_{\boldsymbol{k},\lambda} a_{\boldsymbol{k},\lambda} \, \boldsymbol{e}_\lambda \cdot \boldsymbol{P} + \text{Hermitian conjugate} \tag{VIII-113}$$

where b is a constant, and $\boldsymbol{P}$ the total momentum of all the electrons:

$$\boldsymbol{P} = \sum_q \boldsymbol{P}_q \tag{VIII-114}$$

In (VIII-113), we can replace $\boldsymbol{P}$ by $\boldsymbol{R}$, the position operator of the electrons' center of mass. This is because, in equality (VIII-110), the only matrix element coming into play is between the atomic states $|J_e, M_e\rangle$ and $|J_f, M_f\rangle$, both belonging to a standard basis. The Wigner–Eckart theorem then states that all the matrix elements of $\boldsymbol{P}$ are proportional to those of $\boldsymbol{R}$. Finally, using (VIII-110) and (VIII-113), and the fact that $\langle 0|a_{\boldsymbol{k},\lambda}|\boldsymbol{k}, \boldsymbol{e}_\lambda\rangle = 1$, leads to:

$$\Gamma\left(\boldsymbol{k}, \boldsymbol{e}_\lambda\right)_{M_e \to M_f} \propto |\langle J_e, M_e|\boldsymbol{e}_\lambda \cdot \boldsymbol{R}|J_f, M_f\rangle|^2 \tag{VIII-115}$$

where the proportionality coefficient is independent of M_e, M_f, and $\boldsymbol{e}_\lambda$.

Choosing $\boldsymbol{e}_\lambda = \boldsymbol{e}_z$ allows computing the emission probability of a photon linearly polarized along Oz (this case corresponds to the so-called "π polarization"); the matrix element of the $Q = 0$ component of the vector operator comes into play:

$$\boldsymbol{e}_\lambda = \boldsymbol{e}_0 = \boldsymbol{e}_z \quad (\pi \text{ polarization}) \Leftrightarrow Q = 0 \tag{VIII-116a}$$

Choosing $\boldsymbol{e}_\lambda = -(\boldsymbol{e}_x + i\boldsymbol{e}_y)/\sqrt{2}$ yields the so-called σ^+ circular polarization, and the $Q = 1$ component:

$$\boldsymbol{e}_\lambda = \boldsymbol{e}_{+1} = -\frac{\boldsymbol{e}_x + i\boldsymbol{e}_y}{\sqrt{2}} \quad (\sigma^+ \text{ polarization}) \Leftrightarrow Q = +1 \tag{VIII-116b}$$

When $\boldsymbol{e}_\lambda = (\boldsymbol{e}_x - i\boldsymbol{e}_y)/\sqrt{2}$, we also get a circular polarization, but in the opposite direction, called σ^-, and the $Q = -1$ component:

$$\boldsymbol{e}_\lambda = \boldsymbol{e}_{-1} = \frac{\boldsymbol{e}_x - i\boldsymbol{e}_y}{\sqrt{2}} \quad (\sigma^- \text{ polarization}) \Leftrightarrow Q = -1 \tag{VIII-116c}$$

Finally, choosing any polarization $\boldsymbol{e}_\lambda$, we take its components on $\boldsymbol{e}_{+1}$, $\boldsymbol{e}_0$, and $\boldsymbol{e}_{-1}$ and get a linear combination of the 3 standard components $Q = +1$, 0, and -1 of $\boldsymbol{R}$. We shall limit our reasoning to the three simple cases where the polarization is σ^+, σ^-, or π. The Wigner–Eckart theorem then states that the transition probability is proportional to:

$$|\langle J_e, M_e|J_f\, 1\, M_f\, Q\rangle|^2$$

The square of a Clebsch–Gordan coefficient thus yields the relative intensities of all the vertical or oblique transitions of figure 2. The following selection rules must be obeyed:

$$\begin{aligned} |J_e - J_f| &\leq 1 \\ |M_e - M_f| &\leq 1 \qquad \text{(dipole rules)} \end{aligned} \tag{VIII-117a}$$

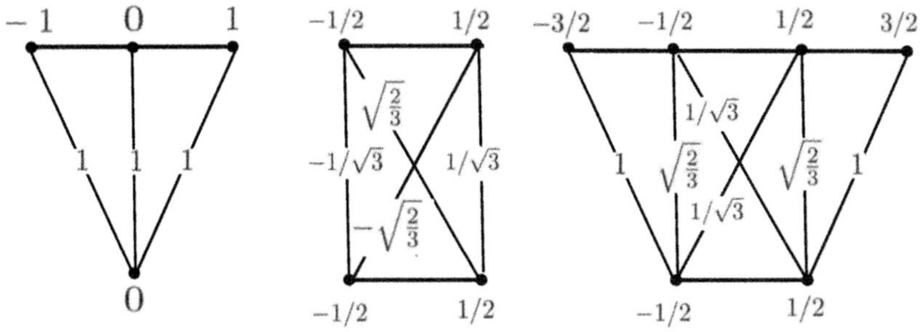

Figure 4: Examples of polarization diagrams. Figure on the left: transition between a lower level with angular momentum $J = 0$ and an upper level with $J = 1$. Middle figure: between two $J = 1/2$ levels. Figure on the right: between a lower level with $J = 1/2$ and an upper level with $J = 3/2$.

and

$$\begin{array}{ll} M_e = M_f & \text{polarization } \pi \\ M_e = M_f + 1 & \text{polarization } \sigma^+ \\ M_e = M_f - 1 & \text{polarization } \sigma^- \\ \text{(conservation of angular momentum)} & \end{array} \qquad \text{(VIII-117b)}$$

The values of the various Clebsch–Gordan coefficients $\langle J_e, M_e | J_f \, 1 \, M_f \, Q \rangle$ often appear in figures such as (VIII-3). Their absolute values are sufficient to evaluate the transition probability when $\boldsymbol{e}_\lambda = \boldsymbol{e}_{+1}$, $\boldsymbol{e}_0$, or $\boldsymbol{e}_{-1}$. However, for an arbitrary polarization $\boldsymbol{e}_\lambda$, which is a linear combination of these vectors, it is useful to know their sign. Figure 4 shows a few examples of such diagrams. In all cases, a π polarization corresponds to a vertical transition, a σ^+ polarization to an oblique transition leaning toward the right, a σ^- polarization to an oblique transition leaning toward the left. These diagrams are sometimes called "polarization diagrams".

Comment:

Since $\boldsymbol{R}$ acts only on the orbital variables of the electrons, we have the following selection rules for an atom where the orbital angular momentum of all the electrons $\boldsymbol{L}$ adds up to the sum of their spins $\boldsymbol{S}$, yielding the total angular momentum $\boldsymbol{J} = \boldsymbol{L} + \boldsymbol{S}$:

$$\begin{aligned} &\Delta L = 0, \pm 1 \\ &\Delta S = 0 \\ &\Delta J = 0, \pm 1 \end{aligned} \qquad \text{(VIII-118)}$$

where ΔL, ΔS, and ΔJ represent the differences between the quantum numbers associated with the excited level and the ground level. In the same way, if we consider an atom with a non-zero nuclear spin $\boldsymbol{I}$, and if we set $\boldsymbol{F} = \boldsymbol{I} + \boldsymbol{J}$, we get:

$$\Delta F = 0, \pm 1 \qquad \text{(VIII-119)}$$

D-3. $K = 2$ tensor operators

For tensor operators of order $K = 2$ (or more), we can make a similar argument as in §§ D-1 and D-2. For example, inside a given subspace $\mathscr{E}(\tau, J)$, all the irreducible tensor operators of a given order K are proportional to each other.

Starting from the vector operator $\boldsymbol{J}$, one can use the Clebsch–Gordan coefficients (VIII-49a) to build the Hermitian operator $\overline{\overline{J}}^{[K=2]}$ with standard components:

$$\begin{aligned} J_{\pm 2}^{[2]} &= (J_{\pm 1})^2 = (J_\pm)^2 /2 \\ J_{\pm 1}^{[2]} &= \frac{1}{\sqrt{2}} (J_{\pm 1}\, J_0 + J_0\, J_{\pm 1}) = \mp (J_\pm\, J_z + J_z\, J_\pm)/2 \\ J_0^{[2]} &= \frac{1}{\sqrt{6}} \left(J_{+1}\, J_{-1} + 2J_0^2 + J_{-1}\, J_{+1} \right) = \left(3J_z^2 - \boldsymbol{J}^2 \right) / \sqrt{6} \end{aligned} \qquad \text{(VIII-120)}$$

Inside $\mathscr{E}(\tau, J)$, all the irreducible tensor operators $T^{[K=2]}$ are proportional to this tensor $\overline{\overline{J}}^{[K=2]}$. The average value $\langle \overline{\overline{J}}^{[2]} \rangle$ of this tensor (i.e. the 5 average values of its components $J_Q^{[2]}$) are called the "alignment" of the physical system (as opposed to the "orientation" which represents the average value of $\boldsymbol{J}$). The Clebsch–Gordan coefficient $\langle J\, 2\, M'\, Q | J\, M \rangle$, which appears thanks to the Wigner–Eckart theorem, shows that the alignment can be different from zero only in a space $\mathscr{E}(\tau, J)$ where $J \geq 1$ (triangle law).

An important example of a $T^{[K=2]}$ operator is the electric quadrupole operator associated in quantum mechanics with an ensemble of charges (complement $\mathrm{C_{VIII}}$).

Comments:

(i) A tensor $\overline{\overline{T}}^{(n=2)}$ is not necessarily proportional to $\overline{\overline{J}}^{[K=2]}$ inside $\mathscr{E}(\tau, J)$. It is only true for its irreducible $K = 2$ part, meaning for its symmetric part with a zero trace. Its antisymmetric part is proportional to the cross product operator $\boldsymbol{J}\times$, i.e. to a $K = 1$ operator (with a coefficient that is, in general, independent of the coefficient for $K = 2$). As for the trace, it is proportional to $\boldsymbol{J}^2$ (with a third independent coefficient), and therefore to $(\mathbb{1})$.

(ii) Expanding the exponential $e^{i\boldsymbol{k}\cdot\boldsymbol{r}}$ of (VIII-112) in powers of a_0/λ introduces, after the electric dipole transitions, multipole transitions of a higher order[15]. Irreducible tensor operators of order $K = 2$, or more, may come into play, which changes the Clebsch–Gordan coefficient written above, and hence, the selection rules.

A well-known example is the green quadrupole line of the oxygen atom ($\lambda = 5\,577$ Å, with a corresponding lifetime of $\tau \simeq 1$ s), whose emission corresponds to the transition:

$$1s^2 2s^2 2p^4\ ,^1\mathrm{S}_0 \rightarrow 1s^2 2s^2 2p^4\ ,^1\mathrm{D}_2$$

It is clear that such a transition where $\Delta J = 2$ (and ΔL equals 2 as well) is forbidden by the dipole selection rules; this is why a quadrupole photon is emitted in the process. As this effect is of a higher order in a_0/λ, the level involved has a long lifetime: it is metastable.

[15]See, for example, complement $\mathrm{A_{XIII}}$ of [9].

Complement A$_{\mathrm{VIII}}$

Short review of classical tensors

This complement gives a short review of the classical properties of tensors. It is obviously out of the question to present a general theory of tensor algebra. We only introduce a few concepts and simple results, as well as the proper notations, in the framework of limited assumptions. Since chapter VIII deals with rotational invariance, we shall limit ourselves to orthonormal basis changes, whereas tensors are generally used for any basis change[1].

1. Vectors

To avoid constant reminders, we briefly summarize a few results of § A-1-a of chapter VIII. We have written the components of a given vector $\boldsymbol{v}$ in two orthonormal reference frames $\{\boldsymbol{e}_1, \boldsymbol{e}_2, \boldsymbol{e}_3\}$ and $\{\boldsymbol{e}'_1, \boldsymbol{e}'_2, \boldsymbol{e}'_3\}$ that are transformed into each other by the rotation $\mathscr{R}_{\boldsymbol{u}}(\varphi)$:

$$\boldsymbol{e}'_{1,2,3} = \mathscr{R}_{\boldsymbol{u}}(\varphi)\ \boldsymbol{e}_{1,2,3} \tag{1a}$$

or:

$$(\boldsymbol{e}'_1\ \boldsymbol{e}'_2\ \boldsymbol{e}'_3) = (\boldsymbol{e}_1\ \boldsymbol{e}_2\ \boldsymbol{e}_3)\ (\mathscr{R}_{\boldsymbol{u}}(\varphi)) \tag{1b}$$

where $(\mathscr{R}_{\boldsymbol{u}}(\varphi))$ is a 3×3 matrix whose elements, in its ith column, are the components of $\boldsymbol{e}'_i$ on the $\boldsymbol{e}_j$ (the i and j indices take the values 1, 2, or 3). This

[1]Considering "oblique" changes of basis may be quite useful in physics, in particular in the study of Lie algebras or relativity theory.

is the matrix associated with the linear (active) rotation operator that brings the e onto the e'.

We call v_x, v_y, v_z the components of an arbitrary vector v in the $\{e_1, e_2, e_3\}$ basis, and v'_x, v'_y, v'_z those in the basis $\{e'_1, e'_2, e'_3\}$:

$$v = v_x e_x + v_y e_y + v_z e_z = v'_x e'_x + v'_y e'_y + v'_z e'_z \tag{2}$$

We have:

$$v'_{x_i} = \sum_k v_{x_k} (\mathscr{R})_{ki} \tag{3}$$

and:

$$\begin{pmatrix} v_x \\ v_y \\ v_z \end{pmatrix} = \left(\mathscr{R}_{\boldsymbol{u}}(\varphi)\right) \begin{pmatrix} v'_x \\ v'_y \\ v'_z \end{pmatrix} \quad \text{or:} \quad \begin{pmatrix} v'_x \\ v'_y \\ v'_z \end{pmatrix} = \left(\mathscr{R}_{\boldsymbol{u}}(\varphi)\right)^{-1} \begin{pmatrix} v_x \\ v_y \\ v_z \end{pmatrix} \tag{4}$$

Since $(\mathscr{R})$ is an orthogonal matrix, this equality can be written:

$$\left(v'_x\, v'_y\, v'_z\right) = \left(v_x\, v_y\, v_z\right) \left(\mathscr{R}_{\boldsymbol{u}}(\varphi)\right) \tag{5'}$$

For an infinitesimal rotation $\mathscr{R}$ by an angle $\delta\varphi$ around the vector $\boldsymbol{u}$, we get:

$$\left(v'_x\, v'_y\, v'_z\right) = \left(v_x\, v_y\, v_z\right) \left[1 + \delta\varphi \begin{pmatrix} 0 & -u_z & u_y \\ u_z & 0 & -u_x \\ -u_y & u_x & 0 \end{pmatrix}\right] \tag{5}$$

where the $\delta\varphi$ terms yield the components of the cross product (vector product) $\boldsymbol{u} \times \boldsymbol{v}$.

2. Tensors

The concept of tensor was used in § B-1 of chapter VIII, in the cases where its components were operators. We use it here in the simple case where they are constants.

Starting from two vectors $\boldsymbol{v}$ and $\boldsymbol{w}$, we can build a table (3 × 3 matrix) including the 9 numbers $v_{x,y,z} \times w_{x,y,z}$:

$$\overline{\overline{T}} = \begin{pmatrix} v_x\, w_x & v_y\, w_x & v_z\, w_x \\ v_x\, w_y & v_y\, w_y & v_z\, w_y \\ v_x\, w_z & v_y\, w_z & v_z\, w_z \end{pmatrix} \tag{6a}$$

This table is called a tensor $\overline{\overline{T}}$ of order two, noted:

$$\overline{\overline{T}} = \boldsymbol{v} \otimes \boldsymbol{w} \tag{6b}$$

A well-known example of a tensor in physics is the moment of inertia of a solid. We assume for simplicity that the origin point O of the solid is immobile (no translation), and that the solid is composed of N particles of identical mass m, located at positions $\boldsymbol{r}_q$ with velocities $\boldsymbol{v}_q$. Since all the particles belong to the same solid body (their relative distances are constant), we must have:

$$\boldsymbol{v}_q = \boldsymbol{\Omega} \times \boldsymbol{r}_q \tag{7}$$

The motion is a pure rotation around O, and we note $\boldsymbol{\Omega}$ the angular velocity of the solid. Its angular momentum is:

$$\boldsymbol{J} = m \sum_{q=1}^{N} \boldsymbol{r}_q \times \boldsymbol{v}_q \tag{8}$$

which, using (7) and the vector triple product formula, can be written as:

$$\boldsymbol{J} = m \sum_{q} r_q^2 \boldsymbol{\Omega} - (\boldsymbol{r}_q \cdot \boldsymbol{\Omega})\, \boldsymbol{r}_q \tag{9}$$

The J_i component of $\boldsymbol{J}$ is given by:

$$J_i = m \sum_{j} \sum_{q} \left[\left(r_q^2 \right) \delta_{ij} - (r_q)_i \, (r_q)_j \right] \Omega_j \tag{10}$$

which can be expressed as a tensor relation:

$$J_i = \sum_j \overline{\overline{T}}_{ij}\, \Omega_j \qquad \text{or:} \qquad \boldsymbol{J} = \overline{\overline{\boldsymbol{T}}}\, \boldsymbol{\Omega} \tag{11}$$

with :

$$\overline{\overline{T}}_{ij} = m \sum_{q} \left[\left(r_q^2 \right) \delta_{ij} - (r_q)_i \, (r_q)_j \right] \tag{12}$$

The 9 quantities define the components of an order two tensor $\overline{\overline{T}}$ (inertia tensor of the solid).

In a more general way, consider n vectors $\boldsymbol{u}, \boldsymbol{v}, \boldsymbol{w}, \ldots, \boldsymbol{m}$. Calling d the space dimension (here $d = 3$), we can form a table with the d^n components of these vectors and obtain a tensor $\boldsymbol{T}^{(n)}$ noted:

$$\boldsymbol{T}^{(n)} = \boldsymbol{u} \otimes \boldsymbol{v} \otimes \boldsymbol{w} \otimes \ldots \otimes \boldsymbol{m} \tag{13}$$

When $n = 1$, we have a simple vector; for $n = 2$, we get an order[2] two tensor, associated with a square matrix; if $n = 3$, we need to use a "cubic matrix" defining an order three tensor, etc. The (Cartesian) components of $\boldsymbol{T}^{(n)}$ are:

$$T^{(n)}_{x_i\, x_j \ldots x_\ell} = u_{x_i}\, v_{x_j}\, \ldots,\, m_{x_\ell} \qquad \text{with:} \quad x_i, x_j, \ldots, x_\ell = x, y, \text{ or } z$$

[2] In the literature, n is sometimes called "rank" instead of "order".

Upon a rotation of the axes (basis change), the components of $\boldsymbol{T}^{(n)}$ become:

$$T^{(n)}_{x'_i\,x'_j\,\ldots\,x'_\ell} = u_{x'_i}\,v_{x'_j}\,\ldots\,m_{x'_\ell} \tag{14}$$

or:

$$T^{(n)}_{x'_i\,x'_j\,\ldots\,x'_\ell} = \sum_{i''j''\ldots\ell''} (\mathscr{R})_{i''i}\,(\mathscr{R})_{j''j}\,\ldots\,(\mathscr{R})_{\ell''\ell}\,T^{(n)}_{x_{i''}\,x_{j''}\,\ldots\,x_{\ell''}} \tag{15}$$

This equality is actually the general definition of an order n Cartesian tensor: a set $\boldsymbol{T}^{(n)}$ of 3^n (orthonormal) Cartesian components is said to form an order n tensor if, upon an orthonormal basis change, the new components are expressed as a function of the old ones according to the linear equalities (15). According to this definition, the $3^n \times 3^n$ matrix that gives the effects of the basis change on the components is equal to n times the tensor product of the matrix $(\mathscr{R})$ by itself[3].

It is clear that definition (15) is broader than (13): the set of tensors $\boldsymbol{T}^{(n)}$ includes not only tensor products such as (13) but all their linear combinations

The set of $T^{(n)}$ form an d^n-dimensional vector space (where $d = 3$ if one started from ordinary space). The tensor products of n vectors $\boldsymbol{e}_i \otimes \boldsymbol{e}_j \otimes \ldots \otimes \boldsymbol{e}_\ell$ (where each of the indices $i, j, \ldots, \ell$ takes d different values) constitute a basis of that space.

Example: We set, in a given basis:

$$\overline{\overline{T}}_{ij} = \delta_{ij} \tag{16}$$

In a new orthonormal basis, the new components are the $\overline{\overline{T}}'_{ij}$ given by:

$$\overline{\overline{T}}'_{ij} = \sum_{i''j''} (\mathscr{R})_{i''i}\,(\mathscr{R})_{j''j}\,T_{x_{i''}\,x_{j''}} = \sum_{i''} (\mathscr{R})_{i''i}\,(\mathscr{R})_{i''j} = \delta_{ij} \tag{17}$$

since $(\mathscr{R})$ is an orthogonal matrix. Equality (16), valid in any basis, defines a particularly simple order two tensor (identity tensor of order two, noted $\overline{\overline{1}}$).

The transformations of order n tensors yield 3^n-dimensional representations of the group $R_{(3)}$ of the rotations around a fixed point. Finding the decomposition of these representations into irreducible representations led to the introduction of irreducible tensors $T^{[K]}$.

[3]If one starts with arbitrary components of $T^{(n)}$ in a single basis, one can always consider that an order n tensor has been defined, and use (14) to compute its components in any other basis. However, if one starts with components of $T^{(n)}$ in all the (orthonormal) bases, one must verify that $T^{(n)}$ is indeed a tensor. This second point of view will be adopted in what follows, where, for example, the definition formulas for a tensor product or a contracted product can be applied simultaneously in all the bases.

Comment:

As already mentioned, limiting ourselves to orthonormal basis changes considerably restricts the problem; it allows ignoring the difference between the contravariant and covariant components of a tensor.

This restriction amounts to considering only the representations of the subgroups SO(3) of GL$(3, R)$. It increases the number of invariant properties of the $T^{(n)}$ in the allowed basis changes; consequently, it allows a further decomposition of the set of $T^{(n)}$ into invariant subsets. For example, the diagonal tensor (16) is by itself an invariant subspace under the action of the basis changes associated with the elements of $R_{(3)}$. However, this would not be the case for GL$(3, R)$ if the δ_{ij} were considered as two times co- or contravariant components (the invariance being preserved for mixed coordinates).

3. Properties

3-a. Tensor addition

The tensor addition of two (or several) tensors of order n is simply obtained by adding the corresponding components (those in the same row and column). It is clear that this operation yields a tensor of the same order.

3-b. Tensor product

Starting from two tensors $T^{(n_1)}$ and $T^{(n_2)}$, we can build a tensor $P^{(n)} = T^{(n_1)} \otimes T^{(n_2)}$ of order $n = n_1 + n_2$, whose components $P^{(n)}_{x_i\, x_j\, \dots\, x_\ell}$ are given by:

$$P_{\underbrace{x_i\, x_j\, \dots\, x_h}_{n_1}\ \underbrace{x_{h+1}\, \dots\, x_\ell}_{n_2}} = T^{(n_1)}_{x_i\, x_j\, \dots\, x_h}\, T^{(n_2)}_{x_{h+1}\, \dots\, x_\ell} \tag{18}$$

It is easy to show that this does indeed yield a tensor of order n.

3-c. Contraction

The index contraction operation consists of transforming a tensor $T^{(n)}$ into a tensor $Q^{(n-2)}$, of order $n-2$, by setting:

$$Q^{(n-2)}_{x_k\, \dots\, x_m} = \sum_{x_i = x, y, z} T^{(n)}_{x_i\, x_i\, x_k\, \dots\, x_m} \tag{19}$$

This amounts to summing all the diagonal elements of $T^{(n)}$ with respect to the first two indices, leaving all the others unchanged. The contraction operation can be easily generalized to any two indices, not necessarily the first ones, or two consecutive ones.

Let us check that this operation does indeed yield a tensor of order $n-2$, by computing the components of $Q^{(n-2)}$ in a new basis:

$$
\begin{aligned}
Q^{(n-2)}_{x'_k \dots x'_m} &= \sum_{i=1,2,3} T_{x'_i x'_i \dots x'_m} \\
&= \sum_{i=1,2,3} \sum_{i''j''\dots m''} (\mathscr{R})_{i''i} (\mathscr{R})_{j''i} \dots (\mathscr{R})_{m''m} \, T_{x_{i''} x_{j''} \dots x_{m''}}
\end{aligned}
\tag{20}
$$

Now:

$$
\sum_i (\mathscr{R})_{i''i} (\mathscr{R})_{j''i} = \delta_{i''j''}
\tag{21}
$$

(product of two rows of an orthogonal matrix) and we get:

$$
\begin{aligned}
Q^{(n-2)}_{x'_k \dots x'_m} &= \sum_{k''\dots m''} (\mathscr{R})_{k''k} \dots (\mathscr{R})_{m''m} \sum_{i''} T_{x_{i''} x_{i''} x_{k''} \dots x_{m''}} \\
&= \sum_{k''\dots m''} (\mathscr{R})_{k''k} \dots (\mathscr{R})_{m''m} \, Q^{(n-2)}_{x_{k''} \dots x_{m''}}
\end{aligned}
\tag{22}
$$

establishing that $Q^{(n-2)}$ is a tensor of order $n-2$.

Examples:

(i) The trace of an order 2 tensor is a scalar (rotation invariant).

(ii) Consider an order 2 tensor $\overline{\overline{\boldsymbol{T}}}$ and an order 1 tensor $\overline{\boldsymbol{V}}$ (meaning a vector). The tensor product

$$
\overline{\overline{\overline{\boldsymbol{P}}}} = \overline{\overline{\boldsymbol{T}}} \otimes \overline{\boldsymbol{V}}
\tag{23}
$$

is a tensor of order 3. By contraction of the last two indices, we get a tensor of order 1, the vector $\overline{\boldsymbol{W}}$ noted:

$$
\overline{\boldsymbol{W}} = \overline{\overline{\boldsymbol{T}}} \cdot \overline{\boldsymbol{V}}
\tag{24}
$$

The W_i component of $\overline{\boldsymbol{W}}$ is given by:

$$
W_i = \sum_j T_{ij} \, V_j
\tag{25}
$$

The index contraction of $\overline{\overline{\overline{\boldsymbol{P}}}}$ does amount to multiplying the rows of $(\overline{\overline{\boldsymbol{T}}})$ by the column of $(\overline{\boldsymbol{V}})$, as written for example in (11).

(iii) In the same way, two second-order tensors $(\overline{\overline{\boldsymbol{T}}})$ and $(\overline{\overline{\boldsymbol{Q}}})$ may be contracted into a scalar:

$$
\boldsymbol{T} : \boldsymbol{Q} = \sum_{ij} T_{ij} \, S_{ij}
\tag{26}
$$

4. Criterium for a tensor

As seen in the previous example, it is also possible to contract the indices of two different tensors, of order n_1 and n_2. This amounts to first consider their tensor product of order $n_1 + n_2$, and then contract two of the indices to obtain a tensor of order $n_1 + n_2 - 2$.

The necessary and sufficient condition for $T^{(n)}$ to be a tensor of order n is that by contraction with any tensor of order $p < n$, it yields a tensor of order $n - p$. We will assume that the condition is sufficient (it is not necessary to consider the ensemble of the a priori possible $n - 1$ values of p: it is already a sufficient condition if one has chosen a single arbitrary value $p < n$).

5. Symmetric and antisymmetric tensors

Consider a tensor $\overline{\overline{T}}$ of order two, such that:

$$T_{x_i\, x_j} = \eta\, T_{x_j\, x_i} \tag{27}$$

If $\eta = +1$, the tensor is said to be symmetric; if $\eta = -1$, it is said to be antisymmetric. This is a property of $\overline{\overline{T}}$, independent of the chosen basis, since:

$$\begin{aligned} T_{x'_i\, x'_j} &= \sum_{i''j''} (\mathscr{R})_{i''i}\, (\mathscr{R})_{j''j}\; T_{x_{i''}\, x_{j''}} = \eta \sum_{i''j''} (\mathscr{R})_{i''i}\, (\mathscr{R})_{j''j}\; T_{x_{j''}\, x_{i''}} \\ &= \eta\, T'_{x'_j\, x'_i} \end{aligned} \tag{28}$$

The inertia tensor defined in (12) is an example of a symmetric tensor.

This can be easily generalized to tensors of higher order: if, in a given basis, a tensor $T^{(n)}$ is symmetric or antisymmetric with respect to two indices (for example, $T_{x_i\, x_j\, x_k} = \eta\, T_{x_j\, x_i\, x_k}$), this property is preserved in any other basis.

This is an example of an invariance property allowing the definition of subspaces in the tensor space that remain globally invariant.

Comment:

For non-orthonormal basis changes, the above property remains valid for covariant or contravariant components with respect to the two indices in question. This is no longer true for mixed components (such as those generally used in the definition of a matrix associated with an operator).

6. Specific tensors

We have already seen the order 2 tensor $\overline{\overline{1}}$ with components δ_{ij}. There is another fundamental tensor, the $\overline{\overline{\overline{\varepsilon}}}_A$ tensor, totally antisymmetric and of order 3. Its components are the ε_{ijk} defined as:

$$\varepsilon_{i,j,k} = \begin{cases} +1 & \text{if } i,j,k \text{ are an even permutation of } 1,2,3 \\ -1 & \text{for an odd permutation} \\ 0 & \text{if 2 of the 3 indices } i,j,k \text{ are equal} \end{cases} \tag{29}$$

Let us check that this definition yields a tensor by computing the components ε'_{ijk} of $\overline{\overline{\overline{\varepsilon}}}_A$ in a new basis:

$$\varepsilon'_{ijk} = \sum_{i''j''k''} (\mathscr{R})_{i''i}\,(\mathscr{R})_{j''j}\,(\mathscr{R})_{k''k}\;\varepsilon_{i''j''k''} \tag{30}$$

The right-hand side of this equality is clearly invariant under the circular permutation of the three indices i, j, k, and changes its sign if two of them are exchanged (it is thus equal to zero if two indices are equal). All the ε'_{ijk} can be readily deduced from $\varepsilon'_{1,2,3}$, which is equal to:

$$\varepsilon'_{1,2,3} = \text{Det}\,(\mathscr{R}) = 1 \tag{31}$$

We thus have:

$$\varepsilon'_{ijk} = \varepsilon_{ijk} \tag{32}$$

meaning $\overline{\overline{\overline{\varepsilon}}}_A$ is indeed an order 3 tensor.

Consider, for example, a tensor $\overline{\overline{T}}$ of order 2, with components $\overline{\overline{T}}_{jk}$. Its tensor product with $\overline{\overline{\overline{\varepsilon}}}_A/2$ yields a tensor of order 5. We then contract it on two pairs of indices (each including one index from $\overline{\overline{T}}$); we are left with a tensor of order $5-2-2=1$. This is a vector, whose components are written:

$$V_{x_i} = \frac{1}{2}\sum_{jk} \varepsilon_{ijk}\; T_{x_j\,x_k} \tag{33}$$

or for instance:

$$V_x = \frac{1}{2}\,(T_{yz} - T_{zy}) \tag{33'}$$

(and the relations obtained by circular permutation of the indices x, y, and z).

Conversely, if a tensor T of order 2 is antisymmetric:

$$\overline{\overline{T}} = \begin{pmatrix} 0 & A & B \\ -A & 0 & C \\ -B & -C & 0 \end{pmatrix} \tag{34}$$

it can be seen as a vector $\boldsymbol{V}$ with components:

$$\begin{cases} V_x = -C \\ V_y = B \\ V_z = -A \end{cases} \tag{35}$$

The action of $\overline{\overline{T}}$ on a vector $\boldsymbol{W}$ (i.e. the contracted product $\overline{\overline{T}} \cdot \overline{W}$) simply yields the cross product (vector product) $\boldsymbol{V} \times \boldsymbol{W}$.

This result can be generalized to any order n. Whenever a tensor of order n is antisymmetric with respect to two of its components, it can be considered to be a "disguised" tensor of order $n-1$. Arranging its components in a different way, one can lower its order by one unity.

7. Irreducible tensors

We shall only give a brief review of the search for irreducible tensors under the action of the $R_{(3)}$ operations. This problem has been treated in detail in chapter VIII using the properties of angular momentum and of the addition of several angular momenta. We will show that a direct approach is possible, and find several results already established in chapter VIII.

Transformation laws (15) for the components of a tensor $T^{(n)}$ yield representations of $R_{(3)}$; the demonstration is the same as for a tensor operator of order 1. Let us focus on the problem stated above: can these laws be restricted to yield irreducible representations of lower dimension? The problem is to find linear combinations of the $T^{(n)}_{x_i x_j \dots x_\ell}$ that always transform into each other under a rotation of the axes, and yield representations of lower dimension.

• We start with the case $n = 2$. Using the results obtained above, we shall decompose the corresponding representation. As it is of dimension 9, numerous options exist: $j = 4$, or $j = 3 \oplus j = 1/2$, etc.

We first notice that the trace of $\overline{\overline{T}}$:

$$\text{Tr}\left\{\overline{\overline{T}}\right\} = \overline{\overline{T}}_{xx} + \overline{\overline{T}}_{yy} + \overline{\overline{T}}_{zz} \tag{36}$$

is invariant under any basis change. Setting:

$$\overline{\overline{T}} = \frac{1}{3}\,\text{Tr}\left\{\overline{\overline{T}}\right\} \times \overline{\overline{1}} + \overline{\overline{K}} \tag{37}$$

we can decompose $\overline{\overline{T}}$ into the sum of two contributions: a scalar part (invariant under a basis change), and a zero trace part (in any basis). The first part is linked to the one-dimensional representation $j = 0$ of the $R_{(3)}$ group.

As for the zero trace tensors, they clearly form an invariant subspace, but not irreducible as we now show. This subspace has a dimension 8. To reduce it, we set:

$$\overline{\overline{K}} = \overline{\overline{S}}_K + \overline{\overline{A}}_K \tag{38}$$

where the components of $\overline{\overline{S}}_K$ are:

$$(S_K)_{ij} = \frac{1}{2}\left[K_{ij} + K_{ji}\right] = \frac{1}{2}\left[T_{ij} + T_{ji}\right] - \frac{1}{3}\delta_{ij}\,\mathrm{Tr}\left\{\overline{\overline{T}}\right\} \tag{39a}$$

and those of $\overline{\overline{A}}_K$:

$$(A_K)_{ij} = \frac{1}{2}\left[K_{ij} - K_{ji}\right] = \frac{1}{2}\left[T_{ij} - T_{ji}\right] \tag{39b}$$

The $\overline{\overline{A}}_K$ tensors belong to the space of the completely antisymmetric tensors. We saw that these can be considered as vectors, and can be associated with the $j = 1$ irreducible 3-dimensional representation of $R_{(3)}$ (*cf.* chapter VII, § A-3-c).

We need now to consider the 5-dimensional space of the $\overline{\overline{S}}_K$ tensors, symmetric and with a zero trace. We assume here[4] that the corresponding representation is irreducible. It explains the presence of the $j = 2$ value in the representation under study.

• The previous resoning can be generalized to tensors of any order n. Taking the trace over any two indices yields tensors of order $n - 2$. In this way, one can subtract from the initial tensor $C_n^2 = n(n-1)/2$ tensors whose order was lowered by two units. One then writes:

$$\begin{aligned} T_{x_i x_j x_k \ldots x_\ell} = \frac{1}{2}&\left[T_{x_i x_j x_k \ldots x_\ell} + T_{x_j x_i x_k \ldots x_\ell}\right] \\ &+ \frac{1}{2}\left[T_{x_i x_j x_k \ldots x_\ell} - T_{x_j x_i x_k \ldots x_\ell}\right] \end{aligned} \tag{40}$$

The second term on the right-hand side, antisymmetric with respect to the first two indices, can be assimilated to a tensor of order $n-1$. It can thus be eliminated, and we keep only the first term, which is symmetric. The same operation can be performed on all the couples of indices.

Finally, after subtracting all the parts of $T^{(n)}$ that are equivalent to orders $n-1$ or $n-2$, we are left with a completely symmetric tensor, with zero traces [the components are invariant under the exchange of any two indices; the $n(n-1)/2$ sums over any two indices are all equal to zero]. The space generated by such tensors is irreducible, and of dimension $2n + 1$.

To check that this dimension is indeed $(2n + 1)$, we try and determine on how many parameters these tensors depend. For a completely symmetric tensor, the value of

[4]The demonstration has been given in § B-4-a of chapter VIII; see also complement B_{VIII}.

a component depends on the label (x, y or z) of the components that it includes, but not on their order:

$x : p$ times $\quad y : q$ times $\quad z : r$ times

if r is given, p can be chosen in $n+1-r$ different ways: $p = 0, 1, 2, \ldots, n-r$. We thus have to compute:

$$\sum_{r=0}^{n} n - r + 1 = n(n+1) - \frac{n(n+1)}{2} + n + 1 = \frac{(n+1)(n+2)}{2}$$

In addition, we have $n(n-1)/2$ conditions for a zero trace. Finally we have:

$$\frac{(n+1)(n+2)}{2} - \frac{n(n-1)}{2} = 2n + 1$$

independent parameters. We simply need to prove the irreducibility of the representations to establish that they correspond to $j = n$.

Comment:

If we consider basis changes that are not necessarily orthonormal, the irreducible representations must be found among the permutation group; the invariant subspaces are then less numerous.

Complement B_{VIII}

Second-order tensor operators

This complement examines in more detail how the equalities (VIII-49) of chapter VIII are used to decompose or build second-order tensors, thus expliciting their properties. Two calculations will be performed. First of all (§ 1), we consider the simple case where $\overline{\overline{T}} = \boldsymbol{V} \otimes \boldsymbol{W}$ and give the expressions of $T_Q^{[K]}$. We then consider (§ 2) an arbitrary tensor $\overline{\overline{T}}$ and write the matrices associated with its decomposition in irreducible parts, as in equality (VIII-55) of chapter VIII. This will explicit the distribution of the various matrix elements associated with each value of K.

1. Tensor product of two vector operators

Assuming that:

$$\overline{\overline{T}} = \boldsymbol{V} \otimes \boldsymbol{W} \tag{1}$$

we compute the components of the tensors $T^{[K]}$ for $K = 0, 1, 2$.

(i) $K = 0$

Equality (VIII-49c) yields:

$$\begin{aligned} T_{(Q=0)}^{[K=0]} &= \frac{1}{\sqrt{3}} \left[V_{+1}W_{-1} - V_0 W_0 + V_{-1}W_{+1}\right] \\ &= -\frac{1}{\sqrt{3}} \left[V_x W_x + V_y W_y + V_z W_z\right] \\ &= -\frac{1}{\sqrt{3}} \boldsymbol{V} \cdot \boldsymbol{W} \end{aligned} \tag{2}$$

The scalar operator we obtain (rotation invariant) is simply proportional to the scalar product $\boldsymbol{V} \cdot \boldsymbol{W}$.

(ii) $K = 1$

Using relation (VIII-49b), we get:

$$T_1^{[K=1]} = \frac{1}{\sqrt{2}}(V_{+1}W_0 - V_0W_{+1}) = -\frac{1}{2}\left[V_xW_z - V_zW_x + i\left(V_yW_z - V_zW_y\right)\right]$$
$$T_0^{[K=1]} = \frac{1}{\sqrt{2}}\left(V_{+1}W_{-1} - V_{-1}W_{+1}\right) = -\frac{i}{\sqrt{2}}\left[V_yW_x - V_xW_y\right]$$
$$T_{-1}^{[K=1]} = \frac{1}{\sqrt{2}}\left(V_0W_{-1} - V_{-1}W_0\right) = \frac{1}{2}\left[V_zW_x - V_xW_z - i\left(V_zW_y - V_yW_z\right)\right] \quad (3)$$

This yields the components of operator $\boldsymbol{K} = \boldsymbol{V} \times \boldsymbol{W}$ and we see that:

$$T_m^{[K=1]} = \frac{i}{\sqrt{2}}[\boldsymbol{V} \times \boldsymbol{W}]_m \qquad (m = +1, 0, -1) \quad (4a)$$

or:

$$T^{[K=1]} = \frac{i}{\sqrt{2}}\,\boldsymbol{K} = \frac{i}{\sqrt{2}}\,\boldsymbol{V} \times \boldsymbol{W} \quad (4b)$$

The commutation relations of this operator with the total angular momentum $\boldsymbol{J}$ are those of a vector operator (pseudovector, parity invariant).

(iii) $K = 2$

Finally, relation (VIII-49a) shows that:

$$T_2^{[2]} = V_{+1}W_{+1}$$
$$T_1^{[2]} = \frac{1}{\sqrt{2}}\left(V_{+1}W_0 + V_0W_{+1}\right)$$
$$T_0^{[2]} = \frac{1}{\sqrt{6}}\left(V_{+1}W_{-1} + 2V_0W_0 + V_{-1}W_{+1}\right)$$
$$T_{-1}^{[2]} = \frac{1}{\sqrt{2}}\left(V_0W_{-1} + V_{-1}W_0\right)$$
$$T_{-2}^{[2]} = V_{-1}W_{-1} \quad (5)$$

The expression for $T_0^{[2]}$ can be written as:

$$T_0^{[2]} = \frac{1}{\sqrt{6}}\left[3V_zW_z - \boldsymbol{V} \cdot \boldsymbol{W}\right] \quad (6)$$

As opposed to the two previous tensors, this tensor cannot be reduced to a tensor of order $n < 2$. Using the Cartesian components, it must be described by a 3×3 matrix, containing 9 operators. We shall see that this matrix is symmetric and has a zero trace.

2. Cartesian components of the tensor in the general case

Consider an arbitrary tensor, not necessarily a tensor product. We will calculate its Cartesian components $\overline{\overline{T}}_{x_i x_j}$ as a function of the $T_{m_i m_j}$, then of the $T_Q^{[K]}$. As an example, relation (VIII-45b) of chapter VIII, as well as equations (VIII-26) for the elements of the (S) matrix, yield for $\overline{\overline{T}}_{xx}$:

$$\overline{\overline{T}}_{xx} = \sum_{i'j'} (S)^{\star}_{1i'} \, (S)^{\star}_{1j'} \, T_{m_{i'} m_{j'}}$$
$$= \frac{1}{2} \left[T_{1\,1} + T_{-1\,-1} - T_{1\,-1} - T_{-1\,1} \right] \tag{7}$$

Using the Clebsch–Gordan coefficients of relation (VIII-49b), we get:

$$\overline{\overline{T}}_{xx} = \frac{1}{2} \left[T_2^{[2]} + T_{-2}^{[2]} - \left(\frac{1}{\sqrt{6}} T_0^{[2]} + \frac{1}{\sqrt{2}} T_0^{[1]} + \frac{1}{\sqrt{3}} T_0^{[0]} \right) \right.$$
$$\left. - \left(\frac{1}{\sqrt{6}} T_0^{[2]} - \frac{1}{\sqrt{2}} T_0^{[1]} + \frac{1}{\sqrt{3}} T_0^{[0]} \right) \right]$$
$$= \frac{1}{2} \left[T_2^{[2]} + T_{-2}^{[2]} \right] - \frac{1}{\sqrt{6}} T_0^{[2]} - \frac{1}{\sqrt{3}} T_0^{[0]} \tag{8}$$

In a similar way:

$$T_{xy} = \frac{i}{2} \left[-T_{1\,1} + T_{-1\,-1} - T_{1\,-1} + T_{-1\,1} \right]$$
$$= \frac{i}{2} \left[T_{-2}^{[2]} - T_2^{[2]} \right] - \frac{i}{\sqrt{2}} T_0^{[1]} \tag{9}$$

as well as:

$$T_{xz} = \frac{1}{\sqrt{2}} \left[-T_{1\,0} + T_{-1\,0} \right] = \frac{1}{2} \left[-T_1^{[2]} + T_{-1}^{[2]} - T_1^{[1]} - T_{-1}^{[1]} \right] \tag{10}$$

We have obtained all the matrix elements of the tensor operator T located on the first row of the matrix. Similar calculations can be carried out for the remaining 6 components, and yield the contribution of $T^{[0]}$, $T^{[1]}$, and $T^{[2]}$ to the Cartesian components of $\overline{\overline{T}}$.

• The contribution of $T^{[0]}$ is:

$$-\frac{1}{\sqrt{3}}\begin{pmatrix} T_0^{[0]} & 0 & 0 \\ 0 & T_0^{[0]} & 0 \\ 0 & 0 & T_0^{[0]} \end{pmatrix} \tag{11}$$

This is a scalar tensor of order $n = 2$. The non-zero components are along the main diagonal; they are equal and invariant under any rotation.

• The contribution of $T^{[1]}$ is written:

$$\frac{i}{\sqrt{2}}\begin{pmatrix} 0 & -T_0^{[1]} & \frac{i}{\sqrt{2}}\left[T_1^{[1]}+T_{-1}^{[1]}\right] \\ T_0^{[1]} & 0 & \frac{1}{\sqrt{2}}\left[T_1^{[1]}-T_{-1}^{[1]}\right] \\ -\frac{i}{\sqrt{2}}\left[T_1^{[1]}+T_{-1}^{[1]}\right] & -\frac{1}{\sqrt{2}}\left[T_1^{[1]}-T_{-1}^{[1]}\right] & 0 \end{pmatrix}$$

$$= \frac{i}{\sqrt{2}}\begin{pmatrix} 0 & -T_z^{[1]} & T_y^{[1]} \\ T_z^{[1]} & 0 & -T_x^{[1]} \\ -T_y^{[1]} & T_x^{[1]} & 0 \end{pmatrix} \tag{12a}$$

We have set, as for any vector operator:

$$T_x^{[1]} = \frac{1}{\sqrt{2}}\left[T_{-1}^{[1]} - T_1^{[1]}\right]$$

$$T_y^{[1]} = \frac{i}{\sqrt{2}}\left[T_1^{[1]} + T_{-1}^{[1]}\right]$$

$$T_z^{[1]} = T_0^{[1]} \tag{12b}$$

The tensor we obtain has a zero trace and is totally antisymmetric. It can be written as a cross product:

$$\frac{i}{\sqrt{2}}\boldsymbol{T}^{[1]} \times \ldots$$

or else, if $\overline{\overline{T}} = \boldsymbol{V} \otimes \boldsymbol{W}$:

$$-\frac{1}{2}\left(\boldsymbol{V} \times \boldsymbol{W}\right) \times \ldots$$

• Finally, the contribution of $T^{[K=2]}$ is written:

$$\begin{pmatrix} \frac{1}{2}\left[T_2^{[2]}+T_{-2}^{[2]}\right]-\frac{1}{\sqrt{6}}T_0^{[2]} & -\frac{i}{2}\left[T_2^{[2]}-T_{-2}^{[2]}\right] & -\frac{1}{2}\left[T_1^{[2]}-T_{-1}^{[2]}\right] \\ -\frac{i}{2}\left[T_2^{[2]}-T_{-2}^{[2]}\right] & -\frac{1}{2}\left[T_2^{[2]}+T_{-2}^{[2]}\right]-\frac{1}{\sqrt{6}}T_0^{[2]} & \frac{i}{2}\left[T_1^{[2]}+T_{-1}^{[2]}\right] \\ -\frac{1}{2}\left[T_1^{[2]}-T_{-1}^{[2]}\right] & \frac{i}{2}\left[T_1^{[2]}+T_{-1}^{[2]}\right] & \frac{2}{\sqrt{6}}T_0^{[2]} \end{pmatrix} \tag{13}$$

This yields, as we said before, a symmetric matrix with a zero trace.

Comments:

(i) Only $T^{[0]}$ contributes to the trace of $\overline{\overline{T}}$.

(ii) A tensor $T^{[K]}$ is Hermitian if its Cartesian components are. Since the (S) matrix is complex, this does not imply the Hermiticity of the spherical components. For the matrix of operators we have just written, $T^{[K=2]}$ and $T^{[K=0]}$ are Hermitian if:

$$\left[T_Q^{[K]}\right]^\dagger = (-1)^Q \, T_{-Q}^{[K]} \tag{14}$$

For $T^{[K=1]}$, this definition ensures that the three operators $T_x^{[1]}$, $T_y^{[1]}$, and $T_z^{[1]}$ defined in (12b) are Hermitian in the usual sense of the word (as components of a vector operator $\boldsymbol{T}$). However, for the contribution to a tensor of order 2, the operators involved are anti-Hermitian, because of the presence of the $i/\sqrt{2}$ factor in (12a).

Complement C_{VIII}

Multipole moments

Consider, in classical physics, a system S of charges and currents localized in a volume $\mathcal{V}_0$. This complement will focus on the electromagnetic interactions of this system with other physical systems localized outside S. In § 1, we shall compute the electric field created by these charges at a point $\boldsymbol{r}$ outside S, starting first with the case where the charges are immobile. This electrostatic field will be expressed as a function of quantities called the "electric multipole moments" Q_l^m of the system. A slightly different problem can also be treated, the interaction of the charges of S with an electric potential created by other charges (all outside S). The result also involves the Q_l^m. In quantum mechanics, the Q_l^m become irreducible tensor operators $T_{Q=m}^{(K=l)}$. The Wigner–Eckart theorem can be applied to them.

We shall continue the discussion in § 2, considering a system of stationary currents instead of motionless charges. This will lead to the introduction of "magnetic multipole moments" M_l^m, whose properties are fairly similar to those of the Q_l^m, except that they have opposite parity. § 3 will provide some examples of this formalism: electric quadrupole of atomic or nuclear levels, selection rules for multipole transitions between different levels.

Comment:

We will remain within the framework of *electrostatics* or *magnetostatics* (density of charges ρ and currents $\boldsymbol{j}$ varying very slowly in time). This restriction is not essential: one can reason in a more general way, within the framework of Maxwell's equations taking into account the effects of *radiation propagation.* This leads to more elaborate computations, and it is useful to introduce the concept of "vector" spherical harmonics (analog to the "scalar" harmonics Y_l^m), which are the eigenfunctions common to $\boldsymbol{J}_2$ and J_z for a vector field (instead of a scalar function, which has only one component). The interested reader may consult chapter 16 of [46], or the Collège de France teaching course of C. Cohen-Tannoudji (1973–1974), available online.

1. Electric multipole moments

1-a. Expanding the potential outside a system of charges

Consider a density of charges $\rho(\boldsymbol{r})$, localized in a volume $\mathcal{V}_0$, and note S_0 a sphere of radius R_0, centered at the origin, containing $\mathcal{V}_0$ (figure 1):

$$\rho(\boldsymbol{r}) = 0 \quad \text{if} \quad r > R_0 \tag{1}$$

We wish to calculate the potential V and the electric field $\boldsymbol{E}$ created by this system of charges outside the sphere S_0. We must look for a solution of the Poisson equation for V:

$$\Delta V(\boldsymbol{r}) = -\rho(\boldsymbol{r})/\varepsilon_0 \tag{2}$$

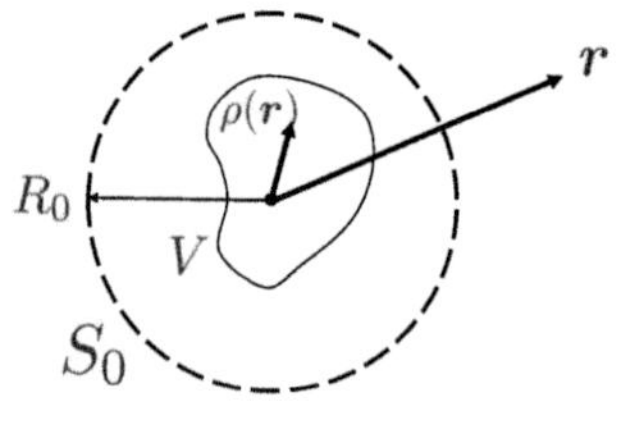

Figure 1: A density of electric charges $\rho(\boldsymbol{r})$ is localized inside a sphere S_0 of radius R_0. The object is to calculate the field at any point $\boldsymbol{r}$ outside the sphere.

that goes to zero at infinity. Operator Δ is the Laplacian:

$$\Delta = \frac{\partial^2}{\partial x^2} + \frac{\partial^2}{\partial y^2} + \frac{\partial^2}{\partial z^2} \tag{3}$$

which is rotation invariant. It thus commutes with L_z and $\boldsymbol{L}^2$ (orbital angular momentum operators) and can be written:

$$\Delta \equiv \frac{1}{r}\frac{\partial^2}{\partial r^2} r - \frac{\boldsymbol{L}^2}{\hbar^2 r^2} \tag{4}$$

We expand the angular dependence of V and ρ on the spherical harmonics, and set[1]:

$$V(\boldsymbol{r}) = \sum_{l=0}^{\infty} \sum_{m=-l}^{+l} v^{\star}_{lm}(r)\, Y_l^m(\theta, \varphi) \tag{5a}$$

$$\rho(\boldsymbol{r}) = \sum_{l=0}^{\infty} \sum_{m=-l}^{+l} \rho^{\star}_{lm}(r)\, Y_l^m(\theta, \varphi) \tag{5b}$$

Since the $Y_l^m(\theta, \varphi)$ are orthonormal as functions of θ and φ, the v_{lm} and ρ_{lm} are given by:

$$v_{lm}(r) = \int_{4\pi} \mathrm{d}\Omega\; Y_l^m(\theta, \varphi)\; V(\boldsymbol{r}) \tag{6a}$$

$$\rho_{lm}(r) = \int_{4\pi} \mathrm{d}\Omega\; Y_l^m(\theta, \varphi)\; \rho(\boldsymbol{r}) \tag{6b}$$

Identifying the expansions of the two sides of (2) and using (4) we get:

$$\left[\frac{1}{r}\frac{\mathrm{d}^2}{\mathrm{d}r^2} r - \frac{l(l+1)}{r^2}\right] v_{lm}(r) = -\frac{\rho_{lm}(r)}{\varepsilon_0} \tag{7}$$

The angular and radial aspects of the problem are now completely separated. For each values of l and m, equation (7) is a differential equation with a single variable.

We now show that its solution is:

$$v_{lm}(r) = \frac{1}{\varepsilon_0}\frac{1}{\sqrt{4\pi(2l+1)}}\frac{Q_l^m}{r^{l+1}} \qquad \text{for } r > R_0 \tag{8}$$

[1] A useful convention is to write these expansions in terms of $v^{\star}_{lm}(r)$ and $\rho^{\star}_{lm}(r)$, instead of v_{lm} and ρ_{lm}. It ensures that these coefficients transform under rotation as Y_l^m, as seen in relations (6). Remember the general relation $Y_l^{m^\star}(\theta, \varphi) = (-1)^m\, Y_l^{-m}(\theta, \varphi)$. As $V(\boldsymbol{r})$ and $\rho(\boldsymbol{r})$ are real, we have $v^{\star}_{lm}(r) = (-1)^m\, v_{l-m}(r)$ and $\rho^{\star}_{lm}(r) = (-1)^m\, \rho_{l-m}(r)$.

where Q_l^m is the constant defined as:

$$\boxed{\begin{aligned} Q_l^m &= \sqrt{\frac{4\pi}{2l+1}} \int_0^{R_0} \mathrm{d}r\ r^{l+2}\ \rho_{lm}(r) \\ &= \sqrt{\frac{4\pi}{2l+1}} \int_{\mathcal{V}_0} \mathrm{d}^3r\ r^l\ Y_l^m(\theta,\varphi)\ \rho(\boldsymbol{r}) \end{aligned}} \tag{9}$$

Q_l^m is called the electric multipole moment of order l, m of the system of charges under study.

Demonstration: For $r > R_0$, the function $\rho_{lm}(r)$ is zero, so that:

$$\left[\frac{1}{r}\frac{\mathrm{d}^2}{\mathrm{d}r^2}r - \frac{l(l+1)}{r^2}\right] v_{lm}(r) = 0 \qquad \text{if} \quad r > R_0 \tag{10}$$

This equation[2] has two independent solutions, r^l and $1/r^{l+1}$. The solutions of equation (7) depend on two parameters, and generally behave as r^l when $r \to \infty$. This diverging behavior must be rejected since the potential created by the charges goes to zero at very large distances. We shall only keep the solution (defined to within a constant multiplicative factor) that has a non-divergent behavior as $1/r^{l+1}$. We set (change of function):

$$v_{lm}(r) = r^l\, x_{lm}(r) \tag{11}$$

This equality defines $x_{lm}(r)$ everywhere (except eventually at $r = 0$). Since $v_{lm}(r)$ remains finite when $r \to \infty$, we expect[3] :

$$x_{lm}(r) \underset{r\to\infty}{\longrightarrow} 0 \tag{12}$$

Inserting (11) into (10) yields:

$$\left[2(l+1)\frac{\mathrm{d}}{\mathrm{d}r} + r\frac{\mathrm{d}^2}{\mathrm{d}r^2}\right] x_{lm}(r) = -\frac{\rho_{lm}(r)}{\varepsilon_0\, r^{l-1}} \tag{13}$$

which can be reduced to a first-order differential equation by setting:

$$y_{lm}(r) = \frac{\mathrm{d}}{\mathrm{d}r} x_{lm}(r) \tag{14}$$

The new function $y_{lm}(r)$ obeys:

$$\left[2(l+1) + r\frac{\mathrm{d}}{\mathrm{d}r}\right] y_{lm}(r) = -\frac{\rho_{lm}(r)}{\varepsilon_0\, r^{l-1}} \tag{15}$$

---

[2]This is a differential equation with real coefficients, obeyed by a complex function of a real variable. Our reasoning applies both to the real and imaginary parts of this function, hence to the function itself.

[3]Actually, if $v_{lm} \sim r^{-(l+1)}$ when $r \to \infty$, we have $x_{lm} \sim r^{-(2l+1)}$ and, as we shall see below, x_{lm} goes rapidly to zero at infinity.

If $r > R_0$, $\rho_{lm}(r)$ is zero, and the solution of this equation is proportional to $1/r^{2(l+1)}$. Setting:

$$y_{lm}(r) = \frac{z_{lm}(r)}{r^{2(l+1)}} \tag{16}$$

we obtain the simpler equation:

$$\frac{\mathrm{d}}{\mathrm{d}r} z_{lm}(r) = -\frac{1}{\varepsilon_0}\, r^{l+2} \rho_{lm}(r) \tag{17}$$

that can be easily integrated to yield:

$$z_{lm}(r) = z_{lm}(0) - \frac{1}{\varepsilon_0} \int_0^r \mathrm{d}r'\, r'^{\,l+2}\, \rho_{lm}(r') \tag{18}$$

The function z_{lm} never has a singular point at the origin. In general, this is not the case for the functions x, y, and v introduced above, since if $z_{lm}(0) \neq 0$ when $r \to 0$:

$$\begin{aligned} y_{lm}(r) &\sim z_{lm}(0)/r^{2l+2} \\ x_{lm}(r) &\sim z_{lm}(0)/r^{2l+1} \\ v_{lm}(r) &\sim z_{lm}(0)/r^{l+1} \end{aligned} \tag{19}$$

This is the result we anticipated above: in general, the solution of (7) diverges at the origin as $r^{-(l+1)}$. We are looking for a solution of (2) that remains valid in all points in space, including the origin, as long as $\rho(\boldsymbol{r})$ is finite everywhere. Consequently, we must set $z_{lm}(0) = 0$ in (18). When $r > R_0$, the function z_{lm} has a constant value given by:

$$z_{lm} = -\frac{1}{\varepsilon_0}\sqrt{\frac{2l+1}{4\pi}}\, Q_l^m \quad \text{if} \quad r > R_0 \tag{20}$$

where we have set:

$$Q_l^m = \sqrt{\frac{4\pi}{2l+1}} \int_0^{R_0} \mathrm{d}r'\; r'^2\, r'^{\,l}\, \rho_{lm}(r') \tag{21}$$

In this definition, the factor $\sqrt{4\pi/(2l+1)}$ has been introduced for convenience, for reasons that will become clear later on; the upper bound R_0 of the integral may be replaced by $+\infty$. We finally get:

$$y_{lm}(r) = -\frac{1}{\varepsilon_0}\sqrt{\frac{2l+1}{4\pi}}\,\frac{Q_l^m}{r^{2(l+1)}} \quad \text{if} \quad r > R_0 \tag{22}$$

yielding the derivative of $x_{lm}(r)$ with respect r.

As we do not know the behavior of x_{lm} at the origin, we cannot integrate (22) from 0 to r; however, condition (12) allows an integration between r and $+\infty$, which yields:

$$x_{lm}(\infty) = 0 = x_{lm}(r) + \int_r^{+\infty} \mathrm{d}r\, y_{lm}(r) \tag{23}$$

or:

$$x_{lm}(r) = \frac{1}{\varepsilon_0}\sqrt{\frac{2l+1}{4\pi}}\frac{Q_l^m}{2l+1}\frac{1}{r^{2l+1}} \tag{24}$$

Consequently, we get:

$$v_{lm}(r) = \frac{1}{\varepsilon_0}\frac{1}{\sqrt{4\pi(2l+1)}}\frac{Q_l^m}{r^{l+1}} \tag{25}$$

establishing equality (8).

Finally, the solution of (2) we have selected is written:

$$\boxed{V(\boldsymbol{r}) = \frac{1}{4\pi\,\varepsilon_0}\sum_{l=0}^{\infty}\sum_{m=-l}^{l}\sqrt{\frac{4\pi}{2l+1}}\,(Q_l^m)^\star\,\frac{Y_l^m(\theta,\varphi)}{r^{l+1}}} \tag{26}$$

The Q_l^m defined in (9) [or (21)] are numbers that allow writing the electric potential outside S_0 as an expansion on spherical harmonics. These numbers are called *electric multipole moments* of the system of charges under study. They completely characterize the electric field $\boldsymbol{E}(\boldsymbol{r})$ created by the charges outside S_0:

$$\boxed{\boldsymbol{E}(\boldsymbol{r}) = -\frac{1}{4\pi\,\varepsilon_0}\sum_{l=0}^{\infty}\sum_{m=-l}^{l}\sqrt{\frac{4\pi}{2l+1}}\,(Q_l^m)^\star\,\boldsymbol{\nabla}\left\{\frac{Y_l^m(\theta,\varphi)}{r^{l+1}}\right\}} \tag{27}$$

1-b. Point charges

If the system consists of a single charge situated at $\boldsymbol{r}_1$, we have:

$$\rho(\boldsymbol{r}') = q\,\delta\left(\boldsymbol{r}' - \boldsymbol{r}_1\right) \tag{28}$$

and $V(\boldsymbol{r})$ is simply given by:

$$V(\boldsymbol{r}) = \frac{1}{4\pi\,\varepsilon_0}\frac{q}{|\boldsymbol{r} - \boldsymbol{r}_1|} \tag{29}$$

In addition, using (28) in equation (9) allows performing the integration to get the value of Q_l^m:

$$Q_l^m = q\sqrt{\frac{4\pi}{2l+1}}(r_1)^l\;Y_l^m(\theta_{\boldsymbol{r}_1},\varphi_{\boldsymbol{r}_1}) \tag{30}$$

Relation (26) then yields, after replacing $\boldsymbol{r}_1$ by $\boldsymbol{r}'$ for the sake of symmetry:

$$\frac{1}{|\boldsymbol{r} - \boldsymbol{r}'|} = \sum_{lm}\frac{4\pi}{2l+1}\,(r')^l\;Y_l^{m\star}(\theta',\varphi')\;\frac{Y_l^m(\theta,\varphi)}{r^{l+1}} \qquad \text{if} \quad r' < r \tag{31}$$

where r', θ', φ' are the spherical coordinates of vector $\boldsymbol{r}'$. When $r' > r$, we simply invert in (31) the role of $\boldsymbol{r}$ and $\boldsymbol{r}'$ to get the spherical harmonic expansion of $1/|\boldsymbol{r} - \boldsymbol{r}'|$. Expansion (31) is frequently used, as we shall see for example in the rest of this complement.

For a system of N point charges $q_1, q_2, q_3, \ldots$ situated at $\boldsymbol{r}_1, \boldsymbol{r}_2, \boldsymbol{r}_3, \ldots$ we have:

$$\rho(\boldsymbol{r}') = \sum_{n=1}^{N} q_n \, \delta\left(\boldsymbol{r}' - \boldsymbol{r}_n\right) \tag{32}$$

and relation (9) leads to:

$$Q_l^m = \sqrt{\frac{4\pi}{2l+1}} \sum_{n=1}^{N} q_n \left(r_n\right)^l Y_l^m\left(\theta_n, \varphi_n\right) \tag{33}$$

where r_n, θ_n, and φ_n are the spherical components of $\boldsymbol{r}_n$.

1-c. Coupling with a field created by external charges

We now consider two systems of charges (figure 2): the first one, identical to the previous one, is entirely located inside a sphere S_0; the second is composed of charges all outside a sphere S_1 centered at the origin like S_0, but with a radius R_1 larger than (or equal to) R_0.

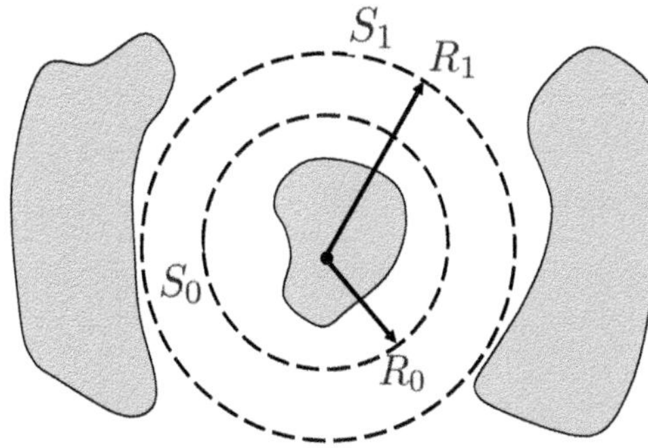

Figure 2: A first system of charges is concentrated around the origin, entirely contained inside a sphere S_0 of radius d R_0. A second system is composed of charges all situated outside a sphere S_1 of radius R_1 larger than R_0; it creates an electrostatic potential $V_{\text{ext}}(\boldsymbol{r})$ that interacts with the first system.

This second system of charges creates an electric potential $V_{\text{ext}}(\boldsymbol{r})$ that interacts with the charges contained in S_0, with an energy written as:

$$W = \int \mathrm{d}^3r' \, \rho(\boldsymbol{r}') \, V_{\text{ext}}(\boldsymbol{r}') \tag{34}$$

where the potential $V_{ext}(\boldsymbol{r}')$ can be expressed as a function of the charge density $\rho_{ext}(\boldsymbol{r}')$ that creates it:

$$V_{ext}(\boldsymbol{r}') = \frac{1}{4\pi\,\varepsilon_0} \int d^3r'' \,\frac{\rho_{ext}(\boldsymbol{r}'')}{|\boldsymbol{r}'' - \boldsymbol{r}'|} \qquad \text{with: } \rho_{ext}(\boldsymbol{r}') = 0 \text{ if } r' < R_1 \tag{35}$$

Using expansion (31) for $1/|\boldsymbol{r}'' - \boldsymbol{r}'|$, we get:

$$V_{ext}(\boldsymbol{r}') = \frac{1}{4\pi\,\varepsilon_0} \sum_{lm} \sqrt{\frac{4\pi}{2l+1}}\; T_l^m \,(r')^l\, Y_l^{m*}\,(\theta', \varphi') \tag{36}$$

with:

$$T_l^m = \sqrt{\frac{4\pi}{2l+1}} \int d^3r'' \,\rho_{ext}(\boldsymbol{r}'')\, \frac{Y_l^m\,(\theta'', \varphi'')}{(r'')^{l+1}} \tag{37}$$

The T_l^m can be called the "interior multipole moments" of the system of charges described by ρ_{ext} (as opposed to the Q_l^m, which are the "exterior multipole moments"). They yield the expansion of $V_{ext}(\boldsymbol{r})$ on the spherical harmonics $r^l\, Y_l^m(\theta, \varphi)$, i.e. the l-degree terms in the expansion of $V_{ext}(\boldsymbol{r})$ in powers of x, y, and z in the neighborhood of the origin.

Inserting (36) in the interaction energy (34), we get:

$$W = \frac{1}{4\pi\,\varepsilon_0} \sum_{l=0}^{\infty} \sum_{m=-l}^{+l} (-1)^m\, T_l^m\, Q_l^{-m} \tag{38}$$

The multipole moments Q_l^m of a system of charges characterize the electric field they create outside the system as well as their interaction energy with an external potential V_{ext}.

1-d. Physical interpretation

We now consider the physical interpretation of the various multipole moments Q_l^m.

α. $l = 0$ *moments; total charge*

We know that Y_0^0 is a constant: $Y_0^0 = 1/\sqrt{4\pi}$. We thus have:

$$Q_0^0 = \int d^3r' \,\rho(r') \tag{39}$$

meaning that Q_0^0 is the total charge of the system under study, that we shall note Q for the sake of simplicity. For point particles with charges q_n, the contribution of the total charge $Q = \sum_n q_n$ to the potential $V(\boldsymbol{r})$ [term $l = m = 0$ of (26)] is:

$$V_0^0(\boldsymbol{r}) = \frac{Q}{4\pi\,\varepsilon_0}\,\frac{1}{r} \tag{40a}$$

and the corresponding electric field is:

$$\boldsymbol{E}_0^0 = \frac{Q}{4\pi\,\varepsilon_0}\,\frac{\boldsymbol{r}}{r^3} \tag{40b}$$

This field is rotation invariant: it is the field that would be created by all the charges if they were all situated at the origin. When studying the field created by an ensemble of charges at a large distance ($r \gg R_0$, characteristic dimension of the system's spatial extension), one often keeps only the $l = m = 0$ term in the expansion (26) of $V(r)$[4].

β. *$l = 1$ moment; electric dipole*

Since:

$$\begin{cases} Y_1^{\pm 1}(\theta,\varphi) = \mp\sqrt{\dfrac{3}{8\pi}}\,\sin\theta\,\mathrm{e}^{\pm i\varphi} = \mp\sqrt{\dfrac{3}{8\pi}}\left(\dfrac{x \pm iy}{r}\right) \\ Y_1^0(\theta,\varphi) = \sqrt{\dfrac{3}{4\pi}}\,\cos\theta = \sqrt{\dfrac{3}{4\pi}}\,\dfrac{z}{r} \end{cases} \tag{41}$$

we get[5]:

$$\begin{aligned} Q_1^{\pm 1} &= \mp\frac{1}{\sqrt{2}}\int \mathrm{d}^3r'\,\rho(\boldsymbol{r}')\,\left[x' \pm iy'\right] \\ Q_1^0 &= \int \mathrm{d}^3r'\,\rho(\boldsymbol{r}')\,z' \end{aligned} \tag{42}$$

Relations (VIII-32) of chapter VIII show how to transform standard components back to Cartesian components:

$$\begin{aligned} Q_{1\,x} &= \frac{1}{\sqrt{2}}\left[Q_1^{-1} - Q_1^1\right] = \int \mathrm{d}^3r'\,\rho(\boldsymbol{r}')\,x' \\ Q_{1\,y} &= \frac{i}{\sqrt{2}}\left[Q_1^{-1} + Q_1^1\right] = \int \mathrm{d}^3r'\,\rho(\boldsymbol{r}')\,y' \\ Q_{1\,z} &= Q_1^0 = \int \mathrm{d}^3r'\,\rho(\boldsymbol{r}')\,z' \end{aligned} \tag{43a}$$

or else, using the vector notation with the traditional notation $\boldsymbol{D}$:

$$\boldsymbol{Q}_1 = \boldsymbol{D} = \int \mathrm{d}^3r'\,\boldsymbol{r}'\,\rho(\boldsymbol{r}') \tag{43b}$$

---

[4]For this approximation to be valid, Q_0^0 must obviously be different from zero (the system of charges cannot be neutral), and we must have $|Q_0| \gg |Q_1^m|/r$, $|Q_2^m|/r^2$, ...

[5]According to the standard notation Y_l^m of the spherical harmonics, we write as a lower index the one corresponding to K, and as an upper index the one corresponding to Q.

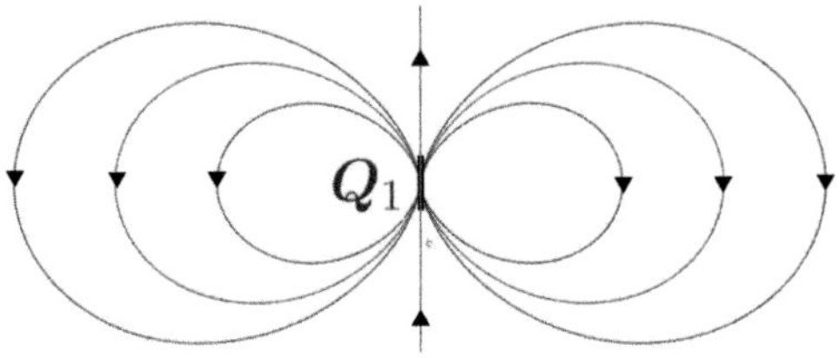

Figure 3: Typical pattern of the field lines created by a dipole $\boldsymbol{Q}_1$ placed at the origin and pointing upwards.

For point charges, we get:

$$\boldsymbol{D} = \sum_n q_n \, \boldsymbol{r}_n \tag{43c}$$

This yields for $\boldsymbol{Q}_1 = \boldsymbol{D}$ the usual electric dipole of an ensemble of charges (thanks to the factor $\sqrt{4\pi(2l+1)}$ introduced above in the definition of Q_l^m). Inserting these equalities in (26) and using again relations (41), we can see that its contribution to $V(\boldsymbol{r})$ is written:

$$\frac{1}{4\pi\,\varepsilon_0} \frac{\boldsymbol{D}\cdot\boldsymbol{r}}{r^3} \tag{44}$$

We get a potential that varies as r^{-2}. As for the electric field, it is proportional to r^{-3} and follows the well-known pattern of a dipole field, shown in figure 3.

We often study in physics electrically neutral systems (atoms or molecules for example). The dipole field is then the main contribution to the electric field they create at large enough distance.

When the system is placed in an external potential, one can isolate in (38) the contribution of the electric dipole to the interaction energy W, a contribution written as:

$$\frac{1}{4\pi\,\varepsilon_0} \sum_m (-1)^m \, Q_1^{-m} \, T_1^m = -\boldsymbol{D}\cdot\boldsymbol{E}_{\text{ext}}(\boldsymbol{0}) \tag{45}$$

where $\boldsymbol{E}_{\text{ext}}(\boldsymbol{0})$ is the electric field created by the external charges at the origin:

$$\boldsymbol{E}_{\text{ext}}(0) = -\boldsymbol{\nabla}\left\{V_{\text{ext}}(\boldsymbol{r})\right\}_{\boldsymbol{r}=0} \tag{46}$$

This relation is easy to establish, as can be seen, for example, in complement E_X of reference [9].

When the spatial extension of the system is very small [compared with the characteristic distance over which the field $\boldsymbol{E}_{\text{ext}}(\boldsymbol{r})$ significantly varies], one often limits the interaction energy W to the $l \leq 1$ terms. Actually, one may even use only expression (45), ignoring the $l = 0$ term, which is written $Q\,V_{\text{ext}}(0)$ and thus yields a constant of no physical consequence on the evolution of the system subjected to the external potential. This situation occurs, for example, when studying the dynamics of an atom subjected to an electric potential created for example by electrodes of macroscopic dimensions, or by a laser of optical frequency that induces a dipole in the atom.

γ. *$l = 2$ moment; electric quadrupole*

Using the expressions for the $Y_2^m(\theta, \varphi)$, one easily shows:

$$\begin{cases} Q_2^{\pm 2} = \dfrac{\sqrt{6}}{4} \displaystyle\int \mathrm{d}^3r'\,\rho(\boldsymbol{r}')\,\left[x' \pm iy'\right]^2 \\ Q_2^{\pm 1} = \mp\dfrac{\sqrt{6}}{2} \displaystyle\int \mathrm{d}^3r'\,\rho(\boldsymbol{r}')\,z'\,\left[x' \pm iy'\right] \\ Q_2^{0} = \dfrac{1}{2} \displaystyle\int \mathrm{d}^3r'\,\rho(\boldsymbol{r}')\,\left[3z'^2 - \boldsymbol{r}'^2\right] \end{cases} \tag{47a}$$

For an ensemble of point charges, these formulas become:

$$\begin{cases} Q_2^{\pm 2} = \dfrac{\sqrt{6}}{4} \displaystyle\sum_n q_n\,\left[x_n \pm iy_n\right]^2 \\ Q_2^{\pm 1} = \mp\dfrac{\sqrt{6}}{2} \displaystyle\sum_n q_n\,z_n\,\left[x_n \pm iy_n\right] \\ Q_2^{0} = \dfrac{1}{2} \displaystyle\sum_n q_n\,\left[3z_n^2 - \boldsymbol{r}_n^2\right] \end{cases} \tag{47b}$$

The corresponding potential varies in r^{-3}, and the field in r^{-4}. Figures showing patterns of such field lines can be found in electromagnetism textbooks. Figure 4 gives two examples of pure electric quadrupole systems, with a zero total charge and a zero electric dipole. A simple way to visualize a pure quadrupole system of charges is to consider an ensemble of two opposite dipoles.

It is easy to show that the contribution to the interaction energy W of the system of charges with the potential $V_{\text{ext}}(\boldsymbol{r})$ involves the product of $\overline{\overline{Q}}_2$ with the vector whose components are the first derivatives of the components of the field $\boldsymbol{E}_{\text{ext}}(\boldsymbol{r})$, taken at the origin.

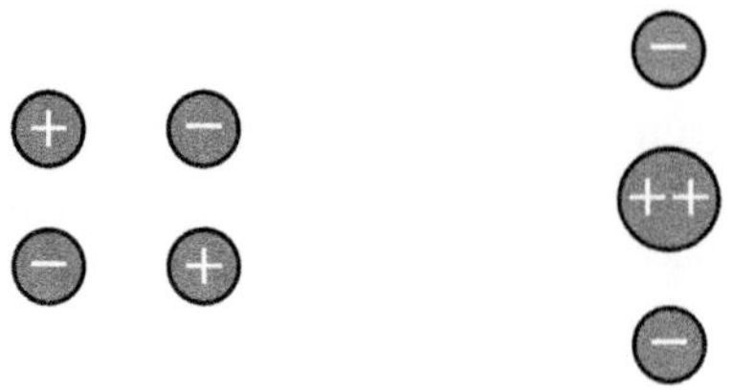

Figure 4: Two systems of charges forming a pure quadrupole, with a zero total charge and a zero electric dipole.

Exercise:

Show that for a system of point charges, the $l = 2$ terms of expansion (38) are written (contraction of two Cartesian tensors of order two):

$$\sum_{i,j} \left\{ \left[\frac{\partial^2 V_{\rm ext}}{\partial x_i \, \partial x_j} \right]_{r=0} \times \sum_n q_n \, x_{in} \, x_{jn} \right\}$$

where $x_{in} = x_n, y_n$ or z_n if $i = 1, 2, 3$

δ. *Generalization; electric l-pole moments*

The higher the value of l considered, the more complex the space variations of the corresponding electric field, and the faster it decreases at infinity [as $r^{-(l+2)}$]; this field depends on the $2l + 1$ components of the moment Q_l. In many cases, this fast decrease allows limiting the summation (26) to a small value of l (for example, $l = 2$), which often considerably simplifies the calculations.

Comment:

One may wonder if the chosen origin of the coordinates has an impact on the values of the Q_l^m. The expressions of all the Q_l^m moments, except Q_{00}, depend a priori on the axes origin used for their calculation. Consider, however, a system with two opposite charges, $+q$ and $-q$. We have:

$$\boldsymbol{Q}_1 = q \, (\boldsymbol{r}_1 - \boldsymbol{r}_2) \tag{48}$$

and it is clear that $\boldsymbol{Q}_1$ does not depend on the chosen origin. This property can be generalized. If all the moments $Q_0, \boldsymbol{Q}_1, \ldots, Q_{l-1\,m}$ of a system are zero in a given reference frame, these moments remain zero under any axes translation. The value of the first non-zero multipole moments, of order l, does not vary in this axes translation.

Exercise:

Demonstrate this property. It will be useful to consider an infinitesimal translation by $\delta \boldsymbol{r}_0$ of the axes origin, and compute the variation δQ_l^m of Q_l^m. This will yield the quantity $\boldsymbol{\nabla}\left\{r'^l\, Y_l^m\,(\theta', \varphi')\right\}$ whose three components are $(l-1)^{\text{th}}$ degree homogeneous harmonic functions of x', y', and z'. This will show that these components are linear combinations of the $r'^{l-1}\, Y_{l-1}^{m'}(\theta', \varphi')$ and that the variation of Q_l^m can be expressed as a function of the $Q_{l-1\,m'}$.

1-e. Operators in quantum mechanics

For an ensemble of N point charges, we obtained in (33) the expression of the Q_l^m multipole:

$$Q_l^m = \sqrt{\frac{4\pi}{2l+1}} \sum_{n=1}^{N} q_n\, (r_n)^l\, Y_l^m\,(\theta_n, \varphi_n) \tag{49}$$

where r_n, θ_n, φ_n are the spherical components of the position $\boldsymbol{r}_n$ of the nth particle, with charge q_n. Expression:

$$(r_n)^l\, Y_l^m\,(\theta_n, \varphi_n)$$

is an nth degree homogeneous (harmonic) functions of the Cartesian components x_n, y_n, z_n of this particle. This function can be noted $F_{lm}(\boldsymbol{r}_n)$:

$$F_{lm}(\boldsymbol{r}_n) = F_{lm}\,(x_n, y_n, z_n) = (r_n)^l\, Y_l^m\,(\theta_n, \varphi_n) \tag{50}$$

The quantum operator $\hat{Q}_l^m$ corresponding to (49) is simply obtained by replacing x_n, y_n, and z_n by the operators X_n, Y_n, and Z_n (no symmetrization is necessary since all these operators commute):

$$\hat{Q}_l^m = \sqrt{\frac{4\pi}{2l+1}} \sum_{n} q_n\, F_{lm}(\boldsymbol{R}_n) \tag{51}$$

Examples:

The total charge ($l = m = 0$) yields an operator that is simply a number (scalar operator), commuting with any other operator. The electric dipole yields:

$$\hat{\boldsymbol{D}} = \sum_{n} q_n\, \boldsymbol{R}_n \tag{52}$$

etc.

Each operator $\hat{Q}_l^m$ is the m component of an irreducible tensor operator of order $K = l$.

Demonstration: We shall reason in the state space $\mathscr{E}$ of a spinless single particle. The generalization to an ensemble of particles is straightforward. We define the quantum operator $\hat{F}_{lm}$ associated with F_{lm}:

$$\hat{F}_{lm}\,|\boldsymbol{r}\rangle = r^l\,Y_l^m\,(\theta,\varphi)\,|\boldsymbol{r}\rangle \tag{53}$$

where θ and φ are the polar angles of $\boldsymbol{r}$. We then compute the action in the $|\boldsymbol{r}\rangle$ representation of the rotation transform of $\hat{F}_{lm}$:

$$\begin{aligned} R_{\boldsymbol{u}}(\varphi)\,\hat{F}_{lm}\,R_{\boldsymbol{u}}^{+}(\varphi)\,|\boldsymbol{r}\rangle &= R_{\boldsymbol{u}}(\varphi)\,\hat{F}_{lm}\,|\boldsymbol{r}'\rangle \\ &= (r')^l\,Y_l^m\,(\theta',\varphi')\;R_{\boldsymbol{u}}(\varphi)\,|\boldsymbol{r}'\rangle = (r')^l\,Y_l^m\,(\theta',\varphi')\;|\boldsymbol{r}\rangle \end{aligned} \tag{54a}$$

where $\boldsymbol{r}'$ is the transform of $\boldsymbol{r}$ under the rotation $(R_{\boldsymbol{u}}(\varphi))^{-1}$ in the usual space, and θ' φ' are the polar angles of $\boldsymbol{r}'$. Now:

$$Y_l^m\,(\theta',\varphi') = \sum_{m'}\left(R^{[j=l]}\,(\boldsymbol{u},\varphi)\right)_{m'm}\,Y_l^{m'}\,(\theta,\varphi) \tag{54b}$$

(rotation transformation of a $j = l$ angular momentum). Equality (54a), valid for any basis ket $|\boldsymbol{r}\rangle$, then yields:

$$R_{\boldsymbol{u}}(\varphi)\,\hat{F}_{lm}\,R_{\boldsymbol{u}}^{+}(\varphi) = \sum_{m'}\left(R^{[j=l]}\,(\boldsymbol{u},\varphi)\right)_{m'm}\,\hat{F}_{lm'} \tag{55}$$

which is the very definition of an irreducible tensor operator.

The parity of operator Q_l^m is $(-1)^l$ since, if I_O represents the parity operation with respect to the axes origin O (complement D_V), we have:

$$\begin{aligned} I_O\,\hat{F}_{lm}\,I_O^{\dagger}\,|\boldsymbol{r}\rangle = I_O\,\hat{F}_{lm}\,|-\boldsymbol{r}\rangle &= I_O\,r^l\,Y_l^m\,(\pi-\theta,\varphi+\pi)\,|-\boldsymbol{r}\rangle \\ &= (-1)^l\,r^l\,Y_l^m\,(\theta,\varphi)\;I_O\,|-\boldsymbol{r}\rangle = (-1)^l\,\hat{F}_{lm}\,|\boldsymbol{r}\rangle \end{aligned} \tag{56}$$

Operator Q_0 (scalar) is thus even, $\boldsymbol{Q}_1$ (dipole) is odd, etc.

2. Magnetic multipole moments

2-a. Expansion of the static magnetic field B outside a system of currents

Consider a system of *stationary* currents, with density $\boldsymbol{j}(\boldsymbol{r})$; this vector necessarily has a zero divergence since:

$$\boldsymbol{\nabla}\cdot\boldsymbol{j}(\boldsymbol{r}) = -\frac{\partial\rho}{\partial t} = 0 \tag{57}$$

We assume these currents are different from zero only inside a sphere of radius R_0 centered at the origin, and we are looking for an expression of the magnetic field they create outside this sphere. This question is reminiscent of the electrostatic problem discussed in the previous § 1, but the sources $\boldsymbol{j}$ now have a vector and not a scalar character. Another difference comes from the zero divergence condition that prevents us from choosing independently the 3 components of $\boldsymbol{j}$ at any point in space, whereas $\rho(\boldsymbol{r})$ could take any value inside S_0.

The magnetic field $\boldsymbol{B}(\boldsymbol{r})$, with zero divergence, is related to the vector potential $\boldsymbol{A}(\boldsymbol{r})$ according to $\boldsymbol{B}(\boldsymbol{r}) = \boldsymbol{\nabla} \times \boldsymbol{A}(\boldsymbol{r})$. Maxwell's equation $\boldsymbol{\nabla} \times \boldsymbol{B}(\boldsymbol{r}) = \mu_0 \boldsymbol{j}(\boldsymbol{r})$ leads to, in the Coulomb gauge where $\boldsymbol{\nabla} \cdot \boldsymbol{A} = 0$:

$$\Delta \boldsymbol{A} = -\mu_0 \, \boldsymbol{j} \tag{58}$$

Taking into account the boundary conditions, the solution of this equation is:

$$\boldsymbol{A}(\boldsymbol{r}) = \frac{\mu_0}{4\pi} \int_{r' \leq R_0} \mathrm{d}^3 r' \, \frac{\boldsymbol{j}(\boldsymbol{r}')}{|\boldsymbol{r} - \boldsymbol{r}'|} \tag{59}$$

Since:

$$\boldsymbol{\nabla}_{\boldsymbol{r}} \times \frac{\boldsymbol{j}(\boldsymbol{r}')}{|\boldsymbol{r} - \boldsymbol{r}'|} = \boldsymbol{\nabla}_{\boldsymbol{r}} \left[\frac{1}{|\boldsymbol{r} - \boldsymbol{r}'|} \right] \times \boldsymbol{j}(\boldsymbol{r}') \tag{60}$$

we get the expression for the magnetic field $\boldsymbol{B}$ (Biot–Savart law):

$$\boldsymbol{B}(\boldsymbol{r}) = \boldsymbol{\nabla}_{\boldsymbol{r}} \times \boldsymbol{A}(\boldsymbol{r}) = \frac{\mu_0}{4\pi} \int_{r' \leq R_0} \mathrm{d}^3 r' \, \frac{\boldsymbol{j}(\boldsymbol{r}') \times [\boldsymbol{r} - \boldsymbol{r}']}{|\boldsymbol{r} - \boldsymbol{r}'|^3} \tag{61}$$

Outside the sphere of radius R_0, the divergence and the curl of field $\boldsymbol{B}(\boldsymbol{r})$ are zero. This field is thus the gradient of a harmonic potential (with a zero Laplacian), as was the case for the electric field $\boldsymbol{E}(\boldsymbol{r})$ in regions where ρ was zero. We can thus apply the results of § 1 and write:

$$\boxed{\boldsymbol{B}(\boldsymbol{r}) = -\frac{\mu_0}{4\pi} \sum_{lm} \sqrt{\frac{4\pi}{2l+1}} \, (M_l^m)^{\star} \, \boldsymbol{\nabla} \left[\frac{Y_l^m(\theta, \varphi)}{r^{l+1}} \right]} \tag{62}$$

where the coefficients M_l^m are yet to be computed. They are called the magnetic multipole moments and we are going to show that they can be expressed as:

$$\boxed{M_l^m = -\frac{1}{l+1} \sqrt{\frac{4\pi}{2l+1}} \int \mathrm{d}^3 r' \, \boldsymbol{\nabla} \cdot [\boldsymbol{r}' \times \boldsymbol{j}(\boldsymbol{r}')] \, r'^l \, Y_l^m(\theta', \varphi')} \tag{63a}$$

or by the equivalent, often useful, relation:

$$\boxed{M_l^m = \frac{1}{l+1} \sqrt{\frac{4\pi}{2l+1}} \int \mathrm{d}^3 r' \, [\boldsymbol{r}' \times \boldsymbol{j}(\boldsymbol{r}')] \cdot \boldsymbol{\nabla} \left[r'^l \, Y_l^m(\theta', \varphi') \right]} \tag{63b}$$

Demonstration:

Following the method proposed in [47], we introduce the auxiliary scalar function $\boldsymbol{r} \cdot \boldsymbol{B}(\boldsymbol{r})$, and compute its Laplacian. Since:

$$\frac{\partial}{\partial x}(xB_x) = B_x + x\frac{\partial B_x}{\partial x} + y\frac{\partial B_y}{\partial x} + z\frac{\partial B_z}{\partial x} \tag{64}$$

and:

$$\frac{\partial^2}{\partial x^2}(xB_x) = 2\frac{\partial B_x}{\partial x} + x\frac{\partial^2 B_x}{\partial x^2} + y\frac{\partial^2 B_y}{\partial x^2} + z\frac{\partial^2 B_z}{\partial x^2} \tag{65}$$

we have:

$$\Delta\big(\boldsymbol{r} \cdot \boldsymbol{B}(\boldsymbol{r})\big) = 2\boldsymbol{\nabla} \cdot \boldsymbol{B}(\boldsymbol{r}) + \boldsymbol{r} \cdot \Delta \boldsymbol{B} \tag{66}$$

Now the divergence of $\boldsymbol{B}$ is zero, and we know that the double-curl formula shows that $\boldsymbol{\nabla} \times \boldsymbol{\nabla} \times \boldsymbol{B} = \boldsymbol{\nabla}(\boldsymbol{\nabla} \cdot \boldsymbol{B}) - \Delta \boldsymbol{B}$. We thus have:

$$\Delta\big(\boldsymbol{r} \cdot \boldsymbol{B}(\boldsymbol{r})\big) = -\boldsymbol{r} \cdot \boldsymbol{\nabla} \times \boldsymbol{\nabla} \times \boldsymbol{B}(\boldsymbol{r}) = -\mu_0[\boldsymbol{r} \cdot \boldsymbol{\nabla} \times \boldsymbol{j}(\boldsymbol{r})] \tag{67}$$

The function $\boldsymbol{r} \cdot \boldsymbol{B}(\boldsymbol{r})$ is thus harmonic when $r > R_0$. It has the same properties as the function $V(\boldsymbol{r})$ studied in § 1, which enables us to write the analog of relation (26) in the form:

$$\boldsymbol{r} \cdot \boldsymbol{B}(\boldsymbol{r}) = \frac{\mu_0}{4\pi} \sum_{l=0}^{\infty} \sum_{m=-l}^{l} \sqrt{\frac{4\pi}{2l+1}} \, (b_l^m)^* \left\{ \frac{Y_l^m(\theta, \varphi)}{r^{l+1}} \right\} \tag{68}$$

where the b_l^m coefficients are obtained by changing the source term in (9):

$$b_l^m = \sqrt{\frac{4\pi}{2l+1}} \int_{\mathcal{V}_0} \mathrm{d}^3 r \; r^l \, Y_l^m(\theta, \varphi) \; [\boldsymbol{r} \cdot \boldsymbol{\nabla} \times \boldsymbol{j}(\boldsymbol{r})] \tag{69}$$

We now compare this expansion with the one written in (62). Taking the scalar product of this relation with the vector $\boldsymbol{r}$, we get under the integral the expression:

$$\boldsymbol{r} \cdot \boldsymbol{\nabla} \left[\frac{Y_l^m(\theta, \varphi)}{r^{l+1}} \right] = r \frac{\partial}{\partial r} \frac{Y_l^m(\theta, \varphi)}{r^{l+1}} = -(l+1) \frac{Y_l^m(\theta, \varphi)}{r^{l+1}} \tag{70a}$$

The divergence under the integral of (63a) can be expressed as:

$$\boldsymbol{\nabla} \cdot (\boldsymbol{r} \times \boldsymbol{j}) = \boldsymbol{j} \cdot (\boldsymbol{\nabla} \times \boldsymbol{r}) - \boldsymbol{r} \cdot (\boldsymbol{\nabla} \times \boldsymbol{j}) = -\boldsymbol{r} \cdot (\boldsymbol{\nabla} \times \boldsymbol{j}) \tag{70b}$$

Identifying both expansions shows that:

$$b_l^m = -(l+1) M_l^m \tag{71}$$

which establishes relation (63a).

To obtain relation (63b), one can use the divergence theorem:

$$\int_{r' \leq R_0} \mathrm{d}^3 r' \, \boldsymbol{\nabla} \cdot \{ r'^l \, Y_l^m(\theta', \varphi') \; \boldsymbol{r}' \times \boldsymbol{j}(\boldsymbol{r}') \} = 0 \tag{72}$$

where the right-hand side is zero as it is the flux across the surface S_0 of an identically zero vector. The expansion of the divergence in this integral introduces the function:

$$r'^{\,l}\, Y_l^m\,(\theta',\varphi')\;\boldsymbol{\nabla}\cdot[\boldsymbol{r}'\times\boldsymbol{j}(\boldsymbol{r}')]+[\boldsymbol{r}'\times\boldsymbol{j}(\boldsymbol{r}')]\cdot\boldsymbol{\nabla}\left[r'^{\,l}\, Y_l^m\,(\theta',\varphi')\right] \tag{73}$$

The integral of the first term of this expression yields the definition (63a) of $-M_l^m$, which establishes relation (63b).

Example: magnetic dipole

Consider a current loop in the plane xOy. The cross product $\boldsymbol{r}'\times\boldsymbol{j}(\boldsymbol{r}')$ of (63b) is parallel to the Oz axis; only the $m=0$ component is different from zero. The l lowest-order component is the dipole component $l=1$, in which case the product rY_1^0 is equal to $(\sqrt{3/4\pi})z$. This yields:

$$M_1^0=\frac{1}{2}\int \mathrm{d}^3r\;[\boldsymbol{r}\times\boldsymbol{j}(\boldsymbol{r})]_z \tag{74}$$

Coming back to the Cartesian components M_x, M_y, and M_z of the dipole, we have:

$$M_z=M_1^0 \qquad M_x=0 \qquad M_y=0 \tag{75}$$

expressed as the more general vector relation:

$$\boldsymbol{M}_1=\frac{1}{2}\int \mathrm{d}^3r\;[\boldsymbol{r}\times\boldsymbol{j}(\boldsymbol{r})] \tag{76}$$

An elementary calculation leads to the well-known result: a circular current loop has a magnetic moment parallel to its axis, with a modulus IS, product of the current I of the loop and its surface S.

Comment:

We only considered the case where the radiation sources, $\rho(\boldsymbol{r})$ or $\boldsymbol{j}(\boldsymbol{r})$, were time-independent. As mentioned in the introduction of this complement, the time dependence can be taken into account in the more general framework of electrodynamics. Time-dependent electric and magnetic dipoles can be defined, and are the sources of Maxwell's spherical waves propagating in space. The expressions for the multipole moments we derived in the present complement are simply the low-frequency limit of these more general expressions. These limits remain valid as long as the volume containing the sources is much smaller than the radiation wavelength.

2-b. Point charges

Consider an ensemble of point charges, whose motion yields a quasi-stationary current density $\boldsymbol{j}$. The particles are supposed to be moving inside a surface S_0. Relation (63b) yields the value of M_l^m for such a system of charges with a quasi-stationary motion:

$$M_l^m = \sum_{n=1}^{N} \left[\frac{1}{l+1}\frac{q_n}{m_n}\left(\boldsymbol{r}_n \times \boldsymbol{p}_n\right)\right] \cdot \boldsymbol{\nabla}\left[r_n^l\, Y_l^m\left(\theta_n, \varphi_n\right)\right] \tag{77}$$

If, in addition, the particles have a spin 1/2 with a dipole magnetic moment $\boldsymbol{\mu}_n$, the moment M_l^m becomes:

$$M_l^m = \sum_{n=1}^{N} \left[\frac{1}{l+1}\frac{q_n}{m_n}\left(\boldsymbol{r}_n \times \boldsymbol{p}_n\right) + \boldsymbol{\mu}_n\right] \cdot \boldsymbol{\nabla}\left[r_n^l\, Y_l^m\left(\theta_n, \varphi_n\right)\right] \tag{78}$$

Exercise: Demonstrate this relation.

2-c. Physical interpretation

The magnetic multipole moments M_l^m are analogous to the electric moments Q_l^m, but their properties are not completely identical. We shall discuss a few simple cases.

α. $l = 0$ moment

Relation (63b) immediately shows that the multipole moment M_0^0 of order zero is equal to zero. There exists no magnetic monopole[6], i.e. free magnetic charge. This is a fundamental difference between the magnetic and electric multipoles.

β. $l = 1$ moment

Taking (76) into account, we can write for a system of point charges moving in a quasi-stationary way and with a spin magnetic moment $\boldsymbol{\mu}_n$:

$$\boldsymbol{M}_1 = \sum_{n} \frac{q_n}{2m_n}\left[\boldsymbol{r}_n \times \boldsymbol{p}_n\right] + \boldsymbol{\mu}_n \tag{79}$$

[6]This non-existence is directly linked to Maxwell's equation $\boldsymbol{\nabla} \cdot \boldsymbol{B} = 0$. One can postulate the existence of free magnetic charges (magnetic monopoles) by modifying Maxwell's theory and stating that $\boldsymbol{\nabla} \cdot \boldsymbol{B}$ is proportional to the density of this new type of charges. One gets an equation completely analogous to the electric equation $\boldsymbol{\nabla} \cdot \boldsymbol{E} = -\rho/\varepsilon_0$, reestablishing a great symmetry between the electric and magnetic phenomena.

Imagine that all the particles under study have the same ratio q/m (they are all electrons, for example). The first term on the right-hand side of (79) yields the orbital magnetic moment:

$$\boldsymbol{M}_1^{(\text{orb.})} = \frac{q}{2m}\,\boldsymbol{l} \tag{80}$$

where $\boldsymbol{l}$ is the total orbital angular momentum:

$$\boldsymbol{l} = \sum_n [\boldsymbol{r}_n \times \boldsymbol{p}_n] \tag{81}$$

As for the spin part, we can write:

$$\boldsymbol{\mu}_n = g_S \frac{q}{2m}\,\boldsymbol{s}_n \tag{82}$$

where $\boldsymbol{s}_n$ is the spin of the nth particle and g_S its gyromagnetic ratio (for electrons, $g_S \simeq 2$). This yields :

$$\boldsymbol{M}_1 = \frac{q}{2m}\,[\boldsymbol{l} + g_S\,\boldsymbol{s}] \tag{83}$$

where $\boldsymbol{s}$ is the total spin $\sum_n \boldsymbol{s}_n$.

The discussion could be extended to magnetic multipole moments of order $l = 2, 3$, etc.

2-d. Quantum operators

In quantum mechanics, one uses the previous expressions but replaces $\boldsymbol{r}_n$ by the operator $\boldsymbol{R}_n$, $\boldsymbol{p}_n$ by the operator $\boldsymbol{P}_n$, and $\boldsymbol{s}_n$ in (82) by the spin operator $\boldsymbol{S}_n$. For example, $\boldsymbol{M}_1$ becomes the operator:

$$\boldsymbol{M}_1 = \sum_{n=1}^{N} \frac{q_n}{2m_n}\,\boldsymbol{L}_n + g_n\,\mu_B\,\frac{\boldsymbol{S}_n}{\hbar} \tag{84}$$

The Bohr magneton μ_B is defined as:

$$\mu_B = \frac{q_{\text{el.}}\hbar}{2m_{\text{el.}}} \tag{85}$$

where $q_{\text{el.}}$ and $m_{\text{el.}}$ are the charge and mass of the electron, and g_n the gyromagnetic ratio of particle n. Like $\boldsymbol{Q}_1$, the magnetic dipole moment $\boldsymbol{M}_1$ is a vector operator; however, it is not an odd but an even operator. This property can be generalized: the M_l^m are irreducible $K = l$, $Q = m$ tensor operators, with parity $(-)^{l+1}$, opposite to that of the electric multipoles of the same order[7].

[7]Such a result is readily understood from relation (77) that includes expressions such as $\boldsymbol{L}\cdot\boldsymbol{\nabla}[\ldots]$; the $\boldsymbol{L}\cdot\boldsymbol{\nabla}$ group is a scalar, and the right-hand side of (77) has the tensor nature of the function within brackets, in this case $R^l\,Y_l^m$. We saw that this function is a $T_{Q=m}^{(K=l)}$ tensor. In addition, the parity of $\boldsymbol{\nabla}\{r^l\,Y_l^m\}$ is equal to $(-1)^{l+1}$, and operator L is even, which explains the parity of M_l^m.

3. Multipole moments of a quantum system with a given angular momentum J

3-a. General considerations

Consider an isolated quantum system (nucleus, atom, molecule, ...) with a given angular momentum J. Its quantum state belongs to a subspace of dimension $(2J+1)$ generated by the kets $|J, M\rangle$. The electric and magnetic properties of this system depend on its multipole moments Q_l^m and M_l^m that are used for computing the system's interactions with external charges and currents. Since the Q_l^m and M_l^m are irreducible tensor operators, we can use the Wigner–Eckart theorem, and the corresponding selection rules (in particular the triangle law). This considerably limits the number of multipole moments to be taken into account. Furthermore, when the states $|J, M\rangle$ have a given parity $\pi = \pm 1$, the average value of the Q_l^m is zero whenever l is odd, and the average value of the M_l^m is zero whenever l is even (the non-zero moments are thus $\boldsymbol{M}_1, \boldsymbol{E}_2, \boldsymbol{M}_3$, etc.). We now discuss the simplest cases.

α. *Case where $J = 0$*

The quantum number J is zero for a certain number of physical systems (α particle, rare gas atoms such as ^{4}He in their ground state, etc.). The Wigner–Eckart theorem then requires $K = 0$: such systems have electromagnetic interactions with the outside that are entirely characterized by a single parameter, their total charge Q_0. Spinless particles have no dipole, or quadrupole, etc. moments.

Comment:

An ^{4}He atom, whose electronic ground state $^1S_0$ has an angular momentum $J = 0$ (and whose nuclear spin I is also zero), has a total charge equal to zero. Consequently, in that state, it is a particle totally insensitive to electromagnetic interactions. Keep in mind, however, that an electric field $\boldsymbol{E}$ (for example) can mix the ground state $^1S_0$ with the $^1P_1$ levels of the atom, which creates an electric dipole proportional to $\boldsymbol{E}$ (calculation of the ground state $^1S_0$ polarisability using first-order perturbation theory). This leads to a coupling energy of the atom with the field $\boldsymbol{E}$, which is of second-order in $\boldsymbol{E}$ (product of the field by the induced dipole). As a result of the distortion of its electronic levels, the helium atom is again sensitive to the electromagnetic interactions, but at a higher order. It is clear that the selection rules we are stating here are no longer valid as soon as the mixing of states $|J, M\rangle$ with other states having different quantum numbers J cannot be neglected.

β. Case where $J = 1/2$

Many particles have a spin 1/2: electron, proton, neutron, $^4He^+$ ion, ^3He nucleus, etc. Their electromagnetic interactions with the outside are characterized by their charge (eventually zero, as for the neutron) and their magnetic dipole moment; because of parity, their electric dipole moment is zero.

Comment:

The parity operation is not considered as a fundamental symmetry law in physics (complement D_V, § 3). This explains the experimental search for an electric dipole of the neutron or of the ^3He nucleus (the ^4He isotope is excluded since the α particle, a spinless nucleus, cannot have a dipole because of its rotational invariance). Such a moment has not been found until now (the experimental upper boundaries are now very low). Note that a positive experimental result would also imply a violation of the time reversal symmetry.

γ. Case where $J = 1$

One can take as an example the 6^3P level of mercury, etc. The $J = 1$ particles are characterized by their charge, their magnetic dipole, and their electric quadrupole.

3-b. Electric quadrupole moment of a nucleus

We discuss in more detail a case important for its application, the electric quadrupole moment of a nucleus with a nuclear spin $I \geq 1$. Considering I as a fixed quantum number in the nucleus ground state amounts to ignoring the existence of its excited levels. This is justified in many cases, because of the high energy of the excited nuclear levels. In atomic physics, for example, this is a very good approximation, since the scale of the energy levels is often on the order of an electron-volt.

We call X_n, Y_n, and Z_n the coordinates of the nucleons, and q_p the charge of the proton. Equalities (47) lead to:

$$Q_2^{\pm 2} = \frac{\sqrt{6}}{4} q_p \sum_{\text{protons}} (X_n \pm iY_n)^2$$

$$Q_2^{\pm 1} = \mp \frac{\sqrt{6}}{2} q_p \sum_{\text{protons}} Z_n (X_n \pm iY_n)$$

$$Q_2^0 = \frac{1}{2} q_p \sum_{\text{protons}} \left(3 Z_n^2 - R_n^2\right) \tag{86}$$

(where $R_n^2 = X_n^2 + Y_n^2 + Z_n^2$). The Wigner–Eckart theorem shows that the electric quadrupole $\overline{\overline{Q}}$ is proportional to the operator [*cf.* (VIII-120)]:

$$\begin{aligned}\overline{\overline{I}}^{\pm 2} &= \frac{1}{2}\,(I_\pm)^2 \\ \overline{\overline{I}}^{\pm 1} &= \mp\frac{1}{2}\,(I_\pm\, I_z + I_z\, I_\pm) \\ \overline{\overline{I}}^{0} &= \frac{1}{\sqrt{6}}\left(3\, I_z^2 - \boldsymbol{I}^2\right)\end{aligned} \tag{87}$$

This leads to:

$$\overline{\overline{Q}} = \alpha\,(\tau, I)\, \overline{\overline{I}} \tag{88}$$

where the proportionality constant is traditionally written as:

$$\alpha\,(\tau, I) = \frac{\sqrt{6}}{2\hbar^2\, I\,(2I-1)}\, q_p\, Q\,(\tau, I) \tag{89}$$

(remember that $I \neq 0$ and $I \neq 1/2$). The quadrupole moment of the nucleus is entirely characterized by the constant $Q\,(\tau, I)$, which has the dimension of a surface; this constant is generally expressed in barns (1 barn $= 10^{-24}$ cm^2). Let us compute the average value of Q_2^0 in the state $|I, I\rangle$ where the eigenvalue of Iz takes its maximum value. The numerical factor in (89) can be simplified and we get:

$$Q\,(\tau, I) = \sum_n \langle I, I|\,(3\, Z_n^2 - R_n^2)\,|I, I\rangle \tag{90}$$

The sign of $Q\,(\tau, I)$ depends on the distribution of the proton charges in the nucleus (figure 5): $Q\,(\tau, I)$ is zero for a spherical distribution, positive for an elongated ellipsoid (cigar), negative for a flattened ellipsoid (pancake).

Comments:

(i) An application of the previous discussion is the determination of the hyperfine Hamiltonian of an atom. When the nuclear spin I is zero, this Hamiltonian is zero. If $I = 1/2$, the hyperfine Hamiltonian is given by the magnetic dipole term (*cf.*, for example, complement A_{XII} in [9]). If $I > 1$, the electric potential "seen" by the electrons depends on the electric quadrupole of the nucleus, which leads to a quadrupole term in the hyperfine Hamiltonian.

(ii) Computing the electromagnetic transition probabilities of quantum systems between levels of different J multiplicities is another important example of multipole moments theory: evaluation of atomic polarizability, emission of quadrupole photons, etc.

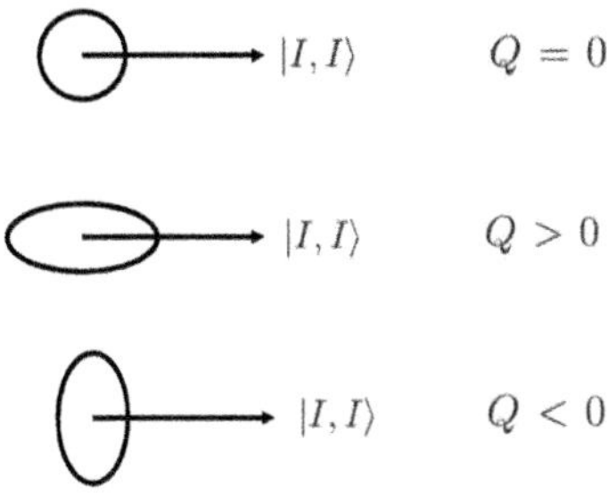

Figure 5: In the state $|I, I\rangle$ with maximum polarization, the quadrupole Q is zero for a spherical distribution of charges in the nucleus, positive for an elongated ellipsoid distribution (similar to a cigar), negative for a flattened ellipsoid distribution (similar to a pancake).

Complement D_{VIII}

Density matrix expansion on irreducible tensor operators

This complement studies a physical system whose state space is restricted to a given irreducible space $\mathscr{E}(\tau, J)$. The $2J+1$ kets $|J, M\rangle$ associated with the values $M = J, J-1, J-2, \ldots, -J+1, -J$, form a basis of the state space $\mathscr{E}$.

1. Liouville space

Consider the ensemble $\mathscr{L}$ of the operators acting in $\mathscr{E}$; in a given basis of $\mathscr{E}$, these operators are represented by $(2J+1) \times (2J+1)$ matrices. It is easy to see that the ensemble $\mathscr{L}$ is a vector space, with dimension $(2J+1)^2$, defined on the set of complex numbers[1]. $\mathscr{L}$ is called a "*Liouville space*". A possible basis for $\mathscr{L}$ is the set of operators $|J, M\rangle\langle J, M'|$ ("dyadic" operators), where M and M' each take $(2J+1)$ values; when these operators are considered as vectors of the Liouville space, we shall note them as "double kets":

$$|J, M\rangle\langle J, M'| \Leftrightarrow |M, M'\rangle\rangle \tag{1}$$

In a similar way, with each operator A will be associated the element of $\mathscr{L}$ noted $|A\rangle\rangle$. A scalar product in the Liouville space between any two elements $|A\rangle\rangle$ and $|B\rangle\rangle$ can be simply defined as:

$$\langle\langle B|A\rangle\rangle = \text{Tr}\left\{B^\dagger A\right\} \tag{2}$$

(where the trace on the right-hand side is to be taken in the usual way in the state space $\mathscr{E}$). Since the traces of two Hermitian conjugate operators are complex conjugates, and since the conjugate of $B^\dagger A$ is $A^\dagger B$, we have:

$$\langle\langle B|A\rangle\rangle = \langle\langle A|B\rangle\rangle^* \tag{3a}$$

[1]It would be the set of real numbers if we only considered Hermitian operators.

As a result, $\langle\langle A|A\rangle\rangle$ is real and we have[2]:

$$\langle\langle A|A\rangle\rangle \geq 0 \tag{3b}$$

(the zero equality occurs only if A is identically zero). The above definition obeys the usual properties of a norm (defined as positive) and a scalar product.

The basis $|M, M'\rangle\rangle$ is orthonormal since:

$$\begin{aligned}\langle\langle M, M'|M'', M'''\rangle\rangle &= \text{Tr}\left\{|J, M'\rangle\langle J, M|J, M''\rangle\langle J, M'''|\right\} \\ &= \delta_{MM''}\,\delta_{MM'''}\end{aligned} \tag{4}$$

Any operator A can thus be decomposed as:

$$\begin{aligned}A &= \sum_{M,M'=-J}^{+J} \langle\langle M, M'|A\rangle\rangle\;|M, M'\rangle\rangle \\ &= \sum_{M,M'} \text{Tr}\left\{|J, M'\rangle\langle J, M|A\right\}\;|J, M\rangle\langle J, M'|\end{aligned} \tag{5}$$

In this equality, the trace is given by:

$$\begin{aligned}\text{Tr}\left\{|J, M'\rangle\langle J, M|A\right\} &= \sum_{M''=-J}^{+J} \langle J, M''|J, M'\rangle\langle J, M|A|J, M''\rangle \\ &= \langle J, M|A|J, M'\rangle\end{aligned} \tag{6}$$

Decomposition (5) is simply equivalent to the introduction of two closure relations on the basis $\{|J, M\rangle\}$, one on each side of the operator A.

In quantum mechanics, a system if often characterized by its density matrix (see, for example, complement E_{III} of reference [9]), which is a Hermitian operator generally called ρ. It allows the computation of the average value of any operator A by the relation:

$$\langle A\rangle = \text{Tr}\{\rho A\} = \langle\langle\rho|A\rangle\rangle \tag{7}$$

---

[2]Any operator A can be written as $A = A_R + iA_I$, where A_R and A_I are the two Hermitian operators (which can thus be diagonalized) given by $A_R = (A + A^\dagger)/2$ and $A_I = (A - A^\dagger)/2i$. We then have:

$$\begin{aligned}\langle\langle A|A\rangle\rangle &= \text{Tr}\left\{A_R^\dagger A_R\right\} + \text{Tr}\left\{A_I^\dagger A_I\right\} + i\;\text{Tr}\left\{A_R^\dagger A_I - A_I^\dagger A_R\right\} \\ &= \text{Tr}\left\{A_R^2\right\} + \text{Tr}\left\{A_I^2\right\}\end{aligned}$$

Each trace is the sum of squared real numbers and is thus positive, which establishes (3b). These sums can be equal to zero only if all the eigenvalues of A_R and A_I are zero, meaning only if A_R and A_I as well as A are equal to zero.

The density matrix ρ of the system can be decomposed as:

$$\rho = \sum_{MM'=-J}^{+J} \rho_{MM'} |M, M'\rangle\rangle = \sum_{MM'} \rho_{MM'} |J, M\rangle \langle J, M'| \tag{8a}$$

where :

$$\rho_{MM'} = \langle J, M|\rho|J, M'\rangle \tag{8b}$$

As ρ is Hermitian, we have:

$$\rho_{M'M} = (\rho_{MM'})^{\star} \tag{8c}$$

2. Rotation transformation

One may wonder how the elements $|A\rangle\rangle$ of the Liouville space are transformed under a rotation. Since the operator A becomes $RAR^{\dagger}$ under a rotation, it is natural to set:

$$|A'\rangle\rangle = |RAR^{\dagger}\rangle\rangle \tag{9a}$$

that can be written as:

$$|A'\rangle\rangle = {}^{\mathscr{L}}\mathscr{R}|A\rangle\rangle \tag{9b}$$

where $|A'\rangle\rangle$ is the rotation transform of $|A\rangle\rangle$, and ${}^{\mathscr{L}}\mathscr{R}$ the linear operator that describes this rotation in the Liouville space.

Let us see in particular the result of the transformation of the basis $|M, M'\rangle\rangle$ introduced above. We have:

$$R\,|J, M\rangle = \sum_{M''=-J}^{+J} \left(R^{[J]}\right)_{M''M} |J, M''\rangle \tag{10a}$$

and thus:

$$\langle J, M'|R^{\dagger} = \sum_{M'''=-J}^{+J} \left(R^{[J]}\right)^{\star}_{M'''M'} \langle J, M'''| \tag{10b}$$

As a result:

$$R\,|J, M\rangle\langle J, M'|R^{\dagger} = \sum_{M''} \sum_{M'''} \left(R^{[J]}\right)_{M''M} \left(R^{[J]}\right)^{\star}_{M'''M'} |J, M''\rangle \langle J, M'''| \tag{11a}$$

or:

$${}^{\mathscr{L}}\mathscr{R}\,|M, M'\rangle\rangle = \sum_{M''\,M'''} \left(R^{[J]}\right)_{M''M} \left(R^{[J]}\right)^{\star}_{M'''M'} |M'', M'''\rangle\rangle \tag{11b}$$

The rotation transformations of the $|M, M'\rangle\rangle$ are thus not that simple, since they involve two rotation matrices.

3. Basis of the $T_Q^{[K]}$ operators

We know the action of a vector operator, the usual angular momentum operator $\boldsymbol{J}$, in the initial state space $\mathscr{E}$. To this operator $T^{[K=1]}$ correspond 3 elements of $\mathscr{L}$ noted $|K,Q\rangle\rangle$ (with $K=1$, $Q=1,\,0,\,-1$).

Taking the tensor product of $\boldsymbol{J}$ with itself, we can construct, as we did in § B-4-a, two new operators $K=0$ (scalar operator proportional to $\boldsymbol{J}^2$) and $K=2$ (alignment, *cf.* § D-3); we are leaving out the $K=1$ operator since, according to the Wigner–Eckart theorem, it is proportional to $\boldsymbol{J}$. We thus introduce in $\mathscr{L}$ one element $|K=0,\,Q=0\rangle\rangle$, and 5 elements $|K=2,\,Q\rangle\rangle$ where Q varies between $+2$ and -2. The process can continue, taking the tensor product of operator $T^{[K=2]}$ with $\boldsymbol{J}$ to obtain $T^{[K=3]}$, etc. It is clear that this procedure cannot go on indefinitely: as soon as $K\geq 2J+1$, the resulting operators are zero, since $\langle J\,K\,M\,Q|J\,M'\rangle$ is zero according to the triangle selection rule.

We have therefore defined $2J+1$ tensor operators of order $K=0,1,2,\ldots,2J$, each having $2K+1$ components. Since:

$$\sum_{K=0}^{2J}(2K+1)=2\frac{2J(2J+1)}{2}+2J+1=(2J+1)^2 \tag{12}$$

the number of these operators is equal to the dimension of the Liouville space $\mathscr{L}$, and the $|K,Q\rangle\rangle$ could, a priori, form a basis; we must however verify that they are independent, proving for example that:

$$\mathrm{Tr}\left\{T_Q^{[K]}\right\}=\delta_{K0}\,\delta_{Q0} \tag{13}$$

Demonstration: We first compute the trace of each operator $T_Q^{[K]}$ we have introduced. If $K=Q=0$, the Wigner–Eckart theorem shows that all the matrix elements are diagonal and equal to each other. Multiplying them by a suitable factor, we can set:

$$T_0^{[K=0]}=\frac{1}{2J+1}\,\mathbb{1} \tag{14}$$

so that:

$$\mathrm{Tr}\left\{T_0^{[K=0]}\right\}=1 \tag{15}$$

As for the traces of all the other operators $T_Q^{[K]}$, they are necessarily equal to zero as we now show. The first of relations (VIII-54) establishes that:

$$\hbar Q\,\mathrm{Tr}\left\{T_Q^{[K]}\right\}=\mathrm{Tr}\left\{\left[J_z,T_Q^{[K]}\right]\right\}=0 \tag{16}$$

(the trace of a commutator is zero). As a result, if $Q\neq 0$, the trace of $T_Q^{[K]}$ is zero. If Q is zero, we use the second equality (VIII-54) to write:

$$\mathrm{Tr}\left\{T_0^{[K]}\right\}\propto\mathrm{Tr}\left\{\left[J_\pm,T_{\mp 1}^{[K]}\right]\right\}=0 \tag{17}$$

where the proportionality constant is zero only if $K = 0$. This establishes equation (13).

We now compute the scalar product of two elements $|K, Q\rangle\rangle$ of $\mathscr{L}$:

$$\langle\langle K, Q|K', Q'\rangle\rangle = \text{Tr}\left\{T_Q^{[K]\dagger}\, T_{Q'}^{(K')}\right\} \tag{18a}$$

We saw in (VIII-98) that operator $T^{[K]}$ is Hermitian if its reduced matrix element in space $\mathscr{E}(\tau, J)$ is real. Multiplying each operator $T^{[K]}$ by a constant can ensure it is Hermitian. Using relation (VIII-67), we can write:

$$\langle\langle K, Q|K', Q'\rangle\rangle = (-1)^Q\, \text{Tr}\left\{T_{-Q}^{[K]}\, T_{Q'}^{(K')}\right\} \tag{18b}$$

To compute this trace, we decompose, as in § B-5-c, the product $T_{-Q}^{[K]}\, T_{Q'}^{[K']}$ into irreducible tensor operators $Z_{Q''}^{[K'']}$:

$$T_{-Q}^{[K]}\, T_{Q'}^{[K']} = \sum_{K''Q''} \langle K\, K' \;-Q\, Q'|K''\, Q''\rangle\, Z_{Q''}^{[K'']} \tag{19}$$

Each operator $Z_{Q''}^{[K'']}$ is proportional to the same order operator $T_{Q''}^{(K'')}$; according to (13), it has a zero trace unless $K'' = Q'' = 0$. Consequently:

$$\langle\langle K, Q|K', Q'\rangle\rangle = \beta(-1)^Q\, \langle K\, K' \;-Q\, Q'|0\, 0\rangle\beta \tag{20}$$

where β is a constant. According to the selection rules on the Clebsch–Gordan coefficients, this expression equals zero unless $K = K'$ and $Q = Q'$, in which case it is independent of Q [*cf.* equality (VII-91)]. The $|K, Q\rangle\rangle$ are thus orthogonal to each other; the multiplication of each $T^{[K]}$ by a real constant yields an orthonormal basis.

We have thus obtained a new orthonormal basis in the Liouville space. The rotation transformation of this new basis is simple since equality (VIII-59) leads to:

$${}^{\mathscr{L}}\mathscr{R}\,|K, Q\rangle\rangle = \sum_{Q'} \left(R^{[K]}\right)_{Q'Q} |K, Q'\rangle\rangle \tag{21}$$

easier to handle than relation (11b).

Any operator A, i.e. any element $|A\rangle\rangle$ of $\mathscr{L}$, can be decomposed onto this new basis. In particular, we can write for the density matrix ρ:

$$|\rho\rangle\rangle = \sum_{K=0}^{2J}\; \sum_{Q=-K}^{+K} \rho_Q^{[K]}\, |K, Q\rangle\rangle \tag{22a}$$

or:

$$\rho = \sum_K \sum_Q \rho_Q^{[K]}\, T_Q^{[K]} \tag{22b}$$

with:

$$\begin{aligned}\rho_Q^{[K]} &= \langle\langle K, Q|\rho\rangle\rangle = \text{Tr}\left\{(T_Q^{[K]})^\dagger\, \rho\right\} \\ &= (-1)^Q\, \text{Tr}\left\{T_{-Q}^{[K]}\, \rho\right\} \end{aligned} \tag{23}$$

This decomposition of ρ onto its "tensor components" is often quite useful. The quantity $\rho_0^{(0)}$ is a rotation invariant constant, the trace of ρ (sometimes called the "total population" of the levels, when this trace is not fixed to the value 1). The three components $\rho_Q^{[1]}$ yield the orientation of the level, the five components of $\rho_Q^{[2]}$ its alignment, etc.

For $J = 1/2$, specifying the three orientation components (and eventually the total population) entirely determines ρ, and hence the state of the system; according to the Wigner–Eckart theorem, all the moments of order 2, 3, ... are always strictly equal to zero.

For $J = 1$, specifying the five $\rho_Q^{(2)}$ (alignment) in additon to the $\rho_Q^{(1)}$ is necessary to completely determine the state of the system. The reasoning can be extended: the higher the angular momentum J of a system, the larger the number of $\rho_Q^{[K]}$ necessary to specify its density matrix.

4. Rotational invariance in a system's evolution

Using Liouville space simplifies the calculations of the evolution of a physical system invariant under rotation.

4-a. Master equation

The time evolution of the density matrix $\rho(t)$ can often be written as:

$$i\hbar \frac{\text{d}}{\text{d}t}\, |\rho(t)\rangle\rangle = \mathscr{C}(t)\, |\rho(t)\rangle\rangle \tag{24}$$

where $\mathscr{C}(t)$ is the so-called linear "liouville operator" acting in $\mathscr{L}$. If the system is isolated, its Hamiltonian H is time independent, and we simply have (von Neumann equation):

$$\mathscr{C}(t)\, |\rho(t)\rangle\rangle = |[H, \rho]\rangle\rangle \tag{25}$$

However, equation (24), which we call the "master equation", is far more general. It includes, for example, possible external interactions (taking into account realistic assumptions for interesting physical cases) that lead to a random character of the Hamiltonian (relaxation theory), etc. It is thus this equation that will be discussed here.

There is a striking analogy between relation (24) and the Schrödinger equation. Here, too, we can introduce an evolution operator $\mathcal{U}(t, t_0)$ that allows writing the solution of (24) in the form:

$$|\rho(t)\rangle\rangle = \mathcal{U}(t, t_0)\ |\rho(t_0)\rangle\rangle \tag{26}$$

This operator must satisfy:

$$i\hbar \frac{\mathrm{d}}{\mathrm{d}t}\,\mathcal{U}(t, t_0) = \mathscr{C}(t)\,\mathcal{U}(t, t_0) \tag{27a}$$

and:

$$\mathcal{U}(t_0, t_0) = \mathbb{1} \tag{27b}$$

If, for example, there exists an operator $U(t, t_0)$ acting in the state space and yielding the following evolution for ρ:

$$\rho(t) = U(t, t_0)\ \rho(t_0)\ U^\dagger(t, t_0) \tag{28}$$

we have simply:

$$\mathcal{U}(t, t_0)\ |\rho\rangle\rangle = |U\,\rho\,U^\dagger\rangle\rangle \tag{29}$$

but, here again, equation (26) is more general.

4-b. Rotational invariance of the master equation

Imagine now that any rotation operation is a symmetry transformation for the evolution of the system under study (as in the study of an atom subjected to collisions with its neighbors in any direction in space with the same probabilities). Going back to the general diagram of figure 5 in chapter I, we can write, just as we did in equation (I-107):

$$^{\mathscr{L}}\mathscr{R}\,\mathcal{U}(t, t_0) = \mathcal{U}(t, t_0)\,^{\mathscr{L}}\mathscr{R} \tag{30a}$$

that is:

$$\left[^{\mathscr{L}}\mathscr{R}\ ,\ \mathcal{U}(t, t_0)\right] = 0 \tag{30b}$$

(where it is understood that the commutator is computed in the Liouville space $\mathscr{L}$). Operators $\mathcal{U}(t, t_0)$ and $\mathscr{C}(t)$ are thus scalar operators acting in $\mathscr{L}$. In addition, equation (21) shows that the $|K, Q\rangle\rangle$ obey, in this space, the definition relations of a standard basis. The Wigner–Eckart theorem can thus be applied, this time in $\mathscr{L}$ rather than in $\mathscr{E}$. As a result, $\mathcal{U}$ and $\mathscr{C}$ have properties analogous to those linked to the presence of the Clebsch–Gordan coefficient (VIII-99) for an operator $T^{(0)}$. We thus have:

$$\langle\langle K, Q|\mathcal{U}(t, t_0)|K', Q'\rangle\rangle = \delta_{KK'}\, \delta_{QQ'}\, C_K(t, t_0) \tag{31}$$

where C is a number independent of Q and Q'. As a result, relation (25) yields:

$$\rho_Q^{[K]}(t) = \rho_Q^{[K]}(t_0) \times C_K(t, t_0) \tag{32}$$

This leads to the following important result, valid anytime the rotational invariance is preserved in the evolution of the system under study: the different tensor components $\rho_Q^{[K]}$ of the density matrix have independent time evolutions; the way they evolve depends on K, but not on Q. It is clear that this property has great practical importance in many physical problems where rotational invariance plays a role.

Example: One often encounters the case where:

$$C_K(t, t_0) = \mathrm{e}^{-(t-t_0)/\tau_K} \tag{33}$$

where τ_K is a positive constant, the relaxation time of the tensor observables of order K.

Comment:

In the particular case where relation (25) yields the evolution of ρ (Hamiltonian evolution), the rotational invariance forces the Hamiltonian H to be a scalar. Operator H is thus proportional to the operator $T_0^{(K=0)}$ written in (14), which commutes with any operator in $\mathscr{E}$ [remember that $\mathscr{E}$ has been restricted to a single subspace $\mathscr{E}(\tau, J)$]. In other words, the physical system does not evolve, and equalities (32) are trivially established.

4-c. Angular momentum in the Liouville space

We can define the action of an angular momentum operator $\boldsymbol{\mathcal{J}}$ in the Liouville space $\mathscr{L}$. We simply consider an infinitesimal rotation of vector $\delta\boldsymbol{a} = \boldsymbol{u}\delta\varphi$ and set:

$$^{\mathscr{L}}\mathscr{R} = 1 - \frac{i\delta\varphi}{\hbar}\, \boldsymbol{u} \cdot \boldsymbol{\mathcal{J}} \tag{34}$$

Using (9), we see that this amounts to defining the action of the $\mathcal{J}_u = \boldsymbol{u} \cdot \boldsymbol{\mathcal{J}}$ component of $\vec{\mathcal{J}}$ as:

$$\mathcal{J}_u|A\rangle\rangle = |\,[J_u, A]\rangle\rangle \tag{35}$$

From the very definition of irreducible tensor operators, it is clear that:

$$\mathcal{J}_z|K,Q\rangle\rangle = Q\,\hbar|K,Q\rangle\rangle \tag{36a}$$

and:

$$\mathcal{J}_\pm|K,Q\rangle\rangle = \hbar\sqrt{K(K+1)-Q(Q\pm1)}\;|K,Q\pm1\rangle\rangle \tag{36b}$$

(where $\mathcal{J}_\pm = \mathcal{J}_x \pm i\mathcal{J}_y$). The $|K,Q\rangle\rangle$ precisely obey the relations of a standard basis in $\mathscr{L}$. The Liouville space can thus be expressed as the direct sum of $2J+1$ irreducible subspaces, invariant under the action of the whole set of rotation operators: the 1-dimensional space generated by $|K=0,Q=0\rangle\rangle$, the 3-dimensional space generated by the $|K=1,Q\rangle\rangle$, the 5-dimensional space generated by the $|K=2,Q\rangle\rangle$, etc. (remember that $K\leq 2J$). Each of these subspace is an eigenspace of the operator:

$$\begin{aligned}\boldsymbol{\mathcal{J}}^2 &= \mathcal{J}_x^2+\mathcal{J}_y^2+\mathcal{J}_z^2\\ &= \frac{1}{2}\left(\mathcal{J}_+\mathcal{J}_- + \mathcal{J}_-\mathcal{J}_+\right)+\mathcal{J}_z^2\end{aligned} \tag{37}$$

with the eigenvalue $K(K+1)\,\hbar^2$:

$$\boldsymbol{\mathcal{J}}^2|K,Q\rangle\rangle = K(K+1)\,\hbar^2\,|K,Q\rangle\rangle \tag{38}$$

In the Liouville space $\mathscr{L}$, a scalar operator (invariant under rotation) is an operator that commutes with any rotation operator ${}^{\mathscr{L}}\mathscr{R}$, i.e. with any component of $\boldsymbol{\mathcal{J}}$. Equality (30b) shows that $\mathcal{U}(t,t_0)$ is a scalar operator. Since $\mathcal{U}(t,t_0)$ commutes with operators $\mathcal{J}_z$ and $\boldsymbol{\mathcal{J}}^2$, its matrix elements between eigenvectors with different eigenvalues are always zero. This readily establishes relation (31) (without reference to the Wigner–Eckart theorem in the Liouville space).

In case the evolution of the physical system under study is invariant only under a rotation around the Oz axis, operateur $\mathcal{U}$ still commutes with $\mathcal{J}_z$ but no longer necessarily with $\boldsymbol{\mathcal{J}}^2$. In this case, the evolutions of the $\rho_Q^{[K]}$ associated with different values of Q are still uncoupled, but not those associated with the same value of Q and different values of K (coupling between the $Q=0$ components of orientation, alignment, etc.).

Chapter IX

Internal symmetries, SU(2) and SU(3) groups

Introduction

In this chapter, we study energy levels of physical systems composed of different types of particles, all of which play a completely symmetric role in the system's Hamiltonian. The object is to obtain a classification of these levels independent, for a given total number of particles, of the distribution of the different types of particles. The reasoning applies independently of the particles nature, all fermions, all bosons, or considered as distinguishable as we shall first assume.

Introduction to Continuous Symmetries: From Space–Time to Quantum Mechanics, First Edition. Franck Laloë.

The results we shall obtain can be applied in many fields of quantum mechanics, such as the classification of energy levels in atomic or molecular physics. They are particularly essential in nuclear physics and elementary particle physics. Historically, the concept of isospin was first introduced by Heisenberg[1] to classify the energy levels of the nuclei in the framework of an SU(2) symmetry. Subsequently, SU(n) symmetries obtained "by substitution of equivalent particles belonging to n different types" have played an increasingly important role, not only in nuclear spectroscopy but also in the physics of elementary particles. These symmetries play an essential role when trying to determine if physical objects should be considered as being an assembly of "more elementary" particles. An important application is the study of the regularities in strong interactions thanks to the SU(3) symmetry. This symmetry comes into play very simply if one assumes that mesons and baryons are made out of elementary "quarks" belonging to three different types characterized by a "flavor" quantum number that can take three different values[2].

Another aim of the following discussion is to familiarize the reader with the use, in some simple cases, of root diagrams and weight diagrams. As its Lie algebra is of rank two, the advantage of the SU(3) group is that the planar representations of the diagrams symbolizing the irreducible representations are particularly easy to read.

In this chapter, the reasoning will be carried out in two steps. In the first (§ A), we introduce symmetry operators K, show that they commute with the Hamiltonian H, and compute the commutation relations between the K (which are those of a Lie algebra). This first step shows the profound analogy existing in quantum mechanics between the nature of a particle (proton, neutron, etc.) and a quantum number that would be associated with an internal variable of the particle, the spin, for example. It will be based on various somewhat technical arguments on the possible symmetries of a state vector, of the products of symmetrization or antisymmetrization operators, etc.

The reader may want to go directly to the physical results and, in a first reading, skip § A to study the second step. Starting from the commutation relations of the K operators, we build the associated irreducible representations via the root diagram. The § B discusses the isospin multiplets of SU(2), the § C the supermultiplets of SU(3) that allow regrouping in diagrams the composite particles of three quarks, baryons, or mesons.

[1]Instead of isospin, one sometimes uses the term "isotopic spin". This term is confusing since the isospin symmetry refers to nuclei having the same total number of nucleons (hence similar masses), whereas the different isotopes of the same element have a fixed number of protons and a variable number of neutrons. The term "isobaric spin", which is sometimes used, would be more adequate.

[2]This number is currently thought to be higher, but the study of SU(3) symmetries is still of great interest!

A. System of distinguishable but equivalent particles

A-1. General hypotheses

Consider a set of n particules, n_α of which belong to a first type α (protons, for example), and n_β to a second type β (neutrons, for example). For the sake of simplicity, we limit the number of types to two, but the reasoning can be generalized to 3, 4, ... different types. Consider an operator H, generally the system's Hamiltonian, and assume that all the particles (whatever their type) play a symmetric role in H. Let us see right away on some particular cases what this hypothesis implies. Consider, for example, the kinetic energy operator:

$$T = \sum_{i=1}^{n_\alpha} \frac{\boldsymbol{P}_i^2}{2m_\alpha} + \sum_{j=n_\alpha+1}^{n_\alpha+n_\beta} \frac{\boldsymbol{P}_j^2}{2m_\beta} \tag{IX-1}$$

where $\boldsymbol{P}$ is the particles' momentum, m_α and m_β the mass of each α and β type of particles. When:

$$m_\alpha = m_\beta = m \tag{IX-2}$$

we can say that all the particles play the same role in T. In this case, we can write:

$$T = \frac{1}{2m} \sum_{i=1}^{n} \boldsymbol{P}_i^2 \tag{IX-3}$$

The mathematical expression of operator T does not change if one (or several) particle α is changed into a particle β (or conversely), though the physical interpretation of T is different as it concerns a different physical system.

If H includes two-particle potential energy terms:

$$H = T + \sum_{i<j} V_{ij}\left(\boldsymbol{R}_i, \boldsymbol{R}_j\right) \tag{IX-4}$$

(the operators $\boldsymbol{R}$ are the particles' position operators), our hypothesis implies that V_{ij} depends neither on i nor on j. Three potentials are supposed to be equal: the one describing the mutual interactions between α particles, the one describing the mutual interactions between β particles, and the one describing "crossed" interactions between α and β particles.

In a general way, regardless of the operator H (including, for example, three-body interactions, spin-spin couplings, etc.) and the number of types α, β, γ... we shall always assume that H *keeps the same mathematical expression if one changes the type of any number of numbered particles.*

This does not exclude the fact that the α or β particles could be fundamentally distinguishable. In this case, one could introduce operators corresponding to other physical quantities and for which the two types of particles play a different role (for instance, such an operator could be N_α, which corresponds to the number of α particles). In other words, even though the α and β particles are in principle distinguishable, we assume they are "not distinguishable" by the operators H we shall consider in what follows[3].

Example: The Hamiltonian H of a nucleus, when one ignores the electromagnetic interactions. It is generally assumed that, in the interactions responsible for the cohesion of nuclei (strong interactions), the proton and the neutron play exactly the same role (charge independence). The reasoning we shall follow allows the classification of the nucleus energy levels[4].

A-2. Operators commuting with H

§ A-2 introduces a number of operators that commute with H, and whose essential characteristic is to change the numbers n_α and n_β of α and β particles (while keeping their sum constant). For the sake of simplicity, we first assume (§ A-2-a) that all the α particles are basically distinguishable from each other (albeit having the same characteristics vis-à-vis the Hamiltonian), the same hypothesis being made for the β particles. This is a fairly elementary case, but useful to establish the notations. In (§ A-2-b), we shall discuss the modifications that occur when all the α particles are identical, as are all the β particles (fermions or bosons).

A-2-a. Distinguishable particles

α. *State space*

When all the particles are distinguishable, the state space of the system composed of n_α particules α and n_β particules β is simply the tensor product of individual state spaces:

$$\mathscr{E}_{n_\alpha,n_\beta} = [\mathscr{D}_\alpha(1) \otimes \mathscr{D}_\alpha(2) \ldots \otimes \mathscr{D}_\alpha(n_\alpha)] \otimes [\mathscr{D}_\beta(n_\alpha + 1) \otimes \ldots \otimes \mathscr{D}_\beta(n)] \quad \text{(IX-5a)}$$

where $\mathscr{D}_\alpha(i)$ is the state space of the particule numbered (i) if it is of the α type, and $\mathscr{D}_\beta(j)$ the state space of the particle numbered (j) of the β type. The total particle number is:

$$n = n_\alpha + n_\beta \quad \text{(IX-5b)}$$

[3]It is not sufficient for the types α and β as a whole to play a symmetric role in H for all the particles, whatever their type, to play a symmetric role in H. As an example, the operator $N_\alpha N_\beta$ meets the first condition but not the second.

[4]In a second step, one can improve the theory to include the electromagnetic repulsion between protons (lifting of degeneracy corresponding to a symmetry breaking).

Since the two kinds of particles play a symmetrical role in the problem, we assume that the individual state spaces $\mathscr{D}_\alpha$ and $\mathscr{D}_\beta$ are isomorphic:

$$\mathscr{D}_\alpha \equiv \mathscr{D}_\beta \tag{IX-5c}$$

which allows dropping the index α or β in the notation of the state spaces $\mathscr{F}$ of a single particle. This space can be decomposed onto an orthonormal basis of individual spaces $\{|\varphi_k\rangle\}$, the index k varying from 1 to $\mathscr{N}$ (the dimension of $\mathscr{E}$, which can eventually be infinite). An orthonormal basis of the space $\mathscr{E}_{n_\alpha,n_\beta}$ is given by the kets:

$$|\overbrace{1:k_1\,;\,2:k_2\,;\,\ldots\,;\,n_\alpha:k_{n_\alpha}}^{n_\alpha}\,;\,\overbrace{n_\alpha+1:k_{n_\alpha+1}\,;\,\ldots\,;\,n:k_n}^{n_\beta}\rangle$$

$$= |\varphi_{k_1}(1)\rangle \otimes |\varphi_{k_2}(2)\rangle \otimes \ldots \otimes |\varphi_{k_n}(n)\rangle \tag{IX-6}$$

To express mathematically that all the particles, regardless of their nature, play a symmetrical role in H, we simply set equal to zero the commutators:

$$[H, P_{ij}] = 0 \tag{IX-7a}$$

where P_{ij} is the transposition operator[5] that exchanges the two (i) and (j) particles:

$$P_{ij}|\varphi_{k_1}(1)\rangle \otimes \ldots |\varphi_{k_i}(i)\rangle \otimes \ldots |\varphi_{k_j}(j)\rangle \otimes \ldots$$
$$= |\varphi_{k_1}(1)\rangle \otimes \ldots |\varphi_{k_j}(i)\rangle \otimes \ldots |\varphi_{k_i}(j)\rangle \otimes \ldots \tag{IX-7b}$$

Relation (IX-7a) must be obeyed for any i and j, whether the i and j numbered particles are of the same nature or not.

Operator P_{ij} does not change either the n_α number of α particles, nor the n_β number of β particules; acting on a ket of the form (IX-6), it exchanges two individual states $|\varphi_k\rangle$. The space $\mathscr{E}_{n_\alpha n_\beta}$ is thus stable under the action of P_{ij}.

β. *Operators changing the nature of the particles*

To introduce the $P_{\beta\to\alpha}$ and $P_{\alpha\to\beta}$ operators that change n_α and n_β, we consider a direct sum of spaces:

$$\mathcal{F}_n = \mathscr{E}_{n_\alpha=n,n_\beta=0} \oplus \mathscr{E}_{n_\alpha=n-1,n_\beta=1} \oplus \ldots \oplus \mathscr{E}_{n_\alpha=0,n_\beta=n} \tag{IX-8}$$

[5]The commutation of H with all the P_{ij} transpositions implies that H commutes with all the $n!$ permutation operators P_λ of the n particles; the symmetric group S_n is a symmetry group for H.

In this larger state space $\mathcal{F}_n$, the numbers n_α and n_β no longer have fixed values, but the value of their sum is fixed[6]. An orthonormal basis of $\mathcal{F}_n$ is given by the kets written in (IX-6), taken for the various values of n_α and $n_\beta = n - n_\alpha$, when n_α goes from 0 to n; by convention, we consider as orthogonal in $\mathcal{F}_n$ two kets coming from two distinct $\mathscr{E}_{n_\alpha,n_\beta}$ spaces. We also must define the action of operator[7] H in $\mathcal{F}_n$ so that H becomes the Hamiltonian of n equivalent particles with non-specified natures (α or β). We must consider an operator that has the same mathematical expression in all the isomorphic spaces $\mathscr{E}_{n_\alpha,n_\beta}$, the only difference being that each particle numbered (i) can, depending on the case, be interpreted physically as an α or β particle. In other words, the action of H in $\mathcal{F}_n$ is the juxtaposition the actions of Hamiltonians associated with different physical systems (n_α and n_β are not fixed), initially defined in each of the spaces $\mathscr{E}_{n_\alpha,n_\beta}$.

We now introduce operator $P_{\beta\to\alpha}$ that transforms an α particle into a β particle:

$$P_{\beta\to\alpha}\,|\overbrace{1:k_1\,;\,\ldots\,;\,n_\alpha:k_{n_\alpha}}^{n_\alpha}\,;\,\overbrace{n_\alpha+1:k_{n_\alpha+1}\,;\,\ldots\,;\,n:k_n}^{n_\beta}\rangle$$
$$=|\overbrace{1:k_1\,;\,\ldots\,;\,n_\alpha+1:k_{n_\alpha+1}}^{n_\alpha+1}\,;\,\overbrace{n_\alpha+2:k_{n_\alpha+2}\,;\,\ldots\,;\,n:k_n}^{n_\beta-1}\rangle \qquad \text{(IX-9a)}$$

No subspace $\mathscr{E}_{n_\alpha,n_\beta}$ remains invariant under the action of $P_{\beta\to\alpha}$, defined on the basis kets (IX-6): acting on any ket of this subspace, $P_{\beta\to\alpha}$ yields a ket of $\mathscr{E}_{n_\alpha+1,n_\beta-1}$, where the only difference is that the $(n_\alpha+1)^{\text{th}}$ particle is of a different nature. As for the action of $P_{\beta\to\alpha}$ on the kets of $\mathscr{E}_{n_\alpha=n,n_\beta=0}$, it can be defined as yielding a zero ket:

$$P_{\beta\to\alpha}\,|\psi\rangle = 0 \qquad \text{if} \quad |\psi\rangle \in \mathscr{E}_{n_\alpha=n,n_\beta=0} \qquad \text{(IX-9b)}$$

In a symmetrical way, one can define an operator $P_{\alpha\to\beta}$ that changes an α particle numbered n_α into a β particle and whose action in $\mathscr{E}_{n_\alpha=0,n_\beta=n}$ yields zero.

γ. *Commutation with H*

Let $|\psi\rangle$ be an eigenket of H, belonging to $\mathscr{E}_{n_\alpha,n_\beta}$ and with the eigenvalue E:

$$\begin{cases} H\,|\psi\rangle = E\,|\psi\rangle \\ |\psi\rangle \in \mathscr{E}_{n_\alpha,n_\beta} \end{cases} \qquad \text{(IX-10)}$$

[6]The space $\mathcal{F}_n$ includes states that are coherent superpositions of states where n_α and n_β have different values. Even though such states do not necessarily have a physical meaning (like a coherent superposition of a proton and a neutron), introducing $\mathcal{F}_n$ is useful for what follows.

[7]For the sake of simplicity, we use the same notation for H as an operator acting either in $\mathscr{E}_{n_\alpha,n_\beta}$ or in $\mathcal{F}_n$.

It then follows that the ket:

$$|\psi'\rangle = P_{\beta\to\alpha}|\psi\rangle \ \in \mathscr{E}_{n_\alpha+1,n_\beta-1} \tag{IX-11}$$

is an eigenvector of H with the same eigenvalue E:

$$H\,|\psi'\rangle = E\,|\psi'\rangle \tag{IX-12}$$

($|\psi'\rangle$ is orthogonal to $|\psi\rangle$). This property comes from the fact that the α and β particles play an equivalent role in H. Take, for example, the case where H is given by expression (IX-4). If we are considering spinless particles, it is clear that the equation for the stationary states:

$$\left\{-\frac{\hbar^2}{2m}\sum_{i=1}^{n}\Delta_i + \sum_{i<j} V(\boldsymbol{r}_i,\boldsymbol{r}_j)\right\}\psi\,(\boldsymbol{r}_1\ldots\boldsymbol{r}_n) = E\,\psi\,(\boldsymbol{r}_1\ldots\boldsymbol{r}_n) \tag{IX-13}$$

will yield the same solution whether the coordinates of the $(n+1)^{\text{th}}$ particle are interpreted as belonging to an α or β particle. If the particles have a spin s, equation (IX-13) is replaced by $(2s+1)^n$ equations associated with the various spinor components, but the conclusion remains the same. In a general way, operators $P_{\beta\to\alpha}$ and $P_{\alpha\to\beta}$ induce a one-to-one correspondence between the isomorphic spaces $\mathscr{E}_{n_\alpha,n_\beta}$ and $\mathscr{E}_{n_\alpha+1,n_\beta-1}$; the matrix elements of H between corresponding states are all equal.

Relations (IX-10) to (IX-12) imply that the two P operators have non-zero matrix elements only between eigenkets of H having the same eigenvalue. It follows:

$$[P_{\alpha\to\beta}, H] = [P_{\beta\to\alpha}, H] = 0 \tag{IX-14}$$

We define operators N_α and N_β, particle number of each species, as (Hermitian) operators acting in $\mathcal{F}_n$, and whose eigenkets are all the kets of each subspace $\mathscr{E}_{n_\alpha,n_\beta}$ with eigenvalues n_α and n_β. It is easy to see that:

$$\begin{cases} [N_\alpha, H] = [N_\beta, H] = 0 & \text{(IX-15a)} \\ [N_\alpha, N_\beta] = 0 & \text{(IX-15b)} \end{cases}$$

Exercises:

- Establish the following commutation relations:

$$[P_{\beta\to\alpha}, N_\alpha] = -P_{\beta\to\alpha} \quad \text{(IX-16a)}$$
$$[P_{\beta\to\alpha}, N_\beta] = P_{\beta\to\alpha} \quad \text{(IX-16b)}$$
$$[P_{\alpha\to\beta}, N_\alpha] = P_{\alpha\to\beta} \quad \text{(IX-16c)}$$
$$[P_{\alpha\to\beta}, N_\beta] = -P_{\alpha\to\beta} \quad \text{(IX-16d)}$$

- Determine the action in $\mathcal{F}_n$ of the commutator between $P_{\beta\to\alpha}$ and $P_{\alpha\to\beta}$ (remember to take into account the kets of $\mathscr{E}_{n_\alpha=0,n_\beta=n}$ and $\mathscr{E}_{n_\alpha=n,n_\beta=0}$; the two operators are neither unitary nor the inverse of each other). It will be convenient to introduce operators:

$$K_{\alpha\beta} = \sqrt{N_\alpha}\, P_{\beta\to\alpha}\, \sqrt{N_\beta}$$
$$K_{\beta\alpha} = \sqrt{N_\beta}\, P_{\alpha\to\beta}\, \sqrt{N_\alpha} \quad \text{(IX-17)}$$

Show that their commutator is given by:

$$[K_{\alpha\beta}, K_{\beta\alpha}] = N_\alpha - N_\beta \quad \text{(IX-18)}$$

and that:

$$[K_{\beta\alpha}]^\dagger = K_{\alpha\beta} \quad \text{(IX-19)}$$

Establish the equalities:

$$[K_{\beta\alpha}, N_\alpha] = -[K_{\beta\alpha}, N_\beta] = K_{\beta\alpha} \quad \text{(IX-20a)}$$
$$[K_{\alpha\beta}, N_\alpha] = -[K_{\alpha\beta}, N_\beta] = -K_{\alpha\beta} \quad \text{(IX-20b)}$$

Deduce from this that the four operators N_α, N_β, $K_{\alpha\beta}$ and $K_{\beta\alpha}$ form a Lie algebra (having operator $N = N_\alpha + N_\beta$ as its center).

Figure 1 summarizes this preliminary study concerning distinguishable particles. The state space $\mathcal{F}_n$ is the direct sum of the subspaces $\mathscr{E}_{n_\alpha,n_\beta}$, $\mathscr{E}_{n_\alpha+1,n_\beta-1}$, etc. represented by vertical oval shapes. These subspaces are isomorphic with each other, and the mathematical calculation of the matrix elements of H is identical in each of them. Consequently, the spectrum of H (and the associated degeneracy degrees) is the same in each of the subspaces $\mathscr{E}_{n_\alpha,n_\beta}$. As schematized in figure 1, operators $P_{\beta\to\alpha}$ and $P_{\alpha\to\beta}$ (or $K_{\beta\alpha}$ and $K_{\alpha\beta}$) establish the correspondence between the eigenstates; they form with N_α and N_β a Lie algebra of operators commuting with the Hamiltonian.

A-2-b. Identical particles

We now assume that the α particles, on the one hand, and the β particles, on the other, are identical particles obeying the same statistics: the two sets are

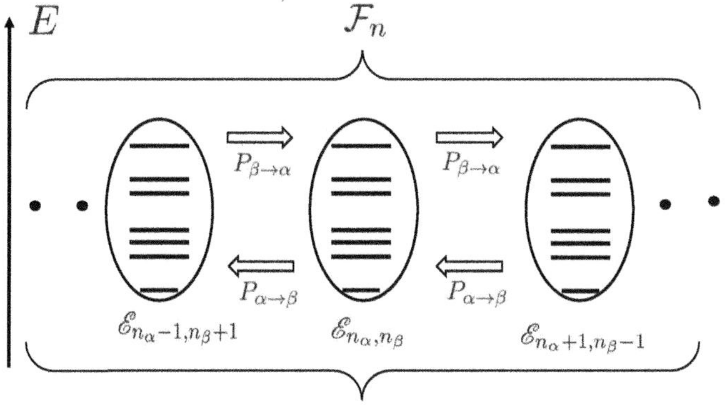

Figure 1: Each vertical oval shape in the figure symbolizes a state space $\mathscr{E}_{n_\alpha,n_\beta}$ for a system of n_α distinguishable particles of type α and n_β distinguishable particles of type β. Their direct sum is a state space $\mathcal{F}_n$ where the total number of particles n remains constant, while their distribution between the two species α and β varies (each population varies between its minimum 0 and its maximum n). The energies E are shown on a vertical axis. Each of these spaces includes the eigenkets of the Hamiltonian H corresponding to the same energy spectrum. Operators $P_{\beta\to\alpha}$ and $P_{\alpha\to\beta}$ allow going from one stationary state to the one with the same energy in the neighboring space state $\mathscr{E}_{n_\alpha,n_\beta}$.

either bosons or fermions. In both cases, we must modify the previous reasoning, since the state space $\mathcal{F}_n$ must be restricted to its part either totally symmetric $\mathcal{F}_n^S$, or totally antisymmetric $\mathcal{F}_n^A$, with respect to each set of α and β particles. For given values of n_α and n_β, the state space of the system combining both sets of particles is the tensor product:

$$\mathscr{E}^{S,A}_{n_\alpha,n_\beta} = \mathscr{E}^{S,A}_{n_\alpha} \otimes \mathscr{E}^{S,A}_{n_\beta} \tag{IX-21}$$

where $\mathscr{E}^{S,A}_{n_\alpha}$ is the space of the (anti) symmetric states of n_α particles α and $\mathscr{E}^{S,A}_{n_\beta}$ the space of the (anti) symmetric states of n_β particles β. The analog of the space $\mathcal{F}_n$ defined in (IX-8), where n_α and n_β vary keeping their sum constant, is now the direct sum:

$$\mathcal{F}_n^{S,A} = \mathscr{E}^{S,A}_{n_\alpha=0,n_\beta=n} \oplus \mathscr{E}^{S,A}_{n_\alpha=1,n_\beta=n-1} \oplus \cdots \mathscr{E}^{S,A}_{n_\alpha=n,n_\beta=0} \tag{IX-22}$$

We can easily see that the diagram of figure 1 will be modified. Each of the subspaces is globally invariant under the action of H (which does not change the number of particles of each species). Any eigenket of H in $\mathscr{E}^{S,A}_{n_\alpha,n_\beta}$ is automatically

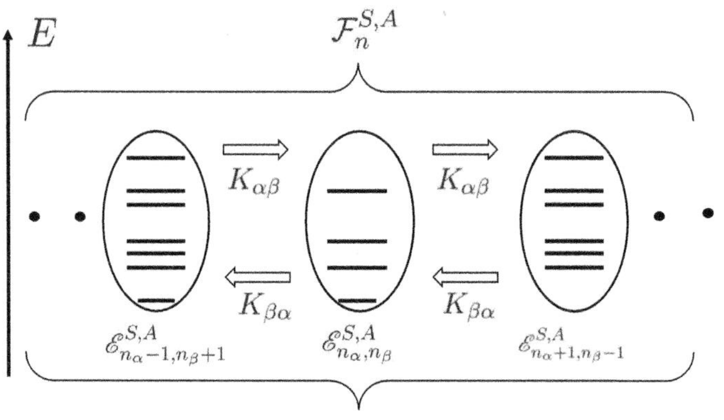

Figure 2: Diagram similar to figure 1, but for the case where the particles α, on the one hand, and β, on the other, are identical. Taking into account the constraints imposed by the symmetrization or antisymmetrization of the state vector (bosons or fermions), only certain levels will be allowed in each of the subspaces. This elimination could be different from one subspace to another: sometimes a given energy is allowed in two neighboring subspaces, and sometimes not. It is now the operators $K_{\alpha\beta}$ and $K_{\beta\alpha}$ defined in (IX-34) that transfer the state vector from one subspace to the next, while preserving the exchange symmetry necessary for identical particles; acting on an eigenstate of the Hamiltonian, they yield a zero ket if the energy level does not exist in the neighboring subspace.

an eigenket of H in the larger space of the distinguishable particles $\mathscr{E}_{n_\alpha,n_\beta}$, but the reverse is not necessarily true. One must eliminate from figure 1 a certain number of energy levels that do not have the exchange symmetry required for identical particles. This leads to a diagram of the type shown in figure 2.

α. *Basis kets, creation, and annihilation operators*

We use the classical notations of chapter XV of reference [9], which we summarize below.

Review and notations: In the state space of n_α distinguishable particles[8] of the α kind, one defines two projectors, S_{n_α} and A_{n_α}, respectively, the symmetrizer and antisymmetrizer:

$$S_{n_\alpha} = \frac{1}{n_\alpha!}\sum_\lambda P_\lambda \qquad \text{and} \qquad A_{n_\alpha} = \frac{1}{n_\alpha!}\sum_\lambda \varepsilon_\lambda P_\lambda \tag{IX-23}$$

[8]The n_α particles α are taken as an example, but the relations below can be easily transposed to n_β particles β with obvious substitutions.

where the P_λ are the $n_\alpha!$ possible permutations of the n_α particles α, and ε_λ is the parity of each of them. The application of the symmetrizer S_{n_α} on the space of n_α distinguishable particles yields the (smaller) boson space, the application of the antisymmetrizer A_{n_α} the (even smaller) fermion space.

For bosons, we introduce the occupation numbers n_k as the number of bosons occupying each individual state $|u_k\rangle$. The basis kets of the space of the completely symmetrical states are then:

$$|n_i,\ n_j, \ldots, n_l, \ldots\rangle$$
$$= \sqrt{\frac{n_\alpha!}{n_i!n_j!\ldots n_l!\ldots}}\ S_{n_\alpha}\,|1 : u_i;\ 2 : u_i; \ldots;\ n_i : u_i;\ \ n_i + 1 : u_j; \ldots;\ n_i + n_j : u_j; \ldots\rangle \tag{IX-24}$$

For fermions, we define in a similar way basis kets of the completely antisymmetrical states by:

$$|u_i, u_j, \ldots, u_l, \ldots\rangle = \sqrt{n_\alpha!}\ A_{n_\alpha}\,|1 : u_i;\ 2 : u_j; \ldots;\ q : u_l; \ldots\rangle \tag{IX-25}$$

where all the individual states must be different, otherwise the ket is zero. In both cases, bosons and fermions, theses states are called "*Fock states*", or sometimes "*occupation number states*".

Taking the direct sum of all the state spaces obtained for all the values of n_α, we get the Fock space, where the particle number is no longer fixed. This enables us to define the action of creation and annihilation operators.

For bosons, we introduce the linear creation operator $a^\dagger_{u_k}$ as:

$$a^\dagger_{u_k}\ |n_1, n_2, \ldots, n_k, \ldots\rangle = \sqrt{n_k + 1}\ |n_1, n_2, \ldots, n_k + 1, \ldots\rangle \tag{IX-26a}$$

This operator adds a particle to the set of α particles. The adjoint operator a_{u_k} removes one, and its action is given by:

$$a_{u_k}\ |n_1, n_2, \ldots, n_k, \ldots\rangle = \sqrt{n_k}\ |n_1, n_2, \ldots, n_k - 1, \ldots\rangle \tag{IX-26b}$$

For fermions, the creation operator $a^\dagger_{u_k}$ is defined by:

$$a^\dagger_{u_k}\ |u_j, \ldots, u_l, \ldots, u_n, \ldots\rangle = |u_k, u_j, \ldots, u_l, \ldots, u_n, \ldots\rangle \tag{IX-27a}$$

It yields a zero ket if it acts on a ket where the individual state u_k is already occupied (Pauli exclusion principle). The corresponding annihilation operator is its adjoint, acting as:

$$a_{u_k}\ |u_k, u_j, u_l, \ldots, u_n, \ldots\rangle = |u_j, u_l, \ldots, u_n, \ldots\rangle \tag{IX-27b}$$

Note that the individual state that is annihilated under the action of a_{u_k} must occupy the first position; if this is not the case, the state should be moved to the first position, with the eventual change of sign imposed by the necessary permutation of states. If the individual state u_k is not occupied in the initial ket, the effect of a_{u_k} is simply to annul the ket it is acting on.

β. *Commutation relations*

We know the commutation relations of the creation and annihilation operators. The commutator of all the operators a (or $a^\dagger$) with all the operators b (or $b^\dagger$) is zero. We simplify the notation $a^\dagger_{u_k}$ into $a^\dagger_k$, and a_{u_k} into a_k. For boson particles, we have:

$$\begin{aligned} \left[a_k, a_{k'}\right] &= \left[a^\dagger_k, a^\dagger_{k'}\right] = 0 \\ \left[a_k, a^\dagger_{k'}\right] &= \delta_{kk'} \end{aligned} \tag{IX-28}$$

(with similar relations for the operators b). For fermions, these relations must be replaced by:

$$\begin{aligned} \left[a_k, a_{k'}\right]_+ &= \left[a^\dagger_k, a^\dagger_{k'}\right]_+ = 0 \\ \left[a_k, a^\dagger_{k'}\right]_+ &= \delta_{kk'} \end{aligned} \tag{IX-29}$$

where $[A, B]_+$ represents the anticommutator $AB + BA$.

In both cases, setting $\eta = +1$ for bosons and $\eta = -1$ for fermions, we have:

$$\begin{aligned} a_k\, a_{k'} &= \eta\, a_{k'}\, a_k \qquad ; \qquad a^\dagger_k\, a^\dagger_{k'} = \eta\, a^\dagger_{k'}\, a^\dagger_k \\ a_k\, a^\dagger_{k'} &= \eta\, a^\dagger_{k'}\, a_k + \delta_{kk'} \qquad ; \qquad a^\dagger_{k'}\, a_k = \eta\left(a_k\, a^\dagger_{k'} - \delta_{kk'}\right) \end{aligned} \tag{IX-30}$$

with similar equalities for b and $b^\dagger$, all the commutators between an operator a (or $a^\dagger$) and an operator b (or $b^\dagger$) being equal to zero.

γ. *Operators changing the nature of the particles*

The products:

$$a^\dagger_k\, a_k \quad , \quad b^\dagger_k\, b_k \qquad \text{and} \qquad a^\dagger_k\, b_k \quad , \quad b^\dagger_k\, a_k \tag{IX-31}$$

do not change the total particle number $n = n_\alpha + n_\beta$ and $\mathcal{F}^{S,A}_n$ is globally invariant under their action. Among these four operators, the first two do not change the nature of the particles and are used to define the operators N_α and N_β, number of α and β particles: :

$$N_\alpha = \sum_k a^\dagger_k\, a_k \qquad\qquad N_\beta = \sum_k b^\dagger_k\, b_k \tag{IX-32}$$

It is easy to see that:

$$[H, N_\alpha] = [H, N_\beta] = 0 \tag{IX-33}$$

The last two operators (IX-31) change the nature of one of the system's particles. We then set:

$$K_{\alpha\beta} = \sum_k a^\dagger_k\, b_k \qquad\qquad K_{\beta\alpha} = \sum_k b^\dagger_k\, a_k \tag{IX-34}$$

and let us show that these operators commute with H. We shall use, as in § A-1, the symmetry properties of the Hamiltonian H, which, when acting on numbered particles, is invariant when the nature of these particles changes. This why we must return to kets containing particles numbered from 1 to n, using relation (IX-24).

• For bosons, the matrix elements of $K_{\alpha\beta}$ between eigenvectors of H with different energies are zero. As a result, operators H and $K_{\alpha\beta}$ commute.

Demonstration: We start from the relation defining $K_{\alpha\beta}$ on the basis of the symmetrized kets (IX-24):

$$K_{\alpha\beta}\,\underbrace{|n_1 \ldots n_k \ldots\rangle}_{\text{particles }\alpha}\;\underbrace{|n'_1 \ldots n'_k \ldots\rangle}_{\text{particles }\beta}$$

$$= \sum_k \sqrt{(n_k+1)n'_k}\;\underbrace{|n_1 \ldots n_k + 1 \ldots\rangle}_{\text{particles }\alpha}\;\underbrace{|\,n'_1 \ldots n'_k - 1 \ldots\rangle}_{\text{particles }\beta} \tag{IX-35}$$

and use several times relation (IX-24) to return to numbered particles. The following expression appears on the left-hand side:

$$\sqrt{\frac{n_\alpha!}{\prod_i n_i!}\frac{n_\beta!}{\prod_j n'_j!}}\;S_{n_\alpha}S_{n_\beta} \tag{IX-36}$$

and on the right-hand side:

$$\sum_k \sqrt{(n_k+1)n'_k}\sqrt{\frac{n_\alpha!}{(n_k+1)\prod_i n_i!}\frac{n'_k\,n_\beta!}{\prod_j n'_j!}}\;S_{n_\alpha+1}S_{n_\beta-1} \tag{IX-37}$$

Several factors cancel out between the two sides of the equation, and we get:

$$\begin{aligned} &K_{\alpha\beta}\;S_{n_\alpha}S_{n_\beta}\;|1:u_i;\ldots;n_\alpha:u_m\rangle|n_\alpha+1:u'_i;\ldots;n:u'_m\rangle \\ &= \sum_k n'_k\;S_{n_\alpha+1}S_{n_\beta-1}|1:u_i;\ldots;n_\alpha:u_m;\;n_\alpha+1:u_k\rangle \\ &\qquad\qquad\qquad\qquad |n_\alpha+2:u'_i;\ldots;n:u'_m\rangle \end{aligned} \tag{IX-38}$$

where, as before, the first ket on each side concerns the α particles, the second the β particles. The sum $\sum_k n'_k$ is simply equivalent to a sum over all the numbered particles included in the initial ket for the β particles.

We now insert on the left-hand side of (IX-38) an eigenvector $|E\rangle$ of the Hamiltonian H, which is a superposition:

$$|E\rangle = \sum x_{ij\ldots q}\;|1:u_i;2:u_j;\ldots;n_\alpha:u_m\rangle|n_\alpha+1:u_l;\ldots;n:u_q\rangle \tag{IX-39}$$

This expression is written in the state space of distinguishable particles, but we assume that it has the appropriate symmetry to describe n_α bosons α and n_β bosons β; in other words, it is invariant under the action of the product $S_{N_\alpha}S_{N_\beta}$. On the right-hand side of (IX-38) we now get, in each term of the sum over the numbered particles, a superposition of states with numbered particles where an additional particle is of the α type, but with the same coefficients $x_{ij...pq}$. When H acts on this right-hand side, and since it commutes with the symmetrizers $S_{N_\alpha+1}S_{N_\beta-1}$, it can act directly on the kets with numbered particles. As we saw in § A-2-a, H is not sensitive to the nature of the particles, and it acts on a superposition of states that is still an eigenvector with the same energy E.

We now simply need to have the product of the two symmetrizers $S_{N_\alpha+1}S_{N_\beta-1}$ act on the ket with numbered particles, multiplied by E. Two cases are possible:

- either the ket is zero;
- or the ket is non-zero, and we have established that the action of operator $K_{\alpha\beta}$ is to yield another eigenket of H with the same energy.

This establishes the announced result.

• For fermions, the reasoning is the same with the restriction that no occupation number n_i or n'_i can be larger than one (otherwise the corresponding ket is zero). In particular, in the final antisymmetrization by the operators $A_{n_\alpha+1}A_{n_\beta-1}$, to avoid obtaining a zero it is imperative that the state which loses a β particle is not already occupied by an α particle. As an example, if we consider two systems of free fermions occupying two Fermi spheres with the same radius, this condition can never be fulfilled, and the action of operator $K_{\alpha\beta}$ yields zero.

δ. *Lie algebra of operators commuting with H*

Let us compute the commutation relations between the four operators N_α, N_β, $K_{\alpha\beta}$, and $K_{\beta\alpha}$ that all commute with H. Relation (IX-15b) shows that N_α and N_β commute with each other. We then determine their commutators with $K_{\alpha\beta}$ and $K_{\beta\alpha}$. Since b_k commutes with all the operators acting in the state space of the α particles, we have:

$$\begin{aligned}\left[a_k^\dagger b_k\,,\,a_{k'}^\dagger a_{k'}\right] &= \left(a_k^\dagger a_{k'}^\dagger a_{k'} - a_{k'}^\dagger a_{k'} a_k^\dagger\right) b_k \\ &= \left(\eta\, a_{k'}^\dagger a_k^\dagger a_{k'} - \eta\, a_{k'}^\dagger a_k^\dagger a_{k'} - \delta_{kk'}\, a_{k'}^\dagger\right) b_k = -\delta_{kk'}\, a_k^\dagger\, b_k \end{aligned} \tag{IX-40a}$$

and, similarly:

$$\begin{aligned}\left[a_k^\dagger\, b_k\,,\,b_{k'}^\dagger\, b_{k'}\right] &= a_k^\dagger \left(b_k\, b_{k'}^\dagger\, b_{k'} - b_{k'}^\dagger\, b_{k'}\, b_k\right) \\ &= a_k^\dagger\left(\eta\, b_{k'}^\dagger\, b_k\, b_{k'} + \delta_{kk'}\, b_{k'} - \eta b_{k'}^\dagger\, b_k\, b_{k'}\right) = \delta_{kk'}\, a_k^\dagger\, b_k \end{aligned} \tag{IX-40b}$$

Summing (IX-40a) and (IX-40b) over k and k', we get:

$$[K_{\alpha\beta}, N_\alpha] = -K_{\alpha\beta} \tag{IX-41a}$$

$$[K_{\alpha\beta}, N_\beta] = K_{\alpha\beta} \tag{IX-41b}$$

A similar reasoning yields[9]:

$$[K_{\beta\alpha}, N_\alpha] = K_{\beta\alpha} \tag{IX-42a}$$
$$[K_{\beta\alpha}, N_\beta] = -K_{\beta\alpha} \tag{IX-42b}$$

We have computed all the commutators involving the operators N. We still need to evaluate the commutator of $K_{\alpha\beta}$ and $K_{\beta\alpha}$. Now:

$$\begin{aligned}\left[a_k^\dagger\, b_k\ ,\ b_{k'}^\dagger\, a_{k'}\right] &= \left(a_k^\dagger\, b_k\, b_{k'}^\dagger\, a_{k'} - b_{k'}^\dagger\, a_{k'}\, a_k^\dagger\, b_k\right)\\ &= \left(\eta\, a_k^\dagger\, b_{k'}^\dagger\, b_k\, a_{k'} + \delta_{kk'}\, a_k^\dagger\, a_{k'} - \eta\, b_{k'}^\dagger\, a_k^\dagger\, a_{k'}\, b_k - \delta_{kk'}\, b_{k'}^\dagger\, b_k\right)\\ &= \delta_{kk'}\left(a_k^\dagger\, a_k - b_k^\dagger\, b_k\right)\end{aligned} \tag{IX-43}$$

Summing over k and k', we get:

$$[K_{\alpha\beta}, K_{\beta\alpha}] = N_\alpha - N_\beta \tag{IX-44}$$

Relations (IX-41) to (IX-44) give the commutation relations of the four operators N_α, N_β, $K_{\beta\alpha}$, and $K_{\alpha\beta}$; they show that they form a 4-dimensional Lie algebra, $\mathscr{L}_4$. It is useful to replace N_α and N_β by the two linear combinations:

$$N = N_\alpha + N_\beta \tag{IX-45a}$$
$$\Delta = \frac{1}{2}\,(N_\alpha - N_\beta) \tag{IX-45b}$$

as it is easy to see that N commutes with all the operators of $\mathscr{L}_4$ (N is the "center" of this Lie algebra). With this new basis of operators, the commutation relations are written:

$$[K_{\alpha\beta}, \Delta] = -K_{\alpha\beta} \tag{IX-46a}$$
$$[K_{\beta\alpha}, \Delta] = K_{\beta\alpha} \tag{IX-46b}$$
$$[K_{\alpha\beta}, K_{\beta\alpha}] = 2\Delta = N_\alpha - N_\beta \tag{IX-46c}$$

These relations are the starting point of all the reasoning proposed in §§ B and C.

A-3. The nature of a particle can be considered as an internal quantum number

At this stage of the reasoning, a question naturally arises: if the physical system under study is invariant under an exchange of particles, whatever their nature, shouldn't we consider them all as identical particles? Their internal

[9]One can also demonstrate relations (IX-41) without computing any commutators, noting from the expressions of $K_{\alpha\beta}$ and $K_{\beta\alpha}$ that they transform any eigenvector of N_α with eigenvalue n_α into another eigenvector with eigenvalue $n_\alpha \pm 1$.

state would then be characterized by two different values of a certain quantum number, corresponding to their nature, and the Hamiltonian would not act on these internal variables (since it does not change the nature of the particles). Conversely, imagine that the Hamiltonian of a given system of identical particles does not act on a certain type of internal variables (like the spin). Can we consider that the system is actually formed of particles of different types, their number being simply the number of distinct values accessible to the quantum number specifying the state of the internal variables? Take, for example, an atom in the non-relativistic approximation where the Hamiltonian does not act on the electron spins. The question becomes whether one can obtain its energy levels as well as their degeneracy by considering that the system is composed of "two sorts of electrons", those whose spin component on a given axis equals $+\hbar/2$ and those whose same spin component equals $-\hbar/2$.

The answer to the previous questions is yes, though in both cases the postulates of quantum mechanics lead to a system's description by very different mathematical kets. In one case, one must symmetrize (or antisymmetrize) the state vector separately with respect to as many groups of particles as there are distinct types. In the second case, the symmetrization must be performed with respect to the ensemble of all the identical particles, regardless of their nature.

A-3-a. Correspondence between Fock states

Consider two physical systems:

- A system $\mathscr{S}$ consisting of two types of particles, α and β, with variable numbers n_α and n_β. The α particles (as well as the β particles) are identical to each other. The Hamiltonian H of the system is supposed to be completely symmetric with respect to all the particles, regardless of their α or β nature; the masses of all the particles (which play a role in the kinetic energy) are the same, the interactions do not depend on the particles' nature.

- A system $\mathscr{S}'$ consisting of an arbitrary number $n = n_\alpha + n_\beta$ of identical particles that can be in an α or β internal isospin state (juxtaposed with a spin state if the particles under study already have a spin). We assume that the Hamiltonian H' of this system does not act on the internal state of the particles.

There is a one-to-one correspondence between the states of $\mathscr{S}$ and those of $\mathscr{S}'$. A Fock state of $\mathscr{S}$ is defined by the knowledge of n_α individual states $|u_k\rangle$ occupied by the α particles, as well as n_β individual states $|u_{k'}\rangle$ occupied by the β particles (for bosons these occupation numbers are any positive integer, for fermions, they can only be equal to one). The one-to-one correspondence between Fock states is due to the fact that, for the system $\mathscr{S}'$, the first n_α states can be attributed to the internal state $|\alpha\rangle$, the other n_β to the internal state $|\beta\rangle$. It can be written as:

$$\begin{array}{l} \mathscr{S}: |\overbrace{\alpha : u_{k_1}, \ldots, u_{k_{n_\alpha}}}^{n_\alpha} \; ; \; \overbrace{\beta : u_{k_{n_\alpha+1}}, \ldots, u_{k_n}}^{n_\beta}\rangle \\ \Updownarrow \qquad\qquad\qquad\qquad \Updownarrow \\ \mathscr{S}': |u_{k_1}, \alpha ; \ldots ; u_{k_{n_\alpha}}, \alpha \; ; \; u_{k_{n_\alpha+1}}, \beta ; \ldots ; u_{k_n}, \beta\rangle \end{array} \tag{IX-47}$$

It can be shown that the matrix elements of H between Fock states of $\mathscr{S}$ are equal to those of H' between the corresponding states of $\mathscr{S}'$. The demonstration is given in complement A_{IX}. It is based on the fact that H, as well as H', commutes with any permutation of the n particles, and it uses a certain number of properties of symmetrizers or antisymmetrizers. It also uses the fact that, when the Hamiltonian does not act in the isospin space, all the permutations of the N particles that change the isospin state of one of them yield matrix elements of H' equal to zero. In a general way, this establishes that in quantum mechanics the nature of a particle can indeed be considered as an internal quantum number without modifying any physical prediction.

If one takes into account interactions in $\mathscr{S}'$ that are not isospin diagonal, the equivalence between the matrix elements disappears, and the properties of the two physical systems can be significantly different.

A-3-b. Creation and annihilation operators

The correspondence between the creation and annihilation operators acting on the Fock states of the two systems $\mathscr{S}$ is $\mathscr{S}'$ is as follows:

$$\begin{array}{ll} a_k \Leftrightarrow a_{k,\alpha} & b_k \Leftrightarrow a_{k,\beta} \\ a_k^\dagger \Leftrightarrow a_{k,\alpha}^\dagger & b_k^\dagger \Leftrightarrow a_{k,\beta}^\dagger \end{array} \tag{IX-48}$$

where the individual states of the system $\mathscr{S}'$ are defined by the quantum number k as well as the α or β isospin state. It is easy to see that the action of these operators preserves the correspondence between Fock states, with the same $\sqrt{n_{\alpha,\beta}}$ or $\sqrt{n_{\alpha,\beta}+1}$ factors, depending on the case.

For the system $\mathscr{S}'$, the operators K defined in (IX-31) and in (IX-34) become:

$$\begin{array}{ll} N_\alpha = \sum_k a_{k,\alpha}^\dagger a_{k,\alpha} & N_\beta = \sum_k a_{k,\beta}^\dagger a_{k,\beta} \\ K_{\alpha\beta} = \sum_k a_{k,\alpha}^\dagger a_{k,\beta} & K_{\beta\alpha} = \sum_k a_{k,\beta}^\dagger a_{k,\alpha} \end{array} \tag{IX-49}$$

We can easily go from one space to another, still using the normal properties of the creation and annihilation operators. We can thus reason indifferently in either space. In what follows, and keeping in mind the correspondence (IX-48), it will often be convenient to use the common notation a_k and b_k in both cases.

We pointed out the difference between the symmetrization of systems $\mathscr{S}$ and $\mathscr{S}'$. This difference has consequences on the commutation or anticommutation properties of the creation and annihilation operators. In the system $\mathscr{S}$, the a_k and $a_k^\dagger$ operators always commute with the $b_{k'}$ and $b_{k'}^\dagger$: they are operators acting on different variables. In the system $\mathscr{S}'$, these operators become $a_{k\alpha}$, $a_{k\alpha}^\dagger$, $a_{k\beta}$, and $a_{k\beta}^\dagger$; for fermion particles, they anticommute. This does not affect the calculations of § A-2-b-δ since, in the first line of (IX-40a), for example, we moved operator b_k over the product of two operators $a_{k'}^\dagger a_{k'}$, which introduces a factor $\eta^2 = 1$ and does not change the result. Similarly, in (IX-40b), operator $a_{k'}^\dagger$ is moved over the product of two operators $b_{k'}^\dagger b_{k'}$, which does not change the sign. Finally, to obtain (IX-43), we had to commute both operators $a_k^\dagger b_{k'}^\dagger$ with $b_k a_{k'}$, which did not change the sign.

B. SU(2) group and isospin symmetry

The previous results are now applied to the isospin symmetry, which involves the Lie algebra of the SU(2) group.

B-1. Lie algebra

The commutation relations (IX-46) are of the same type as those of the components J_z, J_+, and J_- of an angular momentum $\boldsymbol{J}$:

$$[J_+, J_z] = -\hbar J_+ \tag{IX-50a}$$

$$[J_-, J_z] = \hbar J_- \tag{IX-50b}$$

$$[J_+, J_-] = 2\hbar\, J_z \tag{IX-50c}$$

The following correspondence is therefore natural:

$$\begin{cases} K_{\alpha\beta} & \leftrightarrow \quad J_+/\hbar \\ K_{\beta\alpha} & \leftrightarrow \quad J_-/\hbar \\ \ \Delta & \leftrightarrow \quad J_z/\hbar \end{cases} \tag{IX-51}$$

We introduce the operator $\overline{K}$ with components[10]:

$$\begin{cases} K_x = \dfrac{1}{2}\left(K_{\alpha\beta} + K_{\beta\alpha}\right) \\ K_y = \dfrac{1}{2i}\left(K_{\alpha\beta} - K_{\beta\alpha}\right) \\ K_z = \Delta = \dfrac{N_\alpha - N_\beta}{2} \end{cases} \tag{IX-52}$$

These operators are Hermitian since relation (IX-34) shows that:

$$K_{\alpha\beta}^{\dagger} = K_{\beta\alpha} \tag{IX-53}$$

They can thus be used as a basis for a Lie algebra whose exponentials are unitary operators. Relations (IX-46) yield:

$$\begin{cases} [K_x, K_y] = i\,K_z \\ [K_y, K_z] = i\,K_x \\ [K_z, K_x] = i\,K_y \end{cases} \tag{IX-54}$$

These are the usual commutation relations of an angular momentum (to within a factor $\hbar$) that appear in the study of the rotation group $R_{(3)}$.

We know a very simple set of 2×2 matrices obeying these commutation relations: the Pauli matrices $\sigma_x/2$, $\sigma_y/2$, and $\sigma_z/2$. We saw, (*cf.* chapter III, § B-3), that they form the Lie algebra of the SU(2) group, generating all its elements $\mathcal{M}(\boldsymbol{a})$ by exponentiation:

$$\mathcal{M}(\boldsymbol{a}) = \mathrm{e}^{-i\,\boldsymbol{a}\cdot\boldsymbol{\sigma}/2} \in \mathrm{SU}(2) \tag{IX-55a}$$

where $\boldsymbol{a}$ represents a set of 3 real parameters (the modulus of $\boldsymbol{a}$ is included between 0 and 2π). The 3 operators K_x, K_y, K_z generate a Lie algebra $\mathscr{L}_3$, the so-called isospin algebra; it is isomorphic to the SU(2) algebra, referred to as $\mathfrak{su}(2)$ in mathematics. By exponentiation we get the isospin unitary operators:

$$\mathcal{Z}(\boldsymbol{a}) = \mathrm{e}^{-i[a_x K_x + a_y K_y + a_z K_z]} = \mathrm{e}^{-i\boldsymbol{a}\cdot\overline{K}} \tag{IX-55b}$$

that form an isospin group $\mathscr{G}_{\text{isospin}}$, whose elements depend on 3 real parameters (a_x, a_y, and a_z).

There is no need to discuss new lines of reasoning in our study of the isospin symmetry: those proposed for the rotation group $R_{(3)}$, or the SU(2) group, can be applied without changes.

---

[10]The notation $\overline{K}$ is convenient to symbolize the ensemble K_x, K_y, K_z; we do not use the notation $\boldsymbol{K}$ to avoid confusion with a vector operator (§ A of chapter VIII) obeying commutation relations (VIII-8) with the angular momentum operator $\boldsymbol{J}$.

Comment:

The action of an isospin rotation operator $\mathcal{Z}(\boldsymbol{a})$ on a ket where n_α particles are of the type α and n_β of the type β yields in general (if a_x and a_y are different from zero) a linear superposition of kets where n_α and n_β take on different values. The physical interpretation of such kets is not obvious. To avoid dealing with those kets, we can ignore in what follows the operators $\mathcal{Z}(\boldsymbol{a})$ and the isospin group. The only essential point from a physical point of view is the existence of the Lie algebra defined by (IX-46), or the equivalent relation (IX-54).

B-2. Similarity between spin and isospin

Operator $\overline{K}$ can be interpreted as the "total isospin angular momentum" of the particles of $\mathscr{S}'$. Expressions (IX-49) are the classical expressions in the Fock space of a one-particle operator, sum of operators acting on all the individual particles:

$$\overline{K} = \sum_{j=1}^{n} \overline{I}_j \tag{IX-56}$$

where the one-particle isospin operators $\overline{I}$ are defined as:

$$\begin{cases} \overline{I}_z \left|\alpha\right\rangle = \frac{1}{2}\left|\alpha\right\rangle & \overline{I}_z \left|\beta\right\rangle = -\frac{1}{2}\left|\beta\right\rangle \\ \overline{I}_+ \left|\alpha\right\rangle = 0 & \overline{I}_+ \left|\beta\right\rangle = \left|\alpha\right\rangle \\ \overline{I}_- \left|\alpha\right\rangle = \left|\beta\right\rangle & \overline{I}_- \left|\beta\right\rangle = 0 \end{cases} \tag{IX-57}$$

We can interpret $\left|\alpha\right\rangle$ and $\left|\beta\right\rangle$ as the two eigenstates $\left|+\right\rangle$ and $\left|-\right\rangle$ of the S_z component of a spin 1/2:

$$\begin{cases} \left|\alpha\right\rangle \leftrightarrow \left|+\right\rangle \\ \left|\beta\right\rangle \leftrightarrow \left|-\right\rangle \end{cases} \tag{IX-58}$$

B-3. Isospin multiplets

We build an "isospin standard basis" in the spaces $\mathcal{F}_n^{S,A}$, where the particle numbers n_α and n_β vary (but keeping their sum constant). The procedure is exactly the same as the one followed previously for building (in chapter VII) a standard basis for the study of the angular momentum. Operator $[\overline{K}]^2$ is defined as:

$$\begin{aligned} [\overline{K}]^2 &= (K_x)^2 + (K_y)^2 + (K_z)^2 \\ &= \frac{1}{4}\left(N_\alpha - N_\beta\right)^2 + \frac{1}{2}\left(K_{\alpha\beta}\, K_{\beta\alpha} + K_{\beta\alpha}\, K_{\alpha\beta}\right) \end{aligned} \tag{IX-59}$$

and its eigenvalues are expressed as:

$$K(K+1) \tag{IX-60}$$

where K is an integer or a half-integer. As for the eigenvalues of K_z, they take on integer or half-integer values M:

$$M_K = K,\ K-1,\ K-2,\ \ldots, -K \tag{IX-61}$$

and fix the value of the population difference:

$$M_K = \frac{n_\alpha - n_\beta}{2} \tag{IX-62}$$

Since H commutes with K_z and $[\overline{K}]^2$, we can build a standard basis separately inside each eigensubspace of H, with eigenvalue E. This yields the kets:

$$|\sigma, E, K, M_K\rangle$$

where E, K, and M_z specify the respective eigenvalues of H, $[\overline{K}]^2$, and K_z; the reunion of these kets for all possible values of these three numbers constitutes a standard basis in the total space. The index σ distinguishes orthogonal kets associated with the same values of E, K, and M_K; it is therefore not necessary when H, K, and K_z form a CSCO (i.e. when no degeneracy occurs from the point of view of the isospin symmetry alone). As for any standard basis, we have:

$$\begin{cases} K_z\,|\sigma, E, K, M_K\rangle = M_K\,|\sigma, E, K, M_K\rangle \\ [\overline{K}]^2\,|\sigma, E, K, M_K\rangle = K(K+1)\,|\sigma, E, K, M_K\rangle \end{cases} \tag{IX-63a}$$

with:

$$M_K = K, K-1, K-2, \ldots, -K \tag{IX-63b}$$

(K and M_K are both either integer or half-integer). Furthermore:

$$K_\pm|\sigma, E, K, M_K\rangle = \sqrt{K(K+1) - M_K(M_K \pm 1)}\,|\sigma, E, K, M_K \pm 1\rangle \tag{IX-64}$$

Finally, for the chosen standard basis, we can write:

$$H\,|\sigma, E, K, M_K\rangle = E\,|\sigma, E, K, M_K\rangle \tag{IX-65}$$

One can regroup the $(2K+1)$ states associated with the same values of σ, E and K; they form a set of degenerate states, called an "isospin multiplet". The corresponding subspace, noted $\mathcal{F}(E, \sigma, K)$, is stable under the action of all the components of $\overline{K}$; these components are represented by $(2K+1) \times (2K+1)$

Hermitian matrices whose commutators obey relations (IX-54). These matrices yield a representation of the Lie algebra $\mathscr{L}_3$ [11], which is irreducible[12] since the space $\mathcal{F}(E, \sigma, K)$ is not the direct sum of two subspaces stable under the actions of all the operators of $\mathscr{L}_3$. A particle, regardless of its nature α or β, has the same energy when it is in the same individual state. For $n = 2$, we assume the existence of a bound state of two α particles, bosons, or fermions. Another bound state must exist for two β particles, with the same energy and the same symmetry. Consequently, there must also exist an intermediate state with the same properties, but a state with $M_K = 0$ including a particle of each kind. These three states form a $K = 1$ isospin triplet. It is also possible that this bound state only exists for particles of a different nature, the symmetrization constraints forbidding its existence when the two particles are identical. This single bound state corresponds to an isospin singlet.

In a similar way, for higher values of n, one can get several isospin multiplets, since changing the nature of the particles modifies the symmetrization constraints.

In figure 2, the Hamiltonian eigenstates in each subspaces $\mathscr{E}_{n_\alpha n_\beta}$ with fixed values of n_α and n_β are schematized by an horizontal line. The eigenstates of H associated with the same isospin multiplet are represented by $(2K+1)$ segments on the same horizontal line (they have the same energy), centered at $M_K = 0$ (i.e. for $n_\alpha = n_\beta$). In each subspace $\mathscr{E}^{A,S}_{n_\alpha, n_\beta}$, the populations are:

$$\begin{array}{lll} n_\alpha = \dfrac{n}{2} + K & n_\beta = \dfrac{n}{2} - K & M_K = K \\[2ex] n_\alpha = \dfrac{n}{2} + K - 1 & n_\beta = \dfrac{n}{2} - K + 1 & M_K = K - 1 \\ \vdots & \vdots & \vdots \\ n_\alpha = \dfrac{n}{2} - K & n_\beta = \dfrac{n}{2} + K & M_K = -K \end{array} \tag{IX-66}$$

Note that K and $n/2$ are necessarily both integers or half-integers.

Shown in figure 2 is a whole series of isospin multiplets with different values of K. In each of them, starting from a stationary state with any given energy associated with a system of n_α particles α and n_β particles β, the action of operators K_+ or K_- yields other non-zero eigenkets of H, unless n_α takes one of the extreme values on the table (IX-66).

The $K = 0$ case corresponds to an isospin singlet on which the action of operators $K_\pm$ always yield zero. An example of isospin singlet is an ensemble of

[11]Remember that the product (internal operation) in the Lie algebra is defined as the commutator of operators – *cf.* chapter III, relation (III-41). The concept of representation for an algebra is thus the same as for a group.

[12]The same concepts also apply to the (isospin) group $\mathscr{G}$: in $\mathcal{F}(E, \sigma, K)$, its elements $\mathcal{Z}(\boldsymbol{a})$ defined in (IX-55b) are represented in an irreducible way by a set of $(2K+1) \times (2K+1)$ unitary matrices.

$n/2$ non-interacting fermions α occupying the same Fermi sphere as an ensemble of $n/2$ fermions β: no modification of the nature of the particles can lead to the same energy. However, as soon as the Fermi spheres have different radii, other values of K can be found.

Comments:

(i) There are in general other standard bases than that of the eigenvectors of H we have introduced. This happens when the index σ takes several distinct values. In any standard basis such as $(|\xi, K, M_K\rangle)$, we can use the Wigner–Eckart theorem (§ C of chapter VIII) to compute the matrix elements of H (which is an "isospin scalar" since it commutes with any component of $\overline{K}$):

$$\langle \xi, K, M_K|H|\xi', K', M'_K\rangle = \delta_{M_K M'_K}\ \delta_{KK'}\ \langle \xi\, K||H||\xi'\, K\rangle \tag{IX-67}$$

The $\delta_{M_K M'_K}$ simply expresses the fact that H does not change the nature of the particles. We have:

$$|\sigma, E, K, M_K\rangle = \sum_{\xi} C_\xi\,(E, K, M_K)\,|\xi, K, M_K\rangle \tag{IX-68}$$

The change of standard basis does not "mix" the different values of K and M_K.

In a general way, the Wigner–Eckart theorem shows that any scalar operator, such as the operator $\overline{K}^2$, has matrix elements only between kets with the same value of M_K; if it commutes with $N = N_\alpha + N_\beta$, it means that such an operator does not change the nature of the particles.

(ii) Going from (IX-45) to (IX-46), we left out the operator N since it commutes not only with H, but also with all the operators of $\mathscr{L}_3$ (it is a central operator). Just as H, the operator N is an isospin scalar, and its matrix elements obey the same selection rules: in any $\mathcal{F}(\sigma, E, K)$, N is represented by a scalar matrix (with only diagonal elements that are all equal).

One can obviously include the operator N in the Lie algebra under study (which becomes $\mathscr{L}_4$); the corresponding group $\mathscr{G}_4$ depends on four parameters and its operators are written:

$$W(\boldsymbol{a}, \theta) = \mathrm{e}^{-i\theta N}\,\mathcal{Z}(\boldsymbol{a}) \tag{IX-69}$$

where $\mathcal{Z}(\boldsymbol{a})$ is defined by (IX-55b). In each $\mathcal{F}(\sigma, K)$, the contribution of e^{-iN} is a phase factor $\mathrm{e}^{-in\theta}$. It is sometimes said that $\mathscr{G}_4$ is the (tensor) product of the isospin group and a "phase group".

B-4. Application examples

Consider an atom and assume that we can ignore all the interactions depending on its electronic spins. The previous discussion shows that its energy levels regroup into multiplets of $(2K+1)$ levels, $2K$ being even or odd depending on whether the electron number is even or odd. In this example, this classification does not bring anything new: K and M_K can be simply identified with the quantum numbers S and M_S associated with the total spin of the whole set of electrons.

A more interesting example is the isospin symmetry in nuclear physics. Protons and neutrons have almost the same mass (the difference is on the order of 0.1%); if we ignore the electromagnetic interactions between nucleons in the nucleus, which is less important than the strong interactions, these interactions become charge-independent, and protons and neutrons play a symmetrical role. In such a case, all the nucleus energy levels are regrouped in isospin multiplets of $(2K+1)$ degenerate energy levels. This gives the number M_K a simple physical signification, directly related to the electric charge Q of the system under study, since:

$$Q = q_p \, n_\alpha = q_p \left(M_K + \frac{n}{2} \right) \tag{IX-70}$$

where $n_\alpha = Z$ is the number of protons, q_p the charge of each one of them, and n the total number of nucleons of the nucleus under study.

Once each multiplet has been identified, one can compute the energy corrections due to the terms in the Hamiltonian that have been neglected. As an example, the repulsion between the charges of the Z protons introduces an energy correction proportional to $Z(Z-1)$, which lifts the degeneracy between the different values of M_K. There are also corrections linked to surface effects in nuclei (liquid drop model). Discussion of these corrections can be found in nuclear physics textbooks [40].

C. SU(3) symmetry

We now generalize the previous discussion to the case where the number of particle species, or of isospin internal states, is equal to three: α, β, and γ. We note a_k, b_k, c_k the annihilation operators of a particle in the individual states $|\varphi_{k,\alpha}\rangle$, $|\varphi_{k,\beta}\rangle$, $|\varphi_{k,\gamma}\rangle$ respectively; the corresponding creation operators are noted $a_k^\dagger$, $b_k^\dagger$, and $c_k^\dagger$. Their commutation relations are of the same type as those written in (IX-29) and (IX-30).

C-1. Symmetry operator

Generalizing (IX-32) and (IX-34), we introduce the three operators:

$$N_\alpha = \sum_k a_k^\dagger a_k \qquad N_\beta = \sum_k b_k^\dagger b_k \qquad N_\gamma = \sum_k c_k^\dagger c_k \tag{IX-71}$$

as well as the six others:

$$\begin{aligned} K_{\alpha\beta} &= \sum_k a_k^\dagger\, b_k & K_{\beta\alpha} &= \sum_k b_k^\dagger\, a_k \\ K_{\beta\gamma} &= \sum_k b_k^\dagger\, c_k & K_{\gamma\beta} &= \sum_k c_k^\dagger\, b_k \\ K_{\gamma\alpha} &= \sum_k c_k^\dagger\, a_k & K_{\alpha\gamma} &= \sum_k a_k^\dagger\, c_k \end{aligned} \tag{IX-72}$$

It is easy to show that:

$$(K_{\alpha\beta})^\dagger = K_{\beta\alpha} \qquad (K_{\beta\gamma})^\dagger = K_{\gamma\beta} \qquad (K_{\gamma\alpha})^\dagger = K_{\alpha\beta} \tag{IX-73}$$

C-1-a. Commutation relations

When all the particles, regardless of their species, play the same role in the operator H, we have as before:

$$[H, N_{\alpha,\beta,\gamma}] = 0 \qquad \text{and} \qquad [H, K] = 0 \tag{IX-74}$$

where K symbolizes any of the six operators defined in (IX-72). Similar calculations to those of § A-2-b-δ yield:

$$[N_\alpha, N_\beta] = 0 \tag{IX-75a}$$

$$[K_{\alpha\beta}, N_\alpha] = -\,[K_{\alpha\beta}, N_\beta] = -K_{\alpha\beta} \tag{IX-75b}$$

$$[K_{\alpha\beta}, K_{\beta\alpha}] = N_\alpha - N_\beta \tag{IX-75c}$$

with, naturally:

$$[K_{\alpha\beta}, N_\gamma] = [K_{\beta\alpha}, N_\gamma] = 0 \tag{IX-75d}$$

To these relations we must obviously add those obtained by circular permutation of α, β, and γ.

The total number of particles corresponds to operator:

$$N = N_\alpha + N_\beta + N_\gamma \tag{IX-76}$$

which, taking into account relation (IX-75b) in particular, commutes with all the others:

$$[N, N_{\alpha,\beta,\gamma}] = 0 \qquad \text{and} \qquad [N, K] = 0 \tag{IX-77}$$

The above relations do not include all the commutators between K operators. Considering operators K including the three isospin values α, β, and γ, we compute the commutators:

$$[K_{\alpha\beta}\,, K_{\alpha\gamma}] \qquad \text{and} \qquad [K_{\beta\alpha}\,, K_{\gamma\beta}]$$

The computation is made from the isospin point of view, when the permutation of two operators a and b, or b and c, etc. introduces a factor η. We get:

$$\begin{aligned}\left[\, a_k^\dagger\, b_k \,,\, a_{k'}^\dagger\, c_{k'} \right] &= \left(a_k^\dagger\, b_k\, a_{k'}^\dagger - a_{k'}^\dagger\, a_k^\dagger\, b_k \right) c_{k'} \\ &= \left(\eta\, a_k^\dagger\, a_{k'}^\dagger - a_{k'}^\dagger\, a_k^\dagger \right) b_k\, c_{k'} = 0 \end{aligned} \tag{IX-78}$$

A summation over k and k' yields:

$$[K_{\alpha\beta}\,, K_{\alpha\gamma}] = 0 \qquad [K_{\alpha\gamma}\,, K_{\alpha\beta}] = 0 \tag{IX-79a}$$

By circular permutation, we show that the commutators between operators K where the first index α, β, or γ is repeated are all equal to zero. Taking the Hermitian conjugate of these relations and using (IX-73), we see that it is also the case for commutators where the second index is repeated.

We still have to compute commutators of the type $[K_{\alpha\beta}, K_{\beta\gamma}]$ or $[K_{\beta\alpha}, K_{\gamma\beta}]$. Since:

$$\begin{aligned}\left[\, a_k^\dagger\, b_k \,,\, b_{k'}^\dagger\, c_{k'} \right] &= \left(a_k^\dagger\, b_k\, b_{k'}^\dagger - b_{k'}^\dagger\, a_k^\dagger\, b_k \right) c_{k'} \\ &= a_k^\dagger \left(b_k\, b_{k'}^\dagger - \eta\, b_{k'}^\dagger\, b_k \right) c_{k'} = \delta_{kk'}\, a_k^\dagger\, c_k \end{aligned}$$

we get, after summing over k and k':

$$[K_{\alpha\beta}\,, K_{\beta\gamma}] = K_{\alpha\gamma} \tag{IX-79b}$$

By Hermitian conjugation, we also obtain:

$$[K_{\beta\alpha}\,, K_{\gamma\beta}] = -K_{\gamma\alpha} \tag{IX-79c}$$

C-1-b. Lie algebra

Continuing the analogy with § B, we introduce Hermitian operators to obtain a basis for a Lie algebra whose exponentials are unitary operators. We set, as in (IX-52):

$$K_z^{(1)} = \frac{1}{2}\left(N_\alpha - N_\beta\right) \; ; \; K_z^{(2)} = \frac{1}{2}\left(N_\beta - N_\gamma\right) \; ; \; K_z^{(3)} = \frac{1}{2}\left(N_\gamma - N_\alpha\right) \tag{IX-80a}$$

$$K_x^{(1)} = \frac{1}{2}\left(K_{\alpha\beta} + K_{\beta\alpha}\right) \; ; \; K_x^{(2)} = \frac{1}{2}\left(K_{\beta\gamma} + K_{\gamma\beta}\right) \; ; \; K_x^{(3)} = \frac{1}{2}\left(K_{\gamma\alpha} + K_{\alpha\gamma}\right) \tag{IX-80b}$$

$$K_y^{(1)} = \frac{1}{2i}\left(K_{\alpha\beta} - K_{\beta\alpha}\right) \; ; \; K_y^{(2)} = \frac{1}{2i}\left(K_{\beta\gamma} - K_{\gamma\beta}\right) \; ; \; K_y^{(3)} = \frac{1}{2i}\left(K_{\gamma\alpha} - K_{\alpha\gamma}\right)$$

(IX-80c)

Operator $\overline{K}^{(1)}$ is simply the isospin operator $\overline{K}$ already introduced for α and β particles; operators $\overline{K}^{(2)}$ and $\overline{K}^{(3)}$ play similar roles for the other particle isospin states[13].

Whereas the six components $K_{x,y}^{(1,2,3)}$ of the $\overline{K}$ are independent linear combinations, this is not the case for the three components K_z since:

$$K_z^{(1)} + K_z^{(2)} + K_z^{(3)} = 0 \tag{IX-81}$$

Three independent operators can be obtained by taking the operator N and any two operators among the three operators $K_z^{(1,2,3)}$.

As usual, we can set:

$$K_\pm^{(i)} = K_x^{(i)} \pm i\, K_y^{(i)} \tag{IX-82}$$

which amounts to returning to operators of the type $K_{\alpha\beta}$:

$$K_+^{(1)} = K_{\alpha\beta} \quad ; \quad K_-^{(1)} = K_{\beta\alpha} \quad ; \quad K_+^{(2)} = K_{\beta\gamma} \quad ; \quad \text{etc.} \tag{IX-83}$$

This yields the classical commutation relations:

$$\left[K_z^{(i)}, K_\pm^{(i)}\right] = \pm K_\pm^{(i)} \tag{IX-84a}$$

$$\left[K_+^{(i)}, K_-^{(i)}\right] = 2\, K_z^{(i)} \tag{IX-84b}$$

$$\left[K_z^{(i)}, K_z^{(j)}\right] = 0 \tag{IX-84c}$$

In general, all the commutation relations obtained in § C-1-a can be written as a function of these new operators. For example, relation (IX-75b) leads to:

$$\left[K_+^{(1)}, K_z^{(2)}\right] = \frac{1}{2}\left[K_{\alpha\beta}\ ,\ N_\beta - N_\gamma\right] = \frac{1}{2} K_{\alpha\beta} = \frac{1}{2} K_+^{(1)} \tag{IX-84d}$$

and, more generally:

$$\left[K_\pm^{(i)}, K_z^{(j)}\right] = \pm\frac{1}{2}\, K_\pm^{(i)} \qquad \text{if } \ i \neq j \tag{IX-84e}$$

These commutation relations are the starting point of the following discussions.

C-1-c. Hyperspin symmetry

All the operators K we have introduced commute with H when the different particles play a symmetrical role in H, regardless of their nature. We have defined nine operators K, but relation (IX-81) shows that only eight of them are independent. Through linear combinations, the operators (IX-80) generate a Lie algebra[14] $\mathscr{L}_8$ with dimension 8, associated with the (hyperspin) group $\mathscr{G}$ of

[13] Index (1) is associated with the $\alpha\beta$ couple, index (2) with the $\beta\alpha$ couple, and index (3) with the $\gamma\alpha$ couple.

[14] $\mathscr{L}_8$ is the space generated by the real linear combinations of the x, y, z components of the operators $K^{(i)}$. This algebra is often noted $\mathfrak{su}(3)$ in mathematics.

symmetry operators:

$$W\left(\boldsymbol{a}, \boldsymbol{b}, \boldsymbol{c}\right) = \mathrm{e}^{-i\left[\boldsymbol{a}\cdot\overline{K}^{(1)} + \boldsymbol{b}\cdot\overline{K}^{(2)} + \boldsymbol{c}\cdot\overline{K}^{(3)}\right]} \tag{IX-85a}$$

where the notation $\boldsymbol{a}\cdot\overline{K}^{(i)}$ symbolizes the expression:

$$\boldsymbol{a}\cdot\overline{K}^{(i)} = a_x\, K_x^{(i)} + a_y\, K_y^{(i)} + a_z\, K_z^{(i)} \tag{IX-85b}$$

Operators W depend on eight paramters, since the sum $a_z + b_z + c_z$ is no longer present in (IX-85a). Remarks concerning the physical meaning of operators W are the same as those we made concerning the isospin operators $Z(\boldsymbol{a})$.

All the eigensubspaces of H are stable under the action of all the elements of either $\mathscr{L}_8$ or $\mathscr{G}$(hyperspin). We can find in these eigensubspaces irreducible representations of the $\mathscr{L}_8$ algebra, perfectly similar to the irreducible representations of the isospin algebra $\mathscr{L}_3$ introduced in § B. Knowledge and classification of the irreducible representations of $\mathscr{L}_8$ and $\mathscr{G}$ (hyperspin) are important when determining the type of degeneracy imposed by the commutation of H with all the K. The results are significantly different from those obtained for the isospin.

Comments:

(i) Commutation relations in $\mathscr{L}_8$ are not the simple juxtaposition of those of the three isospin operators K: the structure of $\mathscr{L}_8$ produces a sort of intricate connection of the three isospin symmetries into each other.

(ii) Although it commutes with H, we have ignored the operator N. As in the case studied in § B, adding N amounts to adding a "phase group", with no physical consequence.

C-2. SU(3) matrices

The simplest case that allows associating matrices to the K operators is a single particle subspace generated by the kets:

$$|u_1\rangle = |\varphi_k\rangle \otimes |\alpha\rangle \quad ; \quad |u_2\rangle = |\varphi_k\rangle \otimes |\beta\rangle \quad ; \quad |u_3\rangle = |\varphi_k\rangle \otimes |\gamma\rangle \tag{IX-86}$$

where only the hyperspin state varies.

C-2-a. Writing the matrices

This subspace is obviously stable under the action of operators K, and, using the basis kets (IX-86) in that order, we get the matrices:

$$\left(K_z^{(1)}\right) = \frac{1}{2}\begin{pmatrix} 1 & 0 & 0 \\ 0 & -1 & 0 \\ 0 & 0 & 0 \end{pmatrix} \; ; \; \left(K_z^{(2)}\right) = \frac{1}{2}\begin{pmatrix} 0 & 0 & 0 \\ 0 & 1 & 0 \\ 0 & 0 & -1 \end{pmatrix} \; ; \; \left(K_z^{(3)}\right) = \frac{1}{2}\begin{pmatrix} -1 & 0 & 0 \\ 0 & 0 & 0 \\ 0 & 0 & 1 \end{pmatrix}$$

(IX-87a)

as well as:

$$\left(K_+^{(1)}\right) = \begin{pmatrix} 0 & 1 & 0 \\ 0 & 0 & 0 \\ 0 & 0 & 0 \end{pmatrix} \qquad \left(K_-^{(1)}\right) = \begin{pmatrix} 0 & 0 & 0 \\ 1 & 0 & 0 \\ 0 & 0 & 0 \end{pmatrix}$$

$$\left(K_+^{(2)}\right) = \begin{pmatrix} 0 & 0 & 0 \\ 0 & 0 & 1 \\ 0 & 0 & 0 \end{pmatrix} \qquad \left(K_-^{(2)}\right) = \begin{pmatrix} 0 & 0 & 0 \\ 0 & 0 & 0 \\ 0 & 1 & 0 \end{pmatrix}$$

$$\left(K_+^{(3)}\right) = \begin{pmatrix} 0 & 0 & 0 \\ 0 & 0 & 0 \\ 1 & 0 & 0 \end{pmatrix} \qquad \left(K_-^{(3)}\right) = \begin{pmatrix} 0 & 0 & 1 \\ 0 & 0 & 0 \\ 0 & 0 & 0 \end{pmatrix} \tag{IX-87b}$$

In an equivalent way, we can write the components on the Ox and Oy axes:

$$\left(K_x^{(1)}\right) = \frac{1}{2}\begin{pmatrix} 0 & 1 & 0 \\ 1 & 0 & 0 \\ 0 & 0 & 0 \end{pmatrix} \qquad \left(K_y^{(1)}\right) = \frac{1}{2}\begin{pmatrix} 0 & -i & 0 \\ i & 0 & 0 \\ 0 & 0 & 0 \end{pmatrix}$$

$$\left(K_x^{(2)}\right) = \frac{1}{2}\begin{pmatrix} 0 & 0 & 0 \\ 0 & 0 & 1 \\ 0 & 1 & 0 \end{pmatrix} \qquad \left(K_y^{(2)}\right) = \frac{1}{2}\begin{pmatrix} 0 & 0 & 0 \\ 0 & 0 & -i \\ 0 & i & 0 \end{pmatrix}$$

$$\left(K_x^{(3)}\right) = \frac{1}{2}\begin{pmatrix} 0 & 0 & 1 \\ 0 & 0 & 0 \\ 1 & 0 & 0 \end{pmatrix} \qquad \left(K_y^{(3)}\right) = \frac{1}{2}\begin{pmatrix} 0 & 0 & i \\ 0 & 0 & 0 \\ -i & 0 & 0 \end{pmatrix} \tag{IX-87c}$$

These matrices are the usual Pauli matrices, but spread over 3 rows and 3 columns, instead of 2: the first two rows and columns for the $K^{(1)}$ matrices, the last two for $K^{(2)}$, and the first and last for $K^{(3)}$. Noting that the three $K_z^{(i)}$ matrices are not independent (since their sum equals zero), we can eliminate $K_z^{(3)}$, for example; this yields a set of eight linearly independent matrices, with a trace equal to zero. This set, isomorphic by construction to $\mathscr{L}_8$, is a "representation" of that algebra.

It is easy to show that linear combinations with real coefficients of these eight matrices generate all the 3×3 Hermitian matrices with a zero trace, hence the Lie algebra[15] of the group SU(3) (group of the 3×3 unitary matrices with a determinant equal to 1). This explains the denomination "SU(3) symmetry" for the kind of three internal states symmetry we are studying here.

[15]To show this, one can use the same reasoning as for SU(2) in § B-3 of chapter III: any unitary matrix of SU(3) can be seen as the exponential of a Hermitian matrix with zero trace; this matrix can be expanded on the real components of the matrices (K).

The irreducibility of the representation can be easily established. If it were reducible, we could find in the three-dimensional space a vector invariant (direction wise) under the action of any Hermitian operator; hence, such a vector does not exist, in particular under any three-dimensional rotation.

Comments:

(i) Maintaining operator N among the K would yield a matrix with a non-zero trace; the Lie algebra would then be that of the $U(3)$ group, with no restriction on the determinant of the unitary matrices, and with dimension 9.

(ii) Instead of eliminating $K_z^{(3)}$ as we did above, one can proceed in a more symmetrical way and express the 9 matrices (IX-87) as a function of the eight 4×4 Gell-Mann matrices λ_i. These are a generalization of the Pauli matrices and obey the orthonormalization relations:

$$\mathrm{tr}\{\lambda_i \lambda_j\} = 2\delta_{ij} \tag{IX-88}$$

For more details, see reference [26].

C-2-b. Subgroups

Matrices (IX-87) show that there are different ways to decompose $\mathscr{L}_8$ into subalgebra, each generating a subgroup of the (hyperspin) group $\mathscr{G}$. As an example, the three components of $\overline{K}^{(1)}$ generate $\mathscr{L}_3$ and the isospin group discussed in § B. Two other subalgebras $\mathscr{L}_3'$ and $\mathscr{L}_3''$ are generated in a similar way by $\overline{K}^{(2)}$ and $\overline{K}^{(3)}$, respectively.

The 3×3 unitary real (orthogonal) matrices with a determinant equal to unity form another subgroup of SU(3); this group is generally noted SO(3), and simply contains all the rotation matrices in ordinary space. The matrices of the corresponding Lie algebra have already been discussed in chapter VI (to within a constant factor):

$$(R_x) = \begin{pmatrix} 0 & 0 & 0 \\ 0 & 0 & -i \\ 0 & i & 0 \end{pmatrix} = 2\left(K_y^{(2)}\right) \tag{IX-89a}$$

$$(R_y) = \begin{pmatrix} 0 & 0 & i \\ 0 & 0 & 0 \\ -i & 0 & 0 \end{pmatrix} = 2\left(K_y^{(3)}\right) \tag{IX-89b}$$

$$(R_z) = \begin{pmatrix} 0 & -i & 0 \\ i & 0 & 0 \\ 0 & 0 & 0 \end{pmatrix} = 2\left(K_y^{(1)}\right) \tag{IX-89c}$$

C-2-c. Rank; root and weight diagram

By definition, the rank of a Lie algebra is the maximum number of linearly independent operators of this algebra that commute with each other. For the $\mathscr{L}_3$ Lie algebra of an angular momentum, this rank is obviously equal to one ($\boldsymbol{u} \cdot \boldsymbol{J}$ and $\boldsymbol{v} \cdot \boldsymbol{J}$ do not commute unless $\boldsymbol{u}$ is parallel to $\boldsymbol{v}$); this explains the "linear" structure of the angular momentum multiplets (figure 3). For SU(3), we

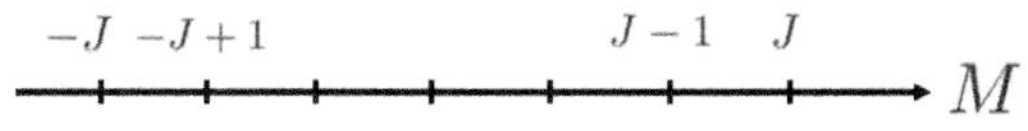

Figure 3: The Lie algebra of an angular momentum is of rank 1, so that an irreducible representation is associated with a segment of a straight line.

shall see that the rank is two, which leads to planar patterns for representing the "supermultiplets" associated with irreducible representations.

It is easy to see that the rank of $\mathscr{L}_8$ must be at least equal to two since $K_z^{(1)}$ commutes with $K_z^{(2)}$. Conversely, this rank cannot be greater than two: otherwise, one could diagonalize simultaneously the corresponding matrices, though we know that three 3×3 diagonal matrices with a zero trace cannot be linearly independent (having a zero trace ensures that these matrices can depend only on two parameters)[16].

In the case of SU(3), two operators play the role of J_z for SO(3), or of σ_z for SU(2). When both these operators are diagonalized, each basis vector has two quantum numbers, each yielding an eigenvalue. This naturally leads to constructing 2-dimensional diagrams, where the two eigenvalues are indicated on two independent axes. Any basis vector will correspond to a given point on the plane.

When $K_z^{(1)}$ and $K_z^{(2)}$ are chosen as the simultaneously diagonalized operators, it is convenient to indicate the corresponding eigenvalues M_K^1 and M_K^2 on two axes at 120 degrees of each other, on which one performs an orthogonal projection (*cf.* figure 4). In that way, the third axis will automatically yield by projection M_K^3:

$$M_K^3 = -M_K^1 - M_K^2 \tag{IX-90}$$

[16]The generalization is straightforward: the rank of the Lie algebra of SU(n) is $(n-1)$.

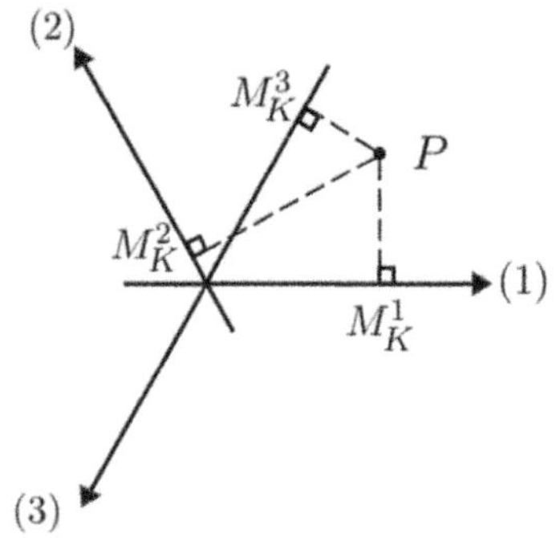

Figure 4: The values of M_K^1, M_K^2, and M_K^3 are indicated on three axes at 120 degrees from each other. For any point P on the plane, the sum of the values of M obtained by orthogonal projections onto the three axes is zero, as required by relation (IX-90).

which, according to (IX-81), is precisely the eigenvalue of $K_z^{(3)}$. This construction preserves the symmetry of the role played by $\overline{K}^{(1)}$, $\overline{K}^{(2)}$, and $\overline{K}^{(3)}$. It is not surprising that the diagrams we shall obtain have a certain 3-fold symmetry, but make no mistake: the $\mathscr{L}_8$ algebra is of rank two, operator $K_z^{(3)}$ being actually redundant.

With each irreducible representation of SU(3) is associated a set of points $P, P', P'', \ldots$ with coordinates M_K^1 and M_K^2. Each point has a certain weight, the number of basis vectors with which it is associated[17]. The ensemble is called the "weight diagram" of the irreducible representation.

Consider now a basis vector $|\xi\,;\,M_K^1, M_K^2\rangle$ associated with one of the points P of the diagram: the number of distinct values of ξ is the weight associated with P. Referring to the general theory of angular momentum, we know that M_K^1 and M_K^2 are either integers or half-integers; in addition, we shall deduce from the commutation relations (IX-84a) that:

$$K_\pm^{(1)}\,|\xi\,;\,M_K^1, M_K^2\rangle$$

is:

- either a zero ket;
- or another eigenket of $K_z^{(1)}$ with eigenvalue $M_K^1 \pm 1$, belonging to the same irreducible subspace.

[17]For SO(3), the weights are always equal to 1, but it is not a general rule for all the groups.

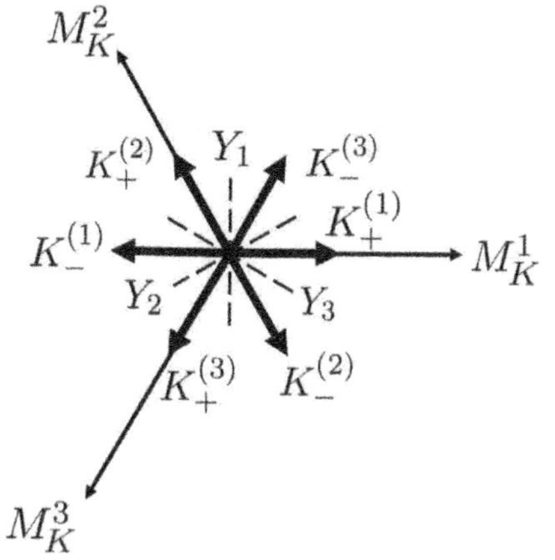

Figure 5: The action of the two operators $K_\pm^{(1)}$ results in a shift of the representative point of plus or minus one unit, parallel to the M_K^1 axis, as shown by the two opposite horizontal arrows. For the two operators $K_\pm^{(2)}$, the shift is along the M_K^2 axis as shown by the two arrows at 120 degrees, and for the two operators $K_\pm^{(3)}$, the shift is along the axis at -120 degrees. The Y_1, Y_2, and Y_3 axes in dashed lines are symmetry axes for any representation of the $SU(3)$ group.

In particular, if $M_K^1 > 0$, this ket can never become zero under the action of $K_-^{(1)}$; if $M_K^1 < 0$, it never becomes zero under the action of $K_+^{(1)}$; if $M_K^1 = 0$, it can become equal to zero depending on the case[18].

Relations (IX-84e) enable us to know how the action of $K_\pm^{(1)}$ changes the eigenvalue of $K_z^{(2)}$ [or of $K_z^{(3)}$]: we see that, as opposed to the eigenvalue of $K_z^{(1)}$, the eigenvalues of $K_z^{(2)}$ and $K_z^{(3)}$ are decreased by 1/2 under the action of $K_+^{(1)}$, and increased by 1/2 under the action of $K_-^{(1)}$.

Finally, the action of $K_+^{(1)}$ is shown in figure 4 by a $+1$ shift parallel to the M_K axis, whereas the action of $K_-^{(1)}$ results in a -1 shift parallel to the same axis. The same reasonings lead to analog results for $K_\pm^{(2)}$ and $K_\pm^{(3)}$. The actions of the six operators $K_\pm^{(i)}$ are symbolized by a diagram of the type shown in figure 5, called a "root diagram".

Comments:

(i) The form of the root diagram suggests a hexagonal-type symmetry for the weight diagram. The fact that the basis kets can be regrouped as those

[18]This property is easily established by expanding $|\xi\,;\,M_K^1, M_K^2\rangle$ on a "standard basis" associated with $\overline{K}^{(1)}$; we know that $J_-|\tau, J, M\rangle$ is never equal to zero if $M > 0$.

of a standard basis associated with $\overline{K}^{(1)}$, where M varies between $+J$ and $-J$, also suggests a symmetry with respect to the Y_1 axis, perpendicular to the M_K^1 axis and going through the origin 0. Similarly, we expect symmetries with respect to the Y_2 and Y_3 axes of figure 5.

(ii) The weight diagram is invariant if we make any basis change in the subspace of the representation we are considering: the eigenvalues of an operator as well as their degeneracy degrees are invariant under a basis change. Two distinct weight diagrams necessarily correspond to different (non-equivalent) representations.

(iii) Our analysis gave a privileged role to the $K_z^{(i)}$ components, in spite of a certain symmetry in the roles played by the x, y, or z components. There are two reasons for this. First of all, the linear dependence relation (IX-81) involves only the $K_z^{(i)}$ components, the other ones remaining independent. Secondly, the operator H we are looking for does not change the number of particles n_α, n_β, or n_γ, just as $K_z^{(i)}$. Consequently, when the $K_z^{(i)}$ are diagonal, the basis vectors of the representations are automatically "physical" eigenvectors of H (belonging to a given subspace $\mathscr{E}_{n_\alpha n_\beta n_\gamma}$).

C-3. Construction of irreducible representations

Let us examine the possible dispositions of the points P forming a weight diagram of SU(3). This will tell us which are the first irreducible representations of SU(3), those with the lowest dimensions.

C-3-a. General comments

Consider an arbitrary point P on the weight diagram, and assume that it is not located at the origin 0. It cannot be located on more than one of the three axes Y_1, Y_2, Y_3 of figure 5. We call $Y_{(i)}$ one of the two axes where it is not located.

If point P has a positive orthogonal projection[19] on the M_K^i axis, the action of $K_-^{(i)}$ on any basis kets associated with P yields another non-zero ket, the eigenvalue of the $K_z^{(i)}$ being lowered by one unit. By recurrence, with any point P is associated a series of points $P', P'', \ldots$ belonging to an isospin multiplet corresponding to the eigenvalue $k^{(i)}\,[k^{(i)}+1]$ of $[\overline{K}^{(i)}]^2$. By the repeated action of $K_\pm^{(i)}$, one can build from the initial point P a series of $(2\,M_K^i+1)$ points. These points are part of the weight diagram, as shown in figure 6 (where we assumed $i=1$).

Note that in such a displacement the values of M_K vary by integer jumps, but those of the other components by half-integer jumps (*cf.* figure 5). It follows

[19] If it is negative, we can obviously follow a symmetric reasoning using K_+.

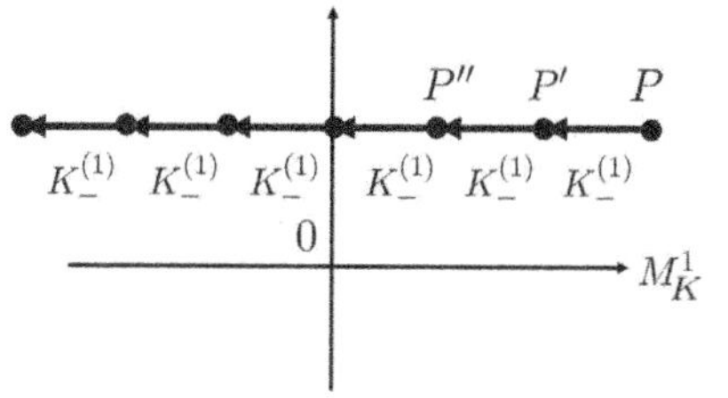

Figure 6: Starting from a point P in the weight diagram, for which we assume $M_K^1 > 0$, we can construct by the repeated action of $K_-^{(1)}$ a whole series of points P', P'', etc. for which M_K^1 decreases by one, two, etc. units. The points are part of $(2K^{(1)} + 1)$ non-zero vectors of the irreducible representation. Analog contructions are also possible along the other two axes M_K^2 and M_K^3, inclined at ± 120 degrees, in figure 5.

that, in general, a weight diagram regroups isospin multiplets corresponding to integer or half-integer values of K_1, K_2, and K_3.

Starting from a point P_1 that is not located on any of the axes Y_1, Y_2, Y_3, we know that there necessarily are five other points $P_2, \ldots, P_6$ of the diagram that are symmetric to the first one with respect to these axes (*cf.* figure 7). The product of two symmetries with respect to two distinct Y_i yields a rotation of ± 120 degrees, so that the six points are at the summit of two equal equilateral triangles. If now P_1 is on one of the axes Y_1, Y_2 or Y_3, the points regroup two by two and the two equilateral triangles merge into into a single one (as an example, P_1 and P_6, P_2 and P_3, P_4 and P_5 become one in figure 7). Two situations can occur for points distinct from the origin 0: they regroup either into sets of six points, or into sets of three points located on the symmetry axes Y_1, Y_2, and Y_3.

Starting from one of the six points $P_1, \ldots, P_6$, one can obtain others by the action of the operators symbolized on the root diagram of figure 5: any ± 1 shift, perpendicular to the Y_i axis, is authorized as long as it yields a point that is not further away from that axis (there exists at least one non-zero ket associated with that additional point). One can thus fill the perimeter $P_1 P_2 \ldots P_6$ of figure 7 with isosceles triangles whose sides equal 1 and are perpendicular to the Y_1, Y_2, and Y_3 axes. (figure 8).

C-3-b. One point diagrams: representation $\{1\}$

The easiest situation is obviously when the weight diagram includes a single point P, which must necessarily be at the origin 0. The only possible eigenvalues of $K_z^{(i)}$ are all equal to zero. The same is true for the eigenvalues of $[\overline{K}^{(i)}]^2$, so

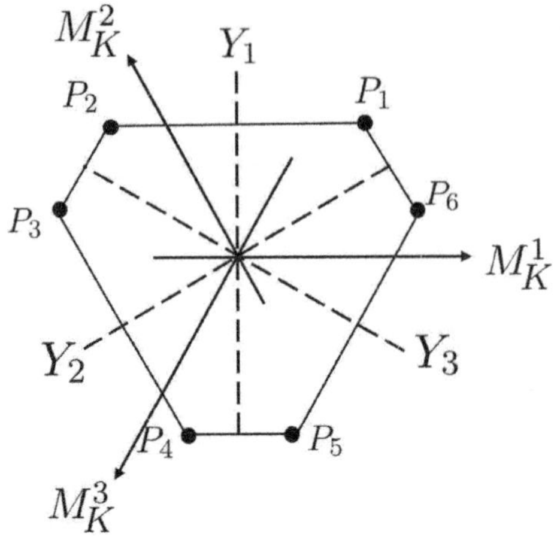

Figure 7: To a point P_1 that is not located on any of the axes Y_1, Y_2, Y_3, correspond five other points $P_2, \ldots, P_6$ of the weight diagram, symmetric to the first one with respect to the axes $Y_{1,2,3}$. If the point P_1 is located on one of the axes Y_1, Y_2, Y_3, the six points reduce to 3.

that the matrices associated with the $K_{x,y,z}$ are all equal to zero. Under these conditions, the only irreducible representation is that of dimension 1 (the weight of point 0 is 1); it is noted $\{1\}$. The (isospin) group $\mathscr{G}$ is represented only by the $(\mathbb{1})$ matrices, which is the so-called "trivial" representation.

C-3-c. Three point diagrams: representations $\{3\}$ and $\{3^*\}$

The minimum number of points different from the origin is three. According to the previous discussion in § C-3-a, two dispositions are possible, as shown in figures 9-a and 9-b (isosceles triangles pointing downwards or upwards).

Such diagrams enable the direct construction of the representation matrices. We first assume that the weight of each point P equals 1: we thus look for representations of dimension 3.

• For the diagram of figure 9-a, we call $|u_1\rangle$, $|u_2\rangle$, and $|u_3\rangle$ the orthonormal basis vectors associated with P_1, P_2, and P_3, respectively. One can read directly on the figure the eigenvalues of the operators $K_z^{(i)}$ associated with $|u_1\rangle$, $|u_2\rangle$, and $|u_3\rangle$. For example, the projection of P_1 onto the M_K^1 axis is $+1/2$, indicating that $|u_1\rangle$ is an eigenvector of $K_z^{(1)}$ with the eigenvalue $+1/2$; the projection of P_2 onto this same axis is $-1/2$, meaning that $|u_2\rangle$ is an eigenvector of $K_z^{(1)}$ with the eigenvalue $-1/2$; finally, $|u_3\rangle$ is an eigenvector with a zero eigenvalue. This leads to the matrix:

$$\left(K_z^{(1)}\right) = \frac{1}{2}\begin{pmatrix} 1 & 0 & 0 \\ 0 & -1 & 0 \\ 0 & 0 & 0 \end{pmatrix}$$

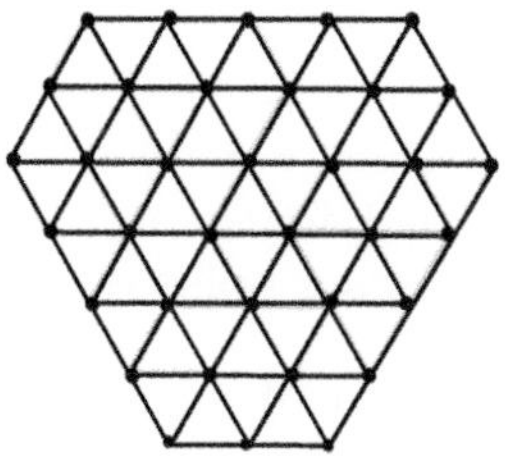

Figure 8: Starting from the points of figure 7 and by the action of the operators symbolized on the root diagram of figure 5, one can obtain a whole series of other points, all falling inside the perimeter drawn in figure 7.

Similar expressions are obtained for $K_z^{(2)}$ and $K_z^{(3)}$, which are the same as the ones already written in (IX-87a).

1/2 −1/2 This root diagram also shows that operator $K_+^{(1)}$ transforms $|u_2\rangle$ into $|u_1\rangle$ (it yields zero when it acts on $|u_1\rangle$ and $|u_3\rangle$); a proper definition of the relative phase of these basis vectors yields the already known expression (IX-87a). The Hermitian conjugate of $K_-^{(1)}$ also yields expression (IX-87a). Defining the

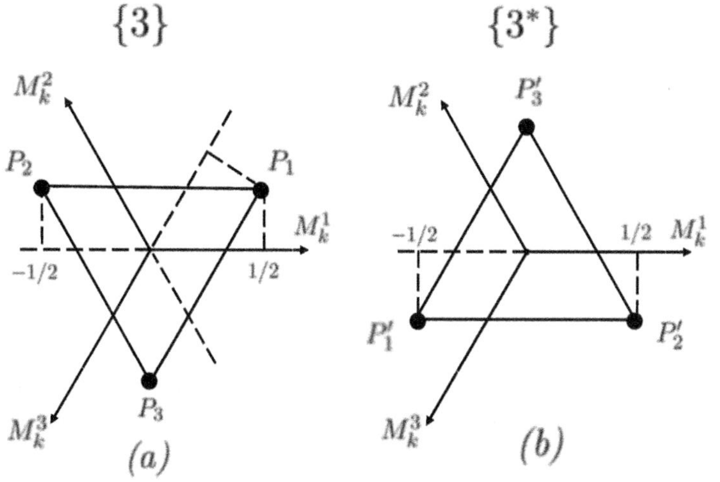

Figure 9: Figure (a) shows the diagram of the three point representation {3} *of the group* SU(3), figure (b) the representation {3*}.

relative phase of $|u_2\rangle$ and $|u_3\rangle$ enables us to find again the expressions of $(K_{\pm}^{(2)})$. Relations (IX-84e) can then be used and show that that the $(K_{\pm}^{(3)})$ matrices can only be those already written in (IX-87a).

We have obtained a unique representation associated with the diagram of figure 9-a, given directly by the matrices of SU(3). It is symbolized by the notation $\{3\}$. Point P_1 is obtained for $n_\alpha = 1$, $n_\beta = n_\gamma = 0$; two circular permutations of the α, β, and γ indices yield the points P_2 and P_3.

• We now consider the diagram of figure 9-b. As it is different from the previous diagram, it may yield a representation, but not one equivalent to the previous one. There is a straightforward method to attribute an irreducible representation to this diagram. Since its triangle and that in figure 9-a are symmetric with respect to the horizontal axis, this amounts to a simple inversion of the role played by $\overline{K}^{(2)}$ and $\overline{K}^{(3)}$. We simply have to exchange in (IX-87a) all the matrices with index $i = 2$ with all those with index $i = 3$. The corresponding representation does exist and will be noted $\{3^*\}$. The point P_1' is obtained for $n_\alpha = 0$, $n_\beta = n_\gamma = 1$; two circular permutations of the α, β, and γ indices yield the points P_2' and P_3'.

Link between the $\{3\}$ and $\{3^*\}$ representations: For the representation $\{3^*\}$, it is convenient to number differently the basis vectors $|u_1\rangle$, $|u_2\rangle$, and $|u_3\rangle$; such a basis change introduces an equivalent representation. The new numbering follows that of the points P_1', P_2', and P_3' of figure 9-b; it is symmetric, with respect to the origin 0, to the numbering chosen in figure 9-a. We now compare representations $\{3\}$ and $\{3^*\}$. In the $\{3^*\}$ symmetry, the operators' eigenvalues change sign, and the associated matrices are the opposite of those of representation $\{3\}$. In the same way, we see that the matrices associated with $K_+^{(i)}$ operators are interchanged with those associated with $K_-^{(i)}$ operators. Attention should be paid to correctly defining the relative phases of the basis vectors, since, in their commutation relations, operators K_+ and K_- play an antisymmetrical role. The reasoning followed for representation $\{3\}$ applies only if the relative phases of the $|u_p\rangle$ have been defined in such a way that $K_{\mp}^{(i)}$ is associated with the opposite of the matrix previously associated with $K_{\pm}^{(i)}$. In this operation, the Hermitian operators $K_{x,y,z}$ undergo the transformations:

$$K_{x,z}^{(i)} \leftrightarrow -K_{x,z}^{(i)} \quad ; \quad K_y^{(i)} \leftrightarrow +K_y^{(i)} \tag{IX-91}$$

Finally, the matrices of representation $\{3^*\}$ can be obtained from equations (IX-87c) by a sign change followed by a complex conjugation. As for the matrices associated with the unitary operators of (hyperspin) $\mathscr{G}$, equality (IX-85a) shows that going from $\{3\}$ to $\{3^*\}$ results in a simple complex conjugation, hence the notation $\{3^*\}$.

Comments:

(i) It is clear that if a set of matrices form a representation of a group $\mathscr{G}$, the set of the complex conjugate matrices will also form a representation

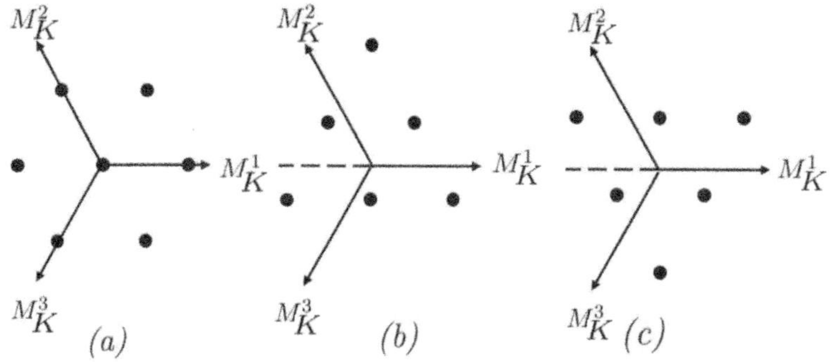

Figure 10: Three other possible diagrams for irreducible representations of the SU(3) Lie algebra. We only represent the positions of the points, not their weight; we shall see below that the point origin of figure (*a*) has a weight equal to two.

(the product of complex conjugate matrices is the conjugate matrix of the product). In the present case, we have conjugate representations that yield different weight diagrams and are thus not equivalent. For SU(2), there also exist complex representations, even if only the SU(2) matrices themselves. However, we saw in chapter VII that the irreducible representations of SU(2) are entirely characterized by the number J, i.e. by the space dimension $2J + 1$. As opposed to SU(3), two conjugate representations in SU(2) are always equivalent.

(ii) Representations $\{3\}$ and $\{3^*\}$ automatically form representations of the SO(3) subgroup. Actually we get twice the same representation ($J = 1$), since the rotation matrices are invariant under complex conjugation.

C-3-d. Diagram with more than three points

Let us try to add the origin to the diagrams of figures 9-a and 9-b. The root diagram cannot connect this point O to any other point (P_1, P_2, and P_3 are not within a unit distance from the origin). The corresponding representations are thus reducible: the weight of point O is equal to the number of times they contain representation $\{1\}$ in addition to one of the $\{3\}$ or $\{3^*\}$ representations.

When looking for diagrams with more than 3 points, two situations may occur:

- If O is a point of the diagram, and if it is not the only one, it is easy to see that the simplest possible disposition of the points is shown in figure 10-a, which has 7 points;

- If O is not a point of the diagram, we must add at least 3 points to figures 9-a or 9-b, and we get the diagrams shown in figures 10-b and 10-c.

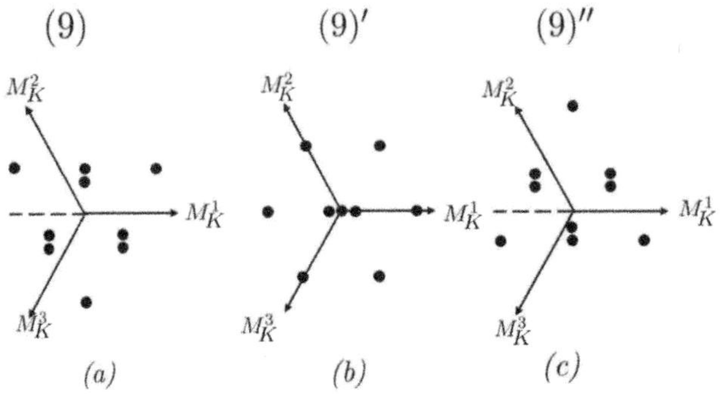

Figure 11: Weight diagram associated with representation (9)*, the product* $\{3\} \otimes \{3\}$*, obtained by combining diagram (a) of figure 9 with itself; with representation* $(9)'$ *, the product* $\{3\} \otimes \{3^*\}$*, obtained by combining diagrams (a) and (b) of figure 9; and with representation* $(9)''$ *, the product* $\{3^*\} \otimes \{3^*\}$*, obtained by combining diagram (b) of figure 9 with itself. The double or triple points indicates a weight of 2 or 3.*

Let us show that each of these patterns is indeed associated with an irreducible representation. We consider the SU(3) representations obtained by a tensor product of the two 3-dimensional representations, which we note (we use a different notation since we shall see that these representations are reducible):

$$(9) \equiv \{3\} \otimes \{3\} \qquad (9)' \equiv \{3\} \otimes \{3^*\} \qquad (9)'' \equiv \{3^*\} \otimes \{3^*\} \tag{IX-92}$$

These representations are of dimension 9, and can be calculated in a ket basis obtained by tensor products of the initial kets. If we associate two kets that are eigenvectors of the same operator of (hyperspin) $\mathscr{G}$, with eigenvalues λ_1 and λ_2, the tensor product ket is still an eigenket, with the eigenvalue $\lambda_1 \times \lambda_2$. For the eigenkets of operators of the $\mathscr{L}_8$ Lie algebra, the situation is similar, but we must now add the eigenvalues (just as when we add two angular momenta).

The weight diagram of the product representation is very simple to obtain. We associate two by two the points P_1 and P_2 of the initial representation, which yields the point P defined by:

$$\overrightarrow{OP} = \overrightarrow{OP}_1 + \overrightarrow{OP}_2$$

The same point P can obviously be obtained from different couples P_1, P_2; this must be reflected in the computation of the weights.

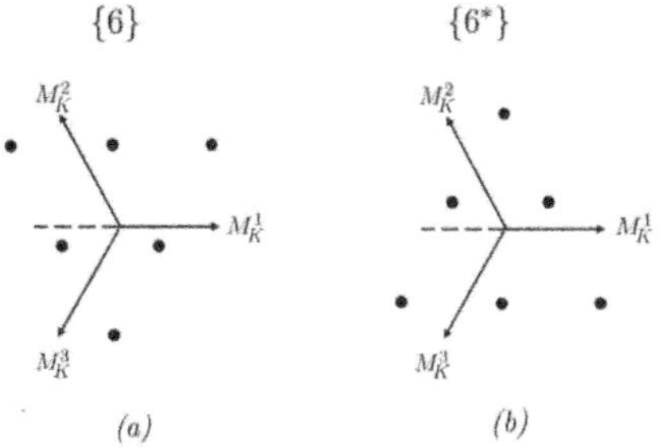

Figure 12: Diagrams of irreducible representations derived from representations in figure 11. The diagram on the left comes from part (a) of this figure, and is noted {6}; the diagram on the right, from part (c), and is noted {6}.*

Figures 11-a, 11-b, and 11-c show the result of this operation for the product representations (9), $(9)'$, and $(9)''$; the double or triple points correspond to weights equal to 2 or 3. Representations (9) and $(9)''$ are actually those associated with a unitary base change for tensors of order two, when we are in a 3-dimensional space (*cf.* complement A_{VIII}) and the two i and j components are both either covariant or contravariant. Now we know two sets of tensors that are invariant in such an operation, the completely symmetric tensors A $[A_{ij} = A_{ji}]$ and the completely antisymmetric tensors $[A_{ij} = -A_{ji}]$. These two sets span spaces of dimension 6 and 3, respectively. Representations (9) and $(9)''$ can thus be decomposed into representations having these dimensions.

Consequently, we must separate the diagrams of figures 11-a and 11-c into two groups, one of 3 points, the other of 6. It can only be achieved in a manner compatible with the rules stated above by identifying in figures 11-a and 11-c the diagrams of figures 9-a and 9-b as well as two others shown in figures 12-a and 12-b. It is easy to show that these diagrams cannot be decomposed any further to obtain a set of points having acceptable properties for a weight diagram. The representations are thus irreducible and are noted $\{6\}$ and $\{6^*\}$. The decomposition can be symbolized by equations:

$$\begin{aligned} \{3\} \otimes \{3\} &= \{3^*\} \oplus \{6\} \\ \{3^*\} \otimes \{3^*\} &= \{3\} \oplus \{6^*\} \end{aligned} \tag{IX-93}$$

We still have to examine representation $(9)'$. It concerns the transformation of a tensor with two indices, one covariant, the other contravariant. We know that the trace $\sum_\mu A^\mu_\mu$ is invariant under any change of basis (unitary or not). This enables us to remove from $(9)'$ a representation, which can only be $\{1\}$. We are left with the diagram of figure 13, which cannot be decomposed in two

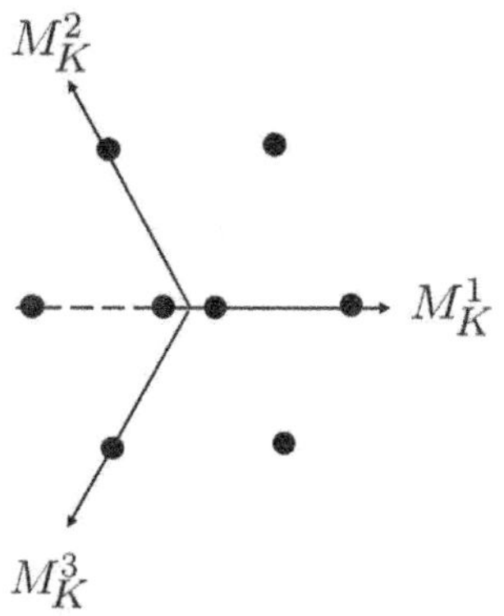

Figure 13: Diagram of an irreducible representation derived from part (b) of figure 11. This representation is noted {8}.

sets of points forming acceptable diagrams. The remaining representation, of dimension 8, is noted {8}, and we write:

$$\{3\} \otimes \{3^*\} = \{1\} \oplus \{8\} \tag{IX-94}$$

Note that {8} is the first irreducible representation we found, whose weights are not all equal to 1 (the weight of the origin equals 2).

Another example: We now take the tensor product:

$$(27) = \{3\} \otimes \{3\} \otimes \{3\}$$

The corresponding diagram, obtained as before, is shown in figure 14; it includes 10 distinct points, some of which (those in the "corners") do not appear in any of the diagrams

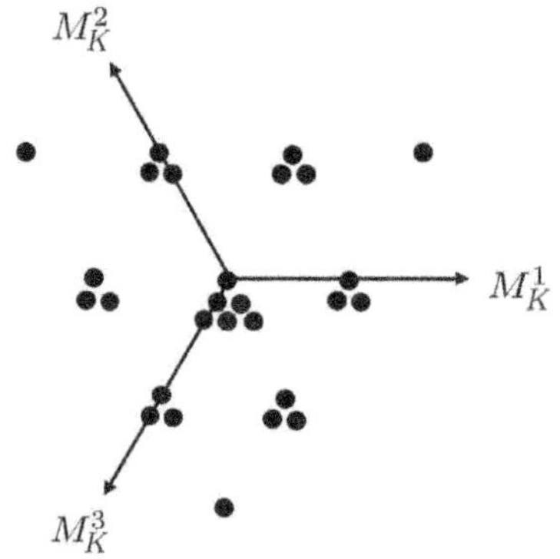

Figure 14: Weight diagram associated with the tensor product representation $(27) = \{3\} \otimes \{3\} \otimes \{3\}$*, which is reducible.*

{10}

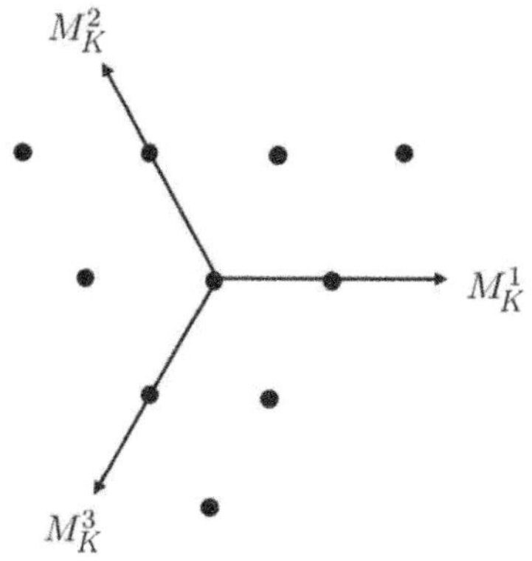

Figure 15: Irreducible representation, noted $\{10\}$.

obtained previously. The (27) representation thus includes a new irreducible representation, of dimension 10 (if all its weights are equal to 1) or more.

In terms of tensors, representation (27) appears when considering the effect of a unitary basis change on the 3 times covariant component μ, μ', and μ'' of a third-order tensor. Now we know that a totally symmetric tensor B:

$$B_{\mu\mu'\mu''} = B_{\mu'\mu\mu''} = B_{\mu''\mu\mu'} = \ldots$$

preserves this property under a basis change: an invariant subspace must exist in the representation space. A totally symmetric tensor depends on 10 parameters[20]. It is also clear that this space includes the vectors associated with the "corners" of figure 14, corresponding to $\mu = \mu' = \mu''$. The diagram of this representation is shown in figure 15; as all the weights are equal to 1, there is no way to subtract from it any of the irreducible representations obtained above. This representation, of dimension 10, is thus irreducible; it is noted $\{10\}$.

In addition, using (IX-93) and (IX-94), we get:

$$\begin{aligned}\{3\} \otimes \{3\} \otimes \{3\} &= \Big[\{3^*\} \oplus \{6\}\Big] \otimes \{3\} \\ &= \{1\} \oplus \{8\} \oplus \Big(\{6\} \otimes \{3\}\Big) \end{aligned} \tag{IX-95}$$

[20]The $B_{\mu\mu'\mu''}$ components depend on the choice of x, y, or z for the indices μ, μ', and μ'', but not on their order because of the symmetry. The number of independent components is thus equal to the number of different ways we can choose x, y, z (including possible repetitions). This number is the sum of:

1	(3 different indices)
$3 \times 2 = 6$	(an index is repeated twice and associated with any of the two others)
3	(3 equal indices).

It is thus equal to 10.

Representation $\{10\}$ must be contained in $\{6\} \otimes \{3\}$, leaving a representation of dimension 8, which can only be $\{8\}$:

$$\{3\} \otimes \{6\} = \{8\} \oplus \{10\} \tag{IX-96}$$

Finally, equation (IX-94) becomes:

$$\{3\} \otimes \{3\} \otimes \{3\} = \{1\} \oplus \{8\} \oplus \{8\} \oplus \{10\} \tag{IX-97}$$

and we can check that this sum yields the weights represented in figure 14.

C-3-e. Applications

An important application[21] of the previous discussion is the quark model: one considers that hadrons (particles subjected to strong interactions: baryons, mésons, ...) are built from more elementary constituents, the quarks. These latter particles are supposed to play the same role in the strong interaction Hamiltonian. When the number of different types of quarks is three (u, d, and s quarks, with different "flavors"), SU(3) symmetry must be obeyed by all the structures they permit building.

The simplest case is when the total particle number n is equal to 1. There are three possible states for the system:

$$|u\rangle \quad |d\rangle \quad |s\rangle$$

that generate an irreducible representation of SU(3) (this group plays the role of a "flavor group"); we assume that it is representation $\{3\}$ of figure 9-a. With the three quarks are associated three "antiquarks" with opposite quantum numbers; the corresponding states are noted:

$$|\overline{u}\rangle \quad |\overline{d}\rangle \quad |\overline{s}\rangle$$

and they must also generate an irreducible representation of SU(3). Changing the sign of the quantum numbers M_K amounts to going from $\{3\}$ to $\{3^*\}$; the antiquarks are associated with the $\{3^*\}$ representation in figure 9-b. This is an advantage of the SU(3) group over SU(2): it has non-equivalent irreducible representations of the same dimension, which allows the introduction of antiparticles.

Quarks and antiquarks are spin 1/2 particles, hence fermions obeying the Pauli exclusion principle (just as leptons: electrons and positrons, muons $\mu^{\pm}$, etc.).

[21] Applications of SU(3) symmetry exist in domains other than particle physics. As an example, this symmetry plays a role in the classification of the energy levels of polyatomic molecules, formed by a certain number of atoms with nuclear spins $I = 1$ (three nuclear spin states m_I are then accessible). One can also assume that the polyatomic atoms are built with different isotopes having almost the same mass (and the same nuclear spin, $I = 0$ for example). SU(3) symmetry is then observed in the vibration-rotation spectrum when one neglects the mass difference between the nuclei. In a second step, one can take into account this mass difference, which leads to a symmetry breaking (lifting of certain degeneracies).

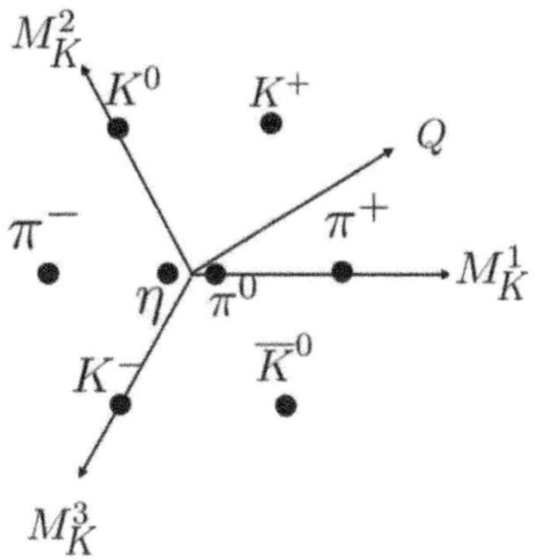

Figure 16: Octet of pseudoscalar mesons. The charges are indicated on the Q axis.

We now discuss the possible symmetries of the various assemblies of quarks and antiquarks. Formula (IX-94) shows which are the supermultiplets obtained by combining a quark and an antiquark: an octet with representation $\{8\}$ and a singlet $\{1\}$. The octet is associated with the so-called pseudoscalar mesons, with zero spin [22] and negative parity, represented in figure 16. There also exists an octet of vector mesons.

If we now combine three quarks, formula (IX-97) shows that we get a singlet, two octets, and one decuplet. An octet of spin 1/2 baryons with positive parity is shown in figure 17. We recognize on the top the proton and neutron isospin doublet. In a general way, on this diagram and those of the other figures, particules belonging to the same isospin multiplet are aligned on the same horizontal segment: moving toward the right by one horizontal unit, we get a particle whose charge is increased by one quantum of charge q_p. The second line of the diagram of figure 17 thus yields an isospin triplet and an isospin singlet. The decuplet of spin 3/2 baryons with positive parity, associated with representation $\{10\}$, is shown in figure 18.

Comments:

(i) If we add to the three quarks $|u\rangle$, $|d\rangle$, $|s\rangle$ a " charm quark" $|c\rangle$, the SU(3) flavor symmetry becomes SU(4) and the supermultiplets must be represented by tridimensional diagrams, whose cross sections are SU(3) diagrams[23]. The "standard model" introduces 6 quarks of dif-

[22] Just like H, the total angular momentum of the quarks is an operator that commutes with all the K operators of the $\mathscr{L}_8$ algebra [it is a scalar of SU(3)]; it thus has a unique value in each super multiplet.

[23] In a similar way, horizontal segments of SU(3) diagrams yield SU(2) diagrams.

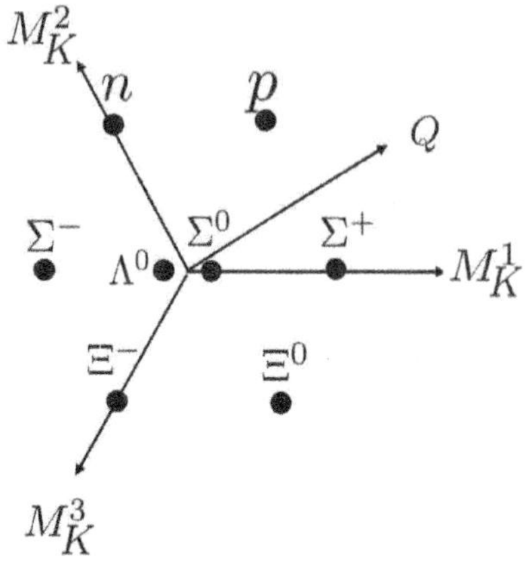

Figure 17: Octet of spin 1/2 baryons with, on the first line, the most usual particles, neutron and proton. The particles on the same horizontal line have the same strangeness quantum number: $S = 0$ for the upper line (neutron and proton), $S = -1$ for the middle line (three Σ particles and one Λ), and $S = -2$ for the bottom line (two Ξ particles). The particles with the same charge are aligned on the same diagonal. All the particles are composed of u, d and s quarks.

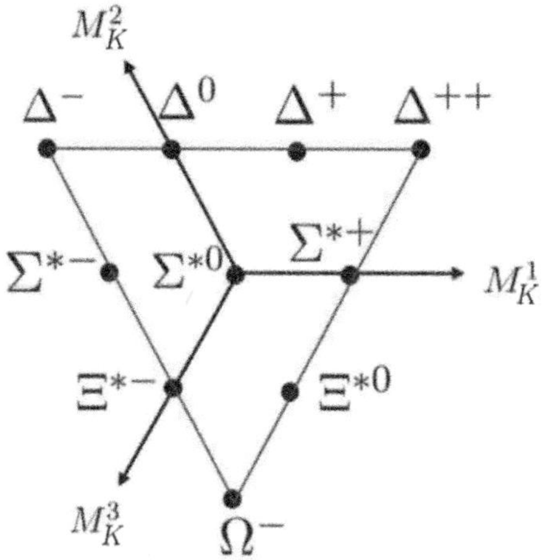

Figure 18: Decuplet of spin 3/2 baryons. The strangeness and charge axes are the same as in figure 17, but the lowest point reaches a strangeness value of $S = -3$ (Ω^- particle). The components of this decuplet are, here also, the u, d, and s quarks.

ferent flavors: u (up), d (down), c (charm), s (strange), t (top), b (bottom), which leads to a SU(6) symmetry[24].

(ii) One should keep in mind that the symmetries we discuss are only an approximation: only the "strongest part" of the strong interactions is invariant under the permutation of quarks of different flavors, meaning that the SU(3) symmetry is only an approximation[25]. In a more refined theory, one introduces a correction that breaks the SU(3) symmetry, and which can be treated by a first-order perturbation theory. This is how Gell-Mann and Okubo derived their mass formula, which is verified with a remarkable precision by experimental results.

(iii) We saw that, in (IX-96), representation $\{10\}$ comes from completely symmetric combinations of the indices associated with the initial representations, i.e. symmetric with respect to the exchange of the quarks internal states. The Pauli principle then requires that the orbital state of the three identical quarks be totally antisymmetric. This is highly unlikely for a set of particles in their ground state: a totally antisymmetric wave function exhibits several "nodes" (more precisely hypersurfaces) where it goes to zero, which implies a high kinetic energy of the particles. This difficulty led to the introduction of an additional quantum number for the quarks, a number that takes on three distinct values, hence defining their "*color*" (this operation multiplies by 3 the total number of distinct quarks).

This leads to the introduction of an "SU(3) color" group, not to be confused with the "SU(3) flavor" group we having been talking about until now. The 3 basis vectors of the $\{3\}$ representation of this color group are associated with 3 quarks of different colors (but of the same flavor). The 3 basis vectors of $\{3\}^*$ are associated with the corresponding antiquarks. The color internal quantum number, often called "color charge", plays a role different from the flavor number, since each observed particle must contain all three colors in equal quantity (it must be a color singlet). This property is linked to the quarks confinement: when trying to separate a quark from the hadron that contains it, the strong coupling constant increases with the separation distance, and more and more energy is needed; at a certain point, this energy is enough to create a quark–antiquark pair, and the initial hadron is reconstituted while a pair of quark–antiquark with a zero color quantum number moves away.

[24]The SU(n) symmetry between n equivalent quarks does not cover all the possible invariances of the theory. We have, for example, a rotation invariance, which involves a symmetry of the SU(2) type. The symmetry group is then the tensor product SU(2) $\otimes$ SU(n), which is simply a subgoup of SU($n+2$).

[25]Nevertheless, the isospin symmetry is considered accurate for strong interactions.

As it name suggests, chromodynamics describes the interactions between color charges, the different colors playing equivalent roles. This is a "gauge theory" presenting some analogies with quantum electrodynamics. In this latter case though, the gauge group is $U(1)$, a very simple group with a single generator, hence commutative. The corresponding gauge field is the electromagnetic field, quantized into photons. In the case of quantum chromodynamics, the color group SU(3) has 8 generators that do not all commute with each other; it is a non-abelian gauge theory. One introduces 8 distinct fields that serve as intermediaries for the interactions of different color quarks (as an example, these interactions are responsible for the bonding of 3 quarks in a proton). With these 8 fields are associated 8 particles, called "gluons".

Exercises

(i) A subgroup $R_{(3)}$ of SU(3) has been introduced in § C-2-b. Compute the values of J associated with the irreducible representations $\{6\}$, $\{8\}$ and $\{10\}$; are these representations irreducible with respect to $R_{(3)}$?

(ii) Redo the calculations of § B-4 (isospin symmetry) in the case of three particles and the SU(3) symmetry. Assume that the three particles have a spin 1/2. Discuss the case where their orbital state is described by a wave function totally symmetric in a particle exchange; do we obtain representation $\{10\}$?

Complement A_{IX}

The nature of a particle is equivalent to an internal quantum number

In § A-3 of chapter IX we mentioned, without a precise demonstration, that when a physical system is invariant under the exchange of particles of different natures, these particles can be considered as identical particles whose internal state is characterized by two different values of a certain quantum number. The Hamiltonian does not act on these internal variables. In this complement, we discuss more precisely why it is equivalent to symmetrize (or antisymmetrize) the state vector either separately with respect to as many sets of particles as there are distinct species, or with respect to all the particles considered to be identical but in different internal states.

1. Partial or complete symmetrization, or antisymmetrization, of a state vector

Consider a set of n_α particles of α nature associated with another set of n_β identical β particles, obeying the same statistics (the two sets are either bosons or fermions). The symmetrizers or antisymmetrizers that must be applied are:

$$\begin{Bmatrix} S_{n_\alpha} \\ A_{n_\alpha} \end{Bmatrix} (1, 2, \ldots, n_\alpha) = \frac{1}{n_\alpha!} \sum_\lambda \begin{Bmatrix} 1 \\ \varepsilon_\lambda \end{Bmatrix} P_\lambda (1, 2, \ldots, n_\alpha)$$

$$\begin{Bmatrix} S_{n_\beta} \\ A_{n_\beta} \end{Bmatrix} (n_\alpha + 1, \ldots, n) = \frac{1}{n_\beta!} \sum_\lambda \begin{Bmatrix} 1 \\ \varepsilon_\lambda \end{Bmatrix} P_\lambda (n_\alpha + 1, \ldots, n) \tag{1a}$$

where the upper line in each of the brackets refers to bosons, the bottom one to fermions (ε_λ is the parity of permutation P_λ). For a set of $n = n_\alpha + n_\beta$ identical particles, the symmetrizer or antisymmetrizer to be applied is the single operator:

$$\begin{Bmatrix} S_n \\ A_n \end{Bmatrix} (1, 2, \ldots, n) = \frac{1}{n!} \sum_\lambda \begin{Bmatrix} 1 \\ \varepsilon_\lambda \end{Bmatrix} P_\lambda (1, 2, \ldots, n) \tag{1b}$$

We are going to express this operator as a function of the partial symmetrizer and antisymmetrizer A and B written in (1a) and show that:

$$\begin{aligned} \begin{Bmatrix} S_n \\ A_n \end{Bmatrix} (1, 2, \ldots, n) = \frac{n_\alpha ! n_\beta !}{n!} \Bigg[1 + \eta \sum_{i=1}^{n_\alpha} \sum_{j=n_\alpha+1}^{n_\beta} P_{i,j} \\ + \sum_{i_1<i_2}^{n_\alpha} \sum_{j_1<j_2}^{n_\beta} P_{i_1,j_1} P_{i_2,j_2} + \ldots \\ +(\eta)^q \sum_{i_1<i_2<\ldots<i_q}^{n_\alpha} \sum_{j_1<j_2<\ldots<j_q}^{n_\beta} P_{i_1,j_1} P_{i_2,j_2} \ldots P_{i_q,j_q} + \ldots \Bigg] \\ \times \begin{Bmatrix} S_{n_\alpha} \\ A_{n_\alpha} \end{Bmatrix} (1, \ldots, n_\alpha) \begin{Bmatrix} S_{n_\beta} \\ A_{n_\beta} \end{Bmatrix} (n_\alpha + 1, \ldots, n) \end{aligned} \tag{2}$$

In this relation, $\eta = \pm 1$ depending on whether the particles are bosons or fermions. The $P_{i,j}$ represent transposition operators of particles i (ranking from 1 to n_α) and j (ranking from $n_\alpha + 1$ to n), and the maximum value of q is the smaller of the numbers n_α and n_β (or one of these numbers if they are equal).

To establish (2) we assume, for example, that $n_\alpha \leq n_\beta$. First of all, we note that any permutation of the n particles can be characterized by:

- the precise numbering of the q particles, among those numbered from 1 to n_α ($q \leq n_\alpha$), that are transferred under this permutation to the group of n_β particles. This is a way of specifying the distribution of particles the permutation puts in the two "boxes", the one ranking from 1 to n_α, and the one ranking from $n_\alpha + 1$ to n;
- the two permutations of particles inside each of the two "boxes".

With each permutation of the n particles, we can associate another intermediary permutation, which puts back the $2q$ particles that changed box into their initial reference box. This operation is performed in a unique way by exchanging the first particle transferred to the second box and numbered i_1 ($1 \leq i_1 \leq n_\alpha$), with the particle numbered j_1 ($n_\alpha + 1 \leq j_1 \leq n$) transferred from the second to the first box; i_1 and j_1 are the smallest particle numbers that must be exchanged between the two groups of particles. One then continues the procedure with the second lowest numbers of particles remaining to be exchanged, etc. until q successive transpositions are performed. The intermediary permutation thus obtained is necessarily the product of two independent permutations, one concerning particles from 1 to n_α, the other particles from $n_\alpha + 1$ to n; it is thus a well-defined permutation among those contained in the product of the symmetrizers S_{n_α} and S_{n_β}, or antisymmetrizers A_{n_α} and A_{n_β}, and each term of that product can be obtained.

The product $S_{n_\alpha}S_{n_\beta}$, or $A_{n_\alpha}A_{n_\beta}$, enables us to obtain the intermediary permutation, which leads to the final permutation using the q transpositions we have just defined. Each permutation of the n particles thus appears on the right-hand side of (2), and the construction we have used shows that each permutation of the n particles appears once and only once. The factor $n_\alpha!n_\beta!/n!$ corresponds to the factorials included in the definition of the symmetrizers S or antisymmetrizers A; the sign $(\eta)^q$ corresponds to the parity change imposed by the q transpositions.

2. Correspondence between the states of two physical systems

From now on, to simplify the notations by avoiding the inclusion of occupation numbers, we shall only consider the case of fermions.

2-a. State spaces

Consider two distinct physical systems:

• a system $\mathscr{S}$ of n_α identical fermions α, and n_β identical fermions β. The state space of each particle is noted $\mathscr{E}$, and the system's Hamiltonian is H;

• a system $\mathscr{S}'$ of $n = n_\alpha + n_\beta$ identical fermions. The one-particle state space is now:

$$\mathscr{E}' = \mathscr{E} \otimes \mathscr{E}_I \tag{3}$$

In this tensor product, $\mathscr{E}_I$ is called the "isospin space": it is the state space associated with the so-called isospin internal variable of each particle that describes its nature, α or β. This two-dimensional space has an orthonormal basis of kets noted $|\alpha\rangle$ and $|\beta\rangle$. Naturally, $\mathscr{E}_I$ would be p-dimensional if $\mathscr{S}$ included particles of p different species $\alpha, \beta, \gamma, \ldots$.

If no other spin variable is already included in the space of internal variables, it is possible to identify spin and isospin by considering $|\alpha\rangle$ and $|\beta\rangle$ as the two orthonormal eigenstates $|+\rangle$ and $|-\rangle$ of a spin component, with respective eigenvalues $+\hbar/2$ and $-\hbar/2$. Otherwise, the internal variable we are discussing is, in a way, an "additional spin".

We call H' the Hamiltonian of $\mathscr{S}'$, and assume it does not act in the $\mathscr{E}_I$ space of the isospin internal variables of each particle.

2-b. One-to-one correspondence

We saw in chapter IX that, under these conditions, there is a one-to-one correspondence between the states of $\mathscr{S}$ and those of $\mathscr{S}'$. This one-to-one correspondence simply comes from the fact that the occupation numbers of the α particles in system $\mathscr{S}$ can be attributed for system $\mathscr{S}'$ to particles in the internal state $|\alpha\rangle$, whereas the other n_β occupation numbers can be attributed to particles in the internal state $|\beta\rangle$. The correspondence is written:

$$\begin{array}{l} \mathscr{S} : |\overbrace{\alpha : k_1, \ldots, k_{n_\alpha}}^{n_\alpha}\rangle \otimes |\overbrace{\beta : k_{n_\alpha+1}, \ldots, k_n}^{n_\beta}\rangle \\ \Updownarrow \qquad\qquad\qquad\qquad \Updownarrow \\ \mathscr{S}' : |k_1, \alpha ; \ldots ; k_{n_\alpha}, \alpha ; k_{n_\alpha+1}, \beta ; \ldots ; k_n, \beta\rangle \end{array} \tag{4}$$

This latter ket of $\mathscr{S}'$ can also be written:

$$\begin{aligned} &|k_1, \alpha ; k_2, \alpha ; \ldots ; k_{n_\alpha}, \alpha ; k_{n_\alpha+1}, \beta ; \ldots ; k_n, \beta\rangle \\ &= \sqrt{n!}\, A(1, \ldots, n) \\ &\quad |1 : k_1, \alpha ; 2 : k_2, \alpha ; \ldots ; n_\alpha : k_{n_\alpha}, \alpha ; n_{\alpha+1} : k_{n_\alpha+1}, \beta ; \ldots ; n : k_n, \beta\rangle \end{aligned} \tag{5}$$

We are going to show that the matrix elements of H between kets of $\mathscr{S}$ are equal to those of H' between the kets of $\mathscr{S}'$ associated via the correspondence (4).

2-c. Equality of the matrix elements

We use the symmetry properties of the Hamiltonian H, as introduced in § A-1 of chapter IX, and in particular the invariance of this Hamiltonian, acting on numbered particles, when the nature of these particles changes. We thus come back to kets including numbered particles. The matrix element of H' between antisymmmetrized kets (5) is the same as the matrix element, between numbered particles, of the operator:

$$n!\, A(1, \ldots, n)\, H'\, A(1, \ldots, n) = n!\, H'\, A(1, \ldots, n) \tag{6}$$

We insert relation (2) in this equality. We see that, in the bracket of (2), which contains the transposition products, all the terms, except the first one equal to 1, yield zero. This is because the transpositions all change the isospin state of one (or several) particles, whereas we assumed that H' does not have any matrix elements modifying the isospin state of any particle. Noting $|\Psi'\rangle$ and $|X'\rangle$ two symmetrized states of $\mathscr{S}'$ like the one written on the left-hand side of (5), the corresponding matrix element of H' is written:

$$\langle X'|H'|\Psi'\rangle = n_\alpha! n_\beta!\, \langle\Theta'|H' A_{n_\alpha} A_{n_\beta}|\Xi'\rangle \tag{7}$$

where $|\Theta'\rangle$ and $|\Xi'\rangle$ are kets of numbered particles. The matrix elements of H' between kets of numbered particles of $\mathscr{S}'$ on the right-hand side of this equality are equal, by hypothesis, to the matrix elements of H between numbered particles of $\mathscr{S}$. We carry out the proper substitution and reintroduce antisymmetrizers on each side of H, which enables us to recover the symmetrized kets of the first line of (4); the coefficient $n_\alpha!n_\beta!$ is necessary for reconstructing the normalization factor for each of the two subsystems, *cf.* (5). We obtain the matrix elements of H between symmetrized and normalized states of $\mathscr{S}$. This establishes the equality between matrix elements of H and H' in the symmetrized bases of $\mathscr{S}$ and $\mathscr{S}'$ written in (4).

3. Physical consequences

From a physical point of view, it is equivalent to speak of particles with different natures, or of identical particles with orthogonal internal quantum states, as long as we only consider operators that do not act on these internal states. It is true that the mathematical structures of the kets in both state spaces are different: in the first case, the antisymmetrization is performed with respect to two distinct subsets of particles, whereas in the second it concerns the ensemble of all the particles. There is, however, a one-to-one correspondence between the state vectors of the two systems, a correspondence that preserves the values of numerous physical quantities (those that do not change the particles nature in the first case, or their isospin internal state in the second).

However, if one considers a state where a particle is described by a linear superposition of different (isospin) internal states (which is perfectly natural for an internal variable), it is associated in the first point of view with a one-particle ket where the nature of that particle is ill-defined. The correspondence between state vectors becomes problematic. It is not essential, however, to give a physical meaning to such a ket, even though it can prove quite useful in the computation.

Complement B_{IX}

Operators changing the symmetry of a state vector by permutation

In § A-2-b of chapter IX, we considered an operation of "resymmetrization" where, once a ket was symmetrized with respect to two groups of n_α and n_β identical particles, its symmetrization was changed as one particle was added to the first group while the second group lost one. We study in this complement the effect of this resymmetrization operation on the state vector. We shall write explicitly the operators Q that allow going from one situation to the other, and we will discuss their properties.

1. Fermions

1-a. Statement of the problem

Starting from a non-symmetrized ket, for example, the product ket written in (IX-6), we can build kets having the proper symmetry by exchanging particles, either when the species of the first n_α particles is α (and the species of the next n_β particles is β), or when n_α is increased by one unit (and n_β is decreased by the same amount). When the first n_α particles are fermions of α nature and the next n_β particles are fermions of β nature, we must use the antisymmetrizers:

$$A\,(1,2,\,\ldots,\,n_\alpha) = \frac{1}{n_\alpha!}\sum_\lambda \varepsilon_\lambda\, P_\lambda\,(1,2,\,\ldots,\,n_\alpha) \tag{1}$$

$$A\,(n_\alpha+1,\,\ldots,\,n) = \frac{1}{n_\beta!}\sum_\lambda \varepsilon_\lambda\, P_\lambda\,(n_\alpha+1,\,\ldots,\,n) \tag{2}$$

where ε_λ is the parity of the permutation P_λ; the summations over λ include all the possible permutations of the particles under study (particles numbered from 1 to n_α in the first case, from $n_\alpha+1$ to n in the second). Using an orthonormal basis of individual states noted $|k\rangle$, the antisymmetrizers yield the kets:

$$\overbrace{|\,\alpha : k_1,\, k_2,\, \ldots,\, k_{n_\alpha}}^{n_\alpha}\ ;\ \overbrace{\beta : k_{n_\alpha+1},\, \ldots,\, k_n\rangle}^{n_\beta}$$

$$= \sqrt{n_\alpha!\, n_\beta!}\ A\,(1,\, \ldots\, n_\alpha)\ A\,(n_\alpha+1,\, \ldots,\, n)$$

$$|1 : k_1\,;\, 2 : k_2\,;\, \ldots\,;\, n_\alpha : k_{n_\alpha}\,;\, n_\alpha+1 : k_{n_\alpha+1};\ldots\,;\, n : k_n\rangle \tag{3}$$

We get an orthonormal basis of the state space of a system of n_α particles α and n_β particles β by taking the whole set of kets (3), for any values of the k indices obeying:

$$k_1 < k_2 < \ldots < k_{n_\alpha} \quad \text{and} \quad k_{n_\alpha+1} < \ldots < k_n \tag{4}$$

(the same individual state can never appear twice, either for the α particles or for the β particles, but the same individual state can be occupied by an α particle and a β particle.)

For a physical system of $n_\alpha+1$ identical α particles and $n_\beta-1$ identical β particles, we must use the antisymmetrizers:

$$\begin{aligned} A\,(1,2,\, \ldots, n_\alpha,\, n_\alpha+1) &= \frac{1}{(n_\alpha+1)!}\sum_\lambda \varepsilon_\lambda\, P_\lambda\,(1,2,\, \ldots,\, n_\alpha, n_\alpha+1) \\ A\,(n_\alpha+2,\, \ldots,\, n) &= \frac{1}{(n_\beta-1)!}\sum_\lambda \varepsilon_\lambda\, P_\lambda\,(n_\alpha+2,\, \ldots,\, n) \end{aligned} \tag{5}$$

which are analog to the ones written in (3), but with a different regrouping of particles.

1-b. Symmetry change of a ket

What happens if we apply operators A written in (5) to kets (3), already antisymmetrized in a different way? We then get the product of operators:

$$\begin{aligned} & A\,(1,\, \ldots,\, n_\alpha+1)\ A\,(n_\alpha+2,\, \ldots,\, n)\ A\,(1, \ldots,\, n_\alpha)\ A\,(n_\alpha+1,\, \ldots,\, n) \\ &= A\,(1,\, \ldots,, n_\alpha+1)\ A\,(1,\, \ldots,\, n_\alpha)\ A\,(n_\alpha+2,\, \ldots,\, n)\ A\,(n_\alpha+1,\, \ldots,\, n) \\ &= A\,(1,\, \ldots,\, n_\alpha+1)\ A\,(n_\alpha+1,\, \ldots,\, n) \end{aligned} \tag{6}$$

where the action of the two successive antisymmetrizers is cumulative on particle $n_\alpha+1$; we have used the fact that the product of the antisymmetrizer of the $n+1$ first particles by the antisymmetrizer of the n first particles reduces to the first antisymmetrizer.

Demonstration: The antisymmetrizers A are projectors. Acting in the space of $n_\alpha+1$ discernible particles, the antisymmetrizer $A(1, \ldots, n_\alpha+1)$ projects onto a subspace $\mathscr{E}^A_{n_\alpha+1}$ totally antisymmetric with respect to the $n_\alpha+1$ particles, hence in particular to the first n_α. This subspace is also a subspace of a larger subspace: the $\mathscr{E}^A_{n_\alpha} \otimes \mathscr{E}(n_\alpha+1)$ subspace onto which $A(1, \ldots, n_\alpha)$ projects, for which no antisymmetry constraint involves the last particle. Consequently, the product of projectors onto nested subspaces reduces to the single projector onto the smallest subspace.

We then use relations:

$$\begin{aligned} A(1, \ldots, n_\alpha+1) &= \frac{1}{n_\alpha+1}\left[1-\sum_{j=1}^{n_\alpha} P_{j,n_{\alpha+1}}\right] A(1, \ldots, n_\alpha) \\ &= \frac{1}{n_\alpha+1} A(1, \ldots, n_\alpha)\left[1-\sum_{j=1}^{n_\alpha} P_{j,n_{\alpha+1}}\right] \end{aligned} \tag{7}$$

where the $P_{j,n_{\alpha+1}}$ are all the transpositions of particle $(n_\alpha+1)$ with the n_α previous ones.

Demonstration: The first equality is a particular case of relation (2) of complement A_{IX}: when $n_\beta = 1$, we have $A_{n_\beta} = 1$, and the summations in the brackets reduce to one. The second equality comes from the first one by noticing that each permutation included in $A(1, \ldots, n_\alpha)$ commutes with the sum over n_α of the permutations $P_{j,n_{\alpha+1}}$, since this latter is symmetric with respect to the n_α indices j. The order of the bracket and the operator A can thus be reversed without changing the product.

Inserting the first equality (7) on the right-hand side of (6) yields:

$$\begin{aligned} &A(1, \ldots, n_\alpha+1)\; A(n_\alpha+1, \ldots, n) \\ &\quad = \frac{1}{n_\alpha+1}\left[1-\sum_{j=1}^{n_\alpha} P_{j,n_\alpha+1}\right] A(1, \ldots, n_\alpha)\; A(n_\alpha+1 \ldots, n) \end{aligned} \tag{8}$$

Consequently, the operator $Q_A\,(n_\alpha \to n_\alpha+1,\, n_\beta \to n_\beta-1)$ that changes the symmetry of the kets is simply written:

$$Q_A\,(n_\alpha \to n_\alpha+1,\, n_\beta \to n_\beta-1) = \frac{1}{n_\alpha+1}\left[1-\sum_{j=1}^{n_\alpha} P_{j,n_\alpha+1}\right] \tag{9}$$

A similar reasoning yields:

$$Q_A\,(n_\alpha \to n_\alpha-1,\, n_\beta \to n_\beta+1) = \frac{1}{n_\beta+1}\left[1-\sum_{j=n_\alpha+1}^{n} P_{n_\alpha,j}\right] \tag{10}$$

Comments:

(i) Applying these symmetry change operators to a normalized ket (3) will generally change its norm. As an example, if all the individual states of the β particles are already occupied by the α particles in the initial ket, the Pauli principle forbids the transfer of the state of a β particle to an α particle: the action of operator Q_A $(n_\alpha \to n_\alpha + 1,\ n_\beta \to n_\beta - 1)$ is to cancel that ket. Conversely, if all the β particles already occupy all the individual states of the α particles, the action of operator Q_A $(n_\alpha \to n_\alpha - 1,\ n_\beta \to n_\beta + 1)$ is also to cancel the initial ket.

(ii) Starting from the second equality (7) instead of the first one, and using it on the antisymmetrizer $A(n_\alpha + 1, \ldots, n)$, we get a relation symmetric to (8):

$$A\,(1, \ldots, n_\alpha + 1)\ A\,(n_\alpha + 1, \ldots, n)$$
$$= \frac{1}{n_\beta}\ A\,(1, \ldots, n_\alpha + 1)\ A\,(n_\alpha + 2 \ldots, n) \left[1 - \sum_{j=n_\alpha+2}^{n} P_{n_\alpha+1,j}\right] \tag{11}$$

This equality can be used to normalize the ket obtained after the symmetrization change.

2. Bosons

If the particles are bosons instead of fermions, a few changes must be made in the previous arguments. Antisymmetrizers A must be replaced by symmetrizers S, which amounts to replacing ε_λ by 1 in equations (1), (2) and the following ones; this eliminates the minus sign in front of the summations over j in relation (7) and the following ones. As an example, (8) becomes:

$$S\,(1, \ldots, n_\alpha + 1)\ S\,(n_\alpha + 1, \ldots, n)$$
$$= \frac{1}{n_\alpha + 1} \left[1 + \sum_{j=1}^{n_\alpha} P_{j,n_\alpha+1}\right] S\,(1, \ldots, n_\alpha)\ S\,(n_\alpha + 1 \ldots, n) \tag{12}$$

As the occupation numbers of the individual states can be larger than 1, the final expression of the normalized ket obtained after a symmetrization change is somewhat more complex. As an exercise, the reader may want to establish it.

Chapter X

Symmetry breaking

This book has focused on the many results that can be deduced from the knowledge that a physical system obeys a certain symmetry. However, when a symmetry no longer applies, very interesting phenomena called "symmetry breaking" may occur. First of all, one can act on a physical system by adding a small perturbation having a symmetry lower than the one of its initial dynamics; this leads to a so-called induced symmetry breaking. As an example, placing an atom in a magnetic field can lift the degeneracy of its energy levels. Another possibility can occur in the absence of any perturbation: a spontaneous phenomenon may lead to a state of the system with a symmetry lower than the one of its dynamics, yielding a so-called spontaneous symmetry breaking. The concept of symmetry breaking has many applications, in condensed matter physics as well as in field theory or particle physics, and it is the subject of many textbooks. Even though this concept is not the subject of this book, we could hardly completely ignore it; we shall simply touch upon that subject, presenting a few simple examples. This chapter will be qualitative, almost without equations, with the results presented in an intuitive way, without demonstrations. For a less elementary discussion of the subject, the reader may consult reference [12].

Introduction to Continuous Symmetries: From Space–Time to Quantum Mechanics, First Edition. Franck Laloë.

Comment: Symmetry breaking is discussed in this chapter from a quantum point of view, but the same concept also exists in classical physics. For example, § 5.2 of reference [5] discusses the spontaneous symmetry breaking occurring when a beam buckles under compression. Similarly, the symmetry breaking of coupled spins that will be discussed in § A-1-b is generally studied by considering the spins as small classical magnetic moments. This has the advantage of simplifying the computations without changing the results in the neighborhood of the critical point.

A. Magnetism, breaking of rotational symmetry

Consider a spherical sample of a magnetic material, modeled by a sphere filled with N spins arranged along a regular network. Each spin has a magnetic dipole, and the total magnetization of the sample is the sum of all the individual magnetic dipoles. Our simplified model only takes into account the spin variables, assuming that all the spins have a quantum number $s = 1/2$, and considers that the position variables are fixed on the regular network. We shall consider the case where the sample is successively in a zero, then non-zero, magnetic field. We shall also discuss the case where the spins are either independent, or coupled with each other.

A-1. Determination of the equilibrium state

The thermal equilibrium at temperature T is obtained by minimizing the free energy thermodynamic potential, $F = U - TS$, where U is the energy and S the Boltzmann entropy (both purely spin related). The density operator of the system at equilibrium is proportional to the exponential e^{-H/k_BT}, where H is the Hamiltonian and k_B the Boltzmann constant.

A-1-a. Independent spins

When the spins are independent, the Hamiltonian H is the sum of the individual spin Hamiltonians h_i, and the thermal equilibrium is particularly simple: it is the juxtaposition of N independent two level systems at thermal equilibrium.

In a zero magnetic field, the two spin states $m_S = \pm 1/2$ have the same energy, and hence the same Boltzmann exponential factor: nothing favors one state over the other (one can choose $H = 0$). The only term left is the entropy, which is maximum when the two states are equally populated. The left-hand side of figure 1 shows the variations of the free energy F (proportional here to the opposite of the entropy) of a spin as a function of its polarization degree m_i; the populations of the two spin levels are $(1 \pm m_i)/2$. The function F is minimal when $m_i = 0$, and maximal (actually, zero) when $m_i = \pm 1$. The equilibrium is obtained when $m_i = 0$, i.e. when the two populations are equal. In this case,

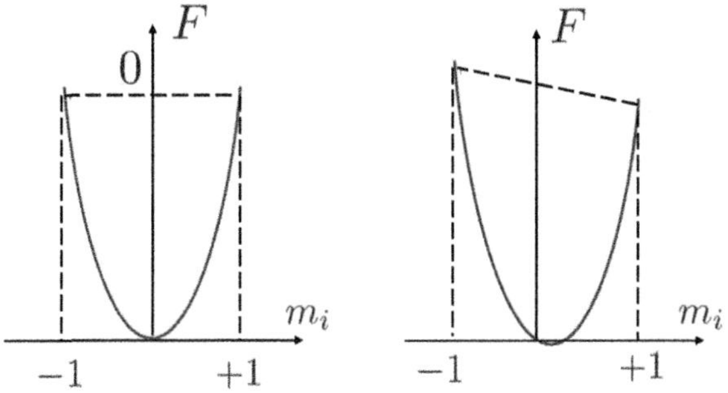

Figure 1: Graph of the free energy $F = U - TS$ of a spin 1/2 as a function of its polarization m_i. The left-hand side of the picture corresponds to the case of a zero magnetic field (and hence $U = 0$): the minimum of the thermodynamic potential $-TS$ occurs for a zero polarization (maximum of the entropy). On the right-hand side, the non-zero magnetic field B is coupled with the spin magnetic moment, which adds to the energy a contribution linear in m_i. The minimum of the thermodynamic potential F is shifted towards a positive value of m_i, proportional to the field B when it is weak.

neither an individual spin nor the total sample have any magnetization, as is to be expected for symmetry reasons.

In a non-zero magnetic field, supposed to be homogeneous over the sample, we must add the contribution of the coupling energy of the spin magnetic moment with the external applied field. This contribution equals $-\mu_S B m_i$, where $\mu_S \boldsymbol{S}/\hbar$ is the magnetic moment of the spin $\boldsymbol{S}$, and B the value of the magnetic field (whose direction determines the quantization axis). The right-hand side of figure 1 shows the function F to be minimized, which is different from the function on the left since a linear function in m_i (a straight line with a slope proportional to B) was added to it. The minimum of this function, which determines the thermal equilibrium, is shifted and each spin has a non-zero polarization. For a weak magnetic field, this polarization is proportional to B, and the proportionality coefficient $\mu_S m_i / B$ is called the spin magnetic susceptibility.

The total magnetization $\mu_S M$ of the sample is obtained by summing the magnetization $\mu_S m_i$ of all the individual spins. One can reinterpret the curves in figure 1 by plotting M instead of m_i on the abscissa axis; the left-hand side of the figure shows the free energy of an ensemble of independent spins when the sum of the m_i is kept constant; the right-hand side shows the variation of F as a function of M in a non-zero magnetic field. M actually characterizes the order

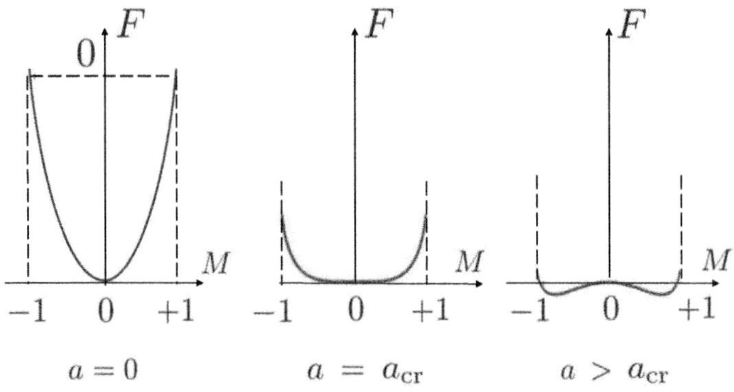

Figure 2: Free energy $F = U - TS$ of a sample, composed of a large number N of spins in a zero magnetic field B, plotted as a function of its magnetization M. The left graph shows the values of F when the spin coupling constant a is zero. In the central graph, a takes the critical value a_{cr} for which the variation of F around $M = 0$ is of the fourth order. The graph on the right shows the variations of F when $a > a_{\mathrm{cr}}$: there are now two minima for non-zero values of M. Even in the absence of an external field, the sample tends to choose one of these two values, exhibiting a symmetry breaking.

of the system: when $M = 1$, the spins are all parallel to the same direction, whereas when $M = 0$ the spins are totally disoriented in the sample; M is called the "order parameter" of the system.

A-1-b. Coupled spins

Imagine now that the nearest neighbor spins are coupled by an interaction of the form $h_{ij} = -a\mathbf{S}_i \cdot \mathbf{S}_j$, where a is a positive coupling constant and where $\mathbf{S}_i$ and $\mathbf{S}_j$ are the two interacting spins. The coupling energy between these two spins is minimal when they are parallel to each other. For the entire sample, we must add to the Hamiltonian H the sum of all the h_{ij} over all the couples of neighboring spins i and j. The result of this coupling is to decrease the energy of states where all the spins are parallel to each other, and increase the energy of states where many pairs have antiparallel spins. It is understandable that such a coupling explains the appearance of ferromagnetism where all the spins point in the same direction. The modulus of the sample magnetization is then at its maximum value.

As before, we first start with the case where the magnetic field B is zero. The equilibrium will be a balance between the average energy U that tends to align the spins in the same direction, and the entropy S that favors a maximal spin disorder. Figure 2 shows the variations as a function of M of the thermodynamic potential $F = U - TS$ of the sample[1]. The graph on the left corresponds to the case where $a = 0$, so that only the entropy plays a role; the minimum of F occurs when $M = 0$. The middle graph shows the variations of F for the so-called "critical" value of $a = a_{\text{cr}}$, where the convex curvature of the graph representing U is exactly balanced by the concave curvature of the curve representing $-TS$. In the vicinity of $M = 0$, the variations of F are now of the fourth order in M, and the "restoring force" that brings M back to its zero value is much weaker. Finally, the graph on the right shows the variations of F for a larger value of the coupling constant a. There are no longer one, but two (opposite) values of the magnetization that minimize the thermodynamic potential, and are hence possible values of M at the equilibrium. The order parameter M spontaneously tends towards one or the other non-zero values, as shown on the figure. At equilibrium, the sample spontaneously exhibits a non-zero magnetization, even in the absence of any external magnetic field, which is the characteristics of a ferromagnetic material. This is an example where equations having a certain symmetry (symmetry with respect to the origin $M = 0$ in this case) yield equilibrium solutions having a lower symmetry. It is a case of spontaneous symmetry breaking.

We now discuss the case of a non-zero magnetic field B. We must add to F the term $-\mu_S MB$, linear in B, and the three graphs of figure 2 become the graphs of figure 3. The graph on the left shows that the minimum of F has been shifted toward a positive value, proportional to B for a weak enough field. The equilibrium value M_{eq} of M is written $M_{\text{eq}} = \chi B$, where χ is the linear magnetic susceptibility of the system. On the central graph, the shift of the minimum is much larger, which is normal since no linear "restoring force" is present to compensate the effect of the magnetic field. In this case, the value of M_{eq} is no longer proportional to the applied field B, and the susceptibility χ is infinite. Finally, the two minima on the graph on the right are slightly shifted proportionally to B, but as M is not linear in B the susceptibility χ can no longer be defined.

A-2. Spontaneous symmetry breaking

In three dimensions, rotational invariance implies that F depends only on M, and not on the space direction of the vector $\mathbf{M}$; this justifies the simple one-dimensional reasoning we have followed until now, and simplifies the graphic representations. It does, however, mask the true nature of rotational invariance,

[1]These curves can be obtained, for example, by a quantum variational computation, where a given total value of the magnetization is imposed by the choice of the variational density operator. However, these curves do not depend on the quantum nature of the computations, and remain the same if the computation is performed in the framework of classical statistical mechanics.

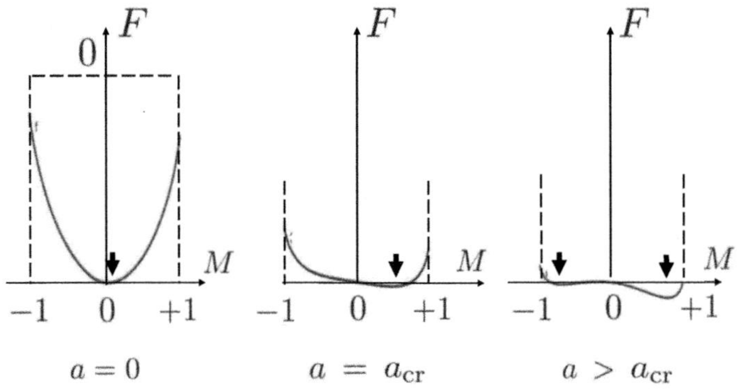

Figure 3: These graphs are similar to those in figure 2, but the magnetic field B is no longer zero. The values of M for which F is minimal are indicated with thick arrows pointing down. On the graph on the left, this value of M is proportional to B, hence small if B is weak. On the middle graph, the constant a takes its critical value, the value of M is much larger, showing the strong response of the system at its critical point (infinite susceptibility). Finally, the graph on the right exhibits symmetry breaking, and the values of M are close to their values in a zero field.

leaving only its symmetry with respect to the origin. We thus come back to the three-dimensional problem of modeling a ferromagnetic object, and discuss the consequences of the true rotational invariance, or SO(3) invariance. We assume that the magnetic field is zero. Since the Hamiltonian H is invariant under rotation, so is the density operator at equilibrium $\rho_{\text{eq}} = Z^{-1}\exp(-H/k_BT)$ (Z is the partition function that insures normalization). We are going to examine the impact of this invariance in the analysis of § A-1-b, in a 3-dimensional framework.

A-2-a. Comparison between density operators, thermodynamic limit

As long as a remains below its critical value a_{cr}, nothing particular happens. The equilibrium value of the order parameter M is zero, which is compatible with the rotational invariance of the density operator at equilibrium ρ_{eq}. On the other hand, if a gets larger than the critical value, it becomes impossible to attribute a non-zero value to M without breaking the rotational invariance, since a particular direction must be chosen for the vector $\mathbf{M}$. How should we solve this contradiction with the rotational invariance of ρ_{eq}?

This contradiction is actually only apparent since it is quite possible to build an equilibrium density operator that is invariant under rotation, even in the presence of this symmetry breaking. To show this, we consider operators

ρ_{M} for which M has a non-zero value, meaning that $\mathbf{M}$ necessarily points in a given direction that we can vary. We then superpose such operators with the same value of M, but different directions of $\mathbf{M}$, so that the sum over all the vectors $\mathbf{M}$ yields zero. This reestablishes the rotational invariance of the matrix ρ_{eq}. Now, aside from the rotational invariance, which is broken in each ρ_{M}, their properties are very close to those of ρ_{eq} in many ways , and they are easier to handle. Concerning the energy U, there is no difference between one of the ρ_{M} and ρ_{eq}, since this latter operator is a superposition of density operators that all have exactly the same energy. The difference only shows up at the entropy level: when taking the average over all the space directions, one increases the domain of the states covered by the distribution, which increases the entropy. This gain is however limited: the average still preserves all the spin correlation properties, and only one variable is averaged, the direction of the total magnetization. In other words, in a system with N degrees of freedom, only one of them varies, so that the relative entropy gain remains very small if N is large, varying as $1/N$ for example. Using the ρ_{M}, one can easily restore the rotational symmetry, gaining that way very little entropy. Furthermore, in the thermodynamic limit where N goes to infinity (keeping the spin spatial density constant), the gain goes to zero: the states with broken symmetry yield the exact minimum for the value of the thermodynamic potential at equilibrium.

Comment:

In the presence of a spontaneous symmetry breaking, this thermodynamic limit is singular, as we now show. Consider, for example, a spin system at the critical point placed in a very low magnetic field. We reduce this field to zero to observe only the spontaneous magnetization, and we also let N go to infinity to get the thermodynamic limit. The result of these operations crucially depends on the order in which these two limits are taken. If we first reduce the magnetic field to zero, N remaining fixed at an eventually very large value, the magnetization will always go to zero even at a residual value of the field (the larger N, the smaller the residual value of the field). But if we first let N go to infinity, in the presence of a constant small magnetic field, the magnetization will remain even after the small field is reduced to zero. This means that the magnetic susceptibility χ of the spin system has become infinite, since M no longer goes to zero when the magnetic field is zero. This shows that the susceptibility takes different values depending on the way the limit is obtained, reflecting the existence of a singular limit.

In conclusion, even though ρ_{eq} is the density operator obtained by the strict application of the statistical mechanics postulates, its corresponding free energy is barely different from that of each $\rho_{\rm M}$, or of any superposition of the $\rho_{\rm M}$ yielding an arbitrary direction for the magnetization. From a mathematical point of view, taking $\rho_{\rm M}$ as the density operator for minimizing the free energy introduces a negligible error when N is large, an error that goes to zero in the thermodynamic limit where N goes to infinity. From a physical point of view, we are going to show that using $\rho_{\rm M}$ is actually more realistic that using $\rho_{\rm eq}$.

A-2-b. Discussion, pertinence of the symmetry breaking

Actually, not only are the $\rho_{\rm M}$ almost as pertinent as the $\rho_{\rm eq}$ in terms of optimization of the thermodynamic potential, but they also provide a better description of the system when taking into account the real conditions of an experiment. Imagine that we start with a sample for which $a < a_{\rm cr}$ or, what amounts to the same thing, whose temperature T is high enough for F to be dominated by the entropy term. The energy term then plays a negligible role, and we are in the situation depicted by the graphs on the left of figures 2 and 3.

In practice, it is impossible to maintain a sample in a strictly zero magnetic field, but this is of no importance if the field remains very small since in that case the figures on the left remain almost the same. If one increases a (or if one lowers the temperature), one necessarily reaches the point where $a = a_{\rm cr}$ where, as we just saw, the sample's magnetic susceptibility becomes infinite. Any residual magnetic field is enough to polarize the sample in its direction, hence breaking the rotational invariance, whatever precautions are taken in the experiment. Continuing the experiment beyond the critical point, the sample exhibits a symmetry breaking, which seems spontaneous even though it has been induced by a very small residual field. Using $\rho_{\rm M}$ is then much closer to reality than using $\rho_{\rm eq}$, which would correspond to the average of observations performed on a sample cooled down in residual fields having different directions.

Note that going through the critical point not only results in an extreme sensitivity to external perturbations, consequence of the divergence of the magnetic susceptibility: the lack of a linear "restoring force" we mentioned above is also responsible for large fluctuations of M, the so-called "critical fluctuations".

We have shown that $\rho_{\rm eq}$ is actually more like a mathematical fiction than a physical density operator: in a real experiment, it is perfectly legitimate to break the symmetry and use $\rho_{\rm M}$. Another practical reason that limits the interest of $\rho_{\rm eq}$ is that, even if we were able to cool down a sample in a strictly zero magnetic field, we would not necessarily obtain the perfectly isotropic density operator. This is because any ferromagnetic material tends to form domains, which are multiple small regions where the magnetization is locally uniform but takes different directions from one region to another. A total magnetization of the sample equal to zero would not necessarily mean that it is well described by $\rho_{\rm eq}$; each domain would actually be well described by its own $\rho_{\rm M}$.

B. A few other examples

We now briefly discuss a few other examples of the symmetry breaking concept.

B-1. Crystallization, translational symmetry breaking

Consider a fluid in equilibrium at a temperature T. We assume it is contained in a cubic box of side L and, to avoid any boundary effects, we use periodic boundary conditions (Born–von Karman conditions). The system is invariant under any spatial translation. Imagine we cool down this fluid so that it solidifies into a crystalline network where each atom or molecule occupies a given position in space. The value of the thermodynamic potential does not depend on the absolute positions but rather on the relative positions of the particles on a regular network: any translation can be applied on the crystalline network without changing the potential or its derivative. As in the case of ferromagnetism, where the spins' relative orientation was important but not their absolute direction, we get an infinite set of system states having all the same relative positions and minimizing the thermodynamic potential. These states are deduced from each other by a translation, but none of them is invariant under a translation (except in the particular case where the whole network is invariant under that particular translation). Here again, the solutions of the equations have a lower symmetry than the equations themselves, and in this case, the translational invariance has been spontaneously broken.

As before, the translational symmetry can be restored by "delocalizing the network", without changing the relative position of the particles. This can be achieved by introducing a linear superposition with equal weights of all the density operators of all the states having a broken symmetry, which will restore the initial symmetry. This operation slightly increases the entropy by the amount associated with the delocalization of a single degree of freedom of the system (the center of mass), which is a negligible gain for a very large number of particles. It is thus more convenient to accept this symmetry breaking and work with the simpler localized states, which have more intuitive physical properties.

B-2. Bose–Einstein condensation and superfluidity, breaking of particle number conservation

Another interesting example of the concept of symmetry breaking is given by Bose–Einstein condensation in a boson gas, and its relation with the emergence of superfluidity.

B-2-a. Simple description of a Bose–Einstein condensate

Quantum statistical mechanics shows that when a gas composed of a macroscopic number of free bosons is at thermal equilibrium, two cases are possible.

Above a certain so-called critical temperature, the fraction of bosons occupying each individual quantum state is infinitesimal: it even goes to zero in the thermodynamic limit (macroscopic system). Below the critical temperature, a finite fraction of the bosons accumulates in the individual quantum sate of lowest energy. The system consists partly of this "condensate", and partly of "excited particles" that populate all the other individual states; one must keep in mind, however, that no spatial separation occurs between the condensate and the excited particles, as the two components generally occupy the same volume. When interactions between bosons come into play, and as long as the gas remains dilute, the previous properties are barely modified; actually, short-range repulsive interactions even tend to stabilize the condensate. In what follows, we shall not go into any detail, and concentrate solely on the condensate without taking into account the excited particles. This is a fairly good approximation when the gas is dilute and at a sufficiently low temperature (compared with the critical temperature).

The simplest description of the condensate is a Fock state $|N : \varphi\rangle$ where N particles occupy the same individual state $|\varphi\rangle$. A completely different state would be a coherent state (*cf.*, for example, complement G_V of [9]), defined by a complex parameter α:

$$|\Psi(\alpha)\rangle_{\varphi} = \mathrm{e}^{-|\alpha|^2/2} \sum_{n=0}^{\infty} \frac{\alpha^n}{\sqrt{n!}} |n : \varphi\rangle \tag{X-1}$$

which is a superposition of Fock states associated with different numbers of particles. As the average number of particles in such a coherent state is $|\alpha|^2$, one can impose the condition $|\alpha|^2 = N$ for the Fock state and the coherent state to correspond to the same average number of particles.

In quantum mechanics, there exists an incertitude relation between the phase and the particle number: the better N is defined, the less the phase is, and vice versa. In a Fock state, the particle number is perfectly defined, but the phase is totally undefined; in a coherent state, the particle number fluctuates (its root mean square equals $|\alpha|$) but for large values of $|\alpha|$, the phase is well defined around a value θ which is the phase of the complex number α. In quantum electrodynamics, coherent states are commonly used to model an electromagnetic field with quasi-classical properties.

B-2-b. Two possible descriptions of interference between two condensates

The question is what happens when two condensates overlap in space, and one measures the particles' position in the overlapping zone. When these condensates are described by a product of coherent states:

$$|\Phi(\alpha_1, \alpha_2)\rangle = |\Psi(\alpha_1)\rangle_{\varphi_1} \otimes |\Psi(\alpha_2)\rangle_{\varphi_2} \tag{X-2}$$

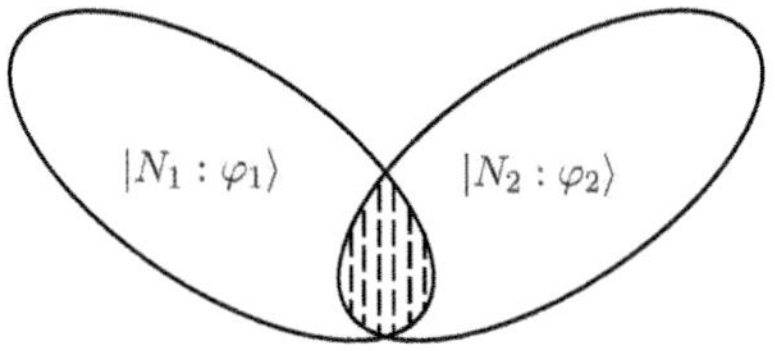

Figure 4: Two condensates with N_1 and N_2 particles, occupying the individual states $|\varphi_1\rangle$ and $|\varphi_2\rangle$ respectively, overlap in a region of space. One measures one by one the positions of the particles in that region. These positions regroup to form an interference pattern with bright and dark fringes (dashed lines on the figure), which can be explained by the existence of a relative phase between the two condensates. When the experiment is repeated several times, the position of the fringes is displaced from one experiment to the other, showing that the relative phase of the two condensates has, each time, a different value.

an optical analogy allows predicting the measurement results: when two quasi-classical fields overlap, they yield an interference pattern, and the particles' detection produces dark and bright fringes (figure 4). The position of these fringes depends on the difference between the phases θ_1 and θ_2 associated with the two coherent states.

However, when the two condensates have well-defined particle numbers, it seems more adequate to describe them as a product of Fock states:

$$|\Phi(N_1, N_2)\rangle = |N_1 : \varphi_1\rangle \otimes |N_2 : \varphi_2\rangle \tag{X-3}$$

Will an interference pattern appear when measuring the particles position with such a state?

The density operator associated with the coherent state plays the same role as the opertor ρ_{M} with the symmetry breaking we discussed above; as for the density operator associated with the product of Fock states, it plays the same role as ρ_{eq}, without any symmetry breaking. As it is generally easier to discuss the evolution of states in the framework of symmetry breaking, we start with the predictions associated with the state $|\Phi(\alpha_1, \alpha_2)\rangle$. We should keep in mind, however, that when two condensates with given particle numbers are prepared separately, they "have never seen each other" before we make them overlap; there is no reason for them to have an absolute or relative phase. We thus consider that θ_1 and θ_2 are random variables evenly distributed between 0 and 2π. For a random choice of θ_1 and θ_2, the difference $\theta_1 - \theta_2$ determines the position of the fringes that will be observed. For another choice, fringes will still be observed but at a different position. This is precisely confirmed by experiments [51]. In a first experiment, a fringe pattern is observed; when a second independent experiment is performed, the fringe pattern is still observed but shifted with respect to the previous one. Using states with symmetry breaking, and with random phases whose values change from one experiment to the next, enables us to easily predict the observations.

Several theoretical approaches [52, 53, 54] show that the same predictions can also be obtained using the state without symmetry breaking (X-3). In a given experiment, the first measured position of a particle is totally random, as in the symmetry breaking approach where nothing about the relative phase is initially known. This first measure creates, however, a first constraint on the relative phase, since that phase cannot have values predicting a dark fringe at that point. Another way to understand this improvement of the phase knowledge is to note that this first measure modifies the state vector because of the von Neuman projection postulate. For the second measure, certain values of the position are now privileged: those corresponding to a bright fringe in a system where the first measure also falls into a bright fringe. The second measure is thus partially correlated with the first one, and continues to modify the state vector towards a better knowledge of the relative phase. The process continues with the following measures, and the relative phase becomes better and better defined. In about ten measurements, the state vector is practically equal to a superposition of states $|\Phi(\alpha_1, \alpha_2)\rangle$ where the difference $\alpha_1 - \alpha_2$ has almost a constant value, as in the analysis without symmetry breaking.

This discussion shows the advantage of directly reasoning in terms of symmetry breaking, which is more direct and intuitive. Nevertheless, from a more fundamental point of view, it is more correct not to break a symmetry when no physical process can justify it[2], as opposed to the magnetism case for example. It is also useful to remember that invoking a symmetry breaking is by no means furnishing a physical explanation of the observed phenomenon: it is not essential to understand the fundamental mechanism, and remains at the level of a very useful computational tool.

B-3. Symmetry breaking in field theory

In field theory, the concept of symmetry breaking is even more important as it is no longer a convenient computational tool, but rather part of the foundations

[2]As opposed to the previous cases, no external breaking is actually realistic for the phase of the condensates: we don't know of any physical process able to create a coherent superposition of states where the particle number varies. In particular, coupling the system to a reservoir of particles, as is done to obtain the grand canonical equilibrium, only introduces an incoherent superposition of various values of N, and does not produce any phase.

of quantum theory.

B-3-a. Standard Lagrangian

The density of the standard Lagrangian $\mathcal{L}$ of a (spinless) scalar field ϕ is written:

$$\mathcal{L} = \frac{1}{2}(\partial_\mu \phi)^\dagger (\partial^\mu \phi) - \frac{1}{2} m^2 \; \phi^\dagger \phi \tag{X-4}$$

where ∂_μ and ∂^μ are respectively the covariant and contravariant derivatives with respect to the space–time components, and where the Einstein convention is used (an upper and lower repeated index μ must be summed over the four space–time components). The constant m is the mass associated with the field, which determines its spatial range. In this Lagrangian density, the first term is a kinetic energy term, the second the opposite of a potential energy. In the field's ground state, all its derivatives are zero, and only the energy term remains:

$$\mathcal{U} = \frac{1}{2} m^2 \; \phi^\dagger \phi \tag{X-5}$$

which is minimal when $|\phi| = 0$. The ground state thus corresponds to a uniformly zero field in all space. Note that, for small variations $\delta\phi$ of ϕ around the ground value $\phi = 0$, the potential variations are:

$$\delta\mathcal{U} = m^2 \; |\phi| \; \delta\phi \tag{X-6}$$

The curvature of the potential (X-5) around the equilibrium value yields the mass associated with the field ϕ.

B-3-b. Lagrangian exhibiting a symmetry breaking

We now assume that to the Lagrangian density we add a term proportional to the fourth power of $|\phi|$, and replace (X-4) by:

$$\mathcal{L} = \frac{1}{2}(\partial_\mu \phi)^\dagger (\partial^\mu \phi) - \frac{1}{2} m^2 \; |\phi|^2 - \frac{\lambda}{4} \, |\phi|^4 \tag{X-7}$$

The potential energy now becomes:

$$\mathcal{U} = \frac{1}{2} m^2 \; |\phi|^2 + \frac{\lambda}{4} \, |\phi|^4 \tag{X-8}$$

To keep the same notations, we shall continue calling m^2 the coefficient of the square term, but it will no longer be interpreted as the square of a mass; consequently, this coefficient has no reason to be forcibly positive. If it is indeed positive, the potential energy is minimal when $|\phi| = 0$, as before. However, if it is negative, the potential energy will be minimal when the square of the field modulus takes the value:

$$|\phi|^2 = \frac{|m^2|}{\lambda} \tag{X-9}$$

Figure 5 shows the variations of the potential energy as a function of $|\phi|$. The situation is similar to that on the right-hand side of figure 2, which exhibits a symmetry breaking.

We now have a situation where, in the ground state of the system, the field is no longer zero but takes the value (X-9). Under these conditions, the variations of the potential around its minimum value are given by the curvature of the potential around the minima of figure 5. This mechanism enables the emergence of a new mass; it is called the "Higgs mechanism".

This mechanism can be applied to the more complex $CU(1) \otimes SU(2)$ symmetry of the electroweak interaction, attributing a non-zero mass to the W and Z gauge bosons, whereas the photon remains a massless particle. The Higgs mechanism plays an essential role in field theory and elementary particle theory. For a simple introduction to this subject, see chapter 6 of reference [5]; more details can be found for example in references [11] and [14].

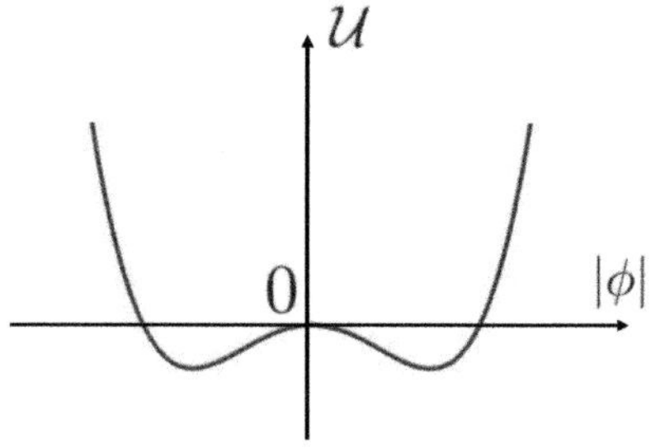

Figure 5: Potential energy $\mathcal{U}$ included in the field Lagrangian, as a function of $|\phi|$ (the phase of ϕ does not play any role in $\mathcal{U}$). Because of the term in $|\phi|^4$, the potential is minimal for a non-zero value of the field $|\phi|$, which leads to a symmetry breaking. If we take into account the phase of the field ϕ, the variable will evolve in a plane, and the surface representing $\mathcal{U}$ will take the form of the so-called "Mexican hat".

Appendix

Time reversal

In complement D_V, we studied space reflections (parity) though the corresponding symmetry is not a continuous symmetry[1]. Our motivation was the natural combination between the displacement group and space parity, which leads to a generalization of the rotation group, transforming SO(3) into $O(3)$. In this appendix, we shall study another non-continuous symmetry, time reversal. It is a discrete operation, which forms, with the identity, a group with two elements for which there exists neither a Lie group nor an associated infinitesimal operator. It is a sort of "time parity", which, in a relativistic framework where time and space are two components of a four-vector, plays a role analog to space parity.

From a quantum mechanics point of view, time reversal presents another interest: whereas all the symmetries we have discussed until now are associated with unitary linear operators acting in the state space, time reversal requires

[1]One could give a continuous character to this symmetry by varying the space point with respect to which the space reflection is performed; this would amount to a simple combination of translation and parity, and would remain somewhat artificial.

Introduction to Continuous Symmetries: From Space–Time to Quantum Mechanics, First Edition. Franck Laloë.

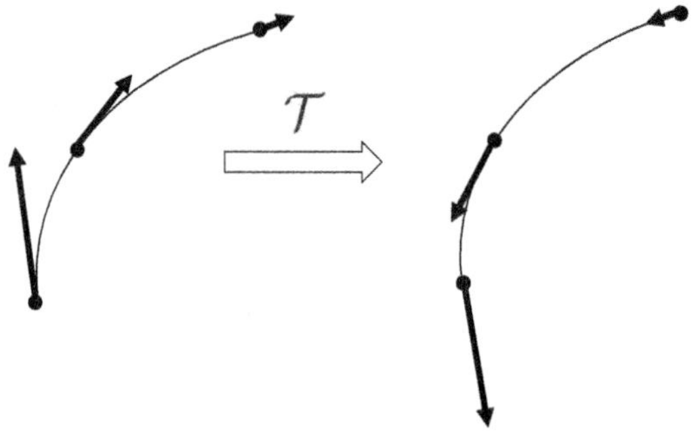

Figure 1: Figure on the left: initial motion (in a case where the particle slows down). Figure on the right: same motion after a time reversal (the particule accelerates).

the introduction of an antilinear, so-called antiunitary, operator. A different formalism must be implemented, which has proved quite useful in many practical problems. That is one more reason to study time reversal.

After a rapid description of time reversal in classical mechanics (§ 1), we shall recall some properties of antilinear operators (§ 2). In § 3, we discuss the conditions a linear or antilinear operator must obey to represent, in quantum mechanics, a time reversal operation. We shall then build explicitly this operator in a few simple cases (§ 4). Finally, § 5 discusses certain important applications as Kramer's theorem and van-Vleck's theorem.

1. Time reversal in classical mechanics

In classsical mechanics, time reversal is the operation $\mathcal{T}$ that associates with any motion of a physical system described by the functions $q_i(t)$, another motion described by the functions $q_i(-t)$. In other words, at any instant t, the values of the generalized coordinates after a time reversal operation are those these same coordinates had at the instant $-t$ in the initial motion:

$$q_i(t) = f_i(t) \stackrel{\mathcal{T}}{\Longrightarrow} q_i(t) = f_i(-t) \tag{1}$$

It is clear that the derivatives $\dot{q}_i$ are transformed according to:

$$\dot{q}_i(t) = \frac{\mathrm{d}}{\mathrm{d}t} f_i(t) \stackrel{\mathcal{T}}{\Longrightarrow} \dot{q}_i(t) = -\frac{\mathrm{d}}{\mathrm{d}t} f_i(-t) \tag{2}$$

To help give an image of time reversal, one sometimes imagines filming the initial motion and projecting the movie backward (starting with the end of the movie); this yields an image of the motion after the action of $\mathcal{T}$ (figure 1).

According to the general definition given in chapter I, operation $\mathcal{T}$ is said to be a symmetry operation if, after the action of $\mathcal{T}$ on any possible motion, there corresponds another possible motion obeying the same evolution laws for the physical system. The diagram shown in figure 1 of chapter I becomes that of figure 2.

Example: Consider a particle with position $\boldsymbol{r}$, subjected to a force $\boldsymbol{F}$ that depends on $\boldsymbol{r}$, $\dot{\boldsymbol{r}}$ (velocity), and time t. The initial equation of motion is:

$$m\ddot{\boldsymbol{r}} = \boldsymbol{F}\left(\boldsymbol{r}, \dot{\boldsymbol{r}}\,;\, t\right) \tag{3}$$

(where m is the particle's mass). In the time reversal operations, each motion described by the function $\boldsymbol{r}(t)$ becomes:

$$\begin{array}{lcl} \boldsymbol{r}(t) & & \boldsymbol{r}(-t) \\ \dot{\boldsymbol{r}}(t) & \stackrel{\mathcal{T}}{\Longrightarrow} & -\dot{\boldsymbol{r}}(-t) \\ \ddot{\boldsymbol{r}}(t) & & \ddot{\boldsymbol{r}}(-t) \end{array} \tag{4}$$

Consequently, the operation $\mathcal{T}$ will be a symmetry operation if:

$$m\ddot{\boldsymbol{r}}(-t) = \boldsymbol{F}\left(\boldsymbol{r}, -\dot{\boldsymbol{r}}\,;\, -t\right) \tag{5}$$

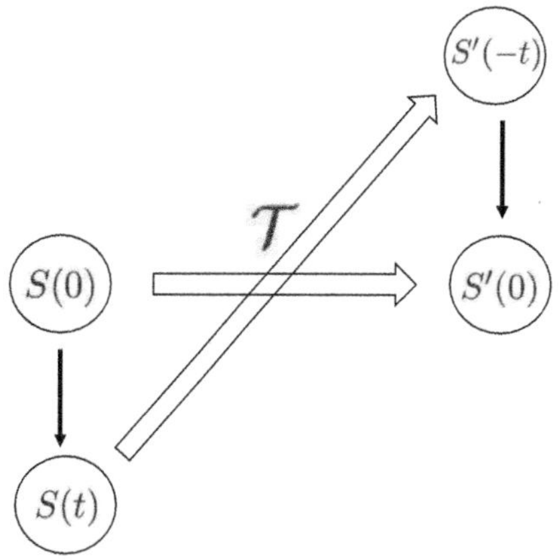

Figure 2: Time-reversal diagram in classical mechanics, in a representation similar to that of figure 1 of chapter I. The left-hand side represents the initial motion; the right-hand side, another motion where the physical system goes, at each instant, through the state associated with the opposite time in the initial motion, and where the velocities have been reversed. The correspondence between states is indicated by double arrows. If the motions on the right-hand and left-hand sides are compatible with the same equations of motion, the dynamics is said to be invariant under time reversal.

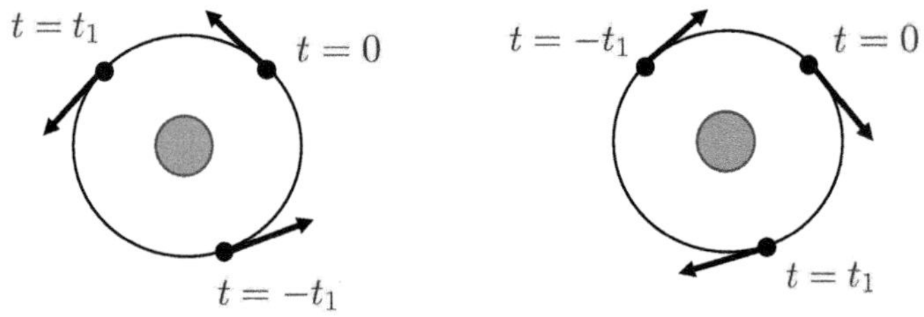

Figure 3: For a particle subjected to a potential $V(\boldsymbol{r})$ that depends only on $\boldsymbol{r}$ (electron rotating around a motionless proton, satellite rotating around the earth, etc.), time reversal is a symmetry operation.

meaning:

$$\boldsymbol{F}\left(\boldsymbol{r}, -\dot{\boldsymbol{r}}\,;\, -t\right) = \boldsymbol{F}\left(\boldsymbol{r}, \dot{\boldsymbol{r}}\,;\, t\right) \tag{6}$$

for any $\boldsymbol{r}$, $\dot{\boldsymbol{r}}$, and t. In particular, if $\boldsymbol{F}$ is the derivative of a potential $V(\boldsymbol{r})$ that only depends on $\boldsymbol{r}$, it is clear that $\mathcal{T}$ is always a symmetry operation. This is the case, for an example, of a charged particle subjected to an electrostatic potential (electron rotating around a motionless proton), or of a satellite rotating around the earth, etc. (*cf.* figure 3).

On the other hand, if magnetic forces are present, the force $\boldsymbol{F}$ acting on a particle with charge q is written (Lorentz force):

$$\boldsymbol{F} = q\left[\boldsymbol{E}(r) + \dot{\boldsymbol{r}} \times \boldsymbol{B}(\boldsymbol{r})\right] \tag{7}$$

Due to the presence of the term in $\dot{\boldsymbol{r}}$, condition (6) is no longer satisfied. Imagine, for example, that $\boldsymbol{E}$ is zero and $\boldsymbol{B}$ uniform, and take a particle with an initial velocity $\boldsymbol{v}_0$ perpendicular to $\boldsymbol{B}$. Its trajectory is a circle, which will be different if the initial velocity $\boldsymbol{v}_0$ is reversed (figure 4). One can easily show that the motion obtained under the action of $\mathcal{T}$ obeys the equations of motion in a field $-\boldsymbol{B}$ opposite to the initial field.

This result can be generalized to a set of charged particles: time reversal is a symmetry operation if these particles are subjected to an external scalar potential (electrostatic field), but this is no longer the case if they are subjected to external magnetic fields.

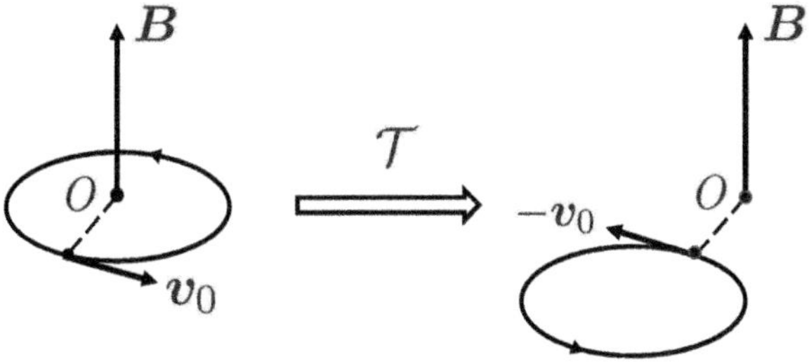

*Figure 4: For a charged particle subjected to a magnetic field **B**, time reversal changes the possible trajectories. In a uniform magnetic field **B**, reversing the initial velocity $\boldsymbol{v}_0$ (assumed to be perpendicular to **B**) of a charged particle changes the trajectory into a different circle: time reversal is not a symmetry operation. However, the initial trajectory is preserved if one changes the sign of both the time and the magnetic field.*

Comments:

(i) When the physical system is isolated (no external applied fields), time reversal is always a symmetry operation when the particle interactions are of electromagnetic nature, even if $\boldsymbol{E}$ and $\boldsymbol{B}$ fields play a role in these interactions: electromagnetic laws are invariant under time reversal. However, it should be noted that time reversal changes the sign of the $\boldsymbol{B}$ fields created by the particles (but not that of $\boldsymbol{E}$):

$$\begin{array}{ccc} \boldsymbol{E}(\boldsymbol{r}\,;t) & \xrightarrow{\mathcal{T}} & \boldsymbol{E}(\boldsymbol{r}\,;-t) \\ \boldsymbol{B}(\boldsymbol{r}\,;t) & & -\boldsymbol{B}(\boldsymbol{r}\,;-t) \end{array} \tag{8}$$

This is consistent with the Biot–Savart law that explicitly involves the velocities, whose sign changes under the action of $\mathcal{T}$, whereas Coulomb's law only depends on the particles' positions.

If the field $\boldsymbol{B}$ is externally applied, time reversal can still be considered as a symmetry operation if it is agreed to change $\boldsymbol{B}$ into $-\boldsymbol{B}$. This is understandable from a physical point of view: $\boldsymbol{B}$ becomes $-\boldsymbol{B}$ if the velocities of the particles creating that field are reversed (for example, reversing the current in the solenoid creating $\boldsymbol{B}$).

(ii) The time reversal operation we have defined depends on the chosen time origin, which was left unchanged. One could also define operations $\mathcal{T}_{t_0}$ centered at any time t_0:

$$q_i(t) \xrightarrow{\mathcal{T}_{t_0}} \tilde{q}_i(t) = q_i\,(2t_0 - t) \tag{9}$$

where it is now the value of time $t = t_0$ that is invariant. Note however that $\mathcal{T}_{t_0}$ is the operation $\mathcal{T}_{t_0=0}$ (i.e. $\mathcal{T}$), followed by a time translation of $2t_0$. If the motion is invariant under a time translation, the study of the single operation $\mathcal{T}$ covers the study of all the operations $\mathcal{T}_{t_0}$ (the situation is similar to the relations between parity and space translations).

(iii) In the Lagrangian point of view of the equations of motion, the Lagrangian L is often the sum of a quadratic function of the $\dot{q}_i$ (without any linear term) and a function that only depends on the q_i and on the time. For example:

$$L = T(\dot{q}_i) - V(q_i\,;\,t) \tag{10}$$

where $T(\dot{q}_i)$ is the kinetic energy, quadratic with respect to the $\dot{q}_i$. Consequently, operation $\mathcal{T}$ is a symmetry operation if:

$$V(q_i\,;\,t) = V(q_i\,;\,-t) \tag{11}$$

(in general, $\mathcal{T}_{t_0}$ will be a symmetry operation for any t_0 if V is time independent). We then have:

$$L(q_i, \dot{q}_i\,;\,t) = L(q_i, -\dot{q}_i\,;\,-t) \tag{12}$$

Since the function L is assumed to be quadratic with respect to the $\dot{q}_i$ (without any linear term), the action of $\mathcal{T}$ changes the momentum p_i:

$$p_i = \frac{\partial L}{\partial \dot{q}_i}(q_i, \dot{q}_i\,;\,t) \tag{13}$$

into its opposite. Consequently:

$$\mathcal{H} = \sum_i p_i\,\dot{q}_i - L \tag{14}$$

is invariant under the operation $\mathcal{T}$. It is easy to show that the Hamilton equations:

$$\begin{cases} \dot{q}_i = \dfrac{\partial \mathcal{H}}{\partial p_i} \\ \dot{p}_i = -\dfrac{\partial \mathcal{H}}{\partial q_i} \end{cases} \tag{15}$$

are invariant under the transformation:

$$\begin{matrix} q_i & & q_i \\ & \stackrel{\mathcal{T}}{\Longrightarrow} & \\ p_i & & -p_i \end{matrix} \tag{16}$$

However, if the Lagrangian does not have a pure quadratic dependence with respect to the $\dot{q}_i$, the situation is more complex. For example, if:

$$L = \frac{1}{2} m\,\dot{\boldsymbol{r}}^2 + q\dot{\boldsymbol{r}} \cdot \boldsymbol{A}(\boldsymbol{r}, t) - q\,U(\boldsymbol{r}\,;\,t) \tag{17}$$

(particle of charge[2] q and mass m in fields $\boldsymbol{E}$ and $\boldsymbol{B}$), the motion is no longer invariant under time reversal; this operation no longer leads to a sign change of $\boldsymbol{p}$, but to:

$$\boldsymbol{p} \stackrel{\mathcal{T}}{\Longrightarrow} \boldsymbol{p}' \tag{18}$$

and a reversal of the velocity expressed as:

$$\boldsymbol{p}' - q\,\boldsymbol{A}(\boldsymbol{r}\,;\,-t) = -\left[\boldsymbol{p} - q\,\boldsymbol{A}(\boldsymbol{r}\,;\,t)\right] \tag{19a}$$

that is:

$$\boldsymbol{p}' = -\boldsymbol{p} + q\left[\boldsymbol{A}(\boldsymbol{r}\,;\,t) + \boldsymbol{A}(\boldsymbol{r}\,;\,-t)\right] \tag{19b}$$

2. Antilinear and antiunitary operators in quantum mechanics

Before continuing with quantum mechanics, it is useful to establish a few mathematical properties of antilinear operators that will be helpful in what follows.

2-a. Antilinear operators

α. *Definition*

An operator A acting in vector space $\mathscr{E}$ is said to be antilinear if, for any kets $|\varphi_1\rangle$ and $|\varphi_2\rangle$ of $\mathscr{E}$ it is acting upon, and for any (complex) constants λ_1 and λ_2, we have:

$$A\Big[\lambda_1|\varphi_1\rangle + \lambda_2|\varphi_2\rangle\Big] = \lambda_1^\star\, A|\varphi_1\rangle + \lambda_2^\star\, A|\varphi_2\rangle \tag{20}$$

In particular, this definition shows that:

$$A\,\lambda|\varphi_1\rangle = \lambda^\star\, A|\varphi_1\rangle \tag{21}$$

which means that the commutator $[A, \lambda]$ of A with a number is different from zero, unless that number is real.

If A is antilinear and $\{|u_n\rangle\}$ a basis in the state space, we have:

$$\begin{aligned} A|\varphi\rangle &= A\left[\sum_n |u_n\rangle\langle u_n|\varphi\rangle\right] = \sum_n \langle u_n|\varphi\rangle^\star\, A|u_n\rangle \\ &= \sum_n \langle\varphi|u_n\rangle\, A|u_n\rangle \end{aligned} \tag{22}$$

[2]The charge q should not be confused with one of the coordinates q_i.

β. Sum and product

Just as for linear operators, the sum $A_1 + A_2$ and product $A_1 A_2$ of two linear or antilinear operators are defined as:

$$\begin{aligned}(A_1 + A_2)|\varphi\rangle &= A_1|\varphi\rangle + A_2|\varphi\rangle \\ (A_1 A_2)|\varphi\rangle &= A_1(A_2|\varphi\rangle)\end{aligned} \tag{23}$$

It is easy to show that if both A_1 and A_2 are antilinear, the product $A_1 A_2$ is linear; if only one of the two operators A_1 and A_2 is antilinear, the other being linear, the product $A_1 A_2$ is antilinear.

γ. Inverse

A and B are, by definition, the inverse of each other if:

$$AB = BA = \mathbb{1} \tag{24}$$

where $\mathbb{1}$ is the identity operator. Let us show that if A is antilinear, so is B. Setting:

$$|\psi\rangle = B|\varphi\rangle \quad \text{and} \quad |\varphi\rangle = A|\psi\rangle \tag{25}$$

we can write:

$$\lambda B|\varphi\rangle = \lambda|\psi\rangle = BA\,\lambda|\psi\rangle = B\left[\lambda^\star A|\psi\rangle\right] \tag{26a}$$

meaning:

$$\lambda B|\varphi\rangle = B\left[\lambda^\star|\varphi\rangle\right] \tag{26b}$$

We can show in a similar way that if we multiply A by a complex constant λ, we must divide the inverse B of A by $\lambda^\star$.

δ. Action of an antilinear operator on a bra

If A is antilinear operator, the scalar product of ket $A|\varphi\rangle$ and bra $|\chi\rangle$:

$$\langle\chi|\Big(A|\varphi\rangle\Big)$$

is, according to (20), an antilinear function of $|\varphi\rangle$. Consequently, the complex conjugate number:

$$\Big[\langle\chi|\Big(A|\varphi\rangle\Big)\Big]^\star$$

is a linear function of $|\varphi\rangle$, whatever the ket $|\varphi\rangle$. We can thus write it as $\langle\chi'|\varphi\rangle$, where $\langle\chi'|$ is a bra (belonging to the dual space of space $\mathscr{E}$) that can be written as:

$$\langle\chi'| = \big(\langle\chi|A\big) \tag{27}$$

Note that the parentheses underline the fact that A acts on the preceding bra. This relation is the definition of the action of A on a bra.

We have thus set:

$$\boxed{\big(\langle\chi|A\big)\,|\varphi\rangle = \big[\langle\chi|\big(A|\varphi\rangle\big)\big]^{\star}} \tag{28}$$

This shows that, when dealing with antilinear operators, one must specify (with parentheses, for example) whether they act on the following ket or on the preceding bra: the results of both choices are complex conjugates of each other.

When A_1 and A_2 are antilinear, their product A_1A_2 is linear and:

$$\begin{aligned}\langle\chi|\big(A_1A_2|\varphi\rangle\big) &= \langle\chi|\big(A_1A_2\big)|\varphi\rangle \\ &= \big[\left(\langle\chi|A_1\right)\left(A_2|\varphi\rangle\right)\big]^{\star} = \big(\langle\chi|A_1A_2\big)|\varphi\rangle\end{aligned} \tag{29}$$

ϵ. *Adjoint operator*

The definition of the adjoint operator $A^\dagger$ of an antilinear operator A is the same as for a linear operator:

$$\langle\chi'| = \big(\langle\chi|A\big) \Longleftrightarrow |\chi'\rangle = A^\dagger|\chi\rangle \qquad \forall\ |\chi\rangle \tag{30}$$

We thus have:

$$\langle\chi'|\varphi\rangle = \big(\langle\chi|A\big)|\varphi\rangle = \langle\varphi|\chi'\rangle^{\star} = \big[\langle\varphi|\left(A^\dagger|\chi\rangle\right)\big]^{\star} \tag{31}$$

Using (28), and taking the complex conjugate, we obtain:

$$\boxed{\langle\chi|\big(A|\varphi\rangle\big) = \langle\varphi|\left(A^\dagger|\chi\rangle\right)} \tag{32}$$

This equality can be used as a definition of the adjoint operator of an antilinear operator A, even though we have not introduced its action on a bra.

Comments:

(i) For a linear operator A, the two sides of equation (32) would not be equal but complex conjugates of each other.

(ii) If A is antilinear, so is $A^\dagger$ as can be seen on (32).

(iii) Replacing A by $A^\dagger$ in (32), we get:

$$\langle\chi|\left(A^\dagger|\varphi\rangle\right) = \langle\varphi|\left(A^{\dagger\dagger}|\chi\rangle\right) \tag{33}$$

Since:

$$\langle\chi|\left(A^\dagger|\varphi\rangle\right) = \langle\varphi|\left(A|\chi\rangle\right) \tag{34}$$

we see that:

$$A^{\dagger\dagger} = A \tag{35}$$

(iv) If A is antilinear and B linear:

$$(AB)^\dagger = B^\dagger A^\dagger \tag{36}$$

as we now show. Setting:

$$|\varphi'\rangle = B|\varphi\rangle \tag{37}$$

we can write, since AB is antilinear:

$$\begin{aligned}\langle\varphi|\left[(AB)^\dagger|\chi\rangle\right] &= \langle\chi|\left[AB|\varphi\rangle\right] = \langle\chi\,|\left[A|\varphi'\rangle\right] \\ &= \langle\varphi'|\left[A^\dagger|\chi\rangle\right] = \left[\langle\varphi|B^\dagger\right]\left[A^\dagger|\chi\rangle\right] = \langle\varphi|\left[B^\dagger A^\dagger|\chi\rangle\right]\end{aligned} \tag{38}$$

Similarly, we can show the relation $(BA)^\dagger = A^\dagger B^\dagger$.

2-b. Antiunitary operators

α. Definition

An antiunitary operator is an antilinear operator such that:

$$A^\dagger = A^{-1} \tag{39a}$$

or else:

$$AA^\dagger = A^\dagger A = \mathbb{1} \tag{39b}$$

With any operator A of this type, we can associate an antiunitary transformation that transforms any ket $|\varphi\rangle$ into $|\overline{\varphi}\rangle$, any bra $\langle\varphi|$ into $\langle\overline{\varphi}|$, and any operator B into $\overline{B}$, such that:

$$\begin{aligned}&|\overline{\varphi}\rangle = A|\varphi\rangle \qquad \langle\overline{\varphi}| = (\langle\varphi|A^\dagger) \\ &\overline{B} = A\,B\,A^\dagger\end{aligned} \tag{40}$$

The following table summarizes the effects of such a transformation:

λ (constant)	$\Longrightarrow$	$\overline{\lambda} = A\,\lambda\,A^\dagger = \lambda^\star$ (complex conjugate)
$\langle\varphi\|\chi\rangle$ (scalar product)	$\Longrightarrow$	$\langle\overline{\varphi}\|\overline{\chi}\rangle = \left(\langle\varphi\|A^\dagger\right)\left(A\|\chi\rangle\right) = \left[\langle\varphi\|\left(A^\dagger A\right)\|\chi\rangle\right]^\star = \langle\varphi\|\chi\rangle^\star$ (complex conjugate)
B (linear operator)	$\Longrightarrow$	$\overline{B} = A\,B\,A^\dagger$ (also linear)
$\langle\varphi\|B\|\chi\rangle$ (matrix element)	$\Longrightarrow$	$\langle\overline{\varphi}\|\overline{B}\|\overline{\chi}\rangle = \left(\langle\varphi\|A^\dagger\right)A\,B\,A^\dagger A\|\chi\rangle = \left\{\langle\varphi\|\left(A^\dagger A\,B\,\|\chi\rangle\right)\right\}^\star = \langle\varphi\|B\|\chi\rangle^\star$ (complex conjugate)
$B\|\varphi\rangle = \lambda\|\varphi\rangle$ (eigenvector)	$\Longrightarrow$	$\overline{B}\|\overline{\varphi}\rangle = A\,B\,A^\dagger A\|\varphi\rangle = A\lambda\|\varphi\rangle = \lambda^\star\|\overline{\varphi}\rangle$ (same eigenvector, complex conjugate eigenvalue)

This means that if B is an observable (hence with real eigenvalues), its spectrum is unchanged in the antiunitary transformation.

In addition:

BC (product of operators)	$\Longrightarrow$	$\overline{BC} = A\,B\,C\,A^\dagger = A\,B\,A^\dagger A\,C\,A^\dagger = \overline{B}\,\overline{C}$
$[X,P] = i\hbar$	$\Longrightarrow$	$\left[\overline{X},\overline{P}\right] = -i\hbar$
$[J_x,J_y] = i\hbar\,J_z$	$\Longrightarrow$	$\left[\overline{J}_x,\overline{J}_y\right] = -i\hbar\,\overline{J}_z$

β. *Complex conjugation operator $K_{\{u\}}$ associated with representation $\{|u\rangle\}$*

Consider a set of orthonormal vectors $\{|u_n\rangle\}$ forming a basis of the state space $\mathscr{E}$. Any ket $|\varphi\rangle$ can be expanded on the $\{|u_n\rangle\}$:

$$|\varphi\rangle = \sum_n c_n\,|u_n\rangle \tag{41}$$

The action of operator $K_{\{u\}}$ can be defined as:

$$|\overline{\varphi}\rangle = K_{\{u\}}|\varphi\rangle = \sum_n c_n^\star \, |u_n\rangle \tag{42}$$

Expanding $|\varphi_1\rangle$ and $|\varphi_2\rangle$ on the $\{|u_n\rangle\}$ basis, it is easy to show that $K_{\{u\}}$ obeys the definition (20) of an antilinear operator. It shall be called "complex conjugation operator in basis $\{|u_n\rangle\}$". Let us compute the adjoint of $K_{\{u\}}$; according to definition (32), we have:

$$\langle u_n| \left(K_{\{u\}}|u_p\rangle\right) = \delta_{np} = \langle u_p| \left(K_{\{u\}}^\dagger|u_n\rangle\right) \tag{43}$$

Since $\delta_{np} = \delta_{pn}$, this simply yields:

$$K_{\{u\}} = K_{\{u\}}^\dagger \tag{44}$$

In addition, we easily see from (42) that:

$$\left(K_{\{u\}}\right)^2 = \mathbb{1} \tag{45}$$

so that:

$$K_{\{u\}} = K_{\{u\}}^{-1} = K_{\{u\}}^\dagger \tag{46}$$

Operator $K_{\{u\}}$ is thus antiunitary. The action of $K_{\{u\}}$ is to change the components (or the wave function) of any ket $|\varphi\rangle$ into their complex conjugates. As for the matrix elements, we saw above that:

$$\langle \overline{u}_n|\overline{B}|\overline{u}_p\rangle = \langle u_n|B|u_p\rangle^\star \tag{47}$$

Since $|\overline{u}_n\rangle = |u_n\rangle$ and $|\overline{u}_p\rangle = |u_p\rangle$, we simply have:

$$\overline{B}_{np} = \langle u_n|\overline{B}|u_p\rangle = \langle u_n|B|u_p\rangle^\star = B_{np}^\star \tag{48}$$

Comments:

(i) In a basis different from $\{|u_n\rangle\}$, the action of $K_{\{u\}}$ on a ket is generally not as simple as a complex conjugation of its components, as we now show. If $\{|v_p\rangle\}$ represents a different basis, we have:

$$\begin{aligned} K_{\{u\}} \sum_p \gamma_p \, |v_p\rangle &= K_{\{u\}} \sum_{pn} \gamma_p \, \langle u_n|v_p\rangle \, |u_n\rangle \\ &= \sum_{pn} \gamma_p^\star \, \langle u_n|v_p\rangle^\star \, |u_n\rangle \end{aligned} \tag{49}$$

In the particular case where all the scalar products $\langle u_n|v_p\rangle$ are real for any n and p (real basis change), a closure relation appears on the right-hand side of this equality:

$$\sum_{pn} \gamma_p^\star \, |u_n\rangle\langle u_n|v_p\rangle = \sum_p \gamma_p^\star \, |v_p\rangle \tag{50}$$

and $K_{\{u\}}$ is simply the complex conjugation operator in basis $|v_p\rangle$. In general, however, $\langle u_n|v_p\rangle$ is not real, and this simplification does not occur.

(ii) Any antilinear operator A can be written:

$$\begin{aligned} A &= A\, K_{\{u\}}^2 = \left(A\, K_{\{u\}}\right) K_{\{u\}} \\ &= K_{\{u\}}^2\, A = K_{\{u\}} \left(K_{\{u\}} A\right) \end{aligned} \tag{51}$$

Any antilinear (antiunitary) operator can thus be expressed as the product of a complex conjugation operator and a linear (unitary) operator: it is easy to show that, if A is unitary, so is the product $AK_{\{u\}}$.

γ. *Complex conjugation operators in position and momentum bases*

- For a spinless particle, we set:

$$|\overline{\psi}\rangle = K_{\{\boldsymbol{r}\}}|\psi\rangle \tag{52}$$

It follows:

$$\overline{\psi}(\boldsymbol{r}) = \langle \boldsymbol{r}|\overline{\psi}\rangle = \psi^\star(\boldsymbol{r}) \tag{53}$$

In addition:

$$\begin{aligned} \overline{\boldsymbol{R}} &= K_{\{\boldsymbol{r}\}}\, \boldsymbol{R}\, K_{\{\boldsymbol{r}\}} = \boldsymbol{R} \\ \overline{\boldsymbol{P}} &= K_{\{\boldsymbol{r}\}}\, \boldsymbol{P}\, K_{\{\boldsymbol{r}\}} = -\boldsymbol{P} \end{aligned} \tag{54}$$

(the matrix elements of $\boldsymbol{R}$ in the $\{|\boldsymbol{r}\rangle\}$ representation are real, the matrix elements of $\boldsymbol{P}$, which acts in this representation as $(\hbar/i)\boldsymbol{\nabla}$, are purely imaginary). Similarly:

$$\overline{\boldsymbol{L}} = K_{\{\boldsymbol{r}\}}\, \boldsymbol{L}\, K_{\{\boldsymbol{r}\}} = K_{\{\boldsymbol{r}\}} \left(\boldsymbol{R} \times \boldsymbol{P}\right) K_{\{\boldsymbol{r}\}} = -\boldsymbol{L} \tag{55}$$

The previous expressions are conveniently expressed as a function of a commutator or of an anticommutator:

$$\begin{aligned} &\left[\boldsymbol{R}, K_{\{\boldsymbol{r}\}}\right] = 0 \\ &\left[\boldsymbol{P}, K_{\{\boldsymbol{r}\}}\right]_+ = \left[\boldsymbol{L}, K_{\{\boldsymbol{r}\}}\right]_+ = 0 \end{aligned} \tag{56}$$

where $[B, C]_+$ designates the anticommutator $BC + CB$.

- Using the complex conjugation operator in the momentum basis:

$$|\tilde{\psi}\rangle = K_{\{p\}} |\psi\rangle \tag{57}$$

we get:

$$\tilde{\psi}_p(\boldsymbol{p}) = \langle \boldsymbol{p}|\tilde{\psi}\rangle = \psi_p^\star(\boldsymbol{p}) \tag{58}$$

Since we know that:

$$\psi(\boldsymbol{r}) = (2\pi\hbar)^{-3/2} \int d^3p \; \mathrm{e}^{i\boldsymbol{p}\cdot\boldsymbol{r}/\hbar} \, \psi_p(\boldsymbol{p}) \tag{59}$$

we have:

$$\tilde{\psi}(\boldsymbol{r}) = \psi^\star(-\boldsymbol{r}) \tag{60}$$

which is different from (53); it was to be expected since $\{|\boldsymbol{r}\rangle\} \Longrightarrow \{|\boldsymbol{p}\rangle\}$ is not a real basis change. It is easy to show that:

$$\left[\boldsymbol{R}, K_{\{p\}}\right]_+ = \left[\boldsymbol{L}, K_{\{p\}}\right]_+ = 0 \tag{61}$$

and that:

$$\left[\boldsymbol{P}, K_{\{p\}}\right] = 0 \tag{62}$$

3. Time reversal and antilinearity

3-a. Need for the introduction of an antilinear operator

We now examine which conditions a transformation in the state space must fulfill in order to be an adequate representation of time reversal. We shall see that a linear operator is not suitable, and that an antiunitary operator is needed. The reader may wish to skip this intermediary step and ignore this § 3-a to directly discuss the form of the time reversal operator.

α. *Simple reasoning: commutators and anticommutators*

Imagine that a unitary operator T represents the time reversal operator in the state space of a spinless free particle, Since the time reversal does not change the position, we must have, for any state vector $|\psi\rangle$:

$$\langle\psi|T\boldsymbol{R}\,T^\dagger|\psi\rangle = \langle\psi|\boldsymbol{R}|\psi\rangle \tag{63}$$

or, since this relation is valid for any $|\psi\rangle$:

$$\boldsymbol{T}\boldsymbol{R}\,T^{\dagger} = \boldsymbol{R} \quad \Rightarrow \quad T\boldsymbol{R} = \boldsymbol{R}T \quad \text{that is:} \quad [T\,,\,\boldsymbol{R}] = 0 \tag{64}$$

Operator T must commute with the position operator.

On the other hand, T must change the momentum sign:

$$\langle\psi|T\boldsymbol{P}\,T^{\dagger}|\psi\rangle = -\langle\psi|\boldsymbol{P}|\psi\rangle \tag{65}$$

so that:

$$T\boldsymbol{P} = -\boldsymbol{P}T \quad \text{or:} \quad [T\,,\,\boldsymbol{P}]_{+} = 0 \tag{66}$$

It is thus the anticommutator of T and the momentum that must be equal to zero.

These two previous conditions are contradictory for a unitary operator, as we now show. If the unitary operator T commutes with the position operator, it is diagonal in the $\{|\boldsymbol{r}\rangle\}$ representation, and we have:

$$T|\boldsymbol{r}\rangle = \mathrm{e}^{-i\alpha(\boldsymbol{r},t)}|\boldsymbol{r}\rangle \tag{67}$$

T simply changes the wave function $\psi(\boldsymbol{r})$ into $\psi'(\boldsymbol{r}) = \mathrm{e}^{i\alpha(\boldsymbol{r},t)}\psi(\boldsymbol{r})$. We can compute the new average value $\langle P\rangle'$ of the momentum as a function of the initial one $\langle P\rangle$:

$$\begin{aligned}\langle P\rangle' &= \frac{\hbar}{i}\int \mathrm{d}^3r\ \mathrm{e}^{-i\alpha(\boldsymbol{r},t)}\psi^*(\boldsymbol{r})\ \boldsymbol{\nabla}[\mathrm{e}^{i\alpha(\boldsymbol{r},t)}\psi(\boldsymbol{r})] \\ &= \langle P\rangle + \hbar\int \mathrm{d}^3r\ \psi^*(\boldsymbol{r})\psi(\boldsymbol{r})\ \boldsymbol{\nabla}\alpha(\boldsymbol{r},t)\end{aligned} \tag{68}$$

The average value of the momentum is thus increased by the average value of the gradient of the phase $\alpha(\boldsymbol{r},t)$, which is not equivalent to a sign change. Even in the very simple case of a spinless free particle, no unitary operator correctly describes the time reversal operation.

β. *More general reasoning*

The reasoning of the previous § 3-a-α, limited to a single spinless particle, is not as general as the one of § C in chapter III, based on the transformations of space–time coordinates. We shall go back to this more general point of view and show that it also concludes that a unitary transformation cannot properly represent a time reversal operation. In chapter III, the commutation relations between the unitary operators associated with the transformations (translations, rotations, parity, change of Galilean reference frame) were direct consequences of the commutation relations between linear transformations acting on space–time coordinates x, y, z, t (increased by one homogeneous coordinate $u = 1$). We shall try and proceed in the same way for the time reversal transformation, but will find again that imposing the unitarity for the time reversal operator leads to contradictions. We shall then show (§ 3-a-γ) that an antiunitary operator is perfectly suitable.

To start with, let's assume we are looking for a unitary operator T whose action in the state space is represented in quantum mechanics by the operator $\mathcal{T}$ schematized in figure 2 (inversion of the velocities without change of the particles' positions). What commutation relations should exist between T and the translation and rotation operators acting in the state space?

Following the same type of reasoning as in § C-1 of chapter III, we can complete relations (III-101), or (III-111) in Einstein's relativity, to include time reversal $\mathcal{T}$. In the space of the space–time coordinates, we now include the velocity to describe the physical system[3]. There are now 8 coordinates: 3 space–time coordinates (x, y, z, and t), 3 components of the velocity ($\dot{x}$, $\dot{y}$, and $\dot{z}$), the time t, and finally the usual homogeneous coordinates $u = 1$. The various translation, rotation, change of Galilean reference frame, etc. operations are described by 8×8 (and no longer 5×5 as in chapter III) matrices. The matrix associated with $\mathcal{T}$ is diagonal, with three elements equal to 1 (positions), three elements equal to -1 (velocity reversal), another one also equal to -1 (time reversal), and a last one equal to 1 (unchanged homogeneous coordinate).

An elementary calculation of matrix products shows that the commutation relations are sometimes replaced by anticommutators. We get:

$$[\mathcal{T}, X_{P_{x,y,z}}] = 0 \tag{69a}$$

$$[\mathcal{T}, X_H]_+ = 0 \tag{69b}$$

$$[\mathcal{T}, X_{K_{x,y,z}}]_+ = 0 \tag{69c}$$

The first commutation relation is natural: we expect that $\mathcal{T}$, which is concerned with the direction of time flow, commutes with all the pure space transformations. The second relation includes an anticommutator. This comes from the fact that an infinitesimal time translation is associated with the differential operator $\partial/\partial t$. Since:

$$\frac{\partial}{\partial t}\left[f(-t)\right] = -\frac{\partial f}{\partial t}(-t) \tag{70a}$$

we have :

$$\frac{\partial}{\partial t} \times (\text{time reversal}) + (\text{time reversal}) \times \frac{\partial}{\partial t} = 0 \tag{70b}$$

[3]The knowledge of the quantum state vector $|\psi\rangle$ of a system corresponds, in classical mechanics, to the knowledge of the positions q_i and their conjugate momenta p_i (related to the velocities $\dot{q}_i$). It is thus not surprising that in the study of operators acting on $|\psi\rangle$, and associated with certain transformations like time reversal, one should take into account the corresponding modifications of the classical velocities $\dot{x}$, $\dot{y}$, and $\dot{z}$.

The third relation also includes an anticommutator as time is directly involved in a change of Galilean reference frame (in the definition of the relative velocity of the two reference frames).

Let us see how these commutation or anticommutation relations in space–time translate into the action of a unitary operator in the quantum state space. The fact that commutator (69a) is zero implies that:

$$[T, \boldsymbol{P}] = 0 \tag{71}$$

Such a relation indicates that a time reversal operation does not change the momentum, which is obviously not true! As for the simple reasoning of § 3-a-α, we arrive at a dead end. The conclusion is that operator T representing time reversal cannot be linear and unitary.

Exercise: Write the eleven 8×8 matrices associated with the Poincaré group completed so as to include time reversal, and discuss their commutation or anticommutation relations.

γ. *Search for an antiunitary operator*

Let us see if a suitable choice could be an antiunitary operator, compatible a priori with Wigner's theorem (chapter IV, § B). How do the commutation relations between symmetry operators in the usual space translate into the state space with an antiunitary operator?

We go back to the reasoning that led to relation (71), to see what is changed by the antiunitarity. The commutation between translations and time reversal yields again:

$$T\,(1 - i\,\delta\boldsymbol{b}\cdot\boldsymbol{P}/\hbar) = (1 - i\,\delta\boldsymbol{b}\cdot\boldsymbol{P}/\hbar)\;T \tag{72}$$

but, since T is antilinear (and δb real), we now have:

$$i\,\delta\boldsymbol{b}\cdot(T\boldsymbol{P}) = -i\,\delta\boldsymbol{b}\cdot\boldsymbol{P}T \tag{73}$$

As this relation is true for any $\delta\boldsymbol{b}$, we must have:

$$[T, \boldsymbol{P}]_+ = 0 \qquad \text{or else:} \qquad [T, i\boldsymbol{P}] = 0 \tag{74}$$

Conversely, these relations show that finite translations also commute, as can be easily shown by expanding the exponential:

$$\mathrm{e}^{-i\boldsymbol{b}\cdot\boldsymbol{P}/\hbar} = \sum_n \frac{1}{n!}\left(\frac{-i\boldsymbol{b}\cdot\boldsymbol{P}}{\hbar}\right)^n \tag{75}$$

Because of the presence of i in (72), which does not commute with T when that operator is antiunitary, it is the anticommutator of $\boldsymbol{P}$ and T (and not the commutator) that must vanish.

Using an antiunitary operator changes commutators into anticommutators when going from the space–time coordinate space to the quantum state space. This will resolve the contradictions associated with the choice of a unitary operator. We now discuss a few additional conditions that T must obey.

Since the time reversal operator does not change the position, we have:

$$\left(\langle\psi|T^{\dagger}\right)\boldsymbol{R}\,T|\psi\rangle = \langle\psi|\boldsymbol{R}|\psi\rangle \tag{76}$$

for any state vector $|\psi\rangle$. According to (28), the left-hand side of this equality is equal to $\langle\psi|\left(T^{\dagger}\boldsymbol{R}\,T\right)|\psi\rangle^{*}$, which must be real for condition (76) to be satisfied. As a result, the diagonal elements between a bra $\langle\psi|$ and a ket $|\psi\rangle$ of the two operators $T^{\dagger}\boldsymbol{R}\,T$ and $\boldsymbol{R}$ must be equal. Since this is true for any $|\psi\rangle$, the two operators themselves are equal:

$$T^{\dagger}\boldsymbol{R}\,T = \boldsymbol{R} \tag{77}$$

or else, multiplying the equality on the left by T:

$$[T\,,\,\boldsymbol{R}] = 0 \tag{78}$$

The time reversal operator must commute with the position operator.

As for the momentum operator, we want a change of sign when time is reversed, meaning:

$$\left(\langle\psi|T^{\dagger}\right)\boldsymbol{P}\,T|\psi\rangle = -\langle\psi|\boldsymbol{P}|\psi\rangle \tag{79}$$

Following the same reasoning, we find again (74). The same reasoning also applies to $\boldsymbol{J}$ and yields:

$$[T, \boldsymbol{J}]_{+} = 0 \tag{80}$$

($\boldsymbol{J}$ can be replaced by $\boldsymbol{L}$ when the system under study is a spinless particle).

Finally, we noted in (69b) that the operations of time reversal and time translation (or change of Galilean reference frame) anticommute. We thus expect:

$$[T, H] = 0 \tag{81}$$

We are going to show that there exists such an antiunitary operator T, adequate from a physical point of view as it obeys these commutation and anti-commutation relations.

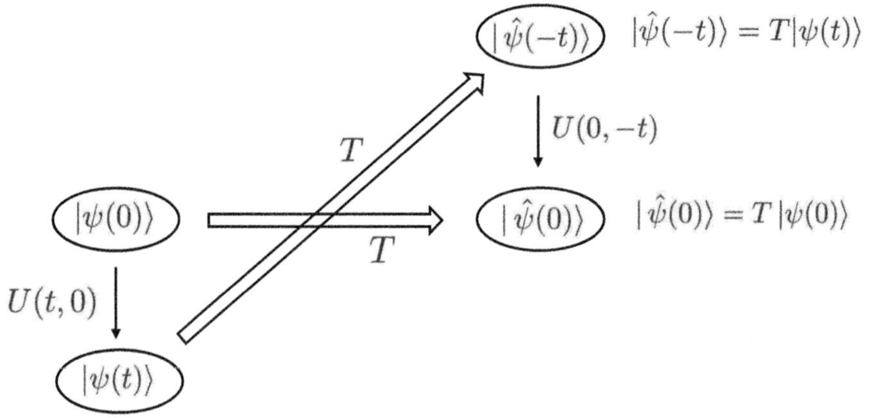

Figure 5: Diagram associated with the time reversal operator T in quantum mechanics, to be compared with the different diagram of figure 5 in chapter I. The time axis is directed downwards. One considers a state vector that, at time $-t$, takes the value $|\hat{\psi}(-t)\rangle = T|\psi(t)\rangle$. Time reversal is a symmetry operation if this state vector obeys the same dynamics as $|\psi(t)\rangle$, meaning if we can close the path on the diagram with the same evolution operator taken between an initial time $-t$ and a final time 0.

3-b. Unitary invariance under time reversal

Now that we established that an antiunitary operator is necessary to represent time reversal, let us go back to the diagram of figure 2; it symbolizes the condition that must be obeyed by the equations of motion for time reversal to be a symmetry operation. In quantum mechanics, this diagram becomes that of figure 5, where two evolution operators U are shown explicitly. Remember that an evolution operator is defined as a function of the Hamiltonian $H(t)$ of the system:

$$\begin{cases} i\hbar \dfrac{\mathrm{d}}{\mathrm{d}t} U(t,t_0) = H(t)\,U(t,t_0) \\ U(t_0,t_0) = \mathbb{1} \end{cases} \tag{82}$$

and that :

$$U^{-1}(t,t') = U^{\dagger}(t,t') = U(t',t) \quad ; \quad U(t,t')\,U(t',t'') = U(t,t'') \tag{83}$$

α. *General case*

We set:

$$|\hat{\psi}(0)\rangle = T\,|\psi(0)\rangle \quad \text{and} \quad |\hat{\psi}(-t)\rangle = T\,|\psi(t)\rangle \tag{84}$$

Time reversal is a symmetry operator if $|\hat{\psi}(t)\rangle$ can be considered as the state vector at time $-t$ of a physical system with the same evolution laws as the initial system. In terms of operators, this condition is written:

$$U(0,-t)\,T\,U(t,0)\,|\psi(0)\rangle = T\,|\psi(0)\rangle \tag{85}$$

or else, since $|\psi(0)\rangle$ can take any value:

$$\boxed{T\,U(t,0) = U(-t,0)\,T} \tag{86a}$$

This equality can be written in a form more symmetric[4] in time:

$$\boxed{U(t,-t) = T^{\dagger}\,U(-t,t)\,T} \tag{86b}$$

Demonstration: Multiplying relation (86a) on the left by $T^{\dagger}$, we get:

$$U(t,0) = T^{\dagger}\,U(-t,0)\,T \tag{87a}$$

Since the equality is valid at any time, we can replace t by $-t$, which yields:

$$U(-t,0) = T^{\dagger}\,U(t,0)\,T \tag{87b}$$

The inverse (or Hermitian conjugate) relation is written:

$$U(0,-t) = U^{\dagger}(-t,0) = T^{\dagger}\,U(0,t)\,T \tag{87c}$$

Multiplying side by side (87a) by this equality, we get:

$$U(t,0)\,U(0,-t) = T^{\dagger}\,U(-t,0)\,T\,T^{\dagger}\,U(0,t)\,T^{\dagger} \tag{88}$$

Since $T\,T^{\dagger} = 1$, and joining together sequential evolution operators, we get relation (86b).

[4] This relation shows that if $|\psi(t)\rangle$ is a solution of Schrödinger equation:

$$i\hbar\frac{\mathrm{d}}{\mathrm{d}t}|\psi(t)\rangle = H(t)|\psi(t)\rangle, \quad \text{the ket} \quad |\varphi(t)\rangle = |\hat{\psi}(-t)\rangle = T|\psi(-t)\rangle$$

is a solution of the same equation (it corresponds to another possible motion with the same Hamiltonian, which can be time-dependent).

β. Conservative system

A conservative physical system is an important particular case (H does not depend on t). We then have:

$$U(t,0) = \exp\{-iHt/\hbar\} = \sum_n \frac{1}{n!}\left(-\frac{iHt}{\hbar}\right)^n \tag{89}$$

Relation (86a) becomes:

$$T\left\{\sum_n \frac{1}{n!}\left(\frac{-iHt}{\hbar}\right)^n\right\} = \left\{\sum_n \frac{1}{n!}\left(\frac{iHt}{\hbar}\right)^n\right\} T \tag{90}$$

It implies, to first-order in t:

$$T\left(\frac{-iHt}{\hbar}\right) = \left(\frac{iHt}{\hbar}\right) T \tag{91}$$

or, since T is antilinear:

$$\left(\frac{it}{\hbar}\right) T\,H = \left(\frac{it}{\hbar}\right) H\,T \tag{92}$$

We thus have:

$$\boxed{[T,H] = 0} \tag{93}$$

Conversely, since $T(-iH) = (iH)T$ this equality yields (90), and hence (86a). Condition (93) establishes the invariance under time reversal of a conservative system (an isolated system for example). Comparison with (69b) shows once again that choosing for T an antilinear operator has switched the roles of commutators and anticommutators.

Comment:

It is easy to see that the diagram of figure 5 does apply to a conservative system. Schrödinger's equation is written:

$$i\hbar\frac{\mathrm{d}}{\mathrm{d}t}|\psi(t)\rangle = H|\psi(t)\rangle \tag{94}$$

Multiplying both sides by the antilinear operator T, we get (since T is not explicitly time-dependent):

$$T\,i\hbar\frac{\mathrm{d}}{\mathrm{d}t}|\psi(t)\rangle = -i\hbar\frac{\mathrm{d}}{\mathrm{d}t}\,T\,|\psi(t)\rangle = T\,H\,|\psi(t)\rangle \tag{95}$$

or, using (93):

$$-i\hbar\frac{\mathrm{d}}{\mathrm{d}t}\,[T\,|\psi(t)\rangle] = H\,[T\,|\psi(t)\rangle] \tag{96}$$

This equation, which has the form of a simple complex conjugate of Schrödinger's equation, directly corresponds to the diagram shown in figure 5.

4. Explicit form of the time reversal operator

4-a. Spinless particle

We already saw in (54) and (55) that $K_{\{r\}}$, the complex conjugation operator in basis $\{|\boldsymbol{r}\rangle\}$, obeys the required commutation relations: commutation with $\boldsymbol{R}$, anticommutation with $\boldsymbol{P}$. We thus set, for a spinless particle:

$$T = K_{\{r\}} \tag{97}$$

Consequently, if $|\psi\rangle$ is the state vector of the particle, and $\psi(\boldsymbol{r}) = \langle\boldsymbol{r}|\psi\rangle$ its wave function, it is easy to get the wave function associated with the state having undergone a time reversal:

$$\langle\boldsymbol{r}|T|\psi\rangle = \psi^\star(\boldsymbol{r}) \tag{98}$$

As for the wave function in the $\boldsymbol{p}$ representation:

$$\psi_p(\boldsymbol{p}) = \langle\boldsymbol{p}|\psi\rangle \tag{99}$$

it becomes, by Fourier transformation:

$$\langle\boldsymbol{p}|T|\psi\rangle = \psi_p^\star(-\boldsymbol{p}) \tag{100}$$

Comment:

We implicitly assumed that the particle under study is not subjected to the action of a vector potential (since we consider that the particle's velocity and momentum are proportional). In the presence of a potential vector $\boldsymbol{A}$, there is in general no operator acting in the state space which, acting alone (without changing $\boldsymbol{A}$), represents the time reversal operator. On the other hand, applying simultaneously $K_{\{r\}}$ and the transformation:

$$\boldsymbol{A}\,(\boldsymbol{r}\,;\,t) \rightarrow -\boldsymbol{A}\,(\boldsymbol{r}\,;\,-t) \tag{101}$$

we get a time reversal transformation [*cf.* equalities (19a) and (19b)]. In the case where $\boldsymbol{A}$ is time-independent, this corresponds to the change of $\boldsymbol{B}$ into $-\boldsymbol{B}$ discussed in the example of § 1.

4-b. Spin 1/2 particle

To the commutation relations of $\boldsymbol{R}$, $\boldsymbol{P}$, and $\boldsymbol{L}$ with T for a free particle, we must add for the spin operator $\boldsymbol{S}$:

$$T\,\boldsymbol{S}\,T = -\boldsymbol{S} \tag{102}$$

This equality ensures that $\boldsymbol{S}$, just as $\boldsymbol{L}$, changes sign under the time reversal operation; since T and $\boldsymbol{L}$ anticommute, it also ensures that the fundamental equality (80) is satisfied.

We set:

$$T = \Theta\, K_{\{\boldsymbol{r},\mu\}} \tag{103}$$

where Θ is a (unitary) operator to be determined, and $K_{\{\boldsymbol{r},\mu\}}$ the (antiunitary) complex conjugation operator in the basis $\{|\boldsymbol{r},\mu\rangle\}$ of the eigenkets common to $\boldsymbol{R}$ and S_z. For the same reasons as before (reality of the matrix elements of $\boldsymbol{R}$ in the chosen basis, purely imaginary character of those of $\boldsymbol{P}$):

$$\begin{aligned} K_{\{\boldsymbol{r},\mu\}}\, \boldsymbol{R}\, K_{\{\boldsymbol{r},\mu\}} &= \boldsymbol{R} \\ K_{\{\boldsymbol{r},\mu\}}\, \boldsymbol{P}\, K_{\{\boldsymbol{r},\mu\}} &= -\boldsymbol{P} \end{aligned} \tag{104}$$

and consequently:

$$K_{\{\boldsymbol{r},\mu\}}\, \boldsymbol{L}\, K_{\{\boldsymbol{r},\mu\}} = K_{\{\boldsymbol{r},\mu\}}\, (\boldsymbol{R} \times \boldsymbol{P})\, K_{\{\boldsymbol{r},\mu\}} = -\boldsymbol{L} \tag{105}$$

Operator T anticommmutes with $\boldsymbol{P}$ so that:

$$\Theta\, K_{\{\boldsymbol{r},\mu\}}\, \boldsymbol{P}\, K_{\{\boldsymbol{r},\mu\}}\, \Theta^\dagger = -\boldsymbol{P} \tag{106}$$

or, taking (104) into account:

$$\Theta\, \boldsymbol{P}\, \Theta^\dagger = \boldsymbol{P} \tag{107a}$$

It follows:

$$[\Theta, \boldsymbol{P}] = 0 \tag{107b}$$

Similarly, one can show that Θ commutes with $\boldsymbol{L}$ (and $\boldsymbol{R}$). The fact that $K_{\{\boldsymbol{r},\mu\}}$ already obeys the required commutation relations with the various orbital operators leads us to take for Θ an operator that does not act in the space of orbital states, i.e. a pure spin operator.

The matrix elements of σ_x and σ_z are real in the $\{|\boldsymbol{r},\mu\rangle\}$ basis; on the other hand, the matrix elements of σ_y are purely imaginary [*cf.* expressions (III-72) of the Pauli matrices]. This means that:

$$\begin{aligned} K_{\{\boldsymbol{r},\mu\}}\, S_x\, K_{\{\boldsymbol{r},\mu\}} &= S_x \\ K_{\{\boldsymbol{r},\mu\}}\, S_z\, K_{\{\boldsymbol{r},\mu\}} &= S_z \end{aligned} \tag{108a}$$

whereas:

$$K_{\{\boldsymbol{r},\mu\}}\, S_y\, K_{\{\boldsymbol{r},\mu\}} = -S_y \tag{108b}$$

As operator T must change the sign of the three components of $\boldsymbol{S}$, the spin operator Θ must change the sign of S_x and S_z without changing that of S_y. In other words, Θ must commute with σ_y, and anticommute with σ_x and σ_z. Matrix σ_y itself satisfies these conditions. One thus chooses for Θ the operator $\sigma_y = 2\,S_y/\hbar$, or more often the matrix $-i\sigma_y$

$$-i\sigma_y = \begin{pmatrix} 0 & -1 \\ 1 & 0 \end{pmatrix} \tag{109}$$

which has the advantage of being real. It is a unitary matrix (like σ_y), and we can set[5]:

$$T = -i\sigma_y\, K_{\{\boldsymbol{r},\mu\}} \tag{110}$$

Under these conditons, if:

$$\begin{pmatrix} \psi_+(\boldsymbol{r}) \\ \psi_-(\boldsymbol{r}) \end{pmatrix} \tag{111a}$$

is a spinor associated with the particle, it becomes after time reversal:

$$\begin{pmatrix} -\psi_-^\star(\boldsymbol{r}) \\ \psi_+^\star(\boldsymbol{r}) \end{pmatrix} \tag{111b}$$

4-c. Particle with arbitrary spin

For an arbitrary spin, the reasoning is the same as in the previous paragraph (we write the matrices expressing the action of S_z, S_+, and S_- in the basis $\{|\boldsymbol{r},\mu\rangle\}$, and show that S_x and S_z are associated with real matrices, whereas S_y is associated with a purely imaginary one).

As for the spin operator Θ, we must choose a unitary operator that changes the sign of S_x and S_z without changing the sign of S_y. This is exactly the action of a rotation by π around the Oy axis. We thus set:

$$T = \mathrm{e}^{-i\pi S_y/\hbar}\, K_{\{\boldsymbol{r},\mu\}} \tag{112}$$

Exercise: If the particle's state is:

$$|\psi\rangle = \sum_{\mu=-S}^{S} \psi_\mu(\boldsymbol{r})\,|\boldsymbol{r},\mu\rangle \tag{113a}$$

[5]This Oy direction for the spin may seem surprising, as a time reversal operation does not favor any direction in space. But keep in mind that this spin operator is associated with a complex conjugation operation $K_{\{\boldsymbol{r},\mu\}}$ in a given basis for which a spin quantization direction has been chosen.

show that:

$$T|\psi\rangle = \sum_{\mu=-S}^{+S} (-1)^{S-\mu}\,\psi_\mu^\star(\boldsymbol{r})\,|\boldsymbol{r},-\mu\rangle \tag{113b}$$

One should start showing that:

$$\mathrm{e}^{-i\pi S_y/\hbar}\,|\mu\rangle = (-1)^{S-\mu}\,|-\mu\rangle \tag{114}$$

by considering that the spin S is obtained by the addition of $2S$ spins 1/2's. In the state space of each of these spins, we have [*cf.* relation (III-76)]:

$$\mathrm{e}^{-i\pi S_y/\hbar} = -i\sigma_y \tag{115}$$

and thus:

$$\mathrm{e}^{-i\pi S_y/\hbar}\,|+\rangle = |-\rangle \quad ; \quad \mathrm{e}^{-i\pi S_y/\hbar}\,|-\rangle = -|+\rangle \tag{116}$$

Comments:

(i) Relation (115) shows that the general formula is equivalent to (110) for a spin 1/2.

(ii) We now compute the square T^2 of the time reversal operator. We get:

$$T^2 = \mathrm{e}^{-i\pi S_y/\hbar}\,K_{\{\boldsymbol{r},\mu\}}\,\mathrm{e}^{-i\pi S_y/\hbar}\,K_{\{\boldsymbol{r},\mu\}} \tag{117}$$

Since the antilinear operator $K_{\{\boldsymbol{r},\mu\}}$ anticommutes with S_y [equality (108b)], it commutes with $-iS_y$ and hence with the rotation operator $\mathrm{e}^{-i\pi S_y/\hbar}$. Consequently:

$$T^2 = \mathrm{e}^{-i\pi S_y/\hbar}\,\mathrm{e}^{-i\pi S_y/\hbar}\left[K_{\{\boldsymbol{r},\mu\}}\right]^2 = \mathrm{e}^{-2i\pi S_y/\hbar} \tag{118}$$

We already discussed in chapter VII the value of a rotation operator by an angle 2π (true or double-valued representation depending on whether the spin is an integer or a half-integer). Consequently:

$$\boxed{T^2 = \begin{cases} 1 \text{ if } S \text{ is an integer} \\ -1 \text{ if } S \text{ is a half-integer}\end{cases}} \tag{119a}$$

These equalities are equivalent to:

$$T^\dagger = \begin{cases} +T \text{ if } S \text{ is an integer} \\ -T \text{ if } S \text{ is a half-integer}\end{cases} \tag{119b}$$

since T is antiunitary and as can be easily shown by multiplying the previous equalities by $T^\dagger$. In all cases, the transformations $(T\,B\,T^\dagger)$ of an observable by T, or $(T^\dagger\,B\,T)$ by $T^\dagger$, are equal.

4-d. System with several particles

The time reversal operator T associated with a system of several numbered particles $1, 2, \ldots, N$ is the tensor product of operators T associated with each particle separately:

$$T = T^{(1)} \otimes T^{(2)} \otimes T^{(3)} \otimes \ldots \otimes T^{(N)} \tag{120}$$

For an ensemble of N fermions (with half-integer spins):

$$\begin{aligned} T^2 &= +1 \text{ if } N \text{ is even} \\ T^2 &= -1 \text{ if } N \text{ is odd} \end{aligned} \tag{121}$$

5. Applications

In this last part, we shall discuss a few situations where time reversal symmetry applies, like microreversibility or the concept of parity of an observable under time reversal. We shall also mention two particularly useful theorems, Kramers' theorem and van Vleck theorem.

5-a. Microreversibility

The "principle of microreversibility in quantum mechanics" is often referred to as an indication that for any *isolated* (hence conservative) physical system, the time-independent Hamiltonian H must commute with the time reversal operator T. This is actually not a real principle generally applicable in physics: whereas electromagnetic and strong interactions are invariant under time reversal, this is not the case for weak interactions. Violation of T symmetry has been observed in some rare disintegrations of mesons (neutral Kaon or B meson). In the vast majority of cases, these processes can be neglected and the microreversibility "principle" can be directly applied.

We multiply on the left equation (86a) by $T^\dagger$ and replace $U(-t,0)$ by $U^\dagger(t,0)$; relation (89) shows that this is possible when H is time-independent. We can write, for any kets $|\varphi\rangle$ and $|\chi\rangle$:

$$\begin{aligned} \langle\chi|U(t,0)|\varphi\rangle &= \langle\chi| \left\{T^\dagger\, U^\dagger(t,0)\, T\right\} |\varphi\rangle = \left\{\left(\left(\langle\chi|T^\dagger\right)\, U^\dagger(t,0)\, \left(T|\varphi\rangle\right)\right)\right\}^* \\ &= \left(\left(\langle\varphi|T^\dagger\right)\, U(t,0)\, \left(T|\chi\rangle\right)\right) \end{aligned} \tag{122}$$

Setting:

$$|\overline{\varphi}\rangle = T\,|\varphi\rangle \quad ; \quad |\overline{\chi}\rangle = T\,|\chi\rangle \tag{123}$$

we get:

$$\boxed{\langle\chi|U(t,0)|\varphi\rangle = \langle\overline{\varphi}|U(t,0)|\overline{\chi}\rangle} \tag{124}$$

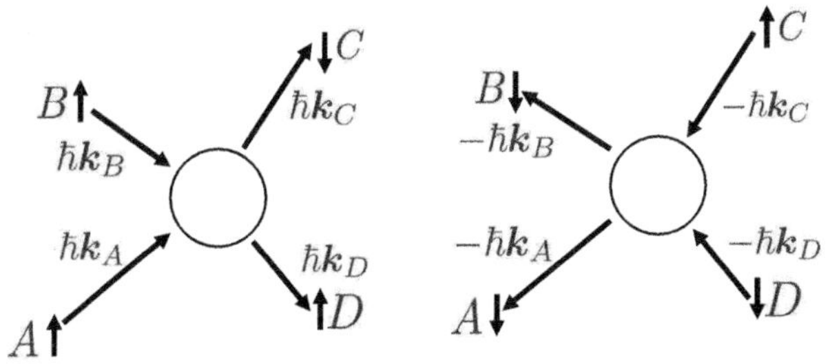

Figure 6: On the left-hand side of the figure is shown a collision where two particles, with initial momenta $\hbar \boldsymbol{k}_A$ and $\hbar \boldsymbol{k}_B$, are transferred to states with final momenta $\hbar \boldsymbol{k}_C$ and $\hbar \boldsymbol{k}_D$ after the collision; the spins are represented by small up or down arrows. The inverse collision is shown on the right-hand side: the states before and after the collision are interchanged, and spin directions are inverted. If the interactions are invariant under time reversal, the probability amplitudes associated with these two collisions are the same.

Consider a system that, at the initial time, is in state $|\varphi\rangle$, and freely evolves until time t. The left-hand side of (124) yields the (complex) probability amplitude of finding that system in state $|\chi\rangle$ if a measurement is performed at that time. The right-hand side concerns a system, initially in state $|\overline{\chi}\rangle$, evolving (with the same dynamical laws) until time t. Equality (124) states that the probability amplitude of finding that system at time t in state $|\overline{\varphi}\rangle$ is the same as the previous probability amplitude (under the condition that the system is invariant under time reversal).

Imagine, for example, a collision (*cf.* figure 6) between two particles A and B with initial momenta $\hbar \boldsymbol{k}_A$ and $\hbar \boldsymbol{k}_B$, and spins pointing both in the Oz direction. We look for the probability amplitude of finding, after the collision, a particle with momentum $\hbar \boldsymbol{k}_C$ and spin in the opposite direction of Oz, and another one with momentum $\hbar \boldsymbol{k}_D$ and spin parallel to Oz[6]. If the collision between particles obeys physical laws invariant under time reversal, we know that this amplitude is the same as that of another process (right-hand side of figure 6): a collision where the incoming particles have momenta $-\hbar \boldsymbol{k}_C$ (with upward spin) and $-\hbar \boldsymbol{k}_D$ (with downward spin) and outgoing particles with momenta $-\boldsymbol{k}_A$ (with downward spin) and $-\boldsymbol{k}_B$ (also with downward spin).

[6]The nature and number of initial and final particles can be different (we could have taken 3 final particles for example): rearrangement collisions in atomic physics, ionizing collisions, fusion or fission in nuclear physics, creation of particles in high energy physics, etc.

Comments:

(i) Remember that time reversal flips the direction of the spins, otherwise this could lead to errors (unless H and hence U do not act in the spin state space, in which case spins do not play any role in the collision).

(ii) In a macroscopic system, invariance under time reversal of all the microscopic processes does not exclude an apparent irreversibility on a macroscopic scale (entropy increase in statistical mechanics).

5-b. Kramers' theorem

We already mentioned a certain analogy between the space parity operator S_0 and the time reversal operator (time parity). When the Hamiltonian H of a physical system is invariant under parity:

$$[S_0, H] = 0 \tag{125}$$

and if $|\varphi\rangle$ is an eigenket of H:

$$H\,|\varphi\rangle = E\,|\varphi\rangle \tag{126}$$

two possibilities exist:

- either $|\varphi\rangle$ is an eigenvector of S_0, even ($S_0|\varphi\rangle = |\varphi\rangle$), or odd ($S_0|\varphi\rangle = -|\varphi\rangle$);
- or $S_0|\varphi\rangle$ is another eigenvector of H with the same eigenvalue E, which is thus degenerate.

These properties are directly transposed to time reversal. If:

$$[T, H] = 0 \quad \text{and if:} \quad H\,|\varphi\rangle = E\,|\varphi\rangle \tag{127}$$

then:

$$H\,T\,|\varphi\rangle = T\,H\,|\varphi\rangle = E\,T\,|\varphi\rangle \tag{128}$$

(since E is real). Consequently, $|\overline{\varphi}\rangle = T\,|\varphi\rangle$ is an eigenvector of H with the same eigenvalue (just as with space parity). The problem is to determine if $|\overline{\varphi}\rangle$ is colinear or not with $|\varphi\rangle$; if it is not, the eigenvalue E is necessarily degenerate.

When $T^2 = -1$ [which is the case for an odd number of fermions, *cf.* (121)], the scalar product $\langle\varphi|\overline{\varphi}\rangle$ is zero; consequently, $|\varphi\rangle$ and $|\overline{\varphi}\rangle$ are not proportional and E is degenerate.

Demonstration: Equality $T^2 = -1$ leads to:

$$-T^\dagger = T^\dagger\,T^2 = \left(T^\dagger\,T\right)\,T \tag{129}$$

and hence $T = -T^\dagger$. It follows:

$$\langle\varphi|\overline{\varphi}\rangle = \langle\overline{\varphi}|\varphi\rangle^\star = \left\{\left(\langle\varphi|T^\dagger\right)|\varphi\rangle\right\}^\star = \langle\varphi|\left(T^\dagger|\varphi\rangle\right)$$
$$= -\langle\varphi|\left(T|\varphi\rangle\right) = -\langle\varphi|\overline{\varphi}\rangle \tag{130}$$

so that $\langle\varphi|\overline{\varphi}\rangle$ is zero.

This leads to the following theorem:

Kramers' theorem: The energy levels of a system invariant under time reversal are all degenerate when the system is composed of an odd number of half-integer spins.

Examples:

(i) Hydrogen atom (the proton spin is ignored): the angular momentum J of the single electron takes on half-integer values ($J = \ell \pm 1/2$) and the $2J+1$ degeneracy is even. All the levels are degenerate. On the other hand, when a magnetic field is applied, the system is no longer invariant under time reversal; Kramers' theorem no longer applies and the magnetic field lifts the degeneracy due to the Zeeman effect.

(ii) Taking into account the proton spin, we get a system with two spin $1/2$ particles; a level such as the hyperfine sub-level $F = 0$ of the ground state is not degenerate. The ground state ${}^1S_0$ of ^{4}He is another example of a non-degenerate system composed of an even number of fermions.

(iii) Kramers's theorem is particularly interesting for cases where rotation invariance does not a priori predict the degeneracy, as was the case for the previous examples. An example is ions in crystals (color centered) subjected to an electrostatic potential possibly of low symmetry and where spin-orbit coupling may play a role. Those ions that have an odd number of electrons are called Kramers ions; Kramer's theorem indicates that, no matter how low the spatial symmetry around these ions, all their energy levels are at least doubly degenerate.

Comment:

When $T^2 = +1$ (even number of fermions), Kramers' theorem does not apply. However, as for the case of space parity, one can draw some conclusions from the invariance of H:

- When $|\overline{\varphi}\rangle$ is proportional to $|\varphi\rangle$, the stationary state we consider is invariant under time reversal (the average values of the particles' velocities are zero everywhere, etc.). Example: ground state ${}^1S_0$ of ^{4}He.

- When $|\overline{\varphi}\rangle$ and $|\varphi\rangle$ are not colinear, they generate a 2-dimensional eigensubspace of H; the energy E is degenerate. We note that:

$$\begin{cases} T\left[\dfrac{1}{\sqrt{2}}\left(|\varphi\rangle + |\overline{\varphi}\rangle\right)\right] = \dfrac{1}{\sqrt{2}}\left(|\varphi\rangle + |\overline{\varphi}\rangle\right) \\ T\left[\dfrac{i}{\sqrt{2}}\left(|\varphi\rangle - |\overline{\varphi}\rangle\right)\right] = \dfrac{i}{\sqrt{2}}\left(|\varphi\rangle - |\overline{\varphi}\rangle\right) \end{cases} \tag{131}$$

The two eigenstates $\frac{1}{\sqrt{2}}(|\varphi\rangle + |\overline{\varphi}\rangle)$ and $\frac{i}{\sqrt{2}}(|\varphi\rangle - |\overline{\varphi}\rangle)$ of H are both invariant under time reversal.

In conclusion, one can always look for eigenstates of H among states that are invariant under time reversal (states that are their own "Kramers conjugate"). For example, when studying the eigenstates of a set of two electrons placed in a real electrostatic potential, on can choose only real wave functions.

5-c. Even or odd observables under time reversal

The definition of an even or odd observable under time reversal is analog to the definition of an even or odd observable under space parity. The time reversal transformation of any observable G is written:

$$\overline{G} = T\,G\,T^{\dagger} \tag{132}$$

If:

$$\overline{G} = \varepsilon\,G \tag{133}$$

and if $\varepsilon = +1$, G is said to be an even observable; however, if $\varepsilon = -1$, the observable is said to be odd.

Since T is antiunitary, relation (133) is equivalent to:

$$T\,G = \varepsilon\,G\,T \tag{134}$$

An *even* observable *commutes* with T, an *odd* obervable *anticommutes* with T. As an example, relation (93) shows that the Hamiltonian of a conservative system invariant under time reversal is an even observable.

Comment:

If G is non-Hermitian, one can always set:

$$G = G_1 + i\,G_2 \tag{135}$$

where:

$$G_1 = \frac{1}{2}\left(G + G^{\dagger}\right) \quad \text{and} \quad G_2 = \frac{1}{2i}\left(G - G^{\dagger}\right) \tag{136}$$

are two Hermitian operators. If each of them has the ε parity under time reversal, we have:

$$\begin{aligned}\overline{G} &= T\,(G_1 + i\,G_2)\,T^\dagger = T\,G_1\,T^\dagger - i\,T\,G_2\,T^\dagger \\ &= \varepsilon(G_1 - i\,G_2) = \varepsilon\,G^\dagger \end{aligned} \tag{137}$$

Consequently, a non-Hermitian operator G will be said to be even ($\varepsilon = +1$) or odd ($\varepsilon = -1$) under time reversal, if:

$$\overline{G} = T\,G\,T^\dagger = \varepsilon\,G^\dagger \tag{138}$$

For example, $\boldsymbol{R}$ is an even operator, $\boldsymbol{P}$, $\boldsymbol{L}$, and $\boldsymbol{J}$ are odd operators; operator J_x^2 is even, but $J_x\,J_y$ does not have a specific parity ε (since $\overline{J_x\,J_y}$ is equal to $J_x\,J_y$, and not to its Hermitian conjugate). According to (138), the evolution operator:

$$U(t,0) = \mathrm{e}^{-iHt/\hbar}$$

is even if H is invariant under time reversal, since [*cf.* § 3-b-β]:

$$\overline{U}(t,0) = U^\dagger(t,0) \tag{139}$$

α. *First property*

Consider an eigenvector $|\varphi_g\rangle$, with a (real) eigenvalue a_g, of an observable G with partiy ε:

$$G\,|\varphi_g\rangle = a_g|\varphi_g\rangle \tag{140}$$

Let us show that $|\overline{\varphi}_g\rangle = T\,|\varphi_g\rangle$ is an eigenvector of G with the eigenvalue $\varepsilon\,a_g$:

$$\begin{aligned} G\,|\overline{\varphi}_g\rangle &= G\,T\,|\varphi_g\rangle = \varepsilon\,T\,G\,|\varphi_g\rangle = \varepsilon\,a_g\,T\,|\varphi_g\rangle \\ &= \varepsilon\,a_g\,|\overline{\varphi}_g\rangle \end{aligned} \tag{141}$$

(since a_g is real).

β. *Second property*

If G is an observable which is even ($\varepsilon = +1$) or odd ($\varepsilon = -1$), we have:

$$\langle\varphi|G|\varphi\rangle = \varepsilon\langle\overline{\varphi}|G|\overline{\varphi}\rangle \tag{142a}$$

and, in a more general way:

$$\boxed{\langle\varphi|G|\chi\rangle = \varepsilon\langle\overline{\varphi}|G|\overline{\chi}\rangle^\star} \tag{142b}$$

This is because:

$$\begin{aligned} \langle\varphi|G|\chi\rangle &= \varepsilon\langle\varphi|T^\dagger\,G\,T\,|\chi\rangle = \varepsilon\left\{\left(\langle\varphi|T^\dagger\right)\left(G\,T|\chi\rangle\right)\right\}^\star \\ &= \varepsilon\langle\overline{\varphi}|G|\overline{\chi}\rangle^\star \end{aligned} \tag{143}$$

γ. *Third properties (matrix elements between two Kramers conjugate states)*

If G is an even operator and $T^2 = -1$, or if G is an odd operator and $T^2 = +1$, we have:

$$\boxed{\langle\varphi|G|\overline{\varphi}\rangle = 0} \tag{144}$$

The diagonal matrix element of G between two states conjugated through time reversal, is always zero, as we now show.

We can write:

$$\begin{aligned}\langle\varphi|G|\overline{\varphi}\rangle &= \langle\varphi|G\,(T|\varphi\rangle) = \left\{\left(\langle\varphi|T^\dagger\right)\,G^\dagger|\varphi\rangle\right\}^* \\ &= \langle\varphi|\left(T^\dagger\,G^\dagger|\varphi\rangle\right)\end{aligned} \tag{145}$$

If $T^2 = \xi$ (with $\xi = \pm 1$), we also have $T^\dagger = \xi T$. Using this equality, we insert $T^\dagger T = 1$, and use (138) to write:

$$\begin{aligned}\langle\varphi|G|\overline{\varphi}\rangle &= \xi\,\langle\varphi|\left(T\,G^\dagger\,(T^\dagger\,T)|\varphi\rangle\right) = \varepsilon\,\xi\,\langle\varphi|\,(G\,T|\varphi\rangle) \\ &= \varepsilon\,\xi\,\langle\varphi|G|\overline{\varphi}\rangle\end{aligned} \tag{146}$$

which shows that $\langle\varphi|G|\overline{\varphi}\rangle$ is zero if the product $\varepsilon\,\xi$ equals -1.

5-d. Van Vleck's theorem

Consider a non-degenerate level of a system whose Hamiltonian is invariant under time reversal. Van Vleck's theorem states that the average value in that level of any operator G, odd under time reversal, is zero.

Since the level is not degenerate, the corresponding ket $|\varphi\rangle$ satisfies:

$$|\overline{\varphi}\rangle = T|\varphi\rangle = \mathrm{e}^{i\zeta}|\varphi\rangle \tag{147}$$

where ζ is a phase a priori unknown. As G is odd:

$$\begin{aligned}\langle\varphi|G|\varphi\rangle &= -\langle\varphi|T^\dagger\,G^\dagger\,T|\varphi\rangle = -\langle\overline{\varphi}|G^\dagger|\overline{\varphi}\rangle^* = -\langle\overline{\varphi}|G|\overline{\varphi}\rangle \\ &= -\mathrm{e}^{i\zeta}\,\mathrm{e}^{-i\zeta}\,\langle\varphi|G|\varphi\rangle = -\langle\varphi|G|\varphi\rangle\end{aligned} \tag{148}$$

Hence:

$$\langle\varphi|G|\varphi\rangle = 0 \tag{149}$$

Van Vleck's theorem is often applied in the particular case where the Hamiltonian H of an ensemble only acts on its orbital variables. This is the case, for example, where an ion is placed in a crystal, and we neglect the relativistic effects responsible for the fine structure Hamiltonian. If H is invariant under time reversal, the average value of the total orbital angular momentum $\boldsymbol{L}$ is zero. This result is a direct consequence of (149) since $\boldsymbol{L}$ is an odd operator. It can also be established by noting that, since H is invariant under time reversal, one can choose real wave functions for the stationary states. Since, on the other hand, $\boldsymbol{L}$ is a purely imaginary operator in the $\{|\boldsymbol{r}\rangle\}$ representation (as an example, for a single particle $\boldsymbol{L}$ corresponds to $-i\hbar\,\boldsymbol{r}\times\boldsymbol{\nabla}$), its average value is purely imaginary. As it must also be real ($\boldsymbol{L}$ is Hermitian), it must necessarily be zero: $\langle\boldsymbol{L}\rangle = 0$.

Let us go back to the ion placed in a crystal, which we discussed above. Imagine that the ion is paramagnetic when it is free ($L \neq 0$) but that in the crystal, the symmetry of the ion orbital Hamiltonian is low enough for the orbital degeneracy to be completely lifted (this is not in contradiction with Kramers' theorem since, in the orbital variables state space, the square of the time reversal operator is equal to $+1$). Consequently, the orbital momentum is zero in the ground state:

$$\langle\boldsymbol{L}\rangle = 0 \tag{150}$$

As a result, the average value of the magnetic orbital momentum of the ion in the crystalline network is zero; the orbital magnetism is said to be "blocked" by the environment (for the free ion, whose ground state is degenerate, the orbital magnetism is different from zero since $L \neq 0$). A physical interpretation of this zero $\langle\boldsymbol{L}\rangle$ is to say that, due to the large complexity of its motion in the potential exerted by the crystal, the ion's orbital momentum $\langle\boldsymbol{L}\rangle$ constantly changes its direction and averages out to zero over time.

Comment:

It is easy to see that adding to the ion's Hamiltonian a fine structure Hamiltonian of the type $\boldsymbol{L}\cdot\boldsymbol{S}$ has no effect to first-order, and will not lift a degeneracy to that order.

We now show that the electric dipole moment of a system, invariant under time reversal and with a given angular momentum, is zero. Consider a conservative system whose Hamiltonian is invariant under both rotation and time reversal.

$$[H, \boldsymbol{J}] = [H, T] = 0 \tag{151}$$

We call E_0 an eigenvalue of H, $(2J+1)$ times degenerate, corresponding to a subspace $\mathscr{E}_0(J)$ generated by the orthonormal kets:

$$|E_0, J, M_J\rangle \quad ; \quad M_J = J, J-1, \ldots, -J \tag{152}$$

We assume the quantum number M_J is enough to distinguish between the different states with the same energy E_J (no accidental degeneracy); this is always the case if $H, \boldsymbol{J}^2$, and J_z form a CSCO.

Consider the ket:

$$T|E_0, J, M_J\rangle$$

It is a normalized ket (T is antiunitary), eigenvector of H and $\boldsymbol{J}^2$ (even operator) with the same eigenvalues E_0 and $J(J+1)\,\hbar^2$, and of J_z with the eigenvalue $-M\hbar$ (since J_z is odd under time reversal). Consequently:

$$T|E_J, J, M_J\rangle = \mathrm{e}^{i\zeta}|E_J, J, -M_J\rangle \tag{153}$$

where ζ is a real phase factor.

The electric dipole operator $\boldsymbol{D}$ is a vector operator, and odd under time reversal:

$$\boldsymbol{D} = T\,\boldsymbol{D}\,T^\dagger = T^\dagger\,\boldsymbol{D}\,T \tag{154}$$

We are going to show that all the matrix elements of $\boldsymbol{D}$ are zero inside $\mathscr{E}_0(J)$. In this subspace, the Wigner–Eckart theorem shows that:

$$\boldsymbol{D} = \alpha\,\boldsymbol{J} \tag{155}$$

and we simply have to prove that α is zero. We take, for example, the component D_z, which has only diagonal elements of the type:

$$\langle E_0, J, M_J|D_z|E_0, J, M_J\rangle$$

According to (153) and (154):

$$\langle E_0, J, M_J|D_z|E_0, J, M_J\rangle = \langle E_0, J, -M_J|D_z|E_0, J, -M_J\rangle \tag{156}$$

meaning:

$$\alpha M_J = -\alpha M_J$$

The coefficient α is thus equal to zero. It was to be expected since (155) shows it is the proportionality coefficient between two operators of opposite time and spatial parity.

Application: A neutron is a spin 1/2 particle. If one discovers it has an electric dipole (colinear with its spin because of the rotational invariance), this will prove that the physical laws at the origin of the neutron's existence are not invariant under time reversal (and space parity).

Bibliography

SIMPLE INTRODUCTORY TEXTS

[1] G. Feinberg and M. Goldbaber, "The conservation laws of physics", *Sci. Am.* **209**, 36 (Oct. 1963).

[2] E.P. Wigner, "Violations of symmetry in physics", *Sci. Am.* **213**, 28 (Dec. 1965).

[3] P. Morrison, "The overthrow of parity", *Sci. Am.* **196**, 34 (Mar. 1963).

[4] E.P. Wigner, "Symmetry and conservation laws", *Physics Today* 34 (Mar. 1964).

[5] J. Iliopoulos, *The origin of mass: elementary particles and fundamental symmetries*, EDP Sciences (2015).

QUANTUM MECHANICS

[6] A. Messiah, *Quantum Mechanics*, Vol. II, 3rd part, Symmmetries and invariances, Dover Publications (2014). Appendix C of this book contains many particularly useful formulas concerning coefficients and matrices related to rotations.

[7] L. Schiff, *Quantum mechanics*, chap. 7, McGraw Hill (1968).

[8] R. Omnès, *Introduction à l'étude des particules élémentaires*, chaps. 2, 3, 4 and 5, Ediscience, Paris (1970). A very simple and clear introduction to the use of continuous groups in particle physics.

[9] C. Cohen-Tannoudji, B. Diu and F. Laloë, *Quantum Mechanics*, Wiley (2020).

[10] R.P Feynman and A.R. Hibbs, *Quantum mechanics and path integrals*, Mc Graw Hill, New York (1968).

[11] S. Weinberg, *The quantum theory of fields*, Vol. I, II, and III, Oxford University Press (1996).

[12] E. Brezin, *Statistical field theory*, Cambridge University Press (2010).

[13] M. Le Bellac, *Quantum Physics*, Vols I and II, Cambridge University Press (2006).

Introduction to Continuous Symmetries: From Space–Time to Quantum Mechanics, First Edition. Franck Laloë.

[14] L. Baulieu, J. Iliopoulos and R. Sénéor, *From classical to quantum fields*, Oxford University Press (2017).

GROUP THEORY AND SYMMETRIES

[15] E.P. Wigner, *Group theory and its applications to quantum mechanics*, Academic Press, New York (1959).

[16] M. Hamermesh, *Group theory and its applications to physical problems*, Addison-Wesley, New Reading (1962).

[17] U. Uhlhorn, “Representation of symmetry transformations in quantum mechanics”, *Arkiv för Fysik* **23**, 307–340 (1962).

[18] P.H.E. Meijer and E. Bauer, *Group theory*, North Holland, Amsterdam (1964).

[19] V. Bargmann, “Note on Wigner's theorem on symmetry operations”, *J. Math. Phys.* **5**, 862–868 (1964).

[20] G. Racah and W. Pauli, lectures in *Ergebnisse der exakten Natürwissenschaften*, Vol. **37**, Springer Verlag (1965).

[21] S. Gasiorowicz, *Elementary particle physics*, Wiley and Sons (1966).

[22] P. Moussa and R. Stora, *Angular analysis of elementary particle reactions*, Lectures given at the Hercegovni International School of Elementary Particle Physics (1966).

[23] A. Zee, *Group theory in a nutshell for physicists*, Princeton University Press (2019).

[24] M.E. Peshkin and D.V. Schroeder, *An introduction to quantum field theory*, CRC Press (2018).

[25] J. Zinn-Justin *Quantum field theory and critical phenomena*, Oxford University Press, fifth edition (2021).

[26] H. Haber, “Properties of the Gell-Mann matrices”, in *Physics 251 Group Theory and Modern Physics*, U.C. Santa Cruz. See also Wikipedia, “Gell-Mann matrices” and “Clebsch–Gordan coefficients for SU(3)”.

[27] J-B. Zuber, “Invariances in physics and group theory”, https://arxiv.org/abs/1307.3970

GALILEO AND POINCARÉ GROUPS

[28] J.-M. Lévy-Leblond, "Une nouvelle limite non relativiste du groupe de Poincaré", *Ann. Inst. H. Poincaré* **3**, 1 (1965).

[29] J.-M. Lévy-Leblond, "Nonrelativistic particles and wave equations", *Commun. Math. Phys.* **6**, 286–311 (1967).

[30] J.-M. Lévy-Leblond, "The pedagogical role and epistemological significance of group theory in quantum mechanics", *Rivista del nuovo cimento* **4**, 99–143 (1974).

[31] O.-M. Bilianuk and E.C.G. Sudarshan, "Particles beyond the light barrier", *Phys. Today* **22**, 43 (May 1969).

[32] J.-M. Lévy-Leblond and J.-P. Provost, "Additivity, rapidity, relativity", *Am. J. Phys.*, **47**, 1045 (1979).

[33] H. Bauke, S. Ahrens, C. H. Keitel and R. Grobe, "What is the relativistic spin operator?", *New J. Phys.* **16**, 043015 (2014).

[34] T. Choi and S.Y Cho, "Spin operators and representations of the Poincaré group", https://arxiv.org/abs/1807.06425.

[35] P. Holland, "Uniqueness of paths in quantum mechanics", *Phys. Rev. A* **60**, 4326–4330 (1999); "Uniqueness of conserved currents in quantum mechanics", *Ann. Phys.* (Leipzig) **12**, 446–462 (2003).

[36] S. Das and D. Dürr, "Arrival times distributions of spin 1/2 particles", *Sci. Rep.* **9**, 2242 (2019). https://doi.org/10.1038/s41598-018-38261-4.

QUANTIZATION

[37] P.A.M. Dirac, *The principles of quantum mechanics*, Oxford University Press (1947), chaps. IV and VI.

[38] H.J. Groenewold, "On the principles of elementary quantum mechanics", *Physica* **XII**, 405–460 (1946).

[39] J.R. Shewell, "On the formation of quantum mechanical operators", *Am. J. Phys.* **27**, 16–21 (1958).

NUCLEAR PHYSICS

[40] J.M. Blatt and V.F. Weisskopf, *Theoretical nuclear physics*, Dover (2010).

EXPERIMENTS

[41] C.S. Wu et al. "Experimental test of parity conservation in beta decay", *Phys. Rev.*, **105**, 1413–1415 (1957).

[42] H. Rauch et al. "Verification of coherent spinor rotation of fermions", *Phys. Lett.* **54**, 425–427 (1975).

[43] S.A. Werner, R. Colella, A.W. Overhauser and C.F. Eagen, "Observation of the phase shift of a neutron due to precession in a magnetic field", *Phys. Rev. Lett.* **35**, 1053 (1975).

ANGULAR MOMENTUM

[44] A.R. Edmonds, *Angular momentum in quantum mechanics*, Princeton University Press (1954).

[45] M. Rotenberg, R. Bivins, N. Metropolis and J.K. Wooten, *The 3j and 6j symbols*, M.I.T. Technology Press (1959).

CLASSICAL ELECTRODYNAMICS

[46] J.D. Jackson, *Classical electrodynamics*, Wiley, second edition (1975).

[47] C.G. Gray, "Simplified derivation of the magnetostatic multipole expansion using the scalar potential", *Am. J. Phys.* **46**, 582–583 (1978).

[48] M. le Bellac and J-M. Lévy-Leblond, "Galilean electromagnetism", *Nuov. Cim.* **14B**, 217–233 (1973).

BROKEN SYMMETRIES

[49] *Basic notions of condensed matter physics*, CRC Press (2018), see in particular chap. 2.

[50] M.R. Andrews, C.G. Townsend, H.J. Miesner, D.S. Durfee, D.M. Kurn and W. Ketterle, "Observation of interference between two Bose condensates," *Science* **275**, 637-641 (1997).

[51] J. Javanainen and S. M. Yoo, "Quantum phase of a Bose–Einstein condensate with an arbitrary number of atoms", *Phys. Rev. Lett.* **76**, 161–164 (1996).

[52] C.J. Pethick and H. Smith, *Bose–Einstein condensates in dilute gases*, Cambridge University Press (2002), chap. 13.

[53] F. Laloë, "The hidden phase of Fock states; quantum non-local effects", *Eur. Phys. J.* **D33**, 87–97 (2005).

Index

Introduction to Continuous Symmetries: From Space–Time to Quantum Mechanics, First Edition. Franck Laloë.